Tree Physiology

Volume 6

More information about this series at http://www.springer.com/series/6644

Guillermo Goldstein · Louis S. Santiago
Editors

Tropical Tree Physiology

Adaptations and Responses in a Changing
Environment

 Springer

Editors
Guillermo Goldstein
Laboratorio de Ecología Funcional,
    Departamento de Ecología Genética y
    Evolución, Instituto IEGEBA
    (CONICET-UBA), Facultad de Ciencias
    Exactas y naturales
Universidad de Buenos Aires
Buenos Aires
Argentina

and

Department of Biology
University of Miami
Coral Gables, FL
USA

Louis S. Santiago
Department of Botany & Plant Sciences
University of California
Riverside, CA
USA

and

Smithsonian Tropical Research Institute
Balboa, Ancon, Panama
Republic of Panama

ISSN 1568-2544
Tree Physiology
ISBN 978-3-319-27420-1   ISBN 978-3-319-27422-5   (eBook)
DOI 10.1007/978-3-319-27422-5

Library of Congress Control Number: 2015957228

Printed on acid-free paper

This Springer imprint is published by SpringerNature
The registered company is Springer International Publishing AG Switzerland

Preface

Our World is changing rapidly, yet, how tropical forests will respond to this change and in turn dampen or accelerate its ripple effects is essentially a physiological question. Addressing important questions regarding the impacts of changes in land utilization, such as deforestation, and effects of global climate change will require specific information on tropical tree physiology. Earth system modeling scientists are clamoring for more physiological data from tropical trees. It seems that the scarcity of information on the physiological responses of trees is the greatest source of uncertainly in predicting how the tropical rain forests will respond to increasing greenhouse gases and in particular increasing atmospheric CO_2. For example, tree species can adjust their physiological behavior to increasing global temperatures or decreases in precipitation, or they can be replaced by other species better adapted to the new environmental conditions. It is also true that the physiology of tropical trees has not been as well-studied as the physiology of trees from temperate regions, leading to major gaps in our understanding of how tropical trees interact with the Earth system over a range of scales.

It is known that the physiological behavior of both tropical and temperate trees is regulated by similar mechanisms. The differences, however, are related to the unique selective pressures to which tropical trees have been subjected during the evolutionary process and its adaptive consequences. The idea put forward by Theodosius Dobzhansky in the 1950s that tropics and temperate zones are areas where selection operates differently, generated fruitful lines of thinking and research. His contention was that in temperate areas mortality was essentially climatically determined, with little or no competition pressure, while in the tropics, where the environment is relatively more constant, at least in terms of seasonal changes in temperature, mortality is the result of the effects of population size and competition. This paradigm of evolutionary pressures has changed substantially but some aspects of it still remain as a guide for understanding differences in patterns of adaptation between temperate and tropical plants. Negative density dependence prevents any single tree species from dominating most tropical forest ecosystems. The reasons for this must be sought not only in ecological and demographic processes but also among the highly diverse physiological characteristics of tropical trees. In the tropics, seasonal temperature variations are relatively small compared to diurnal temperature changes and if soil water is available, growth and metabolic activities can be maintained throughout the entire year. Heavy herbivore pressure is continuous and the adaptive responses of tropical plants to herbivory are impressive. The physiological implications of various types of mutualisms found among tropical trees are also important. Many trees have a relatively short life span of less than 200 years in the wet tropics compared to more than a 1000 years in some temperate-zone trees.

There has been a substantial increase in the number of studies of tropical tree physiology during the last few decades. The reason for this is not only that trees are the dominant growth form in most tropical ecosystems, but also because of increasing availability and refinement of equipment such as portable photosynthesis systems and instruments for studying water relations of plants. Furthermore, a substantially larger number of tropical biologists are now involved in more mechanistic studies. The use of tower cranes during the last 25 years has allowed scientists to reach the canopy of tropical forests, one of the ultimate frontiers for unveiling not only new organisms but also new processes that were unthinkable just a few years ago.

A distinct feature of tropical trees is not only their high species diversity but also the large variety of life history traits and growth forms that are mostly unique to the tropics such as hemiepiphytic trees, stem succulent trees such as baobab trees, mangroves, palms and other arborescent monocots, and unusual arborescent plants near tree line that are not traditional trees. The wide range of shade tolerance from rapidly growing pioneer trees during gap-phase regeneration to species that can survive by growing slowly in deep shade contribute to this diversity.

Tropical trees tend to grow in habitats where soil water availability is high all year round or at least in habitats were it is seasonally available. They grow in arid environments were they access deep soil water such as in the case of phreatophytic trees. They also occur along altitudinal gradients within the tropics up to the upper tree line and in some cases, such as the caulescent giant rosette plants, they can grow above the continuous forest line. They extend to the subtropics, and in many cases they share close phylogenetic relationships with subtropical tree species, and the structure and function of subtropical forests are in many cases similar to tropical forests, in regards to gap-phase regeneration and the high abundance of lianas.

Through 20 chapters authored by 55 people, this book captures the current state of knowledge of the main physiological characteristics of tropical trees. The book was as a way to not only to provide information gathered during the last few years across the world, but also for laying the foundation for discussing controversial paradigms and new hypothesis of physiological process and mechanisms of trees. Thus this book will surely capture the attention not only of tropical biologists but also of biologists working in many different types of environments around the globe. Physiological consequences of global environmental change will permeate most book chapters, as it provides a dynamic arena for tropical trees to respond. The book is organized in six main parts. The first one is on the physiology of unique tropical growth forms. This group of conspicuous plants is extremely important for understanding the structure, function, and dynamics of tropical forests, as well as understanding why certain species live where they do and not elsewhere. Hemiepiphytic trees with an unusual progression of life stages and obligate epiphytes with a unique photosynthetic pathway, are plants that capture the environmental and demographic wonder of forest ecosystems: Do they start their life cycle (as in the case of hemiepiphytic trees) or spend their entire life cycle (as in the case of obligate epiphytes) in the upper canopy to utilize higher levels of incoming solar radiation or to escape the shady understory with high chances of damage by falling debris and exclusion by competition with other plants? Stem succulents trees, such as the fat-stemmed baobabs, which have captured the imagination and attention of writers, such as in the case of "the little prince" by Antoine de Saint-Exupéry, are fascinating trees occurring mostly in seasonally dry forests. The enlarged stem with photosynthetic surfaces, leaves that drop during the dry season and with little biomass allocation to root systems, appear to have a combination of physiological and anatomical traits that at first sight is difficult to understand. Do their low wood density trunks represent conspicuous water storage? Is stored stem water used for new leaf growth near the end of the dry season or to maintain stem conductance during the rainy season? Does the large size of the stem serve a biomechanical role for providing stability to tall mature trees? Palms are another unique growth form, nearly always associated with tropical environments. The hydraulic architectures of these monocots have long intrigued physiologists working on plant water relations and hydraulic architecture. How can trees that do not have secondary growth, and thus cannot produce new xylem tissue after the plants start growing in height, cope with intensive droughts and the dysfunction of cavitated xylem vessels? Finally lianas, which have solved the problem of reaching the upper canopy without investment in a large diameter stem, can move large amounts of water to transpiring leaves. How did these plants solve this important water economy constraint imposed by a relatively narrow stem?

The second part of the book deals with adaptive responses of trees growing in habitats that are unique to the tropics. Mangrove trees occur in coasts across all tropical regions with roots taping seawater. Floodplain trees tolerate freshwater inundation for several months in inland tropical regions and in some cases are completely covered by water. At high elevation, tropical giant rosette plants represent one of the most fascinating cases of evolutionary convergence among tropical alpine climates that are characterized as "summer during the day and winter at night."

The third part of the book discusses emergent paradigms on hydraulic architecture and water relations. The high diversity of tropical tree species allows the use of a wide array of physiological and morphological traits. This provides fertile ground for testing new hypothesis on the adaptive significance of physiological mechanisms for how trees cope with drought, and how they may avoid or repair cavitated xylem vessels, or how close they are to their physiological limits of water availability in the face of extreme climatic events.

The fourth part of the book deals with important responses of trees to a limitation common in tropical soils: low amounts of available nutrients. How valid is the paradigm of widespread phosphorous limitation? What can be inferred from litter manipulation and fertilization experiments in tropical forests? What are the dynamics and the roles of litter accumulation and decomposition? What is the main distinctive characteristic of nutrient cycling in tropical ecosystems dominated by trees? Do nutrients limit the ability of tropical trees to respond to climate change, or can trees adjust and adapt to nutrient limitations to carry on the process of photosynthetic carbon assimilation? Do N and P equally limit the photosynthetic process in tropical trees?

Carbon economy and biomass allocation patterns in tropical trees and forests are the focus of the fifth part of the book. Important issues related to carbon cycling and the strength of carbon sinks across terrestrial ecosystems worldwide are analyzed. In particular, the continuum of physiological traits from high light requiring pioneer trees to slow growing shade-tolerant trees is discussed in terms of primary and secondary succession in tropical forests and gap-phase regeneration. These two groups of trees are considered as the extremes of a gradient of species requiring high light levels for photosynthesis and rapid growth and in species tolerant to diffuse light and the use of light flecks for carbon assimilation. Are there trade-offs in carbon allocation between defense against herbivores and growth? Compared to tropical and temperate forests, subtropical forests have received little attention until now, and the contribution of this region to the global carbon cycle has not been fully assessed. In this part the carbon balance of subtropical forests at different spatial and temporal scales will be analyzed. The reader will be surprised to know that many subtropical forests are strong carbon sinks, and perhaps the strongest when compared to other terrestrial ecosystem.

In the last part, ecophysiological processes at different spatial and temporal scales are analyzed. Until recently the trunks of tropical trees in lowland areas were assumed not to have tree rings. Several studies have found that this is not the case, particularly in seasonally dry environments, which opens a window of opportunities for using tree rings to acquire insights into the ecology and climate sensitivity of tropical trees as well as the possibility of obtaining the age of individual trees. This part also addresses biomechanical characteristics of tree, with special references to the constraints of being a tropical tree. Do tropical trees adhere to the same biomechanical laws as temperate trees?

In the past 20 years since we began working together, we have seen tropical ecophysiology evolve from single species studies to large comparative works that embrace the high diversity of tropical forests. We have seen a transition from descriptive and natural history studies, which provided an important foundation, to advanced quantitative and modeling approaches that reveal broader patterns in space and time. Plant ecophysiology in the tropics has also developed strong linkages to disciplines that focus on larger spatial scales, including community ecology, ecosystem ecology, and landscape ecology, as well as smaller spatial scales such as molecular biology, stable isotope ecology, and genomics. This book represents the work of a community of leading tropical ecophysiologists, many of whom are colleagues and collaborators. We hope that it will provide a useful resource for understanding, conserving, and sustainably managing tropical forests, as well as predicting how these ecosystems will respond to future climate change.

Buenos Aires   Guillermo Goldstein
January 2015   Louis S. Santiago

Contents

Contributors

Mauricio E. Arias Department of Organismic and Evolutionary Biology, Cambridge, MA, USA

Gerardo Avalos Escuela de Biología, Universidad de Costa Rica, San Pedro, San José, Costa Rica; The School for Field Studies, Center for Sustainable Development Studies, Beverly, MA, USA

Eleinis Ávila-Lovera Department of Botany & Plant Sciences, University of California, Riverside, CA, USA

Marilyn C. Ball Research School of Biology, The Australian National University, Canberra, ACT, Australia

L.F. Banin Centre for Ecology and Hydrology, Bush Estate, Midlothian, Scotland

Damien Bonal INRA, UMR EEF—Université de Lorraine/INRA, Champenoux, France

Roel J.W. Brienen School of Geography, Leeds University, Leeds, UK

Sandra J. Bucci Grupo de Estudios Biofísicos y Ecofisiológicos, Facultad de Ciencias Naturales, Universidad Nacional de la Patagonia San Juan Bosco, Chubut, Argentina; Consejo Nacional de Investigaciones Científicas y Técnicas (CONICET), Buenos Aires, Argentina

Paula I. Campanello Laboratorio de Ecología Forestal y Ecofisiología, Instituto de Biología Subtropical, CONICET, FCF, Universidad Nacional de Misiones, Puerto Iguazú, Argentina

Kun-Fang Cao State Key Laboratory for Conservation and Utilization of Subtropical Agro-Bioresources, and College of Forestry, School of Forestry, Guangxi University, Nanning, China

Ya-Jun Chen Key Laboratory of Tropical Forest Ecology, Xishuangbana Tropical Botanical Garden, Chinese Academy of Sciences, Mengla, Yunnan, China

Piedad M. Cristiano Laboratorio de Ecología Funcional, Departamento de Ecología Genética y Evolución, Instituto IEGEBA (CONICET-UBA), Facultad de Ciencias Exactas y Naturales, Universidad de Buenos Aires, Buenos Aires, Argentina

Cátia Nunes da Cunha Departamento de Botânica e Ecologia, Universidade Federal de Mato Grosso, Cuiabá, Mato Grosso, Brazil

James W. Dalling Department of Plant Biology, University of Illinois at Urbana-Champaign, Urbana, IL, USA; Smithsonian Tropical Research Institute, Panama, Republic of Panama

Mark E. De Guzman Department of Botany & Plant Sciences, University of California, Riverside, CA, USA

Débora di Francescantonio Laboratorio de Ecología Forestal y Ecofisiología, Instituto de Biología Subtropical, CONICET, FCF, Universidad Nacional de Misiones, Puerto Iguazú, Argentina

Exequiel Ezcurra Department of Botany and Plant Sciences, University of California, Riverside, CA, USA

Leandro V. Ferreira Museu Paraense Emílio Goeldi, Belém, Pará, Brazil

Guillermo Goldstein Laboratorio de Ecología Funcional, Departamento de Ecología Genética y Evolución, Instituto IEGEBA (CONICET-UBA), Facultad de Ciencias Exactas y naturales, Universidad de Buenos Aires, Buenos Aires, Argentina; Department of Biology, University of Miami, Coral Gables, FL, USA

Guang-You Hao State Key Laboratory of Forest and Soil Ecology, Institute of Applied Ecology, Chinese Academy of Sciences, Shenyang, China

Katherine Heineman Program in Ecology, Evolution and Conservation Biology, University of Illinois at Urbana-Champaign, Urbana, IL, USA

Kaoru Kitajima Graduate School of Agriculture, Kyoto University, Kyoto, Japan; Smithsonian Tropical Research Institute, Balboa, Panama

Ken W. Krauss U.S. Geological Survey, Wetland and Aquatic Research Center, Lafayette, LA, USA

Eloisa Lasso Smithsonian Tropical Research Institute, Balboa, Ancón, Republic of Panama; Departamento de Ciencias Biológicas, Universidad de Los Andes, Bogotá, Colombia

Omar R. Lopez Smithsonian Tropical Research Institute, Panama, Republic of Panama; Instituto de Investigaciones Científicas y Servicios de Alta Tecnología, Ciudad de Saber, Panama, Republic of Panama

Catherine E. Lovelock School of Biological Sciences, The University of Queensland, Brisbane St. Lucia, QLD, Australia

Eric Manzané Smithsonian Tropical Research Institute, Panama, Republic of Panama

Frederick C. Meinzer USDA Forest Service, Forestry Sciences Laboratory, Corvallis, OR, USA

Karl J. Niklas Plant Biology Section, School of Integrative Plant Science Cornell University, Ithaca, NY, USA

Michael J. Osland U.S. Geological Survey, Wetland and Aquatic Research Center, Lafayette, LA, USA

Adela M. Panizza Laboratorio de Ecología Forestal y Ecofisiología, Instituto de Biología Subtropical, CONICET, FCF, Universidad Nacional de Misiones, Puerto Iguazú, Argentina

Pia Parolin Biocentre Klein Flottbek and Botanical Garden, University of Hamburg, Hamburg, Germany; INRA French National Institute for Agricultural Research, Univ. Nice Sophia Antipolis, CNRS, UMR 1355-7254 Institut Sophia Agrobiotech, Sophia Antipolis, France

Nathan G. Phillips Department of Earth and Environment, Boston University, Boston, MA, USA

Maria Teresa F. Piedade INPA, Manaus, Amazônia, Brazil

Fermín Rada Instituto de Ciencias Ambientales y Ecológicas de los Andes Tropicales (ICAE) Facultad de Ciencias, Universidad de Los Andes, Mérida, Venezuela

Ruth Reef School of Biological Sciences, The University of Queensland, Brisbane St. Lucia, QLD, Australia

Heidi J. Renninger Department of Forestry, Mississippi State University, Mississippi, MS, USA

Sabrina A. Rodriguez Laboratorio de Ecología Forestal y Ecofisiología, Instituto de Biología Subtropical, CONICET, FCF, Universidad Nacional de Misiones, Puerto Iguazú, Argentina

Sabrina E. Russo School of Biological Sciences, University of Nebraska, Lincoln, USA

Louis S. Santiago Department of Botany & Plant Sciences, University of California, Riverside, CA, USA; Smithsonian Tropical Research Institute, Balboa, Ancon, Panama, Republic of Panama

E.J. Sayer Lancaster Environment Centre, Lancaster University, Lancaster, England

Fabian G. Scholz Grupo de Estudios Biofísicos y Ecofisiológicos, Facultad de Ciencias Naturales, Universidad Nacional de la Patagonia San Juan Bosco, Chubut, Argentina; Consejo Nacional de Investigaciones Científicas y Técnicas (CONICET), Buenos Aires, Argentina

Jochen Schöngart Instituto Nacional de Pesquisas da Amazônia (INPA), Manaus, AM, Brazil

Katia Silvera Smithsonian Tropical Research Institute, Balboa, Ancón, Republic of Panama; Department of Botany and Plant Sciences, University of California Riverside, Riverside, CA, USA

Martijn Slot Smithsonian Tropical Research Institute, Balboa, Ancón, Republic of Panama

Zheng-Hong Tan Key Laboratory of Tropical Forest Ecology, Xishuangbanna Tropical Botanical Garden, Chinese Academy of Sciences, Mengla, Yunnan, China

Benjamin L. Turner Smithsonian Tropical Research Institute, Balboa, Panama, Republic of Panama

Mariana Villagra Laboratorio de Ecología Forestal y Ecofisiología, Instituto de Biología Subtropical, CONICET, FCF, Universidad Nacional de Misiones, Puerto Iguazú, Argentina

Klaus Winter Smithsonian Tropical Research Institute, Balboa, Ancón, Republic of Panama

Florian Wittmann Department of Biogeochemistry, Max Planck Institute for Limnology, Mainz, Germany

S. Joseph Wright Smithsonian Tropical Research Institute, Balboa, Panama, Republic of Panama

Yi-Ping Zhang Key Laboratory of Tropical Forest Ecology, Xishuangbanna Tropical Botanical Garden, Chinese Academy of Sciences, Mengla, Yunnan, China

Yong-Fei Zhang Department of Geological Sciences, John A. and Katherine G. Jackson School of Geosciences, University of Texas at Austin, Austin, Texas, USA

Yong-Jiang Zhang Department of Organismic and Evolutionary Biology, Harvard University, Cambridge, MA, USA; Key Laboratory of Tropical Forest Ecology, Xishuangbana Tropical Botanical Garden, Chinese Academy of Sciences, Mengla, Yunnan, China

Pieter A. Zuidema Forest Ecology and Forest Management, Centre for Ecosystems, Wageningen University, Wageningen, The Netherlands

Part I
Physiology and Life History Traits of Unique Tropical Growth Forms

Hemiepiphytic Trees: *Ficus* as a Model System for Understanding Hemiepiphytism

Guang-You Hao, Kun-Fang Cao and Guillermo Goldstein

Abstract Woody hemiepiphytes that have an epiphytic juvenile growth stage differ crucially in physiology and ecology from common trees. A relatively high degree of ontogenetic plasticity confers these plants stress tolerance during the epiphytic stage and sufficient competitiveness later as independent trees. The genus *Ficus* consists of about 500 hemiepiphytic and about 300 non-hemiepiphytic woody species. Ecophysiological comparative studies between hemiepiphytic (Hs) and non-hemiepiphytic (NHs) *Ficus* tree species reveal that the existence of an epiphytic growth habit even only for a part of their life cycle involves profound changes that persist to a large degree in their terrestrial growth stage. When growing under similar conditions, both as saplings and mature trees, the Hs have physiological traits resulting in conservative water use and drought tolerance contrasting with more prodigal water use and drought sensitivity in NHs. Divergence in water related functional traits between the two groups are centrally associated with a trade-off between xylem water flux capacity and drought tolerance. Two distinct groups of life history traits for Hs and NHs have evolved—epiphytic regeneration with a slow starting growth rate but enhanced ability to tolerate water deficits in the upper canopy environment and regeneration in the forest understory with an initial

G.-Y. Hao (✉)
State Key Laboratory of Forest and Soil Ecology, Institute of Applied Ecology, Chinese Academy of Sciences, 110016 Shenyang, China
e-mail: haogy@iae.ac.cn

K.-F. Cao
School of Forestry, Guangxi University, Nanning, Guangxi, China
e-mail: caokf@xtbg.ac.cn

G. Goldstein
Laboratorio de Ecología Funcional, Departamento de Ecología Genética y Evolución, Instituto IEGEBA (CONICET-UBA), Facultad de Ciencias Exactas y naturales, Universidad de Buenos Aires, Buenos Aires, Argentina
e-mail: goldstein@ege.fcen.uba.ar; gold@bio.miami.edu

G. Goldstein
Department of Biology, University of Miami, Coral Gables, FL 33146, USA

© Springer International Publishing Switzerland 2016
G. Goldstein and L.S. Santiago (eds.), *Tropical Tree Physiology*,
Tree Physiology 6, DOI 10.1007/978-3-319-27422-5_1

3

burst of growth to rapidly gain a relatively large seedling size that can better survive risks related to terrestrial regeneration. Evidence shows that the underlying physiology distinguishing these two growth forms mostly involves divergences in adapting to contrasting water regimes but not light conditions, contrary to the conventional hypothesis that hemiepiphytism evolved for gaining access to higher irradiance in the canopy than on the forest floor.

Keywords Drought tolerance · Hydraulic architecture · Plant water relations · Regeneration · Shade tolerance

Introduction

Hemiepiphytes are plants that grow epiphytically for a portion, but not all, of their life cycle. They are customarily subdivided into primary and secondary hemiepiphytes depending on which part of their life cycle has root connections with the ground. Primary hemiepiphytes normally germinate and grow on other plants but later establish substantial and permanent connections with the ground via aerial roots (Kress 1986; Putz and Holbrook 1986). Secondary hemiepiphytes germinate on the ground, climb up their host plants, and then lose stem connections with the soil (Kress 1986; Putz and Holbrook 1986). Primary hemiepiphytes have a true epiphytic stage, during which their ecophysiological traits are very similar to those of the true epiphytes (Zotz and Winter 1994), while secondary hemiepiphytes differ fundamentally from epiphytes, but are rather functionally similar to vines even after the lower part of their stems die back (Holbrook and Putz 1996b; Moffett 2000). For example, many aroid species that are called secondary hemiepiphytes can establish connections with the soil by producing adventitious roots after severing the stem connections with the ground and regaining access to soil resources. The current use of the term hemiepiphyte thus confounds two radically different life cycle characteristics. Zotz (2013a) suggested to entirely discard the term "secondary hemiepiphyte" and instead use Moffett's (2000) term "nomadic vine" for climbing plants that germinate on the ground and lose the lower part of their stem later during ontogeny. By doing this, the term "hemiepiphyte" is reserved exclusively for species that were formerly called primary hemiepiphytes. We adhere to this definition hereafter in this chapter.

Hemiepiphytes are an important plant component of tropical vegetation. According to a recent census by Zotz (2013b), there are 19 families and 28 genera that contain more than 800 hemiepiphytic species. Among these taxa, *Ficus* (Moraceae) and *Clusia* (Clusiaceae) are the two most important genera that are composed of more than 600 woody hemiepiphytic species combined. In the tropics many of these species are well known as strangers that germinate on tops of other trees and have the potential to strangle their hosts and become structurally independent trees (Fig. 1a–d). In tropical forests of Panama and Zimbabwe 9.8 and

Fig. 1 **a** The epiphytic growth phase of *Ficus altissima* growing in the canopy of a host palm tree showing multiple dangling aerial roots; **b** *Ficus altissima* during its terrestrial growth stage showing multiple "pseudostems" formed by fused aerial roots; **c** a *Ficus concinna* tree strangling a tree; **d** a free-standing stage *Ficus curtipes* tree showing tangled architecture of aerial roots defining the space occupied by a host tree that has been strangled and decomposed. All plants are growing in the Xishuangbanna Tropical Botanical Garden (XTBG), Yunnan, China *Photo* credits: GYH

12.6 % of trees, respectively, were found bearing hemiepiphytic *Ficus* (Guy 1977; Todzia 1986); in Venezuela 13 % of trees >10 cm DBH carried *Ficus* or *Clusia* (Putz 1983). Hemiepiphytic *Ficus* species are also very commonly found on architectural structures especially in tropical and subtropical regions, such as ancient temples in India and old stone walls in urban Hong Kong (Sitaramam et al. 2009; Jim 2014).

It is commonly considered that there are several potential advantages to start the life cycle as an epiphyte in tropical forests. One of the most important advantages is that the forest canopy offers higher light availability than the forest understory. It has been suggested that hemiepiphytism evolved in plants that colonized rocky areas as an adaptation to access high light environments in the forest canopy (Dobzhansky and Murea-Pires 1954; Ramirez 1977; Putz and Holbrook 1986; Todzia 1986; Laman 1995; Williams-Linera and Lawton 1995). They may also benefit from minimizing risks of fire, flood, terrestrial herbivores and damage or coverage by falling debris. The advantages of spending the initial part of their life cycle as an epiphyte, however, can be offset by the potential limitations of water and nutrient availability (Benzing 1990; Coxson and Nadkarni 1995; Holbrook and Putz 1996a, b, c; Swagel et al. 1997).

The hemiepiphytic *Ficus* (Moraceae) is the most conspicuous group of species with such life history in terms of habitat breath, species richness, abundance and dominance in forest ecosystems (Dobzhansky and Murca-Pires 1954; Putz and Holbrook 1986; Holbrook and Putz 1996b). Species in the genus *Ficus* are among the most important components of tropical lowland rainforests throughout the world (Harrison 2005) and are ecologically important due to their interactions with many frugivorous animals and other plant species (Shanahan et al. 2001). This genus consists of about 500 hemiepiphytic species, including stranglers and banyans, and about 300 non-hemiepiphytic woody species (Putz and Holbrook 1986; Harrison 2005). In *Ficus*, the hemiepiphytic habit most likely evolved four times in the subgenera Urostigma, Sycidium, Pharmacosycea and in a closely related group comprised of the subgenera Conosycea, Galoglychia, Americana and Malvanthera (Harrison 2005). From the point of view of evolution, it is important to consider the main environmental factors that selected for this specialized growing habit. Comparative studies in ecophysiology between hemiepiphytic and non-hemiepiphytic *Ficus* species provide valuable information about the main differences between these two groups in environmental adaptation of the seedling/sapling and adult stages, which allow us to infer major selective pressures for the evolution of hemiepiphytism.

Because of the radical changes in rooting environment between the two growth phases, developmental and physiological plasticity is important for hemiepiphytes and enables them first to survive the harshness of the epiphytic habitat and then to compete successfully with other trees when they are later rooted in the ground (Holbrook and Putz 1996b). The change from functional epiphyte to tree is accompanied by a shift in rooting volume and characteristics of the rooting zone. Previous studies have found that nutrient availability does not exert a major limitation to the epiphytic phase of hemiepiphytic *Ficus* (Putz and Holbrook 1989), whereas measurement of stomatal conductance and leaf phenology indicate that water availability is frequently a major constraint as compared to terrestrially rooted trees of the same species (Holbrook and Putz 1996a, b, c). Epiphytes face frequent and severe water deficits even in areas with very humid climate (Benzing 1990). In seasonally dry climates, hemiepiphytes in the epiphytic stage can experience even more severe drought during the dry season.

In hemiepiphytes, the two different growth phases with contrasting environmental conditions make it interesting and convenient for physiological comparative studies. Some hemiepiphytes, such as species of the genus *Clusia*, switch to the more water efficient facultative CAM metabolism when stressed by drought, but all of the species of *Ficus* studied thus far exhibit only C_3 photosynthesis regardless of the life stage (Ting et al. 1987). Strangler *Ficus* in the epiphytic stage avoids water deficit mainly through strong stomatal control to maintain relatively high leaf water potentials (Holbrook and Putz 1996c). During both the rainy season and the dry season, stomatal conductance of epiphytic stage strangler figs is lower than conspecific trees. Throughout the dry season, epiphytic stage strangler figs only open their stomata in the early morning (Holbrook and Putz 1996c). Furthermore, epiphytic phase *Ficus* can better control water loss from leaf surfaces than conspecific tree-phase plants after stomata are closed (Holbrook and Putz 1996a). The smaller guard cell surface area due to lower stomata density enables the epiphytic stage *Ficus* to lose water more slowly compared to tree phase plants of the same species. Due to these water conservation traits, leaf water potentials of epiphytic stage *Ficus* plants are found to be similar or even less negative than conspecific tree-phase individuals (Holbrook and Putz 1996b). Even when both epiphytic phase and tree phase *Ficus* are well supplied with water during manipulative experiments, the leaves of epiphytic plants still exhibit significantly lower stomatal conductance and much lower epidermal conductance, indicating strong developmental changes from the epiphytic to the terrestrial phase (Holbrook and Putz 1996a).

Physiology related to epiphytic-terrestrial phase transition in *Ficus* has been well studied and reviewed by Holbrook and Putz (1996b). This chapter focuses on ecophysiological comparisons between hemiepiphytic (Hs) and non-hemiepiphytic (NHs) *Ficus* tree species grown under similar environmental conditions to better understand intrinsic differences between these two functional groups.

Comparison of Hydraulics and Water Balance

Stem Hydraulic Conductivity

Only few studies have compared hydraulic architecture between woody hemiepiphytic and free-standing tree species (Patiño et al. 1995; Zotz et al. 1997; Hao et al. 2011). These studies demonstrate that hemiepiphytic plants have stems that are less conductive as shown by relatively low leaf-specific hydraulic conductivity (K_1). According to the studies by Patiño et al. (1995) and Zotz et al. (1997), the lower K_1

found in hemiepiphytic *Ficus* and *Clusia* shoots are largely due to their low investment in water conducting tissue, implying a lower wood cross-sectional area per unit leaf area (Huber value; Hv), rather than less conductive sapwood tissues. Their stem hydraulic conductivity values expressed per wood cross-sectional area are even larger than in tropical and temperate angiosperm trees (Zotz et al. 1997). The more recent study by Hao et al. (2011) comparing mature trees of Hs and NHs grown in a common garden, however, found no significant difference in leaf to sapwood area ratio between the two functional groups but substantially lower specific hydraulic conductivity (K_s, i.e. an intrinsic measure of water transport efficiency of the xylem) in Hs than in NHs.

The discrepancies between different studies, however, may largely be due to methodological differences in the way K_s and Huber values are estimated. A close examination of the Patiño et al. (1995) and Zotz et al. (1997) studies reveal that in both studies K_s and Hv were calculated on a whole stem cross sectional area basis rather than the sapwood area basis. This may strongly affect the interpretation of the results of such comparative studies because non-hemiepiphytic *Ficus* species usually have large pith in their stem, but the pith is negligible in the stems of hemiepiphytic species. In the stems of non-hemiepiphytic *Ficus* species, the pith can account for up to 70 % (on average 45 %) of the "wood" cross-sectional area (G.-Y. Hao unpublished). By using only the sapwood area, excluding the pith, for K_s calculation, we find that Hs have sapwoods that are far less efficient in conducting water even when both types of species are growing as independent trees in a common garden (Hao et al. 2011). Consistent with the differences in stem hydraulic conductivity, Hs and NHs show significant differences in leaf properties related to water transport, water conservation and drought tolerance (Hao et al. 2010). These findings can be parsimoniously explained by the hypothesis that Hs are adapted to drought conditions associated with an epiphytic growth stage in the canopy, whereas NHs are selected for strong competitive ability given the higher water availability that they experience during their first growth stages.

Water Flux Through the Leaf

Hemiepiphytic *Ficus* species have low leaf water flux capacity that parallels their relatively low stem-level hydraulic conductivity. Compared with congeneric terrestrial species, Hs have significantly narrower vessels in their leaf petioles and lower theoretical leaf area adjusted hydraulic conductance calculated from petiole xylem vessel dimension measurements (Fig. 2a, b; $P < 0.05$, t-tests). The NHs have vessel lumen diameters that are on average 30 % larger than Hs but the number of vessels per petiole standardized by leaf area does not differ between the two growth forms. The resultant theoretical xylem hydraulic conductance of NHs averages 104 % higher than that of Hs (Hao et al. 2010).

The lower leaf water flux capacity in Hs is consistent with their lower transpirational water requirement per unit leaf area. Compared to NHs grown under similar

Fig. 2 **a** Leaf petiole average vessel diameter (D_v) and; **b** theoretical hydraulic conductance (K_t) of five hemiepiphytic and five non-hemiepiphytic *Ficus* species. Mean values ± SE for each species are reported (n = 5–6). Species name abbreviations: *Ficus benjamina, BE; F. concinna, CO; F. curtipes, CU; F. religosa, RE; F. tinctoria, TI; F. auriculata, AU; F. esquiroliana, ES; F. hispida, HI; F. racemosa, RA; F. semicordata, SE* (data from Hao et al. 2010)

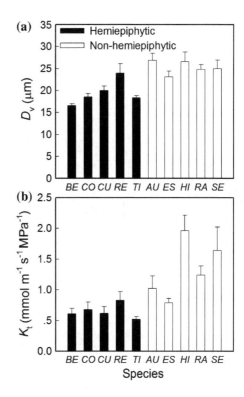

conditions, Hs have significantly lower stomatal conductance and higher intrinsic water use efficiency measured both on mature trees well rooted in the soil and saplings growing in pots supplied with sufficient water (Hao et al. 2010, 2013). Differences in water use between Hs and NHs are most significant when diurnal courses of leaf gas exchange are compared. Epiphytic stage Hs in their native habitats have been found to open stomata only in the early morning during the dry season (Holbrook and Putz 1996c). When saplings are grown in pots and are well watered, Hs still have lower stomatal conductance and a shorter duration of active CO_2 assimilation on sunny days (Fig. 3a, b). In both groups of species, photosynthetic net assimilation rates reach maximum values around 11:00 h, but rates start to decline in Hs there after until the end of the day, whereas in NHs rates remain at high levels until 14:00 h with an afternoon peak following a slight midday depression (Fig. 3b). The different diurnal patterns of photosynthesis between the two growth forms of *Ficus* are likely associated with the intrinsically low xylem hydraulic conductivity in Hs compared to NHs. Higher water use efficiency in Hs suggests a constitutively conservative water use strategy (Table 1), consistent with adaptations to cope with drought-prone canopy habitats.

In juvenile plants of Hs, more conservative water use may contribute to the protection of their xylem vascular systems from catastrophic cavitation when facing unpredictable drought conditions associated with canopy growth in their natural

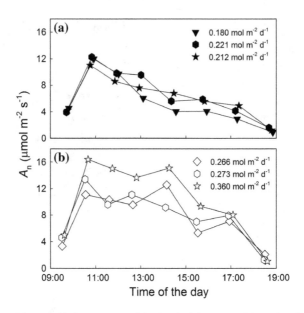

Fig. 3 Leaf net CO_2 assimilation rate (A_n) of (**a**) hemiepiphytic and (**b**) non-hemiepiphytic *Ficus* saplings grown under full sunlight. Numbers in panel **a** and **b** indicate daily cumulative net assimilation. *Ficus concinna* (▼), *F. tinctoria* (●), *F. virens* (★), *F. hispida* (◇), *F. racemosa* (⬡), *F. semicordata* (☆). *Filled* and *open symbols* indicate hemiepiphytic and non-hemiepiphytic species, respectively (Hao et al. 2013)

habitats. Higher leaf xylem hydraulic conductance as found in NHs indicates adaptation of the leaf vascular system to a more prodigal water use, which is only advantageous under conditions of reliable water sources. Considering the frequently occurring drought stress related to a canopy growth during their early ontogeny, high water flux capacity does not appear to be a beneficial trait for Hs.

Traits Conferring Drought Tolerance

Hemiepiphytic *Ficus* species exhibit traits related to greater drought tolerance compared to NHs (Table 1). They have higher leaf mass per area and lower leaf osmotic potential at turgor loss point (π^0), typical adaptations to drought-prone environments (Hao et al. 2010). The Hs almost completely close their stomata at turgor loss point but all of the NHs maintain relatively high stomatal conductance, resulting in a much larger "safety margin" between stomatal closure and turgor loss point in Hs compared to NHs (Fig. 4). Narrow safety margins can benefit some plants by allowing the maintenance of gas exchange and thus optimizing returns on xylem investment (Brodribb and Holbrook 2004). A prerequisite for this set of functional traits is a relatively reliable water supply to the leaves that may not be

Table 1 Ecophysiological traits of seven hemiepiphytic (H) and seven non-hemiepiphytic (NH) *Ficus* species (values are means ± 1 SE). Means of the two growth forms were compared using one-way ANOVAs

Functional traits	Prediction	H mean	NH mean	P-value (ANOVA)
Specific hydraulic conductivity ($kg\ m^{-1}\ s^{-1}\ MPa^{-1}$)	H < NH	2.00 ± 0.22	7.06 ± 1.28	**0.001**
Leaf-specific hydraulic conductivity ($\times\ 10^{-4}\ kg\ m^{-1}\ s^{-1}\ MPa^{-1}$)	H < NH	2.28 ± 0.45	9.55 ± 2.71	**0.007**
Sapwood density ($g\ cm^{-3}$)	H > NH	0.50 ± 0.03	0.45 ± 0.03	0.100
Leaf to sapwood area ratio ($cm^2\ mm^{-2}$)	H < NH	128.7 ± 18.0	115.2 ± 18.5	0.291
Leaf size (cm^2)	H < NH	84.8 ± 26.2	296.7 ± 106.6	**0.030**
Leaf mass per area ($g\ m^{-2}$)	H > NH	107.0 ± 12.5	69.1 ± 6.5	**0.007**
Leaf saturated water content ($g\ g^{-1}$)	H < NH	1.88 ± 0.14	2.30 ± 0.15	**0.025**
Maximum net CO_2 assimilation rate on leaf area basis ($\mu mol\ m^{-2}\ s^{-1}$)	H < NH	13.0 ± 0.8	13.7 ± 0.8	0.274
Maximum net CO_2 assimilation rate on leaf mass basis ($\mu mol\ g^{-1}\ s^{-1}$)	H < NH	0.126 ± 0.016	0.210 ± 0.028	**0.008**
Maximum stomatal conductance ($mol\ m^{-2}\ s^{-1}$)	H < NH	0.301 ± 0.028	0.408 ± 0.021	**0.003**
Intercellular CO_2 concentration ($\mu mol\ mol^{-1}$)	H < NH	290.1 ± 3.5	305.2 ± 2.2	**0.001**
Intrinsic water-use efficiency ($\mu mol\ mol^{-1}$)	H > NH	44.6 ± 2.2	34.3 ± 1.2	**0.001**
Leaf nitrogen content (%)	H > NH	2.11 ± 0.18	1.69 ± 0.12	**0.033**
Photosynthetic nitrogen use efficiency ($\mu mol\ CO_2\ s^{-1}\ mol^{-1}\ N$)	H < NH	83.0 ± 5.9	181.6 ± 31.1	**0.003**

We specified the predictions in comparison between growth forms to allow one-tailed significance testing. *P*-values smaller than 0.05 are shown in bold face (Modified from Hao et al. 2011)

met by Hs with intrinsically low xylem water transport efficiency. Effective stomatal closure in Hs may thus be important in avoiding too large of a water potential gradient across the plant and hence catastrophic hydraulic failure.

The Hs do not only have earlier stomatal closure in response to leaf desiccation but also exhibit more effective water retaining ability after stomata closure (Fig. 5a, b). Average cuticle conductances are 2.1 and 10.7 mmol $m^{-2}\ s^{-1}$ in Hs and NHs ($P < 0.05$, t-test), respectively. Consequently, after excision Hs can keep their relative water content above 70 %, a threshold for physiological damage to occur in many higher plants, for a period on average 10 times that of the NHs. This may confer Hs a greater ability to persist under severe drought. Leaf desiccation

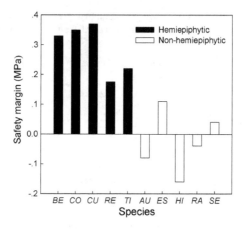

Fig. 4 Safety margin measured as the difference between leaf water potential at 50 % of maximum stomatal conductance (Ψg_{ss50} %) and leaf osmotic potential at turgor loss (π^0) in five hemiepiphytic and five non-hemiepiphytic *Ficus* species. Species name abbreviations are as in Fig. 2

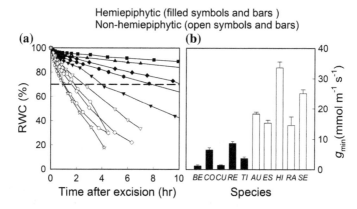

Fig. 5 **a** Relative water content (RWC) changes of water-saturated leaves during bench drying after excision (n = 6). The horizontal dashed line marks 70 % RWC. *Ficus benjamina* (▲), *F. concinna* (▼), *F. curtipes* (■), *F. religosa* (◆), *F. tinctoria* (◗), *F. auriculata* (○), *F. esquiroliana* (▽), *F. hispida* (◇), *F. racemosa* (◖), *F. semicordata* (☆); **b** leaf epidermal conductance. Species name abbreviations are as in Fig. 2 (data from Hao et al. 2010)

avoidance is determined by both the stomatal closure in response to water deficits and water retention after stomata closure, which relates to the resistance to water loss through the epidermal cuticle (Muchow and Sinclair 1989; Holbrook and Putz 1996c). Differences in stomatal control and cuticle conductance may explain the commonly observed leaf wilting and plant dieback in NHs but not in Hs under drought stress both in the field and during drought treatments in pots. The NHs on

Fig. 6 Percentage of leaf loss due to simultaneous drought treatments of the same degrees to seedlings of hemiepiphytic and non-hemiepiphytic *Ficus* species grown under (**a**) full sunlight and (**b**) 5 % sunlight (Hao et al. 2013)

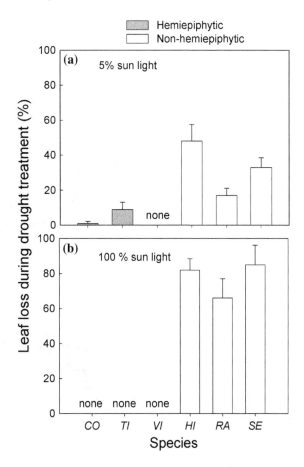

average lost 30 and 70 % of their leaves during a water withholding experiment, whereas Hs did not shed a substantial amount of leaves for a similar degree of drought except in *F. tinctoria* under full daylight (Fig. 6a, b).

When photosynthetic net assimilation rates are plotted against stomatal conductance following a drought treatment in potted plants, contrastingly different patterns are seen between Hs and NHs. The Hs appear to have higher net assimilation rate for a given stomatal conductance (Fig. 7a, b). Furthermore, the Hs show a substantial increase in water use efficiency in response to the drought stress, as shown by the increase in the slopes of the fitted curves, indicating physiological adjustments towards a more efficient water use under drought, whereas this adjustment is lacking in NHs (Fig. 7a, b).

Fig. 7 Leaf net assimilation rate (A_n) vs. stomatal conductance (g_s) following a drought treatment in plants grown under (**a**) full sunlight and (**b**) 5 % sunlight. Symbols for each species are as in Fig. 3

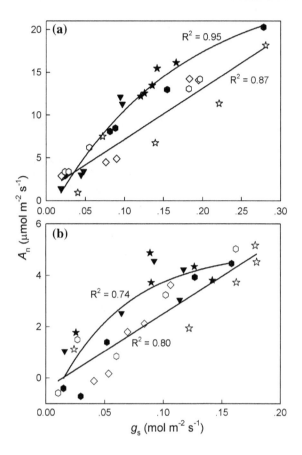

Contrasting Water Use Strategies in Hs and NHs

Although conservative water use in Hs contributes to their ability to survive drought-stressed environments typical of epiphytic habitats during their early ontogeny, it negatively affects their photosynthetic carbon assimilation. While light-saturated net CO_2 assimilation rates per unit leaf area do not differ between the two growth forms, assimilation rates per dry mass are significantly lower in Hs (Table 1). The Hs also have lower photosynthetic nitrogen and phosphorous use efficiencies than NHs (Hao et al. 2011). Differences between Hs and NHs extend the finding that species adapted to different environments tend to differ in a cluster of water flux-related traits, including stem hydraulic conductivity, leaf hydraulic conductance and stomatal conductance (Sack et al. 2003, 2005; Dunbar-Co et al. 2009). These traits probably contribute to NHs having a higher photosynthetic capacity per investment in leaf dry mass and a higher growth capacity.

The contrasts between Hs and NHs in maximum flux-related traits and drought tolerance traits are consistent with the hypothesized trade-off between high hydraulic conductivity and high resistance to hydraulic failure (*e.g.* Martinez-Vilalta et al. 2002; Wheeler et al. 2005). Such a trade-off may affect the distribution of species; species distributed in relatively wet environments are usually more vulnerable to xylem cavitation than species adapted to dry environments (Maherali et al. 2004). The divergences in hydraulics between Hs and NHs are analogous to the differences between xeric and humid-grown species (Bhaskar et al. 2007), indicating that sympatric species within a single genus, can diverge significantly in numerous aspects of leaf structure and function, according to their microhabitat for regeneration. The Hs have leaf traits conferring conservative water use, of particular advantage to plants growing under dry conditions, whereas the NHs exhibit leaf traits conferring prodigal water use that would be advantageous under competitive situations.

It seems that the existence of an epiphytic habit during the juvenile stage in Hs involves a suite of leaf water flux and drought tolerance traits that are different from those of congeneric NHs. In Hs, an epiphytic growth stage that is frequently exposed to water deficits appears to have selected for xylem composed of smaller vessels with lower hydraulic transport efficiency, which carries on to the terrestrial growth stage. These divergences between Hs and NHs persist to a large degree in mature trees that have similar root water access.

Comparison of Adaptations to Irradiance Level

Light is a critical factor affecting the regeneration of forest plants especially in the tropics (Chazdon 1988; Clark et al. 1996). Many tropical forest species have evolved epiphytic growth habits that are thought to allow escape from the very low irradiance level in the forest understory. Hemiepiphytism is commonly hypothesized as an adaptation to exploit the higher irradiance in the forest canopy (Dobzhansky and Murea-Pires 1954; Ramirez 1977; Putz and Holbrook 1986; Laman 1995; Williams-Linera and Lawton 1995). One way to test this hypothesis is by comparing closely related hemiepiphytic and non-hemiepiphytic species in their functional traits related to shade tolerance and their ecophysiological responses to growth irradiance levels. Seeds of both Hs and NHs species were sown at the same time and later grown under four irradiance levels (100, 50, 25 and 5 % sunlight). Different from the expectation of a conventional "light" hypothesis for hemiepiphytism evolution, our results show that differences in light requirement or shade tolerance may not be the major divergence between Hs and NHs in determining their environmental adaptation in tropical rainforests.

Plastic Response to Growth Irradiance Levels

Seedlings of Hs and NHs species do not show contrasting photosynthetic responses to irradiance as found in light-demanding and shade-tolerant species. Under full sun and sufficient water supply, both Hs and NHs have photosynthetic rates comparable to seedlings of typical light-demanding pioneer tree species (*e.g.* Oberbauer and Strain 1984; Turnbull 1991; Davies 1998). The Hs species even show lower average photosynthetic rates than NHs, which is opposite from what is expected if Hs were better adapted to higher irradiance for regeneration. It has also been shown that early-successional or pioneer species that are light demanding during regeneration have high plasticity in physiological traits in response to growth irradiance (Bazzaz and Pickett 1980). However, such a contrast in plasticity to light is lacking in the two groups of *Ficus*. The high plasticity to changes in both Hs and NHs traits under different growth irradiance may thus indicate that both groups of *Ficus* species are adapted to relatively high irradiance for regeneration.

Species that are more shade tolerant are expected to allocate more biomass to leaves to gain maximum light and CO_2 capture with a larger leaf area (Poorter 1999, 2001; Markesteijn et al. 2007). Moreover, plants in the shade tend to have a taller stem per unit stem biomass to escape from low irradiance in the understory (Sasaki and Mori 1981). Such a plastic response is substantial in light demanding species regenerating in gaps (Bazzaz 1979; Bazzaz and Pickett 1980; Bazzaz and Wayne 1994). However, neither obvious differences in allometry nor differences in stem height growth plasticity in response to growth irradiance have been found between Hs and NHs (Hao et al. 2013). Both types of *Ficus* seedlings show a high degree of plasticity in stem growth with height to mass ratio becoming substantially higher under low irradiance, indicating acclimation for enhancing light capture. Although Hs have substantially higher height to mass ratio than NHs under all light treatments, this difference may not suggest adaptation to higher irradiance in Hs, but rather a lower requirement for physical support of stems compared to NHs. The Hs use host trees for support especially at the early stages of the life cycle and thus thick and strong stems or branches are not required from a biomechanical point of view.

Seedling Growth Rate

The *Ficus* species all have tiny seeds and initial small seedlings and thus a rapid initial growth would be critical to them if regenerating terrestrially since tiny seedlings are highly susceptible to risks when growing on the forest floor, such as coverage by falling debris and shading by other fast growing plants. Larger seedlings in forest gaps are more likely to survive and eventually reach the canopy (Brown and Whitmore 1992; Boot 1996; Zagt and Werger 1998). By growing in the canopy, however, Hs seedlings can largely avoid these terrestrial risks and thus surviving environmental stresses experienced at the epiphytic stage (*e.g.* low water availability) would be more important than fast growth and greater competitiveness. Despite the huge amount of seed production and numerous dispersers, hemiepiphytic fig trees do not appear to saturate potential epiphytic micro-sites (Laman 1996a). *Ficus* population densities may be limited by both biotic and abiotic factors reducing early recruitment success (Laman 1995). In a stand of tropical dry forest of Australia, about 95 % of all juvenile individuals of *F. macrophylla* and *F. superba* do not proceed to the strangling phase of the life cycle (Doyle 2000). The struggle to succeed under strong interspecific competition in NHs seedlings may have shifted to a struggle to survive the harsh abiotic stresses in Hs, with selection against a fast growth rate associated with prodigal resource use.

Seedlings of Hs show consistently lower growth rates than those of NHs germinated at the same time under all four irradiance levels, with NHs seedlings having 3.3 and 13.3-fold greater height and biomass, respectively, at the end of a five-month growth period (Fig. 8; Hao et al. 2013). The substantially higher growth rates and larger seedling sizes in NHs during the first few months may be critical for their success, considering that small seedlings at initial stages of the life cycle are

Fig. 8 Seedlings of hemiepiphytic (upper panels) and non-hemiepiphytic (lower panels) *Ficus* species growing in the nursery in XTBG. Seeds of all species were germinated at the same time in petri dishes in an illuminated growth chamber and then transplanted at the same time to pots

expected to be most vulnerable to the risks of terrestrial regeneration. The difference in seedling biomass between the two growth forms, however, decreases as the plants grow older. This is likely related to longer leaf lifespans in Hs seedlings, which allow them to gain an increasingly higher leaf area ratio and leaf mass fraction over time relative to NHs. Despite this, 1-year old Hs saplings still have considerably smaller plant biomass than those of NHs (Hao et al. 2013).

It seems that contrasting "strategies" have evolved in two growth forms of *Ficus* that are crucial for avoiding high risks during regeneration due to their tiny seed sizes: (1) slow growing epiphytic form that escapes the terrestrial related risks at the cost of low water availability, and (2) fast growing terrestrial form that gains a relatively large size in a short period to avoid those risks. Correspondingly, conservative and prodigal water uses appear to have been selected in these two forms of *Ficus* species, which are natively associated with contrasting photosynthetic characteristics and responses to drought. Strong drought tolerance in Hs enables them to regenerate in the drought-prone canopy of dense forests independent of gap formation despite their many traits of pioneer species, such as tiny seed sizes. But such an adaptation is at the cost of a strong competitiveness required for many terrestrially regenerating species.

This finding suggests that light may have played a less important role in driving the adaptive divergence between the two groups of *Ficus* than expected by the conventional "light" hypothesis, whereas naturally contrasting water availability in the juvenile stage likely exerted the strongest selective pressure in shaping distinct physiology of hemiepiphytic and non-hemiepiphytic *Ficus*. The NHs usually regenerate in forest gaps or edges and are most commonly found in forests of early successional stages. In contrasts, the Hs can occupy different microhabitats along the vertical environmental gradient in forest canopies of later succession, depending on the species' ability to reach and survive the potential regeneration sites on their host trees (Laman 1996b). Such divergences may have set one of the most important foundations for different adaptations between these closely related species, which contributes to the high diversity of this tropical genus (Table 2).

Hemiepiphytism and *Ficus* Regeneration in Tropical Forests

The *Ficus* genus has evolved to encompass a variety of species with different life histories including hemiepiphytes and non-hemiepiphytic tree species, but this genus as a group possesses typical traits of pioneer species—very small seeds, high fecundity, high photosynthetic rates and growth rates (Harrison 2005). For example, mature trees of neotropical *Ficus*, including species of both functional groups, appear to be light demanding and have high photosynthetic rates compared to other tree species. The photosynthetic rate of a freestanding species, *Ficus insipida* Willd.

Table 2 Presence (+) or absence of epiphytic phase and aerial roots, growth form and leaf texture of *Ficus* species that naturally occur in Xishuangbanna, China

Species name	Epiphytic phase	Adult growth form	Aerial roots	Leaf texture
Ficus abelii		shrub		papery
F. altissima	+	large tree	+	leathery
F. annulata	+	large tree	+	thinly leathery
F. asperiuscula		shrub		thinly leathery
F. auriculata		small tree	+	thickly papery
F. benjamina	+	tree	+	thinly leathery
F. benjamina var. *nuda*	+	tree	+	thinly leathery
F. callosa	+	large tree		thickly leathery
F. caulocarpa		tree		thinly leathery
F. chapaensis		small tree		papery
F. chrysocarpa		small tree		papery
F. concinna	+	large tree	+	leathery
F. concinna var. *subsessilis*	+	large tree	+	leathery
F. curtipes	+	tree	+	thickly leathery
F. cyrtophylla		small tree		papery
F. drupacea		small tree	+	leathery
F. esquiroliana		tree		papery
F. fistulosa		small tree	+	papery
F. gasparriniana var. *lacerate-folia*		shrub		thinly leathery
F. gasparrinana var. *viridescens*		shrub		thinly leathery
F. glaberrima		tree	+	thinly leathery
F. glaberrima var. *pubescens*		tree	+	thinly leathery
F. hederacea	+	liana	+	thickly leathery
F. heterophylla		shrub		papery
F. hirta	+	shrub		papery
F. hirta var. *brevipila*		shrub		papery
F. hirta var. *imberbis*		shrub		papery
F. hirta var. *roxburghii*		shrub		papery
F. hispida		tree		papery
F. hookeri		large tree		leathery
F. ischnopoda		shrub		papery
F. kurzii		large tree		thinly leathery
F. laevis	+	liana		papery
F. langkokensis	+	small tree		papery
F. maclellandii	+	tree		leathery

(continued)

Table 2 (continued)

Species name	Epiphytic phase	Adult growth form	Aerial roots	Leaf texture
F. macleliandi var. *rhododendrifolia*	+	tree		leathery
F. microcarpa	+	tree	+	thinly leathery
F. neriifolia		small tree		papery
F. nervosa		tree		leathery
F. oligodon		tree		papery
F. orthoneura		tree		leathery
F. pisocarpa	+	tree		leathery
F. pubigera		liana		leathery
F. pubigera var. *anserine*		small tree		leathery
F. pubigera var. *maliformis*		small tree		leathery
F. pubigera var. *reticulate*		small tree		leathery
F. pyrformis		shrub		leathery
F. racemosa		large tree	+	thinly leathery
F. racemosa var. *miquelli*		large tree	+	thinly leathery
F. religiosa	+	large tree	+	leathery
F. sagittata	+	tree	+	leathery
F. sarmentosa var. *lacrymens*		liana	+	leathery
F. semicordata		small tree		papery
F. squamosa		shrub		papery
F. stenophylla		shrub		papery
F. stricta	+	tree	+	thinly leathery
F. subincisa		shrub		papery
F.subinicisa var. *paucidentata*		shrub		papery
F. subulata	+	small tree	+	papery
F. superba var. *japonica*		small tree	+	papery
F. tinctoria ssp. *Gibbosa*	+	tree	+	leathery
F. tinctoria ssp. *Parasatica*	+	tree	+	thinly leathery
F. variegata var. *chlorocarpa*		tree		papery
F. variolosa		small tree		leathery
F. vasculosa		small tree	+	leathery
F. virens	+	large tree	+	leathery
F. virens var. *sublanceolata*	+	large tree	+	leathery

In this region, there are 46 *Ficus* species that naturally occur in the forests as well as two subspecies and 19 varieties. Among the 67 *Ficus* taxa, the 48 tree species, including 23 hemiepiphytic and 25 non-hemiepiphytic tree species are most important to the local ecosystem in terms of abundance and dominance (Adapted from Zhu et al. 1996)

(33.1 μmol m^{-2} s^{-1}) was found to be among the highest of any C$_3$ trees measured under natural conditions (Zotz et al. 1995).

The similarities in responses to growth irradiance between Hs and NHs seedlings found by growth experiments do not support the conventional hypothesis that the hemiepiphytic habit is an adaptation to exploit higher light environment of the canopy (Hao et al. 2013). In tropical forests, besides the vertical irradiance change in light intensity there are also considerable horizontal variations with much higher irradiance in gaps compared to forest understory (*e.g.* Nicotra et al. 1999). In relatively large gaps, pioneer or light-demanding tree species with higher relative growth rates and ecophysiological plasticity have competitive advantages over shade tolerant tree species (Denslow 1980). The seedlings of most NHs are not found in deep shade in the understory but rather in forest gaps or clearings with relatively strong irradiance; Hs seedlings and saplings on host trees are usually partially shaded by the host canopy before extending to high irradiance. Thus, in their natural habitats, the light conditions for seedlings of *Ficus* species of the two growth forms may not be very different, consistent with their lack of contrasting adaptations in physiology and morphology with respect to light requirements.

The epiphytic growth habit of Hs during early life stages may mainly be an adaptation to avoid risks related to the initial forest floor growth, such as strong interspecific competition, flooding, terrestrial herbivores and damage or coverage by falling debris. These risks are common for terrestrial regenerating plants in tropical forests (Holbrook and Putz 1996a), but may be especially high for *Ficus* considering their tiny seeds and small size of young seedlings. To minimize these risks, two contrasting types of life history traits have been selected for in *Ficus*— terrestrial regeneration in gaps with an initial burst of growth to rapidly gain relatively large seedling sizes (NHs) and epiphytic regeneration with slow growth but enhanced ability for tolerating drought-prone canopy environments (Hs).

By having epiphytic growth during their early life stage, Hs seedlings can establish in the canopy of dense tropical rainforests (Harrison et al. 2003). Thus the regeneration of Hs does not heavily depend on forest gap formation as most late successional species do. The high diversity and abundance of hemiepiphytic *Ficus* species may thus be attributable, at least in part, to their canopy regeneration habit, which permits these species to inhabit late succesional forests despite their many traits of typical pioneer species (*e.g.* small seed sizes). In contrast, the regeneration of non-hemiepiphytic species in dense rainforests depends on the formation of forest gaps; for these species a high xylem water transport efficiency would enable relatively high rates of carbon assimilation and growth rates and consequently greater competitiveness in resource acquisition, given a reliable water supply and favorable irradiance levels (Brodribb et al. 2002, 2005; Santiago et al. 2004a, b; Campanello et al. 2008; Zhang and Cao 2009). In this sense, the divergence in growth form and hence ecophysiological traits between hemiepiphytic and non-hemiepiphytic *Ficus* species may have contributed to species co-existence and the diversification of this genus in tropical rainforests.

Acknowledgements GYH thanks colleagues past and present for their support and encouragement. This work was made possible by a joint fellowship granted to GYH by XTBG and University of Miami supporting oversea graduate studies. This work is partially supported by a 100-talents research grant from the Chinese Academy of Sciences awarded to GYH.

References

Bazzaz FA (1979) The physiological ecology of plant succession. Annu Rev Ecol Syst 10:351–371

Bazzaz FA, Pickett STA (1980) The physiological ecology of tropical succession: a comparative review. Annu Rev Ecol Syst 11:287–310

Bazzaz FA, Wayne PM (1994) Coping with environmental heterogeneity: the physiological ecology of tree seedling regeneration across the gap-understory continuum. In: Caldwel MM, Pearcy RW (eds) Exploitation of environmental heterogeneity by plants. Academic Press, San Diego, California, USA, pp 349–390

Benzing DH (1990) Vascular Epiphytes. Cambridge Univ. Press, Cambridge, UK

Bhaskar R, Valiente-Banuet A, Arckerly DD (2007) Evolution of hydraulic traits in closely related species paris from mediterranean and nonmediterranean environments of North America. New Phytol 176:718–726

Boot RGA (1996) The significance of seedling size and growth rate of tropical rain forest tree seedlings for regeneration in canopy openings. In: Swaine MD (ed) The Ecology of Tropical Forest Tree Seedlings. MAB UNESCO Series, Parthenon, Paris, 17: 267–284

Brodribb TJ, Holbrook NM (2004) Stomatal protection against hydraulic failure: a comparison of coexisting ferns and angiosperms. New Phytol 162:663–670

Brodribb TJ, Holbrook NM, Gutiérrez MV (2002) Hydraulic and photosynthetic co-ordination in seasonally dry tropical forest trees. Plant, Cell Environ 25:1435–1444

Brodribb TJ, Holbrook NM, Zwieniecki MA, Palma B (2005) Leaf hydraulic capacity in ferns, conifers and angiosperms: impacts on photosynthetic maxima. New Phytol 165:839–846

Brown ND, Whitmore TC (1992) Do Dipterocarp seedlings really partition tropical rain forest gaps? Philos Trans R Soc Lond B335:369–378

Campanello PI, Gatti MG, Goldstein G (2008) Coordination between water-transport efficiency and photosynthetic capacity in canopy tree species at different growth irradiances. Tree Physiol 28:85–94

Chazdon RL (1988) Sunflecks in the forest understory. Adv Ecol Res 18:1–63

Clark DB, Clark DA, Rich PM, Weiss S, Oberbauer SF (1996) Landscape-scale evaluation of understory light and canopy structure: methods and application in a neotropical lowland rain forest. Can J For Res 26:747–757

Coxson DF, Nadkarni NM (1995) Ecological roles of epiphytes in nutrient cycles of forest systems. In: Lowman MD, Nadkarni NM (eds) *Forest Canopies*. Academic Press, San Diego, pp 495–543

Davies SJ (1998) Photosynthesis of nine pioneer *Macaranga* species from Borneo in relation to life history. Ecology 79:2292–2308

Denslow JS (1980) Gap partitioning among tropical rain forest trees. Biotropica 12:47–55

Dobzhansky T, Murca-Pires BJ (1954) Strangler trees. Sci Am 190:78–80

Doyle G (2000) Strangler figs in a stand of dry rainforest in the lower Hunter Valley, NSW. Aust Geogr 31:251–264

Dunbar-Co S, Sporck MJ, Sack L (2009) Leaf trait diversification and design in seven rare taxa of the Hawaiian *Plantago* radiation. Int J Plant Sci 170:61–75

Guy PR (1977) Notes on the host species of epiphytic figs (*Ficus* spp.) on the flood-plain of the Mana Pools Game Reserve. Rhodesia Kirkia 10:559–562

Hao G-Y, Sack L, Wang A-Y, Cao K-F, Goldstein G (2010) Differentiation of leaf water flux and drought tolerance traits in hemiepiphytic and non-hemiepiphytic *Ficus* tree species. Funct Ecol 24:731–740

Hao G-Y, Goldstein G, Sack L, Holbrook NM, Liu Z-H, Wang A-Y, Harrison RD, Su Z-H, Cao K-F (2011) Ecology of hemiepiphytism in fig species is based on evolutionary correlation of hydraulics and carbon economy. Ecology 92:2117–2130

Hao G-Y, Wang A-Y, Sack L, Goldstein G, Cao K-F (2013) Is hemiepiphytism an adaptation to high irradiance? Testing seedling responses to light levels and drought in hemiepiphytic and non-hemiepiphytic *Ficus*. Physiol Plant 148:74–86

Harrison RD (2005) Figs and the diversity of tropical rainforests. Bioscience 55:1053–1064

Harrison RD, Hamid AA, Kenta T, Lafrankie J, Lee H-S, Nagamasu H, Nakashizuka T, Palmiotto P (2003) The diversity of hemi-epiphytic figs (*Ficus*, Moraceae) in a Bornean lowland rain forest. Biol J Linn Soc 78:439–455

Holbrook NM, Putz FE (1996a) From epiphyte to tree: differences in leaf structure and leaf water relations associated with the transition in growth form in eight species of hemiepiphytes. Plant, Cell Environ 19:631–642

Holbrook NM, Putz FE (1996b) Physiology of tropical vines and hemiepiphytes: plants that climb up and plants that climb down. In: Mulkey SS, Chazdon RL, Smith AP (eds) *Tropical forest plant ecophysiology*. Chapman & Hall, New York, USA, pp 363–393

Holbrook NM, Putz FE (1996c) Water relations of epiphytic and terrestrially-rooted strangler figs in a Venezuelan palm savanna. Oecologia 106:424–431

Jim CY (2014) Ecology and conservation of strangler figs in urban wall habitats. Urban Ecosyst 17:405–426

Kress WJ (1986) The systematic distribution of vascular epiphytes: an update. Selbyana 9:2–22

Laman TG (1995) *Ficus stupenda* germination and seedling establishment in a Bornean rain forest canopy. Ecology 76:2617–2626

Laman TG (1996a) *Ficus* seed shadows in a Bornean rain forest. Oecologia 107:347–355

Laman TG (1996b) Specialization for canopy position by hemiepiphytic *Ficus* species in a Bornean rain forest. J Trop Ecol 12:789–803

Maherali H, Pockman WT, Jackson RB (2004) Adaptive variation in the vulnerability of woody plants to xylem cavitation. Ecology 85:2184–2199

Markesteijn L, Poorter L, Bongers F (2007) Light-dependent leaf trait variation in 43 tropical dry forest tree species. Am J Bot 94:515–525

Martinez-Vilalta J, Prat E, Oliveras I, Josep P (2002) Xylem hydraulic properties of roots and stems of nine Mediterranean woody species. Oecologia 133:19–29

Moffett MW (2000) What's "up"? a critical look at the basic terms of canopy biology. Biotropica 32:569–596

Muchow RC, Sinclair TR (1989) Epidermal conductance, stomatal density and stomatal size among genotypes of *Sorghum bicolour* (L.) Moench. Plant, Cell Environ 12:425–432

Nicotra AB, Chazdon RL, Iriarte VB (1999) Spatial heterogeneity of light and woody seedling regeneration in tropical wet forests. Ecology 80:1908–1926

Oberbauer SF, Strain BR (1984) Photosynthesis and successional status of Costa Rican rain forest trees. Photosynth Res 5:227–232

Patiño S, Tyree MT, Herre EA (1995) Comparison of hydraulic architecture of woody plants of differing phylogeny and growth form with special reference to free-standing and hemi-epiphytic *Ficus* species from Panama. New Phytol 129:125–134

Poorter L (1999) Growth responses of 15 rain-forest tree species to a light gradient: the relative importance of morphological and physiological traits. Funct Ecol 13:396–410

Poorter L (2001) Light-dependent changes in biomass allocation and their importance for growth of rain forest tree species. Funct Ecol 15:113–123

Putz FE (1983) Liana biomass and leaf area of a "Tierra Firme" forest in the Rio Negro basin, Venezuela. Biotropica 15:185–189

Putz FE, Holbrook NM (1986) Notes on the natural history of hemiepiphytes. Selbyana 9:61–69

Putz FE, Holbrook NM (1989) Strangler fig rooting habits and nutrient relations in the llanos of Venezuela. Am J Bot 76:781–788

Ramirez BW (1977) Evolution of the strangling habit in *Ficus* L. subgenus Urostigma (Moraceae). Brenesia 12(13):11–19

Sack L, Cowan PD, Jaikumar N, Holbrook NM (2003) The 'hydrology' of leaves: co-ordination of structure and function in temperate woody species. Plant, Cell Environ 26:1343–1356

Sack L, Tyree MT, Holbrook NM (2005) Leaf hydraulic architecture correlates with regeneration irradiance in tropical rainforest trees. New Phytol 167:403–413

Santiago LS, Goldstein G, Meinzer FC, Fisher JB, Machado K, Woodruff D, Jones T (2004a) Leaf photosynthetic traits scale with hydraulic conductivity and wood density in Panamanian forest canopy trees. Oecologia 140:543–550

Santiago LS, Kitajima K, Wright SJ, Mulkey SS (2004b) Coordinated changes in photosynthesis, water relations and leaf nutritional traits of canopy trees along a precipitation gradient in lowland tropical forest. Oecologia 139:495–502

Sasaki S, Mori T (1981) Responses of Dipterocarp seedlings to light. Malayan Forester 44:319–345

Shanahan M, So S, Compton SG, Corlett R (2001) Fig-eating by vertebrate frugivores: a global review. Biol Rev 76:529–572

Sitaramam V, Jog SR, Tetali P (2009) Ecology of *Ficus religiosa* accounts for its association with religion. Curr Sci 97:636–639

Swagel EN, Bernhard AVH, Ellmore GS (1997) Substrate water potential constraints on germination of the strangler fig, *Ficus aurea* (Moraceae). Am J Bot 84:716–722

Ting IP, Hann J, Holbrook NM, Putz FE, Sternberg L da SL, Price D, Goldstein G (1987) Photosynthesis in hemiepiphytic species of *Clusia* and *Ficus*. *Oecologia* 74:339–346

Todzia C (1986) Growth habits, host tree species, and density of hemiepiphytes on Barro Colorado Island, Panama. Biotropica 18:22–27

Turnbull MH (1991) The effect of light quantity and quality during development on the photosynthetic characteristics of six Australian rainforest tree species. Oecologia 87:110–117

Wheeler JK, Sperry JS, Hacke UG, Hoang N (2005) Inter-vessel pitting and cavitation in woody Rosaceae and other vesselled plants: a basis for a safety versus efficiency trade-off in xylem transport. Plant, Cell Environ 28:800–812

Williams-Linera G, Lawton R (1995) The ecology of hemiepiphytes in forest canopies. In: Lowman MD, Nadkarni NM (eds) forest canopies. Academic Press, San Diego, pp 255–282

Zagt RJ, Werger MJA (1998) Community structure and demography of primary species in tropical rain forest. In: Newbery DM, Brown N, Prins HT (eds) Population and Community Dynamics in the Tropics. Blackwell Scientific Publishers, Oxhord, pp 193–220

Zhang J-L, Cao K-F (2009) Stem hydraulics mediates leaf water status, carbon gain, nutrient use efficiencies and plant growth rates across dipterocarp species. Funct Ecol 23:658–667

Zhu H, Wang H, Xu Z-F, Li B-G (1996) Figs in tropical rainforest of Xishuangbanna and their bio-ecological characteristics. 37:7–14 (in Chinese)

Zotz G (2013a) 'Hemiepiphyte': a confusing term and its history. Ann Bot 111:1015–1020

Zotz G (2013b) The systematic distribution of vascular epiphytes—a critical update. Bot J Linn Soc 171:453–481

Zotz G, Winter K (1994) A one-year study on carbon, water and nutrient relationships in a tropical C_3-CAM hemi-epiphyte, *Clusia uvitana* Pittier. New Phytol 127:45–60

Zotz G, Harris G, Koeniger M, Winter K (1995) High rates of photosynthesis in the tropical pioneer tree, *Ficus insipida* Will. Flora (Jena) 190:265–272

Zotz G, Patiño S, Tyree TT (1997) Water relations and hydraulic architecture of woody hemiepiphytes. J Exp Bot 48:1825–1833

Ecophysiology and Crassulacean Acid Metabolism of Tropical Epiphytes

Katia Silvera and Eloisa Lasso

Abstract Epiphytes are plants that germinate and grow upon other plants without contact with mineral soil and without parasitizing their host plant. Therefore, they derive nutrients and water from the environment. Epiphytes are primarily tropical in distribution and may be the most species-rich life form in very wet rainforest sites, constituting about 10 % of all vascular plants. Nearly 80 % of all vascular epiphytes belong to one of three families: Orchidaceae (orchids), Bromeliaceae (bromeliads), and Polypodiaceae (ferns). Orchids in particular, are the most species rich in epiphytes. In this review, information on the ecophysiology of vascular epiphytes is presented, in an attempt to find patterns that explain the ecophysiological adaptations of canopy living. We highlight the ecophysiology of orchids and bromeliads, and whenever possible, provide insight into other epiphytic families. We discuss morphological, anatomical and physiological novelties that epiphytes have evolved to face the challenges of living in the canopy, including adaptations to increase water capture, to facilitate water storage or to reduce water loss. Because epiphytes are particularly susceptible to climate change, and can be monitored as a component of forest health, we also consider their distribution and physiological responses to climate change as a key aspect of conservation programs.

Keywords Epiphytes · Crassulacean acid metabolism (CAM) · Bromeliaceae · Orchidaceae · Conservation

K. Silvera (✉) · E. Lasso
Smithsonian Tropical Research Institute, P.O. Box 0843-03092, Balboa, Ancón, Republic of Panama
e-mail: katia.silvera@ucr.edu

E. Lasso
e-mail: e.lasso@uniandes.edu.co

K. Silvera
Department of Botany and Plant Sciences, University of California Riverside, Riverside, CA 92521, USA

E. Lasso
Departamento de Ciencias Biológicas, Universidad de Los Andes, Bogotá, Colombia

© Springer International Publishing Switzerland 2016
G. Goldstein and L.S. Santiago (eds.), *Tropical Tree Physiology*,
Tree Physiology 6, DOI 10.1007/978-3-319-27422-5_2

25

Introduction

Epiphytes are plants that germinate and grow all or most of their lives attached to other plants without contact with mineral soil (Benzing 1990), and can grow attached to trunks or branches of host tress and shrubs from the understory floor all the way to tree crowns. Unlike mistletoes, epiphytes do not parasite their hosts, and instead take nutrients and water from the environment (Benzing 1990). They occur on the branches of trees in many biomes, but vascular epiphytes are particularly abundant in tropical forests and can grow so profusely that they substantially increase the mass that tropical tree branches must support (Fig. 1a; Niklas, this volume); epiphytic biomass can be as much as 50 % of the tree leaf biomass (Lüttge 1989), especially in primary montane forests (Nadkarni et al. 2004). In tropical forests, epiphytes form conspicuous masses that are known to support large amounts of animal life including ant nests, arthropods and amphibians (Stuntz et al. 2002), in addition to contributing to the hydrology and nutrient cycling of the ecosystem.

The diversity of epiphytes in tropical forest canopies is also astonishing, constituting about 10 % of all vascular plants. Epiphytes are a conspicuous element of humid tropical forests and may be the most species-rich life form in very wet rainforest sites (Gentry and Dodson 1987). Within this diversity, there are similarities between different species of epiphytes and certain characteristics that are particular to the epiphytic syndrome allowing these species to live in the canopy. With few exceptions, epiphyte species are generally long-lived perennials, have small body size, tend to have small seeds that can be dispersed by wind, and have relatively slow growth. Wind-dispersed seeds allow epiphytes to colonize new trees rapidly. To face the challenges of access to water supply and nutrients, many epiphytes have evolved several morphological adaptions such as leaf succulence, water absorbing trichomes and presence of water absorbing spongy tissue (Benzing 1990; Hietz et al. 1999). Thus the particular growth form and habitat of epiphytes leads to certain physiological outcomes, especially in vascular epiphytes, that appear to facilitate their proclivity for growth in tropical forest canopies.

In this review, information on the ecophysiology of epiphytes is presented, with emphasis on vascular epiphyte literature, in an attempt to find patterns that explain the ecophysiological adaptions of living in the canopy. We emphasize vascular epiphytes because they are the most conspicuous and unique forms of epiphytes in tropical forest, and are particularly abundant in the tropics compared to temperate and boreal forests where non-vascular epiphytes prevail. We also constrain our discussion to epiphytes that germinate in the canopy (Fig. 1a, sometimes called primary epiphytes), and do not consider "nomadic vines", which refer to climbing plants that germinate on the ground and later become epiphytes by losing contact with the ground while climbing host plants (Hao, this volume; Zotz 2013a). Nearly 80 % of all vascular epiphytes belong to one of three families: Orchidaceae (orchids),

Fig. 1 Epiphytic communities from tropical rainforests. **a** Epiphytic community of bromeliads and orchids in a tropical montane cloud forest of Panama. Photo by Gaspar Silvera. **b** *Anthurium* and *Phylodendron* epiphytes in a seasonal tropical forest of Panama. Photo by Gaspar Silvera. **c** Orchid epiphyte in a seasonal lowland forest of Panama. Photo by Katia Silvera

Bromeliaceae (bromeliads), and Polypodiaceae (ferns), and about 85 % of these are orchids and 8 % are bromeliads (Zotz 2013b). Thus, the ecophysiology of orchids and bromeliads will be the main focus of this review, and whenever possible, we will provide insight into other epiphytic families. We will discuss morphological, anatomical and physiological novelties that epiphytes have evolved to face the challenges of living in the canopy. Because epiphytes are particularly susceptible to climate change, and can be monitored as a component of forest health, we also consider their distribution and physiological responses to climate change as a key aspect of conservation programs.

Taxonomic and Ecological Distribution of Epiphytes

Vascular epiphytes are present in 27,614 species, 913 genera and 73 families (Zotz 2013b), and are found worldwide in almost all kinds of ecosystems, especially in tropical and subtropical regions. The real diversity, however, is in tropical rain-forests and cloud forests where they can represent more than 50 % of the total vascular flora of plant species in a given area (Benzing 1990; Gentry and Dodson 1987). Neotropical montane forests, in particular, are exceptionally rich in epi-phytes (Benzing 1989, 1990).

Microclimatic gradients within the canopy of individual host trees provide a variety of niches that can be exploited by different species (Hietz and Briones 1998). Distribution of epiphytes varies greatly within different forests, and has distinct patterns both across environmental gradients, as well as vertically within host tress. To understand the distribution of epiphytes, Burns and Zotz (2010) proposed examining epiphyte assemblages that form different meta-communities as species interaction networks between epiphytes and their host tree. In "classical" meta-communities, epiphytes are restricted to the canopy and are referred to as obligate epiphytes, as in the case of orchids and bromeliads of tropical regions. These meta-communities of epiphytes from different host trees are then linked together by seed dispersal. This differs from temperate zones, where epiphytes are composed of species that can live on the forest floor and occasionally grow as epiphytes.

Distribution of epiphytes also varies horizontally, and this variation occurs between different host trees and among different forest types. Studies that address species composition along altitudinal gradients have revealed that species richness patterns are influenced by the mid-domain-effect (Cardelús et al. 2006), a peak of abundance at mid-elevation. A study focusing on the distribution of CAM in orchids of Panama and Costa Rica found that species richness was highest at sites between 1,000 to 1,500 m altitude, consistent with a peak of abundance at mid-elevation known as the "mid-altitudinal bulge" (Silvera et al. 2009). Almost all epiphytic taxa show hump-shaped curves of distribution with altitude, but their relative contributions change with elevation (Cardelús et al. 2006; Krömer et al. 2005). The mechanism behind this distribution has been explained by the interaction between temperature and water gradients caused by increasing altitude (Peet 1978; Silvera et al. 2009; Whittaker and Niering 1975). Mid elevation sites of around 1,000 m in tropical forests commonly have the highest incidence of cloud cover.

Vertical stratification of epiphytes in tropical forests has been linked to physiological strategies related to microclimate constraints (Graham and Andrade 2004; Reyes-Garcia et al. 2008). Vertical distribution of epiphytes is influenced by ecophysiological adaptations, such as the ability to acclimate to varying levels of photon flux density (PFD) and humidity (Benzing 1990). For example, species of *Phylodendron* tend to clump at lower sites within branches or trunks of host trees (Fig. 1b) where humidity levels are higher and PFD levels are lower, and because part of the life cycle may be dependent on nutrient uptake from roots that are in contact with mineral soil (Putz and Holbrook 1986). On the other hand, epiphytic species that utilize Crassulacean Acid Metabolism (CAM), a water-saving mode of photosynthesis (Ting 1985), tend to clump in areas of higher PFD and exposed sites within branches of the host tree. For example, a study of vertical stratification of epiphytes in a lowland forest of Panama found that >50 % of all epiphytes clustered at intermediate canopy heights (Zotz and Schultz 2008). In a lowland forest of central Panama, the percentage of CAM epiphytes increases with canopy height (Fig. 2; Zotz and Ziegler 1997). In a lowland forest of the Atlantic slope of Panama, total biomass was largest at intermediate heights within the forest, but the relative contribution of CAM species was higher in the upper canopy (Zotz 2004). In another study by Hietz et al. (1999) the proportion of epiphytic species showing CAM decreased with increasing elevation and precipitation along an altitudinal gradient from tropical dry forest to humid montane forest of eastern Mexico.

Fig. 2 Proportion of epiphytic CAM species in different strata of a lowland forest of Central Panama. The percentage of CAM epiphytes increases with canopy height. Percent CAM species was extrapolated from published work (Zotz and Ziegler 1997)

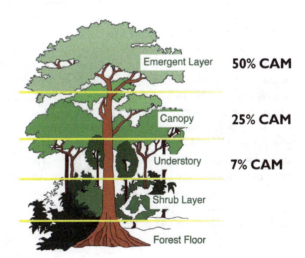

Physiological Adaptations of Epiphytes

There are also numerous ecophysiological adaptations that appear to contribute to epiphyte success in the canopy. Epiphytes often grow in the canopies of very wet tropical forests, with abundant precipitation and high soil water availability, but even in such humid forests, the canopy can dry out quickly and therefore has been described as "the desert in the rainforest". Many epiphytic species use CAM, one of three metabolic pathways found in vascular plants for the assimilation of atmospheric CO_2. CAM plants take up CO_2 from the atmosphere predominantly at night, improving the ability of plants to acquire carbon in water-limited and CO_2-limited environments (Winter and Smith 1996a). Thick leaf cuticles are also observed in many epiphyte species. Many epiphytic species also exhibit less negative osmotic potential compared to their terrestrial congeners (Hao et al., this volume), consistent with careful management of internal water resources. Finally, varying degrees of tolerance to shade and desiccation have been observed in epiphytes compared to related terrestrial species. These characteristics allow epiphytes to survive long periods of drought stress, especially in tropical forests with prolonged dry seasons.

Why not Rooting in Soil Leads to Certain Physiologies

The epiphytic habitat represents a highly dynamic environment, subject to temporal and spatial variations in light, nutrient and water supply. Among these abiotic factors, sporadic or seasonal periods of water shortage are perhaps some of the most common challenges even for epiphytes occurring in humid tropics (Zotz and Thomas 1999). Epiphytes do not take water or nutrients from the host plants. The vast majority of epiphytic plants do not have contact with mineral soil, and therefore must obtain nutrients and water from the microenvironment where they live. Because water availability is a limiting factor for growth and survival of epiphytic species living in the canopy (Lüttge 1989; Zotz and Hietz 2001), and represents a strong selective pressure, epiphytes have evolved myriad physiological and morphological adaptations (Benzing 1990; Hietz et al. 1999), in addition to acquiring necessary nutrients from debris or organic soil from the canopy. These adaptations can be divided into adaptations to increase water capture, to facilitate water storage or to reduce water loss.

To maximize water uptake from rain pulses, dew, or moisture in the air and to survive long periods of drought stress, especially in tropical forests with prolonged dry seasons (Benzing 1990; Hietz et al. 1999), epiphytes have evolved specialized absorptive tissues and organs, and a diversity of growth forms. Bromeliads, for example, possess specialized absorptive trichomes—a key innovation associated with the transition from terrestrial to epiphytic life form (Crayn et al. 2004). In orchids, the functional equivalent to bromeliad trichomes is the "velamen", a multilayered spongy epidermis that allows orchids to absorb atmospheric moisture and nutrients through the roots (Benzing 1987).

Adaptations related to water storage include leaf succulence, composed of abundant hydrenchyma, an enlarged specialized spongy tissue that allows plants to store water. Often hydrenchyma is the main reservoir for maintaining a stable water status in photosynthetic tissues (Freschi et al. 2010). A thick leaf cuticle, on the other hand, is thought to minimize cuticle conductance and therefore water loss during dry periods. A study of 15 vascular epiphytic species evaluated for water permeability of astomatous isolated cuticular membranes showed a lower leaf cuticular permeability to water in comparison to terrestrial forms (Helbsing et al. 2000). Additionally, epiphytes tend to have lower stomatal densities compared to terrestrial plants, thus reducing water loss. Stomatal density can vary from 1–30 stomata/mm^2 in some bromeliads (Martin 1994) to 30–62 stomata/mm^2 in aroids (Lorenzo et al. 2010), to more than 180 stomata/mm^2 in epiphytic *Clusia* species (Holbrook and Putz 1996). In terrestrial plants, stomata densities can reach 600 stomata/mm^2 (Larcher 2003). Finally, many epiphytic species utilize CAM. This metabolic pathway is present in ~ 19 % of epiphytic species in a lowland forest (Zotz 2004), ~ 50 % of tropical epiphytic orchid species (Silvera et al. 2005; Winter and Smith 1996b) and approximately 25–38 % of epiphytic bromeliad species (data estimated from Zotz 2004 and Crayn et al. 2004). Many epiphytic species use combinations of these adaptive traits, which allow them to tolerate limited water supply in the canopy. Below we highlight key adaptations from two of the largest epiphytic plant families: Orchidaceae and Bromeliaceae, and summarize generalities found in other epiphytic families.

Epiphytic Orchids

Of all epiphytic families, the Orchidaceae has been the most successful in colonizing tree crowns and exposed sites (Fig. 1c). One important characteristic that allows epiphytic orchids to colonize the canopy is the presence of roots that serve several key functions. Orchid roots have the capability of photosynthesizing due to presence of chlorophyll, orchids roots are surrounded by spongy velamen that serves to quickly absorb and accumulate water and nutrients, and orchid roots enable the plant to anchor itself securely in branches, thus providing mechanical protection and stability (Benzing 1990). Orchids also use a different strategy during the dry season by having access to stored water and nutrients from the pseudobulb, a storage organ derived from a thickening of the stem between leaf nodes. Orchids generally have long-lived thicker leaves with lower photosynthetic capacity per unit area compared to ferns and bromeliads (Cardelús and Mack 2010). Although thick leaves in orchids are typically associated with $\delta^{13}C$ values in the CAM range, some thin-leaved orchids are also capable of CAM photosynthesis, as demonstrated by acid titration (Silvera et al. 2005).

CAM Photosynthesis in Epiphytic Orchids

In contrast to C_3 and C_4 plants, CAM has evolved multiple independent times within vascular plants, and is widely distributed throughout semi-arid tropical and subtropical environments, including epiphytic habitats in tropical regions (Silvera et al. 2009). Several studies indicate that CAM photosynthesis may be widespread among tropical epiphytic plants, particularly in the largest family of vascular plants, the Orchidaceae with approximately 30,000 species (Earnshaw et al. 1987; Silvera et al. 2009, 2005; Winter et al. 1983). It is estimated that up to 50 % of tropical epiphytic orchid species exhibit different degrees of CAM (Silvera et al. 2005) with a gradient of species showing C_3 to weak CAM to strong CAM modes. The largest proportion of CAM epiphytic orchid species are usually distributed at lower altitudinal sites (Silvera et al. 2009). Additionally, the multiple origins of CAM and its strong positive correlation with epiphytism has been linked to the rapid degree of speciation in the Orchidaceae (Silvera et al. 2009). This is particularly true within the Subfamily Epidendroideae, which represents the largest epiphytic clade within flowering plants (Fig. 3)

Epiphytic Bromeliads

The Bromeliaceae family has approximately 3350 species that cover a wide range of habitats; from very humid forest to extremely xeric sites, and from lowland sites to near 5000 m altitude (Crayn et al. 2015). About half of the species are epiphytes. The epiphytic bromeliads show a morphological progression from tank life forms, to atmospheric or nebulophyte life forms (Zotz 2013b). The term "nebulophyte" has been used in the literature to characterize epiphytic bromeliads with long thin leaves and low leaf succulence, a morphology that maximizes fog-catching (Martorell and Ezcurra 2007; Reyes-Garcia et al. 2012). Tank species have large rosette type leaves that are arranged in such way that can hold large amounts of water after rain events. Tank bromeliads receive nutrient inputs from rain, mist, dust, and nutrient released from ground-rooted plants through leaching or decomposition, as well as from remains of animals and/or organic matter (Benzing 1990; Romero et al. 2010) These "continuously supplied" (Benzing 1990) tank epiphytes tend to have thick leaf cuticles, high stomatal control, and are more restricted to the upper canopy, all characteristics that allow them to resist photoinhibition and avoid dessication (Reyes-Garcia et al. 2012). Species with large tanks, however, can be subjected to long periods of desiccation. Accumulated water can dry out after 1–2 weeks (Zotz and Thomas 1999), and thus tank species rely on the ability to switch from C_3 to CAM in order to survive these water shortage episodes (Benzing 2000). The tank structure or "phytotelmata" can intercept and retain debris and water, and form a nice microcosm where a diverse community of invertebrate and vertebrate animals thrive (Benzing 2000). Organism debris living inside of the tank can provide

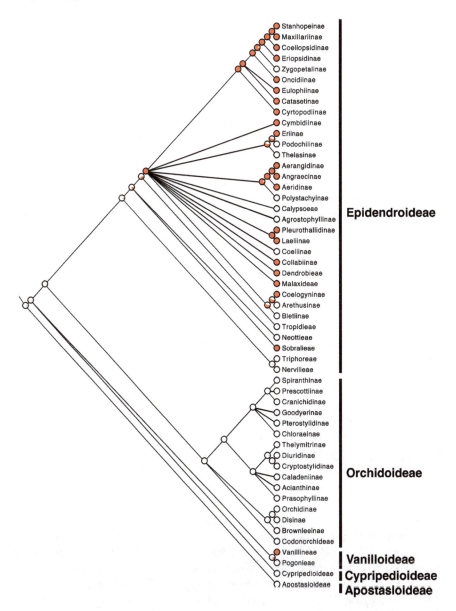

Fig. 3 Cladogram of the relationship among 53 Orchidaceae subtribes. The five orchid subfamilies are represented on the right side. Presence and absence of CAM were mapped onto the cladogram showing lineages with CAM in *red*. The *red* area within the *pie chart* indicates relative support for different ancestor states. This analysis supports the multiple, independent origins of CAM in the epiphytic Subfamily Epidendroideae

nutrition to the bromeliad, as shown by experiments with stable isotopes of nitrogen. Romero et al. (2010) demonstrated that treefrog feces from *Scinax hayii* contributed 27.7 % of the total N to the plant, whereas dead termites contributed 49.6 % of the total N to the plant. They also showed that plants receiving additional N through frog feces and termites had higher photosynthetic rates compared to control plants, but only during the rainy season. Epiphyte nutrient availability is generally less critical than water availability (Laube and Zotz 2003), but can still be a limiting factor when water is available.

In contrast to tank bromeliads, atmospheric life forms or "pulse supplied" epiphytes only have access to water after a precipitation event, and maximize water absorption through abundant leaf trichomes (Benzing 1990). Atmospheric bromeliads receive nutrient inputs from dust, mist and rain through dense surface trichomes. Indeed, the main adaptation of atmospheric bromeliads for living in the canopy is the presence of specialized trichomes for both minerals and water absorption (Benzing et al. 1985), especially because bromeliad roots do not engage in nutrient acquisition nor have mycorrhizal fungi associations (Lesica and Antibus 1990). Trichomes, however, vary in their absorptive capabilities. For example, some tank bromeliads within the subfamily Tillandsioideae have trichomes capable of absorbing large organic molecules such as amino acids (Endres and Mercier 2003), and derive nutrients from predator feces and arthropod carcasses living in their tanks (Romero et al. 2010). Species from the subfamilies Pitcairnioideae and Bromelioideae have trichomes that absorb smaller molecules.

CAM Photosynthesis in Bromeliad Species

The appearance of CAM is considered one of the key innovations associated with the transition from terrestrial to epiphytic life form. In Bromeliads, the epiphytic habit has evolved a minimum of three times (Crayn et al. 2004), and bromeliad epiphytes show great plasticity in the expression of CAM with a range of expressions from C_3 to facultative CAM to full CAM (Andrade et al. 2007; Griffiths and Smith 1983). Some species under stress have shown high levels of recycling of respiratory CO_2 via CAM (Martin 1994; Maxwell et al. 1994). Griffiths and Smith (1983) were able to relate the ecology of different bromeliad species to the occurrence of CAM and C_3 on the island of Trinidad. The authors found that tank bromeliads, which include light demanding C_3 species and CAM species, show a wide range of distribution across rainfall gradients, but are always found in the uppermost part of the canopy. The more xerophytic bromeliads all show CAM and are restricted to sites with the lowest rainfall, whereas the shade-tolerant bromeliads are all C_3 species and are found in a wide geographical range but on sites with high humidity. This study highlights that it is not possible to distinguish C_3 and CAM bromeliad species based only on their stratification within the forest.

Bromeliads can also be heteroblastic, changing their growth forms during their lifetime by starting as an atmospheric life form and developing water holding tanks

Fig. 4 Bromeliad species *Werauhia sintenisii.* **a** Adult or tank form and **b** juvenile or atmospheric form of the heteroblastic bromeliad species *Werauhia sintenisii*. Production of anthocyanins for photoprotection in the adult form is common in epiphytic bromeliads. Photos by Eloisa Lasso

as adults (Fig. 4, Benzing 2000; Zotz et al. 2011). A study by Beltrán et al. (2013) demonstrated that heteroblastic species shift anatomically and physiologically from C_3 to CAM during their development to increase the survival of juveniles that are more susceptible to drought stress given their lack of tank. In other words, species were able to display features of facultative CAM when exposed to drought stress irrespective of developmental stage. The only obvious difference between the adult and the juvenile was the size of the hydrenchyma, in which juveniles presented thicker hydrenchyma with higher water storage (Beltrán et al. 2013). Zotz et al. (2011) reported that the juvenile atmospheric form of *Werauhia sanguinolenta* showed higher trichome density compared to the adult form. Recent work by Meisner et al. (2013) reported similar size-related variations when studying the functional relevance of water storage tissue, stomata and trichome densities, and transpiration and nutrient uptake rates in seventeen homoblastic and heteroblastic species from Panama, thus questioning the functional relevance of heteroblasty in bromeliads.

Epiphytic Ferns and Aroids

Few species of epiphytic ferns have been studied regarding their adaptation to water scarcity. Even in humid cloud forest, fern species have xeric adaptations including coriaceous leaves, low rates of water loss, leaf scales and high cell wall elasticity. Some ferns species are poikilohydric, and can tolerate high water deficit and recover without any physiological damage (Hietz and Briones 1998), whereas others display a variety of strategies for coping with the xeric conditions in the tree crown, including succulence of leaves and rhizomes, early closure of stomata upon water loss, osmotic adjustment, low rates of cuticular water loss, water-absorbing leaf scales and low cell-wall elasticity (Hietz and Briones 1998). Fern distribution in the canopy is correlated with the relative water content at which stomata close, leaf thickness, and stomatal density and size (Hietz and Briones 1998).

Canopy species of *Anthurium* and *Philodendron* present morpho-physiological adaptations to the canopy environment, such as litter-trap growth, thick cuticles, and ramification (Jacome et al. 2004). Ontogenetic and size-related changes in anatomy and ecophysiology relate to performance and survival of *Anthurium scandens* epiphytes in the field (Lorenzo et al. 2010). For example, stomatal conductance did not change with ontogenetic phase and was highly dependent on water availability at the root level. Nevertheless, adult epiphytes had lower epidermal conductance to water loss as well as higher values of leaf succulence and sclerophylly, and showed higher retranslocation rates of leaf nitrogen during senescence, all traits that confer higher resistance to low resources in comparison to seedlings living in the shaded humid understory (Lorenzo et al. 2010).

To CAM or not to CAM? That Is the Question

CAM is an ecophysiological adaptation to water limitation influencing the distribution of plants within their natural environments. One way to predict extinction threats is to understand the evolutionary history of drought resistance mechanisms in plants. CAM photosynthesis is one of the best examples of convergent evolution in vascular plants; estimates report that the CAM pathway is found in approximately 7 % of vascular plants from 35 families and 343 genera (Holtum et al. 2007; Silvera et al. 2010; Smith and Winter 1996). It is estimated that over 57 % of all epiphytes are capable of performing CAM (Benzing 1987; Lüttge 2004). Epiphytes that utilize different levels of CAM photosynthesis are plastic in their responses to environmental cues, and with elevated CO_2, CAM plants have an average increase in biomass of 35 % without the downward acclimatization of photosynthesis that is typically observed in many C_3 species (Drennan and Nobel 2000). CAM evolution is accompanied by a series of anatomical and biochemical modifications from common C_3 photosynthesis that allows CAM species to have a competitive advantage in drought prone epiphytic habitats (Silvera et al. 2010). As a consequence, CAM species have a higher water use efficiency (WUE) when compared to C_3 species due to reduced amount of water lost per unit carbon assimilated. The possibility that the habitats and growth potential for selected CAM species may be favored, relative to C_3 species, is important considering currents trends of increased atmospheric CO_2 and associated climate change patterns, along with predicted increases in arid and semi-arid land area (Cushman and Borland 2002; Yang et al. 2015).

CAM has contributed to the exploitation of epiphytic habitats through mid elevation tropical humid environments, which is especially true for the Orchidaceae (Silvera et al. 2009). Studies using leaf carbon isotopic composition ($\delta^{13}C$) of epiphytic orchids have shown a bimodal distribution with the majority of species exhibiting values of approximately –28‰, typically associated with the C_3

pathway, and a second peak of distribution at values around −15‰, which is associated with strong CAM. The presence of weak CAM, as evidenced by acid titrations, can also contribute to a second peak of abundance within the C_3 photosynthesis isotopic cluster (Silvera et al. 2005). These data suggest that among CAM orchids, there may be preferential selection for species to exhibit strong CAM or weak CAM, rather than intermediate metabolism (Silvera et al. 2005). The role of weak CAM as an intermediate state between C_3 and strong CAM, or as an evolutionary reservoir for CAM adaptive radiations is an unknown and interesting question for further studies.

Nutrient Limitation

Even though roots of epiphytes are not in touch with mineral soil, nutrient alterations of mineral soil can have an effect on epiphytic communities by altering the edaphic factors of soils used by the host trees. Experimental evidence in support of the hypothesis that epiphytes are nutritionally linked to their host tree is, however, contradictory to date. Nutritional links between host trees and epiphytes in high mountain trees are expected to be more nutrient limited than those in lowland forests (Grubb 1977). Cardelús and Mack (2010) explored epiphytic nutrient status along an elevation gradient in Costa Rica and found no relationship between foliar chemistry and elevation for any of the epiphytic groups studied, which included ferns, orchids and bromeliads. In contrast, a long-term fertilization experiment in the elfin forest of Puerto Rico showed that nutrient addition to soil stimulated tree leaf litter production (Walker et al. 1996), and increased nutrient content of litterfall (Yang et al. 2007), which in turn affected epiphytes growing on those trees. Epiphytic bromeliads growing on fertilized trees clearly responded to nutrient augmentation received by their host trees, and accumulated more nutrients in their vegetative body thus increasing their growth rate, fruit production and lifetime fitness (Lasso and Ackerman 2013). These studies together with Benner and Vitousek (2007), and Boelter et al. (2014), suggest that epiphytes are nutrient limited, and that a nutritional link exists between host trees and their epiphytes. This may be especially true for tank bromeliads since their impounding leaf rosette traps litter and leachates from their host trees. However, epiphytes with different nutrient acquisition mechanisms may not be strongly linked to their host tree. Different epiphyte groups use diverse nutrient uptake mechanisms, which ultimately provide an explanation for the coexistence and maintenance of species richness in tropical forest canopies (Cardelús and Mack 2010).

Vascular epiphytes of lowland tropical forests are nutrient limited, mainly by P and sometimes by N, or P and N simultaneously (Wanek and Zotz 2011; Zotz and Asshoff 2010; Zotz and Richter 2006). Epiphytes living in tropical montane cloud forests are more likely nutrient limited because there is constant input of water and excessive rain that may wash out nutrients from the tank, leaf surface or even the layer of canopy soil where epiphytes grow. Additionally, lower temperatures in tropical mountain cloud forest reduce decomposition rate, potentially decreasing the release of nutrients from canopy soil. Benner and Vitousek (2007) documented a dramatic increase in the abundance and species richness of canopy epiphytes in a forest fertilized annually with phosphorus (P) for 15 years, but found no response in forest fertilized with nitrogen (N) or other nutrients, suggesting that P is also the main limitation in highland forests. Studies of two bromeliad species, *Werauhia sanguinolenta* from the lowlands of Panama (Zotz and Richter 2006) and *Werauhia sintenissi* from the cloud mountain forest in Puerto Rico (Lasso and Ackerman 2013) also point to phosphorous limitation for growth and reproduction at all altitudes. The accumulation of P, Zn, Al, Fe, Na in the foliar tissue of *Werauhia sintenissi* growing in plots fertilized regularly for 10 years with macro and micro nutrients (Lasso and Ackerman 2013), suggest luxury consumption and storage not only of P, as reported before for *Vriesea sanguinolenta* (Wanek and Zotz 2011), but also for other nutrients.

Epiphytes and Climate Change

Potential Responses or Indicators of Climate Change

Atmospheric CO_2 levels are predicted to double that of pre-industrial levels by 2050 (IPCC Assessment Report). With increasing levels of atmospheric CO_2, tropical forests are expected to increase carbon uptake, which may lead to faster growth. In the case of epiphytes, this may translate into higher biomass and subsequently higher loads on tree branches, which might lead to increased structural damage of host trees. Tropical species, including epiphytes, are likely to shift upslope rather than latitudinally with climate change mostly because of the very steep temperature gradient found per meter elevation as compared to temperate zones (5.2–6.5 °C decrease per 1000 m elevation, Colwell et al. 2008). To estimate the potential responses of epiphytes to climate change, it is crucial to estimate the effects of increasing levels of CO_2 in epiphytic CAM species, and on species with facultative CAM. Whether these species would have an advantage with increased photosynthetic flexibility remains unknown.

Are Epiphytes Really Indicators of Climate Change?

Epiphytes are one of the most sensitive plant growth forms to disturbance and microclimate changes due to their lack of access to permanent water sources and nutrients, which forces them to be tightly dependent on seasonal precipitation, fog or dew. For this reason, epiphytes are thought to be exceptionally sensitive to air quality and climate, and considered indicators of climate change and forest health. In addition, epiphytes are an important component of watersheds because of their ability to intercept cloud and fog water thus increasing net water inputs. Managing and implementing conservation programs for epiphytes is highly dependent on understanding which factors influence epiphytic diversity in species-rich regions, and the physiological responses of epiphytes to environmental change. Climate change, through its effect on temperature, is likely to have an effect, albeit small, on evaporation rates, which in turn can have an effect on communities of animals living in water reservoirs inside epiphytic bromeliads. Examining patterns of epiphytic community distribution among different climate areas may provide an early indication of vegetation responses to climate change.

Undoubtedly, the major threat to epiphytic diversity is deforestation (Fig. 5). It is estimated that about 21,000 square km of forest is deforested annually in South America, which harbors one of the richest epiphytic communities in the world, most of it in the Amazonian Basin. Conservation of tropical rainforest is key to conservation of epiphytes.

Fig. 5 Epiphytic habitat loss due to deforestation. **a** Deforestation for agricultural plantations. Photo by Katia Silvera, and **b** Deforestation due to slash-and-burn agricultural practices. Photo by Gaspar Silvera

Concluding Remarks

Because most studies on epiphytes are species or genus specific, we need an array of experiments using broad species samples, in order to estimate how communities will change over time. As Zotz (2004) pointed out, when studying epiphytic communities it is important to distinguish prevalence of vascular epiphytes in terms of species numbers or biomass, because many orchid species occur in very low numbers and their contribution to community biomass can be low. Because of this inherit bias, interpretations of whether CAM is prevalent in tropical moist forests compared to semiarid regions remain biased by number of species (Zotz 2004). Most of the ecophysiological studies conducted on epiphytic plants have focused on water relations of leaves, which provide a skewed view of plant function across communities. Integration of studies using different organs is lacking and necessary to understand how species will respond to climate change.

Epiphytes, as indicators of climate change, may be the first to respond to global changes; therefore how we implement conservation strategies to forests is key to mitigate changes to tropical forests.

Acknowledgments We dedicate this chapter to the memory of Mónica Mejía-Chang, a great friend and passionate epiphyte and bromeliad researcher in Panama. Her infectious personality and enthusiasm for life was reflected in her work and radiated throughout her life. We kindly acknowledge John Cushman and artist Kirah Van Sickle for providing figure 2. This work is supported in part by funding from the National Science Foundation (Award DEB 1442199 to KS), the Smithsonian Tropical Research Institute Tupper Fellowship (to KS), and the Panamanian Secretaría Nacional de Ciencia, Tecnología e Innovación (SENACYT to KS). This publication was also made possible by the Colombian Fondo de Apoyo a Profesores Asistentes (FAPA to EL) from the Universidad de Los Andes, Bogotá.

References

Andrade JL, de la Barrera E, Reyes-Garcia C, Ricalde MF, Vargas-Soto G, Cervera JC (2007) Crassulacean acid metabolism: diversity, environmental physiology and productivity. Boletín de la Sociedad Botánica de México 81:37–50

Beltrán JD, Lasso E, Madriñán S, Virgo A, Winter K (2013) Juvenile tank-bromeliads lacking tanks: do they engage in CAM photosynthesis? Photosynthetica 51:55–62

Benner JW, Vitousek PM (2007) Development of a diverse epiphyte community in response to phosphorus fertilization. Ecol Lett 10:628–636

Benzing DH (1987) Vascular epiphytism: taxonomic participation and adaptive diversity. Ann Mo Bot Gard 74:183–204

Benzing DH (1989) The evolution of epiphytism. In: Lüttge U (ed) Vascular plants as epiphytes: evolution and ecophysiology, vol 76. Springer, Berlin, Heidelberg, New York, pp 15–41

Benzing DH (1990) Vascular epiphytes. General biology and related biota. Cambridge University Press, Cambridge

Benzing DH (2000) Bromeliaceae-profile of an adaptive radiation. Cambridge University Press, Cambridge

Benzing DH, Givnish TJ, Bermudes D (1985) Absorptive trichomes in *Brocchinia reducta* (Bromeliaceae) and their evolutionary and systematic significance. Syst Bot 10:81–91

Boelter CR, Dambros CS, Nascimento HEM, Zartman CE (2014) A tangled web in tropical tree-tops: effects of edaphic variation, neighbourhood phorophyte composition and bark characteristics on epiphytes in a central Amazonian forest. J Veg Sci 25:1090–1099

Burns KC, Zotz G (2010) A hierarchical framework for investigating epiphyte assemblages: networks, meta-communities, and scale. Ecology 91:377–385

Cardelús CL, Colwell RK, Watkins JE (2006) Vascular epiphyte distribution patterns: explaining the mid-elevation richness peak. J Ecol 94:144–156

Cardelús CL, Mack MC (2010) The nutrient status of epiphytes and their host trees along an elevational gradient in Costa Rica. Plant Ecol 207:25–37

Colwell RK, Brehm G, Cardelús CL, Gilman AC, Longino JT (2008) Global warming, elevational range shifts, and lowland biotic attrition in the wet tropics. Science 322:258–261

Crayn DM, Winter K, Schulte K, Smith JAC (2015) Photosynthetic pathways in Bromeliaceae: phylogenetic and ecological significance of CAM and C_3 based on carbon isotope ratios for 1893 species. Bot J Linn Soc 178:169–221

Crayn DM, Winter K, Smith JAC (2004) Multiple origins of crassulacean acid metabolism and the epiphytic habit in the Neotropical family Bromeliaceae. Proc Natl Acad Sci USA 101:3703–3708

Cushman JC, Borland AM (2002) Induction of crassulacean acid metabolism by water limitation. Plant, Cell Environ 25:295–310

Drennan PM, Nobel PS (2000) Responses of CAM species to increasing atmospheric CO_2 concentrations. Plant, Cell Environ 23:767–781

Earnshaw MJ et al (1987) Altitudinal changes in the incidence of crassulacean acid metabolism in vascular epiphytes and related life forms in Papua New Guinea. Oecologia 73:566–572

Endres L, Mercier H (2003) Amino acid uptake and profile in bromeliads with different habits cultivated in vitro. Plant Physiol Biochem 41:181–187

Freschi L et al (2010) Specific leaf areas of the tank bromeliad *Guzmania monostachia* perform distinct functions in response to water shortage. J Plant Physiol 167:526–533

Gentry AH, Dodson C (1987) Contribution of nontrees to species richness of a tropical rain forest. Biotropica 19:149–156

Graham EA, Andrade JL (2004) Drought tolerance associated with vertical stratification of two co-occurring epiphytic bromeliads in a tropical dry forest. Am J Bot 91:699–706

Griffiths H, Smith JAC (1983) Photosynthetic pathways in the Bromeliaceae of Trinidad: Relations between life-forms, habitat preference and the occurrence of CAM. Oecologia 60:176–184

Grubb PJ (1977) Control of forest growth and distribution on wet tropical mountains. Annu Rev Ecol Syst 8:83–107

Helbsing S, Riederer M, Zotz G (2000) Cuticles of vascular epiphytes: Efficient barriers for water loss after stomatal closure? Ann Bot 86:765–769

Hietz P, Briones O (1998) Correlation between water relations and within-canopy distribution of epiphytic ferns in a Mexican cloud forest. Oecologia 114:305–316

Hietz P, Wanek W, Popp M (1999) Stable isotopic composition of carbon and nitrogen and nitrogen content in vascular epiphytes along an altitudinal transect. Plant, Cell Environ 22:1435–1443

Holbrook NM, Putz FE (1996) From epiphyte to tree: Differences in leaf structure and leaf water relations associated with the transition in growth form in eight species of hemiepiphytes. Plant, Cell Environ 19:631–642

Holtum JAM, Winter K, Weeks MA, Sexton TR (2007) Crassulacean acid metabolism of the ZZ plant, *Zamioculcas zamiifolia* (Araceae). Am J Bot 94:1670–1676

Jacome J, Galeano G, Amaya M, Mora M (2004) Vertical distribution of epiphytic and hemiepiphytic Araceae in a tropical rain forest in Chocó, Colombia. Selbyana 25:118–125

Krömer T, Kessler M, Robbert Gradstein S, Acebey A (2005) Diversity patterns of vascular epiphytes along an elevational gradient in the Andes. J Biogeogr 32:1799–1809

Larcher W (2003) Gas exchange in plants. In: Larcher W (ed) Physiological ecology: ecophysiology and stress physiology of functional groups, 4th edn. Springer, Berlin, pp 91–139

Lasso E, Ackerman J (2013) Nutrient limitation restricts growth and reproductive output in a tropical montane cloud forest bromeliad: findings from a long-term forest fertilization experiment. Oecologia 171:165–174

Laube S, Zotz G (2003) Which abiotic factors limit vegetative growth in a vascular epiphyte? Funct Ecol 17:598–604

Lesica P, Antibus RK (1990) The occurrence of mycorrhizae in vascular epiphytes of two Costa Rican rainforests. Biotropica 22:250–258

Lorenzo N, Mantuano DG, Mantovani A (2010) Comparative leaf ecophysiology and anatomy of seedlings, young and adult individuals of the epiphytic aroid *Anthurium scandens* (Aubl.) Engl. Environ Exp Bot 68:314–322

Lüttge U (1989) Vascular plants as epiphytes. Evolution and ecophysiology. Springer, Berlin

Lüttge U (2004) Ecophysiology of crassulacean acid metabolism (CAM). Ann Bot 93:629–652

Martin EG (1994) Physiological ecology of the Bromeliaceae. Bot Rev 60:1–82

Martorell C, Ezcurra E (2007) The narrow-leaf syndrome: a functional and evolutionary approach to the form of fog-harvesting rosette plants. Oecologia 151:561–573

Maxwell C, Griffiths H, Young AJ (1994) Photosynthetic acclimation to light regime and water stress by the C_3-CAM epiphyte *Guzmania monostachia*: gas exchange characteristics, photochemical efficiency and the xanthophyll cycle. Funct Ecol 8:746–754

Meisner K, Winkler U, Zotz G (2013) Heteroblasty in bromeliads – anatomical, morphological and physiological changes in ontogeny are not related to the change from atmospheric to tank form. Funct Plant Biol 40:251–262

Nadkarni NM, Schaefer D, Matelson TJ, Solano R (2004) Biomass and nutrient pools of canopy and terrestrial components in a primary and a secondary montane cloud forest, Costa Rica. For Ecol Manage 198:223–236

Peet RK (1978) Forest vegetation of the colorado front range: patterns of species diversity. Vegetatio 37:65–78

Putz FE, Holbrook NM (1986) Notes on the natural history of hemiepiphytes. Selbyana 9:61–69

Reyes-Garcia C, Mejia-Chang M, Griffiths H (2012) High but not dry: diverse epiphytic bromeliad adaptations to exposure within a seasonally dry tropical forest community. New Phytol 193:745–754

Reyes-Garcia C, Mejia-Chang M, Jones GD, Griffiths H (2008) Water vapour isotopic exchange by epiphytic bromeliads in tropical dry forests reflects niche differentiation and climatic signals. Plant, Cell Environ 31:828–841

Romero GQ et al (2010) Nitrogen fluxes from treefrogs to tank epiphytic bromeliads: an isotopic and physiological approach. Oecologia 162:941–949

Silvera K, Neubig KM, Whitten WM, Williams NH, Winter K, Cushman JC (2010) Evolution along the crassulacean acid metabolism continuum. Funct Plant Biol 37:995–1010

Silvera K, Santiago LS, Cushman JC, Winter K (2009) Crassulacean acid metabolism and epiphytism linked to adaptive radiations in the Orchidaceae. Plant Physiol 149:1838–1847

Silvera K, Santiago LS, Winter K (2005) Distribution of crassulacean acid metabolism in orchids of Panama: evidence of selection for weak and strong modes. Funct Plant Biol 32:397–407

Smith JAC, Winter K (1996) Taxonomic distribution of crassulacean acid metabolism. In: Winter K, Smith JAC (eds) Crassulacean acid metabolism: biochemistry, ecophysiology and evolution. Springer, Berlin Heidelberg, pp 427–436

Stuntz S, Simon U, Zotz G (2002) Rainforest air-conditioning: the moderating influence of epiphytes on the microclimate in tropical tree crowns. Int J Biometeorol 46:53–59

Ting IP (1985) Crassulacean acid metabolism. Ann Rev Plant Physiol 36:595–622

Walker LR, Zimmerman JK, Lodge DJ, Guzman-Grajales S (1996) An altitudinal comparison of growth and species composition in hurricane-damaged forests in Puerto Rico. J Ecol 84:877–889

Wanek W, Zotz G (2011) Are vascular epiphytes nitrogen or phosphorus limited? A study of plant ^{15}N fractionation and foliar N : P stoichiometry with the tank bromeliad *Vriesea sanguinolenta*. New Phytol 192:462–470

Whittaker RH, Niering WA (1975) Vegetation of Santa Catalina Mountains, Arizona. V. Biomass, production and diversity along elevation gradient. Ecology 56:771–790

Winter K, Smith JAC (1996a) Crassulacean acid metabolism: current status and perspectives. In: Winter K, Smith JAC (eds) Crassulacean acid metabolism: biochemistry, ecophysiology and evolution. Springer, Berlin Heidelberg, pp 389–426

Winter K, Smith JAC (1996b) An introduction to crassulacean acid metabolism: biochemical principles and ecological diversity. In: Winter K, Smith JAC (eds) crassulacean acid metabolism. Springer, Berlin, pp 1–13

Winter K, Wallace BJ, Stocker GC, Roksandic Z (1983) Crassulacean acid metabolism in Australian vascular epiphytes and some related species. Oecologia 57:129–141

Yang X et al (2015) A roadmap for research on crassulacean acid metabolism (CAM) to enhance sustainable food and bioenergy production in a hotter, drier world. New Phytol 207:491–504

Yang X, Warren M, Zou X (2007) Fertilization responses of soil litter fauna and litter quantity, quality, and turnover in low and high elevation forests of Puerto Rico. Appl Soil Ecol 37:63–71

Zotz G (2004) How prevalent is crassulacean acid metabolism among vascular epiphytes? Oecologia 138:184–192

Zotz G (2013a) 'Hemiepiphyte': a confusing term and its history. Ann Bot 111:1015–1020

Zotz G (2013b) The systematic distribution of vascular epiphytes – a critical update. Bot J Linn Soc 171:453–481

Zotz G, Asshoff R (2010) Growth in epiphytic bromeliads: response to the relative supply of phosphorus and nitrogen. Plant biology (Stuttgart, Germany) 12:108–113

Zotz G, Hietz P (2001) The physiological ecology of vascular epiphytes: current knowledge, open questions. J Exp Bot 52:2067–2078

Zotz G, Richter A (2006) Changes in carbohydrate and nutrient contents throughout a reproductive cycle indicate that phosphorus is a limiting nutrient in the epiphytic bromeliad, *Werauhia sanguinolenta*. Ann Bot 97:745–754

Zotz G, Schultz S (2008) The vascular epiphytes of a lowland forest in Panama-species composition and spatial structure. Plant Ecol 195:131–141

Zotz G, Thomas V (1999) How much water is in the tank? Model calculations for two epiphytic bromeliads. Ann Bot 83:183–192

Zotz G, Wilhelm K, Becker A (2011) Heteroblasty—A review. Bot Rev 77:109–151

Zotz G, Ziegler H (1997) The occurrence of crassulacean acid metabolism among vascular epiphytes from Central Panama. New Phytol 137:223–229

Stem-Succulent Trees from the Old and New World Tropics

Eleinis Ávila-Lovera and Exequiel Ezcurra

Abstract Stem-succulent trees are common in tropical drylands. Besides their ability to store water, these trees also possess photosynthetic bark, which can re-assimilate internally respired CO_2 at virtually no water cost. Both of these traits are advantageous in seasonally dry ecosystems, where plants are exposed to periods of limited water availability and, consequently, carbon gain. In most species, plants do not use the stored water in stems to buffer daily water deficits; they use this water to flush new leaves before the onset of rains. This gives an extra advantage to stem-succulent trees over other functional groups because leaves are already present when the first rain falls. Having succulent stems does not pose a mechanical constraint in these plants, rather the succulence of the tree stem can act as hydrostatic pressure against the bark, contributing to the biomechanical support of tall trees. Stem-succulent trees are also able to maintain physiological processes and growth during drought, making them good candidates to be used in reforestation of degraded arid lands.

Keywords Photosynthetic bark · Retamoid · Sarcocaulescent · Stem photosynthesis · Tropical dry forest · Water use efficiency

Introduction

Stem-succulent trees with photosynthetic bark have evolved in association with tropical drylands. The large mid-latitude coastal deserts of the world such as the Namib in Southern Africa, Atacama in Chile, the Saharan Atlantic Coastal Desert,

E. Ávila-Lovera (✉) · E. Ezcurra
Department of Botany and Plant Sciences, University of California,
2150 Batchelor Hall, Riverside, CA 92521, USA
e-mail: eleinis.avilalovera@email.ucr.edu

E. Ezcurra
e-mail: exequiel.ezcurra@ucr.edu

© Springer International Publishing Switzerland 2016
G. Goldstein and L.S. Santiago (eds.), *Tropical Tree Physiology*,
Tree Physiology 6, DOI 10.1007/978-3-319-27422-5_3

and Baja California in Mexico, are found on the west side of the African and American continents associated with cold coastal currents that move towards the equator along the eastern fringe of the Atlantic and Pacific oceans. North and south, these deserts are flanked by semiarid regions: Mediterranean sclerophyllous scrubs in their high-latitude boundary, and tropical dry scrubs towards the equator. The seasonality of precipitation in these drylands changes dramatically from the tropics towards higher latitudes: whereas Mediterranean shrubs survive mostly with winter rains brought in from oceanic westerly winds, moisture in the dryland tropics is almost entirely provided by summer rains delivered by summer monsoons and, secondarily, by late summer hurricanes and tropical storms (García-Oliva et al. 1991; Douglas et al. 1993; Stensrud et al. 1995). Even within a single ecological region, the transition from winter to summer rains can be marked. Most coastal deserts receive winter rains in their high-latitude, temperate reaches, but are fed predominantly by summer monsoon rains at their tropical edge, where they transition into tropical thorn scrub and seasonally dry forest (Douglas et al. 1993). Winter- or summer-dominated seasonality generates different types of drylands. Winter-rain drylands are dominated by evergreen shrubs with small and/or tough leaves (e.g., the South African fynbos, California chaparral, Chilean *matorral*, and other sclerophyllous scrubs; Dimmit 2000; Ezcurra et al. 2006), while the tropical summer-rain drylands are dominated by drought-deciduous trees and shrubs (Bullock et al. 1995; Gordon et al. 2004; Becerra 2005).

In the tropics, dry forests (TDF, Tropical Dry Forest) are diverse ecosystems in terms of plant life-forms (Medina 1995)—or what has been known more recently as plant functional groups (PFG). The understory of these forests is dominated by herbs and woody shrubs, while the canopy is dominated by trees from different PFG: evergreen, brevi-deciduous (leaf-exchanging), deciduous, stem-succulent, and lianas (Schnitzer and Bongers 2002). At the same time, TDFs are one of the most endangered ecosystems worldwide (Rodríguez et al. 2010), since they are in favorable areas for agriculture, cattle, and human settling. Currently, some efforts are underway to recover these ecosystems. An important step to achieve this is reforestation with key plant species, for example, stem-succulent Baobab trees (*Adansonia*, Fig. 1) have been used in reforestation practices in Africa, Madagascar and Australia where they are one of the most representative trees (Wickens and Lowe 2008). The amount of fauna associated with these forests is high, and attempts to increase populations of endangered animal species usually start with re-planting trees that support wildlife. One example is the yellow-headed parrot (*Amazona barbadensis*) associated with one specific type of TDF on Margarita Island, Venezuela (Rodríguez and Rojas-Suárez 2008).

Global Climate Change has prompted many researchers to investigate possible effects of longer, more intense, and/or more frequent drought periods on plant physiology (Chaves and Pereira 1992; Chaves et al. 2002; McDowell et al. 2008; Tezara et al. 2010). Attention has turned to the prevention or remediation of the detrimental effects of these droughts worldwide. We need to know how climate change affects vegetation and how vegetation can feed back on climate. The ecophysiology of many plant species is known due to applied research in particular crops,

Fig. 1 Baobab (*Adansonia sp.*) in Tanzania, East Africa. Photo by Pedro Piqueras

biofuel plants, timber trees, or useful dryland species. The study of natural tropical dry forests has lagged behind but there is now a growing number of studies being published with interest in stem-succulent trees and their physiological responses to drought. Because plants "belong" to different PFG, they likely respond differently to climate change. The focus of this chapter is to describe the physiology and emphasize the role of stem-succulent trees in the seasonally dry ecosystems they inhabit.

Plants have evolved a number of different strategies to cope with drought in arid regions. In many species, drought tolerance has led to the evolution of a reduced leaf area or to a drought-deciduous habit, both of which contribute to reduced water loss during critically dry periods. Drought-deciduous and leafless plants in hot deserts and tropical drylands frequently occur in the form of shrubby or arborescent species with photosynthetic stems. These plants can have either stem net photo-synthesis (SNP) or stem recycling photosynthesis (SRP) (Ávila et al. 2014), both of which positively affect the carbon economy of plants. When plants have SNP, their stems are usually more efficient in the use of water than leaves, i.e., they have higher photosynthetic water use efficiency (WUE; photosynthetic carbon gain divided by water loss from transpiration) (Ehleringer et al. 1987; Osmond et al. 1987; Smith and Osmond 1987; Nilsen and Sharifi 1997). When plants have SRP, their stems do not lose water because the photosynthetic bark re-assimilates internally respired CO_2. Within plants that have photosynthetic stems one important group is the *sarcocaulescent* group, which has large-sized stems with translucent exfoliating bark, a large amount of parenchymatous tissue that serves as a water reservoir, and non-succulent, drought-deciduous leaves (Franco-Vizcaino et al. 1990). Another group, the *retamoid* group, comprises leafless or almost leafless woody plants that have stomata in the stem's epidermis or other structures such as lenticels in the bark surface that permit gas exchange (Schaedle 1975). A third group, the *cactoid* group, is composed of succulents with Crassulacean Acid Metabolism (CAM), such as the New World cacti or the African cactoid euphorbs. In this chapter we will discuss the physiology, ecology, and biogeographic distri-bution of sarcocaulescent and retamoid species, with an emphasis on sarcocauls, trees with photosynthetic bark and possessing the ability to store large amounts of water in their stems. The cactoid group will not be discussed here as CAM phys-iology is addressed in an earlier chapter of this book (Silvera and Lasso, this volume). As we will see in the following sections, the retamoid growth form tends to be more frequent in the pole-ward edge of deserts in temperate drylands with winter rains, whereas sarcocauls tend to be more frequent in the equator-ward edges of deserts in tropical drylands with monsoon summer rains.

Sarcocaulescent Trees

Sarcocaulescent, or fleshy-stemmed trees, also known as "pachycauls" (thick stems), are plants with a disproportionately thick trunk for their height and canopy. They are often referred to as "bottle trees" for their abnormally swollen stems. In contrast with cactoid and retamoid species, which are frequently protected by dense spines, sarcocaulescent plants show massive trunks tapering upward into relatively small branches. The most distinctive traits of sarcocauls are: (1) the large amount of undifferentiated parenchyma both in the stem rays and central axis that serves as a water reservoir for the plant, (2) the presence of smooth, translucent exfoliating bark with photosynthetic cells, and (3) non-succulent drought-deciduous leaves. Although the bark of sarcocaulescent trees normally has no lenticels or stomata, limiting the diffusion of CO_2 to photosynthetic tissue, SRP, previously known as bark or corticular photosynthesis (see Ávila et al. 2014), helps to maintain adequate carbohydrate supplies during leafless periods by re-fixing and recycling respiratory CO_2 (Franco-Vizcaino et al. 1990).

Sarcocauls often coexist in nature with a related morphology: caudiciform plants. Caudiciform species generally have an enlarged basal caudex or stem axis, a thick, tuber-like structure at ground level from which the stems and roots arise (Rowley 1987). The caudex may extend below the ground and often gives rise to deciduous twining stems. This adaptation is well developed in species of the gourd family (Cucurbitaceae), such as *Marah macrocarpa* in California, USA or *Ibervillea sonorae* in Mexico.

Plants with giant fleshy stems occur in a number of families in the dry regions of the Americas, Africa, and Australia (Wickens and Lowe 2008, see Table 1), and sometimes in montane forests (Carlquist 1962, 2001). Remarkably, fleshy-stemmed trees are particularly dominant and diverse in some highly isolated insular environments such as the island of Socotra, that lies in the northwestern Indian Ocean near the mouth of the Red Sea between the Arabian Peninsula and the Horn of Africa (Brown and Mies 2012), or Madagascar (Fischer and Theisen 2000). The third hotspot of sarcocaulescent plants is the Peninsula of Baja California in Mexico (Franco-Vizcaino et al. 1990), which is not a true island now but has evolved as an island ecosystem for most of the last 6 million years. The reason for this extraordinary diversity of sarcocaulescent trees in islands is still a matter of debate. Mabberley (1974) attributed the extraordinary concentration of pachycauly in some islands to lineages of herbaceous ancestors evolving arborescent life-forms in isolation from competition with other trees. Alternatively, the absence of large herbivores in many of these insular environments during most of the Pliocene–Pleistocene could have played an important role in the evolution of fleshy stems.

Apart from the role of isolation in the evolution of the syndrome, it seems also clear that the sarcocaulescent life-form is particularly advantageous in hot dry environments. In contrast with retamoid species, trees with giant fleshy stems tend to occur in the equator-ward edge of the warm deserts and in tropical dry environments, in places such as the Horn of Africa, the Kaokoveld and Succulent Karoo in Namibia, the Tehuacán desert in southern Mexico, the Caatinga in Brazil, or the Dry Chaco in Paraguay. Even within a region, the association with aridity is evident. In Baja California, for example, the largest concentration of sarcocaulescent growth forms occurs in the dry central deserts of the peninsula (Franco-Vizcaino et al. 1990; Perea et al. 2005). Similarly, in Madagascar, the distribution of sarcocauls is chiefly restricted to the TDFs and thickets of the western and southwestern regions (Fischer and Theisen 2000; Wickens and Lowe 2008).

There are some species that seem to lie somewhere in the middle of the retamoid-to-sarcocaulescent gradient. In the American Continent, all species within the genus *Parkinsonia* ('paloverde', formerly in the genus *Cercidium*; Fabaceae, Caesalpinioideae) possess thick stems with green photosynthetic bark, but lack the

Table 1 Some noteworthy sarcocaulescent genera and their geographic distribution

Plant family	Genus	Location	References
Aizoaceae	*Psilocaulon*	South Africa	Adie and Yeaton (2013)
Anacardiaceae	*Cyrtocarpa*	Mexico	Wiggins (1980), Medina-Lemos and Fonseca (2009)
	Pachycormus	Mexico	Nilsen et al. (1990), Franco-Vizcaino et al. (1990)
	Spondias	Costa Rica, Panama, Mexico	Borchert (1994, 1996), Goldstein et al. (1998), Medina-Lemos and Fonseca (2009)
Apocynaceae	*Adenium*	Tropical Africa, Madagascar	Wickens and Lowe (2008)
	Frerea		
	Pachypodium	Madagascar	Wickens and Lowe (2008)
	Plumeria	Costa Rica, Mexico, Nigeria, Puerto Rico	Borchert and Rivera (2001), Alvarado-Cárdenas (2004), Sloan et al. (2006)
Asteraceae	*Dendrosenecio*	Montane Tropical Africa	Wickens and Lowe (2008)
Bixaceae	*Cochlospermum*	Costa Rica	Borchert (1996)
Burseraceae	*Bursera*	Mexico, Tropical America	Nilsen et al. (1990), Borchert (1996), Medina-Lemos (2008), Wickens and Lowe (2008)
	Commiphora	Africa, Mexico	Medina-Lemos (2008), Wickens and Lowe (2008)
Cactaceae	*Pereskia*	Tropical America	Britton and Rose (1963), Arias et al. (2004)
Campanulaceae	*Cyanea*	Hawaii	Wickens and Lowe (2008)
	Brighamia	Hawaii	
	Lobelia	Montane S. America, Africa, India	
Crassulaceae	*Crassula*	Tropical America	
Cucurbitaceae	*Dendrosicyos*	Socotra, Southern Arabia	Wickens and Lowe (2008)
Didiereaceae	*Portulacaria*	South Africa	Cowling and Mills (2011), Adie and Yeaton (2013)
Euphorbiaceae	*Euphorbia*	Africa, Tropical America	
	Givotia	Madagascar	Wickens and Lowe (2008)
	Jatropha	Mexico, Central America	Maes et al. (2009), Díaz-López et al. (2012), Wickens and Lowe (2008)
	Ricinus		Simbo et al. (2013)
Fabaceae	*Enterolobium*	Costa Rica	Borchert (1994)

(continued)

Table 1 (continued)

Plant family	Genus	Location	References
	Dalbergia	Costa Rica, Mexico	Borchert (1994), Olvera-Luna et al. (2012)
	Delonix	Madagascar	Wickens and Lowe (2008)
	Gliricidia	Costa Rica	Borchert (1994)
Fouquieriaceae	*Fouquieria*	Baja California, Sonoran Desert (Arizona)	Franco-Vizcaíno et al. (1990), Nilsen et al. (1990), Pockman and Sperry (2000)
Malvaceae: Bombacoideae	*Adansonia*	Tropical Africa, Madagascar. NE Australia	Wickens and Lowe (2008)
	Bombacopsis	Central America	Borchert and Rivera (2001), Borchert and Pockman (2005), Wickens and Lowe (2008)
	Bombax	Old World Tropics	Coster (1923), Wickens and Lowe (2008)
	Cavanillesia	Tropical America	
	Ceiba (+ *Chorisia*)	Tropical America	Borchert and Rivera (2001), Wickens and Lowe (2008)
	Ochroma	Costa Rica, Panama	Borchert (1994), Machado and Tyree (1994)
	Pseudobombax	Argentina, Brazil, Costa Rica, Panama	Machado and Tyree (1994), Borchert (1996), Borchert and Rivera (2001), Schöngart et al. (2002)
Malvaceae: Sterculioideae	*Brachychiton*	Australia	Wickens and Lowe (2008)
	Hildegardia	Nigeria	Borchert and Rivera (2001)
Moraceae	*Dorstenia*	Tropical Africa, Socotra, India, Mexico	Wickens and Lowe (2008), González-Castañeda and Ibarra-Manríquez (2012)
Moringaceae	*Moringa*	Madagascar	Wickens and Lowe (2008)
Vitaceae	*Cyphostemma*	Africa, Madagascar	Wickens and Lowe (2008)

fleshy parenchymatic tissues or the ability to store water of the true sarcocauls (Fig. 3 of Santiago et al. this volume). Roots of these species can also be green if they grow exposed to sunlight (Fig. 2). A similar intermediate case is posed by the arid-zone tree *Geoffroea decorticans* (Fabaceae, Caesalpinoidea) in Argentina and Chile, which has drought-deciduous leaves and large stems with exfoliating green bark, but lacks fleshy parenchyma and the ability to store water.

Fig. 2 Roots of *Parkinsonia praecox* when growing exposed to sunlight. Tropical dry forest in Margarita Island, Venezuela. Photo by Wilmer Tezara

Retamoid Shrubs

The retamoid syndrome (Zohary 1962; Shmida and Whittaker 1981) is common in certain Old World Mediterranean legumes in genera such as *Retama* (from where the syndrome takes its name), *Calcycotome*, *Cytissus*, *Genista*, *Spartium*, and *Ulex*, all common in southern Europe, northern Africa, and the Near East. Retamoid species are mostly shrubs with highly reduced leaves, photosynthetic stems (Table 2), and, often, spinescent shoots that give them a characteristic "crown of thorns" appearance. In North American drylands, a diverse array of species with similar morphology occur in different families. The most outstanding example is that of the largely leafless green spiny shrubs which resemble *Castela emoryi*, a member of the Simaroubaceae. The convergent forms are *Koeberlinia spinosa* in the Koeberliniaceae, *Canotia holacantha* in the Celastraceae, *Thamnosma montana* in the Rutaceae, *Adolphia californica* in the Rhamnaceae, and *Glossopetalon spinescens* in the Crossosomataceae, all having leaves reduced to scales and persistently green stems that carry out photosynthesis. Compared to North America, the retamoid habit is more common, and taxonomically more diverse in South America, where retamoid forms frequently occur outside of strict drylands, in environments such as the Puna of Chile, the Patagonian steppe, or the Chaco forests (Johnston 1940). The genus *Colletia*, in the family Rhamnaceae, forms the most common set of retamoid plants in South America, including five species widely dispersed in Chile, Argentina, Bolivia, and Uruguay. Other notable retamoid species in South America include *Cassia aphylla* (Caesalpinioideae), *Prosopidastrum globosum* (Fabaceae-Mimosoideae), *Retanilla ephedra* (Rhamnaceae), and *Bulnesia retama*, the only retamoid species within the New World Zygophyllaceae. Following the name of the dominant genera with this characteristic green-stem morphology, retamoid species have also been referred to as "holacanthoid" plants in North America (Muller 1941), or "colletoid" species in South America (Johnston 1940).

The Evolution of Stem Succulence and Photosynthetic Bark

In the primary shoot of dicotyledons the vascular bundles that run along the stem perform the basic function of connecting the leaves to the rest of the plant (Gibson 1978; Tomlinson and Wheat 1979). When the stem's secondary growth begins, a continuous cambial layer develops and the vascular bundles give way to a continuous woody tissue made of xylem, inside the cambial layer, and surrounded by an external layer of phloem covered externally by bark. In short, the main difference between primary and secondary stems lies in the arrangement of conductive tissues in the form of vascular bundles surrounded by large amounts of undifferentiated parenchyma in primary shoots as opposed to a continuous cambium in secondary stems, and the presence of a photosynthetic epidermis in the former as opposed to a suberous bark in the latter.

The trunks of both retamoid and sarcocaulescent growth forms show one or both of these juvenile traits: retamoid plants are characterized by their green photosynthetic stems with a stomata-bearing epidermis, while sarcocauls are characterized by their fleshy stems with large parenchymatic radii, and often by the presence of photosynthetic chlorenchyma in their externally smooth, non-suberous bark. At an anatomical level, Carlquist (1962, 2001) described ancestral juvenile traits in the wood anatomy of sarcocauls, and hypothesized that the evolutionary mechanism for the development of the sarcocaulescent growth form was the retention of juvenile characteristics in the adult trees, a phenomenon he called paedomorphism. Carlquist's hypothesis has been challenged by other authors, such as Mabberley (1982), who believes that pachycauly in islands has evolved from ancestral herbaceous plants becoming larger and larger individuals in the absence of tree competitors, and Olson (2003) who showed that the main morphologic traits in sarcocauls, namely wide parenchymatic rays and abundant axial parenchyma, are present also in vines and lianas. Olson suggests that pachycauly in many taxa evolved repeatedly from lianas in the core eudicots.

So while Carlquist hypothesizes that sarcocaulescence evolved from woody trees with non-fleshy stems, Mabberly believes the syndrome evolved from herbaceous plants that in the absence of competitors became large and tree-like, while Olson supports the idea that giant fleshy stems evolved from vines and lianas. In practice, because the evolution of sarcocaulescence is polyphyletic, all three models could have operated independently in different taxa. Mabberly's model seems a likely hypothesis for the origin of pachycauly in the case of some arborescent Asteraceae (*Dendrosenecio*) and Campanulaceae (*Cyanea*, *Brighamia*, and *Lobelia*), all plants whose nearest relatives are herbaceous, and Olson's model seems plausible in the case of sarcocaulescent Cucurbitaceae (*Dendrosicyos*) and Vitaceae (*Cyphostemma*), two families dominated by vines and creepers. In many other woody taxa [e.g., Anacardiaceae, Burseraceae (Fig. 3), or bombacoid Malvaceae (Fig. 4)], however, the evolutionary pathway is less clearly defined and Carlquist's hypothesis cannot be ruled out. Beyond the details of the discussion, these three models jointly constitute an appealing and evolutionarily parsimonious idea: paedomorphism, the retention of primary shoot traits in enlarged adult stems, could be a simple mechanism of evolution in response to selective forces favoring either reduced leaves and photosynthetic shoots, or a succulent stem with a large proportion of parenchymatic cells

Fig. 3 *Bursera simaruba* in an early successional forest in Gamboa, Panama. Note the photosynthetic bark. Photo by Eleinis Ávila-Lovera

Fig. 4 *Ceiba speciosa* in Riverside, California, USA. Photo by Louis S. Santiago

capable of storing water. It is important to note that both of these characteristics are advantageous in seasonally dry ecosystems, where water is scarce during at least one period of the year. One way or another, these three evolutionary models involve heterochrony, the evolution of changes in the timing of morphologic development events in one taxon relative to another, as their driving mechanism. Conceptually, they might explain why the retamoid and the sarcoculescent syndromes are so common in drylands throughout the world, and why they have arisen independently in so many taxonomically unrelated families.

Ecophysiology of Trees with Succulent Stems

When we think of trees with succulent stems the first that come to mind are the Baobabs (Fig. 1). These trees might be the most famous among stem-succulent trees, and Wickens and Lowe (2008) have described them as "grotesque trees dominating the landscape"; the landscape usually being African savannas. These singular trees usually store water in their stems during the rainy season and it is thought that the water is used during the dry season. In this section we will describe the use of water by this type of tree and how they physiologically respond to environmental stresses such as drought.

Physiology of Succulent Trees

Plants with succulent stems are found in families spread among the Angiosperms (Table 1). Representative genera mostly belong to the family Malvaceae-Bombacoideae, followed by Apocynaceae, Euphorbiaceae and Fabaceae. A common characteristic is that they all inhabit seasonally dry environments.

Table 2 Ratio of stem net photosynthesis to leaf photosynthesis in retamoid species from North American Deserts (taken and modified from Ávila et al. 2014)

Species	Family	Location	Stem-to-leaf A ratio[a]	References
Bebbia juncea	Asteraceae	Sonoran Desert	0.52	Ehleringer et al. (1987)
Chrysothamnus paniculatus	Asteraceae	Sonoran Desert	0.64	
Dyssodia porophylloides	Asteraceae	Sonoran Desert	0.83	
Gutierrezia microcephala	Asteraceae	Sonoran Desert	0.85	
G. sarothrae	Asteraceae	Sonoran Desert	0.26	
Hymenoclea salsola	Asteraceae	Sonoran Desert	0.67	
Lepidium fremontii	Brassicaceae	Sonoran Desert	0.70	
Porophyllum gracile	Asteraceae	Sonoran Desert	0.63	
Psilostrophe cooperi	Asteraceae	Sonoran Desert	1.01	
Salazaria mexicana	Lammiaceae	Sonoran Desert	1.11	
Senecio douglasii	Asteraceae	Sonoran Desert	0.06	
Sphaeralcea parvifolia	Malvaceae	Sonoran Desert	0.61	
Stephanomeria pauciflora	Asteraceae	Sonoran Desert	1.04	
Thamnosma montana	Rutaceae	Sonoran Desert	0.48	
Eriogonum inflatum	Polygonaceae	Mojave Desert	0.50	Osmond et al. (1987)
Hymenoclea salsola	Asteraceae	Arizona	0.60	Comstock and Ehleringer (1988)
Spartium junceum	Fabaceae	California	0.38	(Nilsen and Bao 1990)
Justicia californica	Acantaceae	Sonoran Desert	1.29	Tinoco-Ojanguren (2008)

[a]Photosynthetic rate of stems were expressed in projected area before calculating the ratio of stem-to-leaf photosynthesis

Usually, succulence in stems is associated with the presence of chlorenchymatic tissue underneath the periderm which can re-assimilate CO_2 released by respiration (Nilsen et al. 1990). Stem recycling photosynthesis has been found in African Baobab (*Adansonia digitata*) and Castor bean (*Ricinus communis*), and effectively contributes to bud development in both plants (Simbo et al. 2013). The stem contribution to bud development was estimated by excluding light from penetrating the stem periderm; a reduction of 50 and 67 % in dry biomass of developed buds was found in drought and watered *Adansonia digitata* plants, respectively, while the reduction was lower in *Ricinus communis* (25 and 40 % in drought and watered plants, respectively) (Simbo et al. 2013).

When comparing leaf and stem photosynthesis, rain was found to be detrimental for CO_2 diffusion in leaves because it can clog stomata, but wetting the stem periderm decreases reflectance which subsequently increases light absorption by the chlorenchyma, electron transport rate (ETR), and photosynthesis in *Quercus coccifera* (Manetas 2004). For plants with SRP, CO_2 diffusion does not decrease with stem wetting because the CO_2 source is derived from respiration and not from the atmosphere.

Similarities and differences have also been found when comparing leaf and SNP. In general terms, stem photosynthesis functions just like leaf photosynthesis. Stem net photosynthesis has the same responses to environmental variables such as photosynthetic photon flux density (PPFD), internal concentration of CO_2 (C_i), temperature and vapor pressure deficit (VPD). Both stem and leaf photosynthesis show C_3 metabolism, and both organs have high stomatal density (Osmond et al. 1987; Nilsen et al. 1989; Nilsen and Sharifi 1994; Aschan and Pfanz 2003). However, WUE has been found to be higher in stems, which can be incredibly valuable in periods of water deficit when most plants with SNP are leafless (Ávila et al. 2014). On the other hand, stem photosynthesis in plants with SRP occurs at no water cost, making these plants very successful during periods of water deficit and/or low temperature, where they can re-assimilate respired CO_2 without losing any water.

Since stem-succulent trees are typically found in seasonally dry ecosystems, one advantage that has been associated with this trait is the use of stored water during the dry season to buffer daily water deficits. However, some studies have evaluated this assumption and have found that daily use of stored water is usually negligible in *Adansonia* species (Chapotin et al. 2006a, b, c). In *Adansonia rubostripa* and *Adansonia za*, stored water does not buffer daily water potential (Ψ) because of the difficulty in withdrawing water from the storage tissue (Chapotin et al. 2006c). Instead, stored water is used to flush new leaves before the onset of the rainy season in *Adansonia* species, giving them an advantage over other species since leaves are already present when the first rain comes, thus maximizing photosynthetic capacity and extending the growing season of the plants (Chapotin et al. 2006b). This phenology also takes place in *Plumeria alba* from Guánica, Puerto Rico, where the peak of leaf flushing occurs 2–4 months before the peak of rainfall when water availability of the soil is low (Sloan et al. 2006).

Other species show daily use of stored water during the lag between leaf transpiration and stem basal sap flow, as demonstrated in five species of tropical canopy trees from a lowland seasonal moist forest in Panama (Goldstein et al. 1998). Here, the use of stored water and recharge of reservoirs is a dynamic process that can even be altered by fluctuating environmental conditions (Goldstein et al. 1998). In drier ecosystems, such as thornscrubs and deserts, seasonal and diurnal variation in leaf Ψ is small due to the buffering capacity of succulent stems as is found in *Fouquieria columnnaris*, *Pachycormus discolor*, and *Bursera microphylla* in Baja California (Nilsen et al. 1990). The SRP found in *Fouquieria columnnaris* and *Pachycormus discolor*, in addition to the capacity to store water in their stems, may ensure survival during extreme drought conditions (Franco-Vizcaino et al. 1990).

Not only leaf-flushing but also flowering has been associated with stored water in stem-succulent trees. In a study performed in a TDF in Costa Rica where five PFG were evaluated (deciduous hardwood, deciduous lightwood, deciduous softwood, evergreen lightwood and evergreen softwood), deciduous lightwood trees were found to have the highest capacity to store water during the rainy season (Borchert 1994). On average, stem water content of stem-succulent trees was 63 % compared to 31 % in deciduous hardwood, 47 % in deciduous softwood, 46 % in evergreen lightwood and 51 % in evergreen softwood (Borchert 1994). This capacity to store water in stem-succulent trees was associated with low wood density (0.40 g cm^{-3}), and the water stored during the wet season was found to be used at the end of the dry season for leaf flushing and flowering (Borchert 1994). It was also found that deciduous lightwood trees experienced less water deficit, and both their leaf and stem Ψ remained high during the dry season after leaf shedding (Borchert 1994).

Water storage capacity of stems is highly correlated to wood anatomy and biochemical support (Borchert and Pockman 2005; Chapotin et al. 2006a). Stem-succulent plants among other plant types have the highest capacity to store water and can maintain higher Ψ than deciduous and leaf-exchanging species during drought (Borchert and Pockman 2005). One might think that succulence is a disadvantage for stability and mechanical support. However, in six species of *Adansonia*, lighter wood and its intrinsic high capacity of water storage acts as hydrostatic pressure against the bark which can contribute to biomechanical stability in tall trees (Chapotin et al. 2006a; Niklas 2016 this volume). It seems that succulent plants are better armed to face drought since water uptake during rehydration and minimum Ψ in the dry season are correlated to water storage capacity (Borchert and Pockman 2005), and they suffer less from collapse when fully hydrated (Chapotin et al. 2006a).

Leaf and root morphology are often indicative of where plants live or can live. Baobab trees can have smaller leaves with higher stomatal density in drier and hotter areas than in wetter areas (Cuni Sanchez et al. 2010). Also, while rooting depth is sometimes deeper in arid than in sub-humid ecosystems, stem succulents have an intermediate depth, with widely spread shallow roots (Schenk and Jackson 2002). Furthermore, pruning has a significant effect on leaf size: Baobab trees growing in the same environment have smaller leaves on pruned branches than on non-pruned branches (Cuni Sanchez et al. 2010). The genetics of different populations and the phenotypic plasticity in physiological traits of a single population found in Baobabs may also play a role in drought responses (Cuni Sanchez et al. 2010).

Plant functional groups (evergreen, deciduous, brevi-deciduous, and stem-succulent trees) were studied in Guanacaste, Costa Rica to determine what trait or suite of traits are part of the strategies tropical trees use to respond to drought periods in terms of water balance, wood traits, and phenological behavior (Worbes et al. 2013). In a Principal Component Analysis (PCA), the two first axes were related to hydraulic conductivity, control of transpiration and water loss (Worbes et al. 2013). In *Cochlospermum vitifolium*, plants flush leaves all year round despite the seasonality of rainfall in Guanacaste (Fallas-Cedeño et al. 2010). Stem succulents also have higher leaf and wood carbon isotopic composition ($\delta^{13}C$) when comparing to deciduous, brevi-deciduous and evergreen species, suggesting that they have tighter control over their stomata, making them water conservative species and successful pioneers in this TDF (Worbes et al. 2013). In another TDF in Belize, stem succulents flush leaves early in the dry season, likely using water stored during the previous rainy season (Sayer and Newbery 2003).

Most stem-succulent trees occur in TDF, yet there is some evidence of stem succulents in Amazonian forests. Schöngart et al. (2002) studied the same PFGs as Worbes et al. (2013) and their phenological and stem-growth responses to flood-pulses and found that *Pseudobombax munguba*, a stem-succulent tree species, flushed new leaves only after the end of the flooded period. The whole reproductive phase was completed within one single aquatic phase in contrast with the others PFGs, and maturation of fruits finished by the end of the aquatic phase or with a little extension intro the terrestrial phase (Schöngart et al. 2002). Stem diameter increment had the highest correlation with monthly precipitation among the PFGs studied (Schöngart et al. 2002). It was argued that the phenological processes are correlated to the flood-period and not to photoperiod (Parolin et al. 2016, this volume), as it has been stated for TDF trees in Costa Rica and Puerto Rico by Borchert and Rivera (2001) and Sloan et al. (2006), respectively.

Physiological Responses to Drought

In terms of drought, numerous experiments have been performed to assess photosynthesis and growth responses to water deficit. Most of these studies are conducted in pots and in greenhouses under controlled conditions using juvenile plants. Despite the lack of realism compared to field-based studies, these data are needed to advance our understanding of the mechanisms driving plant processes. Some studies have directly compared adult populations of the same species or closely related species in contrasting environments. Some of the plant species listed in Table 1 have been used in drought experiments and the most relevant results are presented below.

Two populations of *Adansonia digitata* from West and Southeast Africa were compared in terms of physiological and morphological responses to drought (De Smedt et al. 2012). Seedlings were used to investigate the mechanisms by which Baobab juveniles cope with soil drought and it was found that the population from West Africa had the strongest drought-avoidance mechanism, making them more water conservative than seedlings from Southeast Africa (De Smedt et al. 2012). In another comparative study, *Adansonia grandidieri*, *Adansonia madagascariensis* and *Adansonia rupostripa* from Madagascar, the species with the lowest drought tolerance came from the highest rainfall ecosystem (Randriamanana et al. 2012). All species cope with drought by reducing stomatal conductance (g_s) and having water stored in the taproot, with a high WUE at the expense of maintaining high photosynthetic rates (Randriamanana et al. 2012).

Seedlings of *Adansonia digitata* have been used to study sap flow and water use in drought experiments, which include a drought + recovery treatment (Van den Bilcke et al. 2013). The mechanisms associated with survival during drought included succulence of the taproot, which represented 17.5 % of total daily water use, and SRP which takes place in the chlorenchyma below the periderm (Van den Bilcke et al. 2013).

Drought can affect different aspects of the physiology of stem-succulent trees, depending on the species. *Jatropha curcas* shows no changes in specific leaf area, Ψ range, relative water content, transpiration efficiency, or aboveground biomass, but phenology (Maes et al. 2009) and biomass production did change (Achten et al. 2010; Díaz-López et al. 2012). Drought-induced production of new leaves with reduced leaf area and higher stomatal density (Maes et al. 2009), as has been found in populations of *Adansonia digitata*, as well as reduced leaf, stem and root growth (Achten et al. 2010), leaf Ψ, pressure potential, photosynthetic rate, g_s, WUE, and maximum quantum efficiency of photosystem II (Díaz-López et al. 2012). These morphological and physiological responses to drought allow plants to have a conservative water use strategy.

Contrary to what is expected, photosynthetic WUE and crop WUE (kg fruits m^{-3} H_2O) in *Jatropa curcas* decrease with drought (Abou Kheira and Atta 2009; Díaz-López et al. 2012). The highest crop WUE was found in the treatment with the highest water availability (Abou Kheira and Atta 2009). On the other hand, most other characteristics, including oilseed quality, do not change with drought treatments, which indicates that *Jatropa curcas* can be used to re-vegetate a wide range of arid lands to exploit its oil without changes in its quality (Abou Kheira and Atta 2009).

Another trait to cope with drought in succulent trees is starch storage during the rainy season (Fallas-Cedeño et al. 2010). As with water, starch is used seasonally and not daily in *Cochlospermum vitifolium*; it is used as a storage reserve for phenological events such as branch extension, leaf flushing, and reproduction that take place during the dry season before the onset of rains (Fallas-Cedeño et al. 2010). However, both starch storage and stem succulence are typical traits that correspond to a drought avoidance strategy.

Use of Stem-Succulent Trees for Conservation and Rehabilitation of Degraded Arid Lands

Stem-succulent plants have great potential for use in restoration of degraded lands due to their exceptional physiological performance and tolerance to drought. However, little has been done to actually address this hypothesis and determine whether these traits are enough to promote effective reforestation practices.

One of the few studies that support this hypothesis was performed in South Africa where *Portulacaria afra* was found to be a nurse plant playing an important role in the regeneration dynamics of arid subtropical thicket vegetation (Adie and Yeaton 2013). This species modifies microhabitats and creates opportunities for plants that are more susceptible to extreme conditions, which are common in this ecosystem (Adie and Yeaton 2013). *Portulacaria afra* clumps comprised approximately 50 % of the studied area and approximately 90 % of tree seedlings were recorded under its canopy (Adie and Yeaton 2013). *Portulacaria afra* may simply provide shade and protection against intense rain events to young seedlings, but a high soil carbon content has been found under its clumps (Cowling and Mills 2011), which is known to have important effects on soil structure and, possibly, on the soil microbial community, enhancing the recruitment of other plant species.

More informal than an experiment is the observation that Baobab trees are now being planted—consciously or unconsciously—in areas where they were not originally present. Some of these areas are even drier than their native range. This provides evidence that stem-succulent trees can cope with severe drought periods, which would be ideal to restore plant communities in arid lands. Furthermore, Baobabs have multifunctional uses as shade and street trees, for water storage, shelter and storage, food, wood, fiber, fertilizer, fuel, insecticide, and as an ornamental (Wickens and Lowe 2008). Recently, Baobab seed oil and fruit pulp have been exported to countries outside of Africa, such as Canada, USA and European countries (Venter and Witkowski 2010). Using them to reforest degraded lands can have significant positive effects on both the health of the land and the economy of nearby villages.

Conclusions and Future Directions

In seasonally dry ecosystems, there are predominantly three plant growth forms with photosynthetic stems, based on morphology, anatomy and physiology of the stems: sarcocaulescent, retamoid, and cactoid. The sarcocaulescent and retamoid forms include plant species with a broad biogeographic distribution and many similar characteristics. Even when the stem does not look green it might have a layer of chlorenchyma beneath the periderm, which can carry either SNP or SRP (Ávila et al. 2014). This feature and the possibility to store water in fleshy stems of sarcocaulescent species, supports the ability to cope with prolonged periods of water deficit, common in deserts and TDF.

There is not a unique hypothesis about the origin of stem succulence, photosynthetic bark, or sarcocauls in general. There are multiple hypotheses that cannot be ruled out because the evolution of the green stem syndrome is polyphyletic, and all possible models of evolution could have operated independently in different taxonomically unrelated families.

Ecophysiological performance of different sarcocaulescent species have been described in tropical countries of the Old and New Worlds. However, more in situ studies under field conditions that take into account all biotic and abiotic factors affecting the physiology of adult populations need to be done if we are to use this information to make decisions about land management and conservation of endangered species. This effort is underway, but we still need the help of new physiological ecologists to work on the still unanswered questions.

References

Abou Kheira AA, Atta NMM (2009) Response of Jatropha curcas L. to water deficits: yield, water use efficiency and oilseed characteristics. Biomass Bioenergy 33:1343–1350

Achten WMJ, Maes WH, Reubens B et al (2010) Biomass production and allocation in Jatropha curcas L. seedlings under different levels of drought stress. Biomass Bioenergy 34:667–676

Adie H, Yeaton RI (2013) Regeneration dynamics in arid subtropical thicket, South Africa. S Afr J Bot 88:80–85

Aschan G, Pfanz H (2003) Non-foliar photosynthesis—a strategy of additional carbon acquisition. Flora—Morphol Distr Funct Ecol Plants 198:81–97

Alvarado-Cárdenas LO (2004) Apocynaceae. In: Medina-Lemos R, Sánchez-Ken JG, García-Mendoza A, Arias-Montes S (eds) Flora del Valle de Tehuacán-Cuicatlán 38: 1–57. Instituto de Biología, Universidad Nacional Autónoma de México, Mexico

Arias S, Gama-López S, Guzmán-Cruz LU, Vázquez-Benítez B (2004) Cactaceae. In: Medina-Lemos R, Sánchez-Ken JG, García-Mendoza A, Arias-Montes S (eds) Flora del Valle de Tehuacán-Cuicatlán 95: 1–235. Instituto de Biología, Universidad Nacional Autónoma de México, Mexico

Ávila E, Herrera A, Tezara W (2014) Contribution of stem CO_2 fixation to whole-plant carbon balance in nonsucculent species. Photosynthetica 52:3–15

Becerra JX (2005) Timing the origin and expansion of the Mexican tropical dry forest. Proc Natl Acad Sci USA 102:10919–10923

Borchert R (1994) Soil and stem water storage determine phenology and distribution of tropical dry forest trees. Ecology 75:1437

Borchert R (1996) Phenology and flowering periodicity of Neotropical dry forest species: evidence from herbarium collections. J Trop Ecol 12:65–80

Borchert R, Pockman WT (2005) Water storage capacitance and xylem tension in isolated branches of temperate and tropical trees. Tree Physiol 25:457–466

Borchert R, Rivera G (2001) Photoperiodic control of seasonal development and dormancy in tropical stem-succulent trees. Tree Physiol 21:213–221

Britton NL, Rose JN (1963) The Cactaceae, vol 1, 2. Revised edition. Dover Publications, New York

Brown G, Mies BA (2012) Vegetation ecology of Socotra. Springer, Dordrecht

Bullock SH, Mooney HA, Medina E (1995) Seasonally dry tropical forests. Cambridge University Press, Cambridge

Carlquist S (1962) Theory of paedomorphosis in dicotyledonous. Phytomorphology 12:30–45

Carlquist S (2001) Comparative wood anatomy. Springer, Berlin

Chapotin SM, Razanameharizaka JH, Holbrook NM (2006a) A biomechanical perspective on the role of large stem volume and high water content in baobab trees (Adansonia spp.; Bombacaceae). Am J Bot 93:1251–1264

Chapotin SM, Razanameharizaka JH, Holbrook NM (2006b) Baobab trees (Adansonia) in Madagascar use stored water to flush new leaves but not to support stomatal opening before the rainy season. New Phytol 169:549–559

Chapotin SM, Razanameharizaka JH, Holbrook NM (2006c) Water relations of baobab trees (Adansonia spp. L.) during the rainy season: does stem water buffer daily water deficits? Plant Cell Environ 29:1021–1032

Chaves MM, Pereira JS (1992) Water stress, CO_2 and climate change. J Exp Bot 43:1131–1139

Chaves MM, Pereira JS, Maroco J et al (2002) How plants cope with water stress in the field? Photosynthesis and growth. Ann Bot 89:907–916

Comstock JP, Ehleringer JR (1988) Contrasting photosynthetic behavior in leaves and twigs of Hymenoclea salsola, a Green-Twigged Warm Desert Shrub. Am J Bot 75:1360–1370

Coster CH (1923). Lauberneuerung und andere periodische Lebensprozesse in dem trockenen MonsungebietOst-Javas. Annales du Jardin botanique de Buitenzorg 33:117–189

Cowling RM, Mills AJ (2011) A preliminary assessment of rain throughfall beneath Portulacaria afra canopy in subtropical thicket and its implications for soil carbon stocks. S Afr J Bot 77:236–240

Cuni Sanchez A, Haq N, Assogbadjo AE (2010) Variation in baobab (Adansonia digitata L.) leaf morphology and its relation to drought tolerance. Genet Resour Crop Evol 57:17–25

De Smedt S, Cuní Sanchez A, Van den Bilcke N et al (2012) Functional responses of baobab (Adansonia digitata L.) seedlings to drought conditions: differences between western and south-eastern Africa. Environ Exp Bot 75:181–187

Díaz-López L, Gimeno V, Simón I et al (2012) Jatropha curcas seedlings show a water conservation strategy under drought conditions based on decreasing leaf growth and stomatal conductance. Agric Water Manag 105:48–56

Dimmit MA (2000) Biomes and communities of the Sonoran Desert region. In: Phillips SJ, Comus PW (eds) A natural history of the Sonoran Desert. ASDM Press/University of California Press

Douglas MW, Maddox RA, Kenneth H (1993) The Mexican monsoon

Ehleringer JR, Comstock JP, Cooper TA (1987) Leaf-twig carbon isotope ratio differences in photosynthetic-twig desert shrubs. Oecologia 71:318–320

Ezcurra E, Mellink E, Wehncke E, González C, Morrison S, Warren A, Dent D, Driessen P (2006) Natural history and evolution of the world's deserts. In: Ezcurra E (ed) Global deserts outlook. United Nations Environment Programme (UNEP), Nairobi

Fallas-Cedeño L, Holbrook NM, Rocha OJ et al (2010) Phenology, lignotubers, and water relations of Cochlospermum vitifolium, a pioneer tropical dry forest tree in Costa Rica. Biotropica 42:104–111

Fischer E, Theisen I (2000) Vegetation of Malagasy Inselbergs. In: Porembski PDS, Barthlott PDW (eds) Inselbergs. Springer, Berlin, pp 259–276

Franco-Vizcaino E, Goldstein G, Ting IP (1990) Comparative gas exchange of leaves and bark in three stem succulents of Baja California. Am J Bot 77:1272–1278

García-Oliva F, Ezcurra E, Galicia L (1991) Pattern of rainfall distribution in the Central Pacific Coast of Mexico. Geografiska Annaler Ser A, Phys Geogr 73:179–186

Gibson AC (1978) Architectural designs of wood skeletons in cacti. Cactus Succulent J Great Britain 40:73–80

Goldstein G, Andrade JL, Meinzer FC et al (1998) Stem water storage and diurnal patterns of water use in tropical forest canopy trees. Plant Cell Environ 21:397–406

González-Castañeda N, Ibarra-Manríquez G (2012) Moraceae. In: Medina-Lemos R, Sánchez-Ken JG, García-Mendoza A, Arias-Montes S (eds) Flora del Valle de Tehuacán-Cuicatlán 96: 1–33. Instituto de Biología, Universidad Nacional Autónoma de México, Mexico

Gordon JE, Hawthorne WD, Reyes-García A, et al (2004) Assessing landscapes: a case study of tree and shrub diversity in the seasonally dry tropical forests of Oaxaca, Mexico and southern Honduras. Biol Conserv 117:429–442

Johnston IM (1940) The Floristic Significance of Shrubs Common to North and South American Deserts. Journalof the Arnold Arboretum 21:356–363

Mabberley DJ (1974) Pachycauly, vessel-elements, islands and the evolution of arborescence in "Herbaceous" families. New Phytol 73:977–984

Mabberley DJ (1982) On Dr Carlquist's defence of paedomorphosis. New Phytol 90:751–755

Machado J-L, Tyree MT (1994) Patterns of hydraulic architecture and water relations of two tropical canopy trees with contrasting leaf phenologies: Ochroma pyramidale and Pseudobombax septenatum. Tree Physiol 14:219–240

Maes WH, Achten WMJ, Reubens B et al (2009) Plant-water relationships and growth strategies of Jatropha curcas L. seedlings under different levels of drought stress. J Arid Environ 73:877–884

Manetas Y (2004) Photosynthesizing in the rain: beneficial effects of twig wetting on corticular photosynthesis through changes in the periderm optical properties. Flora-Morphol Distr Funct Ecol Plants 199:334–341

McDowell N, Pockman WT, Allen CD et al (2008) Mechanisms of plant survival and mortality during drought: why do some plants survive while others succumb to drought? New Phytol 178:719–739

Medina E (1995) Diversity of life forms of higher plants in neotropical dry forests. In: Bullock SH, Mooney HA, Medina E (eds) Seasonally dry tropical forests. Cambridge University Press, New York

Medina-Lemos R (2008) Burseraceae. In: Medina-Lemos R, Sánchez-Ken JG, García-Mendoza A, Arias-Montes S (eds) Flora del Valle de Tehuacán-Cuicatlán 66: 1-76. Instituto de Biología, Universidad Nacional Autónoma de México, Mexico

Medina-Lemos R, Fonseca RM (2009) Anacardiaceae. In: Medina-Lemos R, Sánchez-Ken JG, García-Mendoza A, Arias-Montes S (eds) Flora del Valle de Tehuacán-Cuicatlán 96: 1-33. Instituto de Biología, Universidad Nacional Autónoma de México, Mexico

Muller CH (1941) The holocanthoid plants of North America. Madroño 6:128–132

Nilsen E, Sharifi M (1997) Carbon isotopic composition of legumes with photosynthetic stems from mediterranean and desert habitats. Am J Bot 84:1707–1713

Nilsen ET, Bao Y (1990) The influence of water stress on stem and leaf photosynthesis in Glycine max and Sparteum junceum (Leguminosae). Am J Bot 77:1007–1015

Nilsen ET, Sharifi MR (1994) Seasonal acclimation of stem photosynthesis in woody legume species from the Mojave and Sonoran deserts of California. Plant Physiol 105:1385–1391

Nilsen ET, Meinzer FC, Rundel PW (1989) Stem photosynthesis in Psorothamnus spinosus (smoke tree) in the Sonoran desert of California. Oecologia 79:193–197

Nilsen ET, Sharifi MR, Rundel PW et al (1990) Water relations of stem succulent trees in north-central Baja California. Oecologia 82:299–303

Olson ME (2003) Stem and leaf anatomy of the arborescent Cucurbitaceae Dendrosicyos socotrana with comments on the evolution of pachycauls from lianas. Plant Syst Evol 239:199–214

Olvera-Luna AR, Gama-López S, Delgado-Salinas A (2012) Fabaceae. In: Medina-Lemos R, Sánchez-Ken JG, García-Mendoza A, Arias-Montes S (eds) Flora del Valle de Tehuacán-Cuicatlán 107: 1–42. Instituto de Biología, Universidad Nacional Autónoma de México, Mexico

Osmond CB, Smith SD, Gui-Ying B, Sharkey TD (1987) Stem photosynthesis in a desert ephemeral, Eriogonum inflatum. Oecologia 72:542–549

Perea MC, Ezcurra E, León de la Luz JL (2005) Functional morphology of a sarcocaulescent desert scrub in the bay of La Paz, Baja California Sur, Mexico. J Arid Environ 62:413–426

Pockman WT, Sperry JS (2000) Vulnerability to xylem cavitation and the distribution of Sonoran desert vegetation. Am J Bot 87:1287–1299

Randriamanana T, Wang F, Lehto T, Aphalo PJ (2012) Water use strategies of seedlings of three Malagasy Adansonia species under drought. S Afr J Bot 81:61–70

Rodríguez JP, Rojas-Suárez F (2008) Libro Rojo de la Fauna Venezolana

Rodríguez JP, Rojas-Suárez F, Giraldo Hernández D (2010) Libro Rojo de Los Ecosistemas Terrestres de Venezuela

Rowley G (1987) Caudiciform and pachycaul succulents: pachycauls, bottle-, barrel-and elephant-trees and their kin, 1st edn. Strawberry Press, Mill Valley, Calif

Sayer EJ, Newbery DM (2003) The role of tree size in the leafing phenology of a seasonally dry tropical forest in Belize, Central America. J Trop Ecol 19:539–548

Schaedle M (1975) Tree photosynthesis. Annu Rev Plant Physiol 26:101–115

Schenk HJ, Jackson RB (2002) Rooting depths, lateral root spreads and below-ground/above-ground allometries of plants in water-limited ecosystems. J Ecol 90:480–494

Schnitzer SA, Bongers F (2002) The ecology of lianas and their role in forests. Trends Ecol Evol 17:223–230

Schöngart J, Piedade MTF, Ludwigshausen S et al (2002) Phenology and stem-growth periodicity of tree species in Amazonian floodplain forests. J Trop Ecol

Shmida A, Whittaker RH (1981) Pattern and biological Microsite effects in two shrub communities, Southern California. Ecology 62:234–251

Simbo DJ, Van den Bilcke N, Samson R (2013) Contribution of corticular photosynthesis to bud development in African baobab (Adansonia digitata L.) and Castor bean (Ricinus communis L.) seedlings. Environ Exp Bot 95:1–5

Sloan SA, Zimmerman JK, Sabat AM (2006) Phenology of Plumeria alba and its herbivores in a tropical dry forest. Biotropica 39:195–201

Smith SD, Osmond CB (1987) Stem photosynthesis in a desert ephemeral, Eriogonum inflatum. Morphology, stomatal conductance and water-use efficiency in field populations. Oecologia 72:533–541

Stensrud DJ, Gall RL, Mullen SL, Howard KW (1995) Model climatology of the Mexican monsoon. J Climate 8:1775–1794

Tezara W, Urich R, Coronel I et al (2010) Asimilación de carbono, eficiencia de uso de agua y actividad fotoquímica en xerófitas de ecosistemas semiáridos de Venezuela. Ecosistemas 19:67–78

Tinoco-Ojanguren C (2008) Diurnal and seasonal patterns of gas exchange and carbon gain contribution of leaves and stems of Justicia californica in the Sonoran Desert. J Arid Environ 72:127–140

Tomlinson PB, Wheat DW (1979) Bijugate phyllotaxis in Rhizophoreae (Rhizophoraceae). Bot J Linn Soc 78:317–321

Van den Bilcke N, De Smedt S, Simbo DJ, Samson R (2013) Sap flow and water use in African baobab (Adansonia digitata L.) seedlings in response to drought stress. S Afr J Bot 88:438–446

Venter SM, Witkowski ETF (2010) Baobab (Adansonia digitata L.) density, size-class distribution and population trends between four land-use types in northern Venda, South Africa. For Ecol Manage 259:294–300

Wickens GE, Lowe P (2008) The baobabs: pachycauls of Africa, Madagascar and Australia: the pachycauls of Africa. Springer, Madagascar and Australia

Wiggins IL (1980) Flora of Baja California. Stanford University Press, Stanford, Calif

Worbes M, Blanchart S, Fichtler E (2013) Relations between water balance, wood traits and phenological behavior of tree species from a tropical dry forest in Costa Rica-a multifactorial study. Tree Physiol 33:527–536

Zohary M (1962) Plant life of Palestine: Israel and Jordan. Ronald Press Company

Palm Physiology and Distribution in Response to Global Environmental Change

Heidi J. Renninger and Nathan G. Phillips

Abstract Palms (Arecaceae) represent one of the oldest surviving monocot families maintaining a presence in tropical rainforest-like biomes throughout history. Comprising a variety of plant growth forms (arborescent, acaulescent, lianoid), palms are one of the few monocots that achieve significant heights. In doing so, they face many of the same environmental and physiological constraints as dicotyledonous trees including long-distance water transport and longevity making them an important, but largely missing, component of comparative tree physiological studies. Palms differ from dicot trees in several key ways including lacking dormancy mechanisms that restrict them to mainly tropical climates. Palms also lack a vascular cambium and the constant addition of new conduits, and instead, rely exclusively on vascular bundles for fluid transport and mechanical stability. The majority of arborescent palm species also possess only one apical meristem complex from which all new leaf and stem growth originates thereby limiting their options for leaf positioning and light acquisition. These differences will likely alter the response of palms to global change compared with dicot species. Temperature increases have the potential to extend palm distributions to higher elevations and latitudes, but could negatively affect individual palm carbon balance. Within the tropics, precipitation has been shown to have the strongest positive effect on palm species richness and future changes in rainfall patterns will likely alter palm distributions. Therefore, global change has the potential to alter both palm distributions and individual physiological functioning, but palms will likely continue to have a considerable presence in many tropical ecosystems.

H.J. Renninger (✉)
Department of Forestry, Mississippi State University,
Box 9681, Mississippi, MS 39762-9601, USA
e-mail: hrenninger@gmail.com

N.G. Phillips
Department of Earth and Environment, Boston University,
685 Commonwealth Avenue, Boston, MA 02215, USA
e-mail: nathan@bu.edu

© Springer International Publishing Switzerland 2016   67
G. Goldstein and L.S. Santiago (eds.), *Tropical Tree Physiology*,
Tree Physiology 6, DOI 10.1007/978-3-319-27422-5_4

Keywords Agriculture · Architecture · Climate change · Global distribution · Leaf life span

Introduction

Rising atmospheric carbon dioxide (CO_2) and temperature are changing the structure and function of trees (Phillips et al. 2008; Way and Oren 2010) and their global distributions, including palms (Walther et al. 2007). The unique monopodial growth form in palms suggests that they may respond to environmental change in categorically different ways than dicotyledonous trees that display secondary growth characteristics. Yet little information is available about climate and environmental change responses of palms. In addition to their economic and ecological importance, palms can also provide a broader value in comparative research of tree physiology, but are absent from some pertinent studies (e.g., Mäkelä and Sievänen 1992; Way and Oren 2010; Stephenson et al. 2014). In this chapter we review physiological ecology of palms, and assess their potential responses to increased temperature and [CO_2] and altered water availability in the context of global change.

The palm family (Arecaceae) exhibits a pan-tropical distribution with individual species found in a wide range of ecosystem types including rainforests, montane regions, dry forests, savannas and desert oases (Tomlinson 2006). Palms are monocots and represent one of the oldest extant families in this clade based on molecular dating (Janssen and Bremer 2004), with the oldest palm fossils found from the Turonian stage in the Late Cretaceous (Berry 1914; Kvacek and Herman 2004; Harley 2006). There is also evidence that palms have existed throughout the entire recorded span of the tropical rainforest-like biome (Couvreur et al. 2011; Baker and Couvreur 2013b) making them a model plant family to study the evolution of tropical rainforests (Couvreur et al. 2011). Palms likely originated in the northern latitudes on the Laurasian supercontinent and subsequently dispersed to equatorial South America, Africa and Southeast Asia before the end of the Cretaceous (Bjorholm et al. 2006; Couvreur et al. 2011; Baker and Couvreur 2013a; Couvreur and Baker 2013). They were widespread during the Early Eocene Climatic Optimum where global temperatures were much warmer than today and boreotropical forests extended well into the northern latitudes (Morley 2000; Bjorholm et al. 2006). Likewise, palms have shown consistent diversification rates and low extinction rates throughout their history (Couvreur et al. 2011). Therefore as a plant family, palms exhibit great resilience as well as evidence of long distance dispersal and the ability to colonize most areas with suitable climate (Baker and Couvreur 2012).

Today, the majority of palm species diversity occurs in tropical Asia, with around 1200 species followed by the Neotropics with about 800 species (Dransfield et al. 2008). In a study of tree species composition and species richness in the

Fig. 1 *Iriartea deltoidea*
(Ruiz & Pav.), the fifth most
abundant tree species in the
Amazon (Ter Steege et al.
2013), growing at Tiputini
Biodiversity Station in eastern
Ecuador

Amazonian rainforest, seven of the 20 most common tree species were palms
(Fig. 1) with the palm family having five times more "hyperdominant" species than
expected by chance and being second only to the Fabaceae in individual tree
abundance throughout the region (Ter Steege et al. 2013). Surprisingly, although
there are about 200 palm species in Madagascar, there are only 65 species in
continental Africa (Dransfield et al. 2008). Research suggests that tropical Asia and
South America have greater species richness due to higher diversification benefit-
ting from plate tectonic activity (Baker and Couvreur 2013a), as opposed to Africa
which has remained relatively isolated from other landmasses since the breakup of
Gondwana (Couvreur and Baker 2013; Couvreur 2014). Historical climate change
and species extinctions may have also played a role in Africa's shortage of palm
species as well as the current climate because Africa tends to be drier than other
tropical locations (Couvreur and Baker 2013; Couvreur 2014).

Although palms have a dominant presence in the tropics and, in particular tropical
rainforest biomes, they lack dormancy mechanisms and are absent from locations
with extensive freezing temperatures (Tomlinson 2006). Palms exhibit a variety of
growth types including understory shrubs-like forms, vine forms (rattans; *Calamus
spp.* in the old world tropics and *Desmoncus spp.* in the Neotropics) and arborescent
tree-like forms. Likewise, palms are one of the few monocot families containing
species that achieve significant heights with the tallest palm being a species that
grows in the Colombian Andes (*Ceroxylon quindiuense* (Karst.) H. Wendl) and
reaches heights of 60 m (Henderson et al. 1995). Despite arborescent palms being as

tall as, or taller than, sympatric dicot trees and displaying long lifespans, they differ in several key ways that affect their physiological functioning within an ecosystem. For example, the majority of palm species lack aerial branching and all leaves and stem growth occurs at a single apical meristem (Tomlinson 1990). This gives palms little flexibility in terms of their aboveground body plan as opposed to many other plant types that have a more modular body plan (Tomlinson 2006). Palms also lack secondary meristems meaning that they do not produce secondary xylem (wood), secondary phloem or bark tissues and all fluid transport occurs through primary vascular bundles that are scattered within a matrix of parenchyma cells throughout the stem. Within tropical ecosystems, these differences could alter the way that palms respond to climate change compared with other plant types.

This chapter will begin with a discussion of the unique structural and physio-logical features of palms compared to dicot tree species. Drawing off the literature and our previous studies, we consider implications of these differences for how physiological functioning and growth in palms responds to environmental variation. Next, because palms are a key component of many tropical agricultural systems, we will review physiological data from these systems and discuss their effect on the broader landscape. Finally, we consider environmental and physiological con-straints on palm growth and distribution and the potential for global change to alter palm abundance and range size.

Key Features of Palms and Physiological Implications

Leaf and Crown Structure

The vast majority of palm species are monopodial (Tomlinson 1961) and, unlike dicot trees, lack a mechanism for branching. All leaves are successively produced by one apical meristem complex that, in caulescent species, also produces new stem tissue associated with each leaf that increases the overall height of the palm. When a leaf falls, it creates a permanent scar or node on the palm's trunk leaving a record of its height growth (Fig. 2). These nodes hold promise for identifying the age of an individual palm (Oyama 1993) as well as acting as a potential "timescale" in the aseasonal tropics to determine the age of gaps and forest disturbances (Martínez-Ramos et al. 1988; Ratsirarson et al. 1996). However, leaf turnover rates are difficult to predict because they are dependent on environmental variation and canopy position (Lugo and Rivera Batlle 1987), male versus female individuals in dioecious species (Ataroff and Schwarzkopf 1992; Cepeda-Cornejo and Dirzo 2010) and ontogeny, increasing with age in some species (Ratsirarson et al. 1996; Renninger and Phillips 2010) and decreasing in others (De Steven et al. 1987; Lugo and Rivera Batlle 1987; Renninger and Phillips 2010). Because palm leaves are produced in a rosette pattern by the apical meristem, new leaves are in a better position for light interception (Corley 1983) as well as being free of epiphylls

Fig. 2 Nodal leaf scars on the stem of the palm *Mauritia flexuosa* (L.f.) leave a record of vertical height growth

(Coley and Kursar 1996) and any other dry deposition compared to older leaves. As palm leaves age, chlorophyll content (Corley 1983), leaf nitrogen content (Hogan 1988) and stomatal conductance tend to decline (Corley 1983; Hogan 1988).

Because all leaves originate from a single apical meristem in most shrub-like and arborescent palms, they exhibit a more compact crown with limited options in terms of canopy architecture compared with dicot trees. This makes leaf area estimations and scaling of physiological parameters more straightforward for palms (Renninger et al. 2009; Renninger and Phillips 2010) because crowns are composed of relatively low numbers of leaves, although individual leaves are some of the largest in the plant kingdom (Tomlinson 2006). Therefore, construction costs and investment of carbon and nutrients into a single leaf is higher for palms than dicot trees and a single palm leaf will experience a variety of light and environmental conditions as it progresses from a newly formed leaf at the top of its canopy to a lower leaf at the base of its canopy (Chazdon 1985). There is also evidence that when these large senescent leaves eventually fall, they affect the understory seedling bank and species composition by damaging seedlings and covering seeds (Peters et al. 2004; Venceslau Aguiar and

Tabarelli 2010). Palms are limited in their options for increasing total leaf areas for light interception with different species either increasing the number of leaves they carry and retaining individual leaves longer (Holbrook and Sinclair 1992b; Avalos and Sylvester 2010; Renninger and Phillips 2010) or increasing the size of individual leaves (Rich et al. 1995; Renninger and Phillips 2010). Pinnately-compound leaves appear to have much greater capacity for increases in individual leaf area through increases in length and three-dimensionality of leaves (Rich et al. 1995) compared with simple and palmately-compound leaves (Renninger and Phillips 2010).

For palms, there are diminishing returns for increasing light interception capacity with increasing leaf biomass because new leaves shade older ones (Chazdon 1985) do to the compact crown structure of palms. However, they do possess some strategies to deal with their compact crown structure including altering the size, width and number of leaves they hold, changing the zenith angle of the petiole, and adjusting the biomass allocations between petioles and the lamina in order to maximize light capture efficiency in a variety of understory conditions (Chazdon 1985; Takenaka et al. 2001). Nevertheless, leaf support costs were shown to outpace increases in leaf and crown size in two understory palms and these costs could not be completely compensated for by increased leaf longevity (Chazdon 1986). Despite the constraint to leaf development due to monopodial growth, some tree palms can adjust leaf demography and biomass allocation under different light levels inside the forest. For example, leaf life span of *Euterpe edulis*, a conspicuous palm in subtropical Atlantic Forests, increases by 100 days with decreasing irradiance while the rate of leaf production decreases (Gatti et al. 2011). At higher light levels, typical of small forest gaps, adjustments in biomass allocation to leaf components allow *E. edulis* to reduce self-shading and increase light interception. At high light, palms allocate more biomass to roots and the plants exhibit small leaf sizes when leaves are compared using an explicit ontogenetic analysis. Ontogeny constrains the maximum size that each consecutive leaf could achieve, while growth irradiance determines the rate of leaf production (Gatti et al. 2011). Thus, although several studies have been performed on leaf and crown properties of understory palms, more detailed research on leaf physiology of tall, overstory palms including light capture efficiency, leaf construction costs, leaf temperatures, boundary layer conductance and wind resistance compared with tropical dicot trees is needed to fully understand the implications of a compact crown structure and large leaf size on global change responses in palms.

Functional Consequences of the Lack of Secondary Growth

As monocots, palms lack a vascular cambium from which secondary growth originates in woody dicot species. Instead of secondary xylem (wood) and secondary phloem performing long distance fluid transport, primary xylem and phloem tissues located within thousands of vascular bundles transport water and photosynthetic sugars throughout the palm. Radially, vascular bundles exhibit the lowest frequencies in the center of the palm stem and increase in frequency in positions closer to the stem periphery (Tomlinson 1961; Rich 1987). Longitudinally, vascular bundles do not remain in the same position within the stem, but instead create a shallow helix as they travel through the stem center before diverging to the stem periphery to connect with the leaves (Zimmermann and Tomlinson 1965, 1972; Zimmermann 1973). In the stem periphery, a vascular bundle splits into elements that connect to the leaf and elements that continue in a vertical direction moving back toward the stem center (Zimmermann and Tomlinson 1965). With global change and anthropogenic activities increasing the occurrence of fire in the tropics (Goldammer and Price 1998; Cochrane 2003, 2009), the lack of a vascular cambium, the occurrence of both xylem and phloem tissues throughout the stem and the cluster of leaf bases that insulates the apical meristem may confer palms with greater fire resistance compared to co-occurring tropical dicot trees (Cochrane 2003). Results are, however, mixed with some studies finding that adult arborescent palms are relatively resistant to fire-induced mortality (Ratsirarson et al. 1996; Souza and Martins 2004; Van Nieuwstadt and Sheil 2005), whereas Williams et al. (1999) found that palms exhibited much lower survival in a wildfire in northern Australia compared with deciduous and evergreen eucalypt species.

In terms of stem longitudinal and diameter growth, tall palms follow two main strategies. In many arborescent palms, all diameter growth occurs underground, and the aboveground stem emerges at its final diameter (Waterhouse et al. 1978). In other palms, a small diameter, aboveground stem occurs relatively early in development and continues to increase in diameter until reaching a final size (Fig. 3a, b, Waterhouse et al. 1978; Alves et al. 2004; Avalos et al. 2005; Avalos and Fernández Otárola 2010; Avalos and Sylvester 2010). In these palms, stems increase in girth through increases in the size and space between parenchyma cells (Fig. 3c, d) within the stem (Waterhouse et al. 1978; Rich 1987). These microscopic changes across the entire stem can lead to four- to six-fold increases in overall stem diameter (Waterhouse et al. 1978; Rich et al. 1986; Renninger and Phillips 2010). Recently, these increases in size and distance between parenchyma cells as well as the helical path of vascular bundles have also been associated with sustained increases in the length of the lower stem region in an Amazonian tropical palm species (Renninger and Phillips 2012). Some palm species that exhibit an aboveground stem in their early growth stages also possess stilt roots (Fig. 4) which may provide additional stability to the aboveground stem (Schatz et al. 1985). Therefore, although palms lack the ability to add new vascular tissue through secondary, radial growth, they still exhibit a large capacity to alter the structure and functioning of their main stem axis.

Because long distance water transport occurs under negative pressures subjecting the water in vessels to a metastable state and the possibility for malfunction (Zimmermann 1983), plants must either efficiently avoid cavitation, replace cavitated vessels with functioning ones, or refill cavitated vessels and reconnect them to the water transport stream. Without a vascular cambium, palms lack the ability to replace vessels in their stem or leaves, and therefore need to either avoid cavitation throughout the lifetime of the palm or refill cavitated vessels. There is evidence that cavitations occur and are refilled in the petioles of palms on a regular basis (Milburn

Fig. 3 Radial stem expansion in the palm *Iriartea deltoidea* (Ruiz & Pav.) which has a small diameter, aboveground stem as a sapling (**a**) that continues to expand until reaching its adult size (**b**). Stem expansion occurs through increases in the size of parenchyma cells as well as the intercellular space between them as seen when comparing a stem cross section from a small diameter stem (**c**) and a large diameter stem (**d**). Scale bars = 0.5 mm

and Davis 1973; Zimmermann and Sperry 1983; Sperry 1986; Renninger and Phillips 2011) but there is little data on if, and how widespread, cavitations occur in the stem vessels. Root-generated, positive pressures have been found in palms (Davis 1961), but likely play a minor role in cavitation refilling (Milburn and Davis 1973). More likely, the overall hydraulic architecture of the palm protects stem xylem from widespread cavitations. Palms possess hydraulic constrictions between the stem and subtending leaves because only the tiny protoxylem vessels within

Fig. 4 Stilt roots in (**a**) *Iriartea deltoidea* (Ruiz & Pav.) and (**b**) *Socratea exorrhiza* (Mart.) H. Wendl., the ninth most abundant tree species in the Amazon (Ter Steege et al. 2013)

bundles connect the leaf to the stem (Zimmermann and Tomlinson 1965; Zimmermann and Sperry 1983). Conductance is highest in the stem tissue, extremely low at the leaf insertion point, and increases distally in the petiole (Zimmermann and Sperry 1983). This anatomical arrangement provides evidence that palms conform to the hydraulic segmentation hypothesis (Zimmermann 1983; Tyree and Ewers 1991; Pivovaroff et al. 2014) in which the integrity of the stem xylem is protected from the large negative water potentials experienced by the leaf and petiole xylem (Zimmermann and Sperry 1983). However, empirical evidence for this hypothesis is lacking, particularly for tall, arborescent palm species that can reach 60 m in height (e.g. *Ceroxylon quindiuense* (Karst.) H. Wendl.; Henderson et al. (1995)) and whose stem xylem in certain species functions for more than 100 years (Tomlinson and Huggett 2012).

Stem and Tissue Longevity

Palms of various species have been found to live for more than a century (Tomlinson and Huggett 2012), achieving significant heights while lacking a vascular cambium and the radial additions of new vascular tissues. Because palms continue to use vessels located at their stem bases and produced while they were younger and smaller in stature, they may be overbuilt hydraulically when young to accommodate increases in water transport distance as they become older and taller. An overbuilt hydraulic architecture in young palms and the lack of a vascular cambium could suggest that long distance water transport may begin to limit stomatal conductance and carbon uptake in tall palms. However leaf-specific transpiration rates determined from sap flux measurements were found to be similar between short, young palms and the tallest individuals of a given species (Renninger et al. 2009; Renninger and Phillips 2010). These results are consistent with leaf gas exchange measurements which showed no difference in stomatal conductance or photosynthesis with palm height in subtropical *Washingtonia robusta* (H. Wendl), from measurements spanning the full range of palm heights in this species on sites in two continents (Renninger et al. 2009). Therefore, our studies have not found a hydraulic limitation to gas exchange (Ryan and Yoder 1997; Ryan et al. 2006) in the tallest individuals of the study species. However, we did find evidence for sharply reduced leaf epidermal sizes in the tallest palms of *Washingtonia robusta* (Renninger et al. 2009), and smaller average frond area, and smaller total leaf area in taller palms, consistent with a turgor-hydraulic limitation to palm height (Woodruff et al. 2004) although in two palm species growing in the humid tropics (*Iriartea deltoidea* Ruiz & Pav. and *Mauritia flexuosa* L.f.), the tallest individuals had the largest leaf areas with similar leaf-specific transpiration rates as shorter individuals (Renninger and Phillips 2010).

Not only do xylem vessels need to function in water transport throughout the lifetime of the palm (see *Functional consequences of the lack of secondary growth*), but individual phloem tissues and parenchyma cells also remain functional throughout the 100+ year lifespan of some palm species (Zimmermann 1973; Bullock and Heath 2006; Tomlinson and Huggett 2012). This condition in palms as arborescent monocots is particularly novel in the plant kingdom, because in dicot tree species, phloem transports solutes for one to a few years before its function is replaced by new phloem tissues produced by the vascular cambium. Likewise, parenchyma cells in the stems of dicot trees remain functional throughout the sapwood but become nonfunctional during the transition to heartwood. Therefore, in many dicot species, stem parenchyma cells remain alive for a few years to several decades with a few long-lived conifer species exhibiting sapwood longevities exceeding 100 years (Spicer 2005). Not only do parenchyma cells in palms remain alive throughout the lifetime of the palm, but they also retain the ability to change size and wall composition exhibiting cell expansion and cell wall thickening over extended periods of time (Waterhouse et al. 1978; Rich 1987).

The longevity of the phloem tissues in palm stems deserves increased attention, however a few studies present data on its structure and functional properties. Using phloem exudation from developing coconut inflorescences, Milburn and Zimmermann (1977) measured specific mass transfer rates as high as 70 g h^{-1} cm^{-1} and phloem pressures as high as 0.76 MPa which is in agreement with estimates of phloem osmotic and xylem water potentials. Parthasarathy and Tomlinson (1967) and Parthasarathy and Klotz (1976) performed anatomical investigations of phloem in palms and found that, unlike dicot species, sieve plates lack the definitive callose material that occludes sieve pores during typical phloem senescence, but possess a small amount of callose material surrounding individual sieve plate pores. They also found that appreciable amounts of p-protein, a substance that seals phloem sieve tube cells in the event of large pressure changes, were absent. Therefore, the majority of palm phloem tissue appears to lack the dormancy mechanisms seen in the majority of dicot tree species with the exception of the phloem supporting fallen palm leaves which become occluded by outgrowths from surrounding parenchyma cells (Parthasarathy and Tomlinson 1967). It is clear that more research is needed on the longevity and functioning of palm phloem and parenchyma tissues as well as their interaction with xylem vessels and the potential for positive stem pressure and cavitation refilling within palm stems (Renninger and Phillips 2011).

A Higher Proportion of Living Cells in Stems

While dicot trees generally have sapwood containing between 5–25 % living parenchyma cells (Spicer and Holbrook 2007), palm stems contain vascular bundles within a matrix of parenchyma cells that can make up more than 50 % of the cross-sectional area of the main central portion of the stem (not including parenchyma cells located within the vascular bundles themselves; Renninger and Phillips (2010)). Therefore, palm stems contain a high capacity for stored water and non-structural carbon reserves within these parenchyma cells that could potentially increase as palm stems increase in size. However, this may also lead to increased levels of maintenance respiration in larger palms which could put them at a disadvantage compared to co-occurring dicot species. Navarro et al. (2008) found that autotrophic respiration in coconut (*Cocos nucifera* L.) palms accounted for about 60 % of total GPP which is slightly higher than an estimate of average autotrophic respiration of trees from a tropical forest site (50 % from Malhi et al. (1999)). In terms of foliar respiration, Cavaleri et al. (2008) found that dicot trees and liana species had higher rates than palms and Henson (2004) found no differences between maintenance respiration in oil palm (*Elaeis guineensis* Jacq.) fronds of differing ages. Per unit stem biomass, maintenance respiration rates were found to decrease with age (Breure 1988; Henson 2004) being highest near the crown and decreasing towards the base (Henson and Chang 2000). Nevertheless, total maintenance respiration rates increased with palm size due to the accumulation of living tissues (Henson 2004) and showed a temperature dependence (Henson 2004) which may increase the respiration costs incurred by palms in a warming climate.

Increased living stem tissue may also mean that stored stem water plays a larger role in daily and seasonal water relations in palms compared with co occurring species. Killmann (1983) found that moisture content increased with stem height in coconut palms. In some species, palm stems also contain a swollen region with Fisher et al. (1996) finding this region in the Cuban belly palm (*Acrocomia crispa* (Kunth) C. Baker ex. Becc.) to have 89 % moisture content which was higher than the upper and lower stem regions (62 and 46 % moisture content respectively). Stem parenchyma tissue has also been found to give up its water relatively freely having a specific capacitance that was 84 times higher than leaf tissue in *Sabal palmetto* (Walt.) Lodd. (Holbrook and Sinclair 1992a). However, this large amount of stored stem water did not prevent large water deficits in the leaves of *Sabal palmetto* and stored stem water reserves were only mobilized during periods of high transpiration due to the large hydraulic constriction at the leaf insertion point (Holbrook and Sinclair 1992b). Therefore, stored stem water may serve the function of protecting stem vessels from cavitation under periods of high transpiration and maintaining leaf turgor during periods of severe drought and stomatal closure (Holbrook and Sinclair 1992b). Diurnally, transpiration in taller palms showed a greater reliance on stored water compared with shorter palms (Holbrook and Sinclair 1992b; Renninger et al. 2009; Renninger and Phillips 2010). Holbrook and Sinclair (1992b) found that about 35 % of daily water use in *Sabal palmetto* was derived from stored stem water with this percentage increasing during a soil dry-down. In subtropical *Washingtonia robusta*, stored stem water contributed about 22 % to daily water use in the tallest individuals (Renninger et al. 2009). Even in the palm *Mauritia flexuosa*, which grows in permanently inundated swamps and therefore faces no soil water deficit, stored stem water in the tallest individuals contributed almost 20 % to daily water use (Renninger and Phillips 2010).

The large proportion of living parenchyma cells within palm stems allows for storage of significant amounts of non-structural carbon reserves in the form of starch and sucrose that serve a physiological function as well as making certain species valuable as an edible food source for indigenous tropical populations (Kahn 1988). Fisher et al. (1996) found that in Cuban belly palms, the largest amounts of starch were located in the upper trunk region with little to no starch located in the middle, swollen region and the lower trunk. In coconut palms, non-structural carbon was stored mainly as soluble sucrose as opposed to insoluble starch (Mailet-Serra et al. 2005), the preferred storage form in dicot trees. In coconut palms, about 60 % of stored non-structural carbon reserves are located in the stem and about 14 % are found in the leaf petioles with little to no storage in the root tissues (Mailet-Serra et al. 2005). Petiole carbon reserves appear to act as a buffer by storing excess photosynthetic carbon in the absence of other sinks and releasing carbon during periods of high demand and/or low photosynthetic assimilation rates thereby maintaining vegetative growth over a range of environmental conditions (Mialet-Serra et al. 2008). In oil palm both glucose and starch are stored in stems with starch concentrations being highest below the crown and increasing when assimilation rates are high and decreasing when fruits are being produced (Legros et al. 2009). The role of the large glucose and sucrose pool in the remainder of the palm stem remains unknown but could become mobilized under more severe environmental stressors and disturbances (Mialet-Serra et al. 2008; Legros et al. 2009). While non-structural carbon reserves have been studied in agriculturally important palm species, they remain largely uncharacterized in the undomesticated palms occurring widely across the tropics where they potentially represent an important source of carbon sequestration.

Stem Biomechanics

A lack of secondary growth also has implications for mechanical stability of palms as they grow taller and may mean that palms have a predetermined critical height that is set by the diameter expansion that occurs well before the palm reaches its maximum stature (Niklas 1992). For dicot trees, wood tissues perform the dual role of long distance water transport and mechanical stability. Vascular bundles in palms allocate the role of water transport to large metaxylem vessels and the role of mechanical stability to fibers, meaning that palms can have a higher hydraulic efficiency for a given carbon investment compared with dicot trees (Renninger et al. 2013). Palms attain similar heights as co-occurring dicot trees reaching the top of the canopy in tropical forests. However, palms tend to have larger height to diameter ratios than dicot trees (Niklas 1994; Niklas et al. 2006) and would be theoretically unstable if they shared similar material properties (Rich 1986; Rich et al. 1986; Alves et al. 2004). In addition, palms cannot maintain a constant slenderness ratio as they grow taller. Due to their lack of secondary growth, they get proportionally thinner as they increase in height (Niklas 1992). However, palms can continue to alter the density and stiffness of their tissues as they age. As palms grow taller, the density, stiffness and strength of the palm base and stem periphery increases (Fig. 5, Rich 1986, 1987; Rüggeberg et al. 2009) through sustained thickening and sclerification of fibers (Rich 1986) allowing palms to achieve higher strength and stiffness values at a given density compared with dicot wood tissues (Rich 1987). The location of the dense stem material in the lower stem periphery puts it in an optimal location to withstand the highest mechanical stresses that the stem experiences (Niklas 1992).

Fig. 5 Decay resistant, peripheral stem tissue of a palm containing highly lignified fibers and sclerenchyma cells

In some species of palms that reach significant heights, the top of the stem remains flexible allowing the stem to bend without breaking under wind loads (Rich 1987; Winter 1993; Alves et al. 2004; Rüggeberg et al. 2009). The leaf sheaths that surround the main stem also provide additional mechanical support (Tomlinson 1962; Niklas 1999) against buckling and torsional deformation. Likewise, palm leaves can reconfigure in the wind, particularly pinnately compound leaves that exhibit a flexing of the rachis and a rotating of the individual leaflets such that they form a more compact, aerodynamic structure with less drag (Niklas 1999). Some palm individuals can grow in the direction of a canopy gap exhibiting a trunk that diverges significantly from vertical (Fig. 6). The small crown and lack of branching in palms decreases their mechanical load (Rich et al. 1986) but also means that they exhibit more dynamic swaying motion in the wind compared to dicot trees whose sideways motion is dampened by the larger branch mass (James et al. 2006). The arrangement of vascular bundles within the palm stem also functions in the transfer of external forces to the ground. The central region of the palm stem contains rigid vascular bundles within a matrix of pliable parenchyma cells and intercellular air spaces that perform the function of energy dissipation of external forces (Rüggeberg et al. 2009). Stress discontinuities between the vascular bundles and the parenchyma matrix are avoided by a pattern of decreasing stiffness of fiber cells as they approach the surrounding parenchyma tissues (Rüggeberg et al. 2008, 2009). Therefore, a single palm stem has been shown to contain the full range of tissue density and tensile strength measured across all woody dicot species (Niklas 1992). These unique mechanical features allow palms to reach significant heights while

Fig. 6 *Washingtonia robusta* (H. Wendl.) palms growing at the Los Angeles County Arboretum and Botanic Garden and exhibiting stem axes that diverge significantly from vertical

withstanding intense wind loads (Winter 1993; Duryea et al. 1996), but may be an additional limitation to the overall crown size and photosynthetic capacity of a palm individual.

Palms in Tropical Agricultural Systems

Palms have been cultivated by humans for thousands of years and continue to be an important component of both subsistence agroforestry systems (Pritchard Miller and Nair 2006; Clement et al. 2009) and industrialized monoculture plantations (Chao and Krueger 2007; Fowler et al. 2011). For South American subsistence cultures, the most useful palms are either tall, have large fruits or large leaflets with these individuals used for both food and construction (Cámara-Leret et al. 2014). In terms of more wide-scale cultivation and production, some Amazonian palm species like peach palm (*Bactris gasipaes* Kunth) face challenges due to the shelf life of its fruits (Clement et al. 2004) while other native palms like açaí palm (*Euterpe oleracea* Mart.) are gaining in global popularity (Van Looy et al. 2008). Increasingly, palm monocultures comprise vast areas of the tropics. Therefore, in order to better quantify and model fluxes of carbon, water vapor and other greenhouse gases across many tropical landscapes, an increased understanding of the physiological responses of these palm crops to current and future environmental factors is necessary. Likewise, environmental change may alter the productivity and profitability of these systems affecting the livelihoods of many in the tropics. In addition, because of their economic importance, these cultivated palms have been the most widely studied in terms of physiological functioning and response to environmental conditions. The following section provides background information and ecological physiology for the three main palm crops; date palm (*Phoenix dactylifera* L.), coconut palm and oil palm.

Date Palm

Palms have been identified as one of the most important plant families in terms of supplying products for human use (Tregear et al. 2011) and the date palm is thought to be one of the oldest fruits in cultivation with its usage dating to ancient Mesopotamia (Chao and Krueger 2007). Date palms differ from many other species of palms in that they thrive in very hot, dry climates with little rainfall, low humidity and abundant sunshine. However, they require abundant soil moisture in the rooting zone (2 m depth) provided from either a shallow water table or constant irrigation (Chao and Krueger 2007). Date palms continue to be grown in arid areas largely in the Middle East and North Africa where they play a significant role in the economies of many countries as well as providing subsistence agriculture for rural desert populations (Chao and Krueger 2007; El-Juhany 2010). Date palms may also prevent desertification in arid areas (El-Juhany 2010) and can provide shade allowing understory crops to be grown (Chao and Krueger 2007). Dates are a rich source of protein, vitamins and mineral salts (El-Juhany 2010) as well as energy with ripe fruits having sugar concentrations exceeding 80 % (Chao and Krueger 2007). Likewise, because ripe fruits have low water contents, they can be stored for prolonged periods of time under ambient conditions (Chao and Krueger 2007). Globally, date palm production exceeded 7.5 million Mg in 2012 with 1.1 million ha in production (FAOSTAT 2014). Egypt, Iran and Saudi Arabia all produced over 1 million Mg of dates in 2012 and Iraq, Iran, Saudi Arabia and Algeria all contain over 100,000 ha of date palms in production (FAOSTAT 2014).

Date palms grow in extreme locations with little rainfall and high radiation, temperatures and vapor pressure deficits. Of the incoming solar radiation absorbed by the leaf, 80 % is dissipated through non-photochemical quenching (Sperling et al. 2014a). Date palms require access to constant soil moisture using about 120 L of water per day and exhibiting immediate decreases in transpiration if irrigation is withheld (Sperling et al. 2012). Because consistent irrigation is important for maintaining crop yield, automated systems including thermal imaging have been developed to detect water stress in date palms (Cohen et al. 2012). Symbiotic relationships with arbuscular mycorrhizae have also been found to aid growth of seedlings during drought conditions (Baslam et al. 2014). Because date palms require abundant irrigation and freshwater is limited in many locations, saltwater is frequently used to irrigate palms. Date palms have been identified as salt tolerant, however high salinity water (EC = 10.5–12 dS m^{-1}) decreases growth and fruit yields (Tripler et al. 2007, 2011; Sperling et al. 2014a). In a long term study, date palms irrigated with high salinity water were half as large and their fruit production was 50 % smaller than palms irrigated with low salinity water (Tripler et al. 2011). High salinity irrigation does not affect the photosynthetic apparatus but instead causes decreases in stomatal conductance and water use due to osmotic stress and more negative soil water potentials (Tripler et al. 2007; Sperling et al. 2014a; b). Decreased stomatal conductance in palms receiving high salinity irrigation also led to increased photorespiration rates compared with control palms (Sperling et al. 2014a). Despite effects of drought stress on date palm productivity, increased rain or humidity with climate change may make certain areas in North Africa and elsewhere unsuitable for date palm because fruits require dry, hot summers to avoid cracking and fungal diseases (Shabani et al. 2012). However, other locations including areas in North and South America may become more suitable for date palms with climate change (Shabani et al. 2012).

Coconut Palm

Coconut palms likely originated from Southeast Asia (Harries 1978) and are currently found throughout the humid tropics in areas that receive ample rainfall. Coconut palms are grown primarily for their fruits which provide products including the white endosperm tissue or coconut "flesh", coconut oil and coconut water from the immature fruits. Both dwarf and tall varieties are grown for coconut production with conflicting information about the water use and resistance to drought stress of each variety (Gomes and Prado 2007). Globally, almost 62 million Mg of coconut were produced in 2013 with over 12 million ha in cultivation (FAOSTAT 2014). Asia, by far, has the largest area in production in 2013 with over 9.7 million ha followed by Africa with almost 1.2 million ha and the Americas with over 600,000 ha (FAOSTAT 2014). Indonesia, the Philippines, and India led coconut production by country with each producing over 10 million Mg and having over 2 million ha in production in 2013 (FAOSTAT 2014). Well-watered palms are capable of producing over 90 kg of fruit dry matter per year (Rees 1961), with low light intensity, drought stress, nutrient deficiencies (Prado et al. 2001; Gomes and Prado 2007) and diseases including lethal yellowing decreasing yield. Lethal yellowing affects coconut palms throughout the Americas (Maust et al. 2003) and is caused by a phytoplasma that affects phloem transport (Maust et al. 2003) and leads to increasing leaf sugar concentrations, decreased root carbohydrate concentrations and permanent stomatal closure (McDonough and Zimmermann 1979; Eskafi et al. 1986; Leon et al. 1996; Martinez et al. 2000; Maust et al. 2003).

Water use and carbon uptake in coconut palms have been found to respond to seasonal differences in soil moisture, radiation and vapor pressure deficit. Well-irrigated dwarf coconut palms in Brazil had evapotranspiration rates ranging from 2.5 mm day^{-1} in the rainy season to 3.2 mm day^{-1} in the hot summer season (de Azevedo et al. 2006). During the dry summer season in India, evapotranspiration rates averaged 5.5 mm day^{-1} and were as high as 7 mm day^{-1} (Rao 1989). In coconut palm fields (LAI = 3) growing in Vanuatu under ample soil moisture, evapotranspiration was about 950 mm year^{-1} with palm transpiration representing about 68 % of this total and averaging 1.3–2.3 mm day^{-1} depending on atmospheric conditions (Roupsard et al. 2006). In locations where rainfall is highly seasonal, carbon uptake and water use is limited by radiation during the wet season and by decreased stomatal conductance during the dry season (Prado et al. 2001), although coconut palms tend to be more water use efficient in the dry season (Rees 1961; Prado et al. 2001) showing increases in epicuticular wax (Kurup et al. 1993) as well as sugar and amino acid concentrations in leaves (Kasturi Bai and Rajagopal 2000). During the dry season, rainfed coconut palms experienced decreased stomatal conductance due to both soil and atmospheric drought compared to irrigated palms (Shivashankar et al. 1991; Repellin et al. 1997) that decreased dry matter production (Rees 1961) and leaf water potentials by about 0.5 MPa (Rees 1961; Kasturibai et al. 1988). Prolonged drought has also been found to cause long-lasting negative effects on the photosynthetic apparatus in coconut palms (Gomes et al. 2008).

Oil Palm

Oil palm is native to West Africa (Dufrene and Saugier 1993) and its cultivation has been steadily increasing globally with over 4 million ha added between 1993 and 2003 and almost 6 million ha added between 2003 and 2013 (FAOSTAT 2014). Worldwide, over 17 million ha were in production in 2013, with the vast majority located in Asia (over 12 million ha) and almost exclusively in Indonesia (7 million ha) and Malaysia (4.5 million ha) followed by Africa with over 3 million ha (with 2 million ha located in Nigeria) and the Americas with over 1 million ha (FAOSTAT 2014). In 2013, over 55 million Mg of palm oil were produced globally (FAOSTAT 2014) and is used as a cooking oil, in a variety of processed foods and other products and, recently, as a biofuel. In terms of aboveground production of temperate and tropical forests, oil palm plantations are second only to fertilized *Eucalytpus* (Dufrene and Saugier 1993). In Borneo, daytime maximum CO_2 uptake was almost two times higher in an oil palm plantation compared with a nearby rainforest site (Fowler et al. 2011) and, in West Africa, oil palm plantations accumulated almost 3 Mg of carbon ha^{-1} yr^{-1} (Thenkabail et al. 2004). However, the drainage of peatswamp forests in Southeast Asia for palm oil production leads to large losses of above- and below-ground carbon as well as the carbon sequestration capacity of further peat accumulation (Germer and Sauerborn 2007; Koh et al. 2011). Therefore, increasing yields per unit area, incentivizing smallholder systems and prioritizing conversion of anthropogenic grasslands should be priorities for future oil palm production (Sayer et al. 2012).

Although oil palms maintain large leaf areas throughout the year reaching a maximum LAI over 7 (Gerritsma and Soebagyo 1999) and maximum assimilation rates of over 14 μmol m^{-2} s^{-1} (Legros et al. 2009), they exhibit large stomatal control over transpiration responding to both soil drought and high vapor pressure deficits (Smith 1989; Dufrene and Saugier 1993). Even though about 43 % of total biomass in oil palm plantations is located in the root systems, the ratio of transpiration to potential evapotranspiration decreased from about 0.7 under well-watered conditions to a minimum of around 0.35 during the dry season (Dufrene et al. 1992). In a comparison of a rainforest and oil palm plantation site in Southeast Asia, the oil palm plantation exhibited about twice the net primary productivity even though both sites had similar LAI of around 6 (Fowler et al. 2011). The oil palm plantation also had larger latent heat fluxes that accounted for a larger portion of the available energy compared to the rainforest (Fowler et al. 2011). However, the oil palm plantation also exhibited about 25 % higher emissions of nitrous oxide (N_2O) likely due to its fertilization and over five times greater total volatile organic compound (VOC) emissions than the rainforest site (Fowler et al. 2011). VOC emissions in the oil palm plantation were largely comprised of isoprene (Fowler et al. 2011; Misztal et al. 2011) and these emissions were correlated with sensible heat flux indicating that VOC emissions could increase with temperature increases due to climate change (Misztal et al. 2011). VOCs have the potential to react with NOx compounds in the atmosphere to form tropospheric ozone which has negative effects on human health, crop yields and climate change. While ozone levels did not differ above the oil palm plantation and an adjacent rainforest site, increased industrialization in the area and increasing release of NOx compounds could increase tropospheric ozone to detrimental levels (Hewitt et al. 2009).

Physiology, Growth and Potential Future Distributions of Palms Under Environmental Change

Environmental Drivers of Palm Distributions

The foreseeable future is one of increasing atmospheric [CO_2], increasing temperatures, and an intensified water cycle, albeit with highly variable and uncertain changes in regional hydrology (Hartmann et al. (2013), pp 201–204). The uncertainty in future regional hydrology, in particular, makes forecasting future palm responses difficult. Globally, precipitation is one of the most important predictors of palm species richness followed by modern day temperature and temperature changes occurring throughout the Quaternary (Kissling et al. 2012). In continental scale studies of palm species distributions, water availability was the most important predictor of palm species richness (Bjorholm et al. 2005; Blach-Overgaard et al. 2010). For the African continent, the majority of palm species distributions responded positively to increasing precipitation and peaked in tropical rainforest climates with a few species preferentially located in low precipitation regimes (Blach-Overgaard et al. 2010). Again, temperature had little effect but spatial constraints were significant and may indicate dispersal limitations from historical refugia (Blach-Overgaard et al. 2010). The low species richness on the African continent compared to the Neotropics and Southeast Asia may also reflect the strong climatic changes that pre-date the Quaternary (Kissling et al. 2012). On islands, palm species richness is most strongly controlled by habitat heterogeneity, area, and geological origin (Kissling et al. 2012).

For the Neotropics, Bjorholm et al. (2005, 2006) report that annual precipitation, number of wet days, latitude and soil fertility were identified as strong predictors of species richness, with temperature and habitat being relatively unimportant. Svenning et al. (2008) also found that high net diversification and species richness in palms was associated with decreasing absolute latitude and increasing water and energy availability with the greatest species richness found in areas that maintained favorable environments for palm diversification throughout the Cenozoic. Kreft et al. (2006) found that precipitation and actual evapotranspiration were the strongest single predictors of palm species richness in addition to latitude whereas potential evapotranspiration and temperature had relatively low importance. It is interesting to note that latitude as a predictor was significant even when environmental variables of rainfall and temperature were included in models (Bjorholm et al. 2005; Kreft et al. 2006; Svenning et al. 2008), which may indicate that energy availability or historical biogeography are also key determinants in palm species richness. The importance of latitude in neotropical palm species richness may also be an area effect with the largest land area available to palms located closest to the Equator due to the geography of the Americas (Chown and Gaston 2000; Svenning et al. 2008). In a comparison of rare versus more widespread palm species, Kreft et al. (2006) found that while climatic factors were most strongly associated with species richness in widespread species, rare species were more strongly predicted by topographic complexity. Therefore across the Neotropics, the highest palm species richness occurred in grid cells with >3000 mm of annual precipitation in both lowland and montane habitats (Kreft et al. 2006). Palm species richness patterns at the subfamily level in the Neotropics also reflects historical dispersal routes of each group with the Arecoideae being most strongly controlled by modern day environmental variables reflecting its long presence throughout South and Central America while Coryphoideae species richness shows strong bias toward North and Central America due to its northern dispersal route and the Calamoideae exhibits bias towards South America due to its long history in this region (Bjorholm et al. 2006). Likewise, long-term climate stability appears to be important as regions with the highest palm species richness have remained wet throughout the climatic changes occurring in the Pleistocene (Bjorholm et al. 2005).

At more local and regional scales, other environmental and habitat factors predicting palm species richness and density become significant. In the western Amazon basin, Vormisto et al. (2004) found that the only environmental determinants of palm species patterns were soil cation concentration and texture while dispersal limitation also played a role with climatically similar locations exhibiting significant differences in species composition. Kristiansen et al. (2011) found that local species diversity in the western Amazon basin was most strongly controlled by precipitation seasonality while regional species diversity was most strongly controlled by long-term habitat stability. In a study of seven palm species at La Selva Biological Station, Costa Rica, abundance of tall palm individuals varied with topography being highest on slope crests and steeper slopes and lowest in areas with shallow slopes or lowland flat terrain (Clark et al. 1995). This pattern is unrelated to small palms of each species available for recruitment but could be inversely related to soil fertility (Clark et al. 1995). Kristiansen et al. (2009) compared factors of palm species abundance at the local, regional and continental scales across the Neotropics and found that topographic niche breadth and stem height were the most important determinants of palm species abundance. Vormisto et al. (2004) also found that tall palm species tended to have wider geographical ranges which may relate to their increased dispersal ability (Kristiansen et al. 2009) and more generalized habitat needs compared with other palm growth forms. Therefore, regionally, climatic history, environmental factors, terrain and palm-specific characteristics all interact to determine species composition in a given area.

Topography and hydrology have also been shown to affect palm species richness and abundance on a regional basis. In Amazonian Ecuador, 17 % of individuals in *terra firme* locations were palms compared to almost twice that percentage in floodplain forests (Fig. 7, Balslev et al. 1987). In Amazonian Brazil, locations with

Fig. 7 A grouping of *Bactris sp.* palms growing along the floodplain of an oxbow lake at Tiputini Biodiversity Station in eastern Ecuador

waterlogged soils had over two times the density of palm individuals compared to *terra firme* locations (Peres 1994). However, these waterlogged locations are typically dominated by only a few species of palms, and therefore, exhibit lower species richness than *terra firme* locations (Kahn and Mejia 1990, 1991; Eiserhardt et al. 2011b). Many permanently inundated swamps in the Amazon basin are dominated by the palm *Mauritia flexuosa* (Fig. 8) which has been estimated to have a population of 1.5 billion stems covering around 3 million hectares across the region (Ter Steege et al. 2013). Likewise, these waterlogged regions in the tropics are significant sources of carbon sequestration (Vegas-Vilarrubia et al. 2010) and their low species richness compared to *terra firme* forests could make ecological modeling studies in these palm swamps more tractable.

Potential Effects of Global Change on Palm Distributions and Individual Physiology

Global increases in temperature have the potential to affect both palm species distributions as well as individual palm functioning within ecosystems. For individuals, the large respiration load from the abundance of living parenchyma cells within stems has the potential to negatively affect carbon balance under warmer temperatures (Henson 2004). Likewise, increases in temperature may also lead to increases in atmospheric vapor pressure deficits (VPD). Stomatal conductance in palms has been shown to be sensitive to increasing VPD (Smith 1989; Shivashankar et al. 1991; Dufrene and Saugier 1993; Repellin et al. 1997; Prado et al. 2001; Renninger et al. 2010) leading to greater stomatal closure and potentially limiting carbon gain as a result of increasing temperatures.

Fig. 8 Permanently inundated palm swamp in Amazonian Ecuador dominated by *Mauritia flexuosa* (L.f.)

In terms of overall palm species distributions, temperature limitations restrict palms to tropical and warm-temperate latitudes. Eiserhardt et al. (2011a) found that the ability of either water or energy to predict palm species richness shifted with latitude within the tropical/subtropical region such that energy became more important at increasing latitudes. Because palms cannot withstand low freezing temperatures (Tomlinson 2006), their presence/absence in the fossil record can be used to track past climates and changes between freezing and non-freezing conditions (Archibald et al. 2014). Notably, of the palm species that extend furthest into the higher latitudes (Table 1), most belong to a single subfamily, the Coryphoideae (Morley 2000; Couvreur et al. 2011; Thomas and De Franceschi 2013). These subtropical palms are considered to be relict taxa with low net diversification rates and are remnants of a time when tropical forests extended into the boreal zone during the Paleocene-Eocene (Svenning et al. 2008). Cooling temperatures in the Cenozoic lead to widespread extinction of palms in regions outside of the tropics (Svenning et al. 2008). Moreover, within this subfamily, most

Table 1 Selection of palms occurring furthest from the tropics

Subfamily	Genus	Species	Height (m)	Leaf form	Maximum latitude
Arecoideae	*Jubaea*	*chilensis*	15	pinnate	36°S
Arecoideae	*Rhopalostyls*	*sapida*	15	pinnate	44°S
Coryphoideae	*Chamaerops*	*humilis*	5	palmate	44°N
Coryphoideae	*Sabal*	*mexicana*	15	costapalmate	30°N
Coryphoideae	*Sabal*	*minor*	2	palmate/ costapalmate	30°N
Coryphoideae	*Sabal*	*palmetto*	20	costapalmate	34°N
Coryphoideae	*Trachycarpus*	*fortnuei*	10	palmate	46°N
Coryphoideae	*Washingtonia*	*filifera*	15	palmate	35°N
Coryphoideae	*Washingtonia*	*robusta*	22	palmate	35°N

genera occurring at latitudinal extremes are members of the tribe Corypheae. General characteristics like maximum palm height or leaf type are not consistent among palms occurring at latitudinal extremes (Table 1). However, a characteristic common of the Coryphoideae is two or more relatively small diameter vessel elements per vascular bundle (as opposed to only one or two larger diameter vessel elements in other subfamilies), and may confer a relatively greater level of freezing and drought tolerance to members in this subfamily (Thomas and De Franceschi 2013). Certain members of the Coryphoideae have, in recent decades, been increasing their latitudinal range, including *Trachycarpus fortunei* (Hook.) H. Wendl. (Walther et al. 2007), *Washingtonia filifera* (Lindl.) H. Wendl. (Cornett 1991), *Chamaerops humilis* L. and species in the genus *Sabal* (Lockett 2004). How species within this subfamily are able to withstand occasional freezing conditions and maintain functionality has not been studied extensively and requires further attention.

Frost resistance and subzero temperature effects on photosynthesis, survival and distribution were studied in *Euterpe edulis*, a tree palm species of the Atlantic Forest, near the southern limit of the species distribution (Gatti et al. 2008). *Euterpe edulis* is absent from forest stands in valley bottoms. Subzero temperatures are observed in the lowest site, but the medium and higher elevation sites never experienced absolute minimum temperatures below 0 °C. This palm species can exhibit substantial supercooling (temperatures can be lowered experimentally down to the ice nucleation temperature without extracellular ice formation). Ice formation was observed at about −4 °C, relatively close to the equilibrium freezing temperature, only after an increase in ambient humidity resulted in dew formation, triggering ice seeding on the plant surface (Gatti et al. 2008). Dew formation is commonly observed in valleys during the mild winter season of the subtropics. This observation gives further support to the hypothesis that strong infrequent frost events could be an important environmental factor determining the spatial distribution pattern of palm species in their latitudinal limit of distribution despite the capacity to avoid low subzero temperatures by supercooling. As extremes of cold continue to move poleward, it seems reasonable to expect that these species may follow suit, as is occurring, for example, with successfully established and reproducing populations of *Trachycarpus fortunei* in southern Switzerland (Walther et al. 2007).

While altered regional water availability is inherently difficult to forecast, its impacts, along with a more predictable regional warming, seem relatively clear on palm physiology and growth. We understand less about potential impacts of other global changes, including increased $[CO_2]$, nitrogen deposition, changes in local or regional wind intensity and patterns (Bichet et al. 2012), and the potential interactive effects among these environmental factors on palm physiology and growth. However, it is possible to speculate on how palms may respond to a combination of changes in the above-mentioned environmental variables, based upon the basic constraints of the monopodial growth form and associated features described earlier. For example, while research is increasingly showing that trees with secondary growth show continuous girth growth through old age and large size (Sillett et al. 2010; Stephenson et al. 2014), and that girth growth in old trees may be responding to global environmental change (Phillips et al. 2008), this primary degree of freedom in secondary girth growth is not available to palms, and continuous addition of stem tissues can only occur primarily along the axial dimension. This creates an uncertain set of potential costs and benefits to palms given environmental change. Our research (Renninger et al. 2009; Renninger and Phillips 2010) has demonstrated that gas exchange in the tallest palms in a species was not hydraulically limited, so frond carbon uptake could be enhanced with an increase in atmospheric $[CO_2]$. This is a necessary, if not sufficient, condition for a growth response to elevated $[CO_2]$, and it may be that height growth is ultimately limited by gravity and its effects on turgor pressure of expanding leaf cells (Woodruff et al. 2004). Nevertheless, even if elevated $[CO_2]$ could promote increased net carbon uptake and further height growth in palms, it is not clear if benefits of enhanced light capture of taller palms in closed canopy forests may outweigh costs to mechanical stability. This mechanical stability may be as much about the impacts of neighboring trees on palm mechanical stability as it is about the intrinsic ability of taller palms to withstand winds (Zimmerman et al. 1994).

In open-grown conditions, palms have little ability to modify crown form to the extent that dicotyledonous branching trees can (Mäkelä and Sievänen 1992), and it is difficult to imagine how a vertical growth response to elevated $[CO_2]$ could be of physiological benefit. Impacts on reproductive structures or respiration components of palm carbon budgets are even less well known, although research has shown substantial effects of elevated $[CO_2]$ on non-structural carbohydrates and secondary compounds in palm seedlings, e.g., Ibrahim and Jafaar (2012). Perhaps even more so than for taller palms in closed-canopy forests, the cost of mechanical stability may limit increases in open-grown palm height. While there appears to be a global tendency toward reduced mean winds associated with climate change (Bichet et al. 2012), extreme weather events containing high gusting winds that can contribute to palm windthrow are expected to increase (Hartmann et al. (2013), pp 216–217).

In spite of the fundamental difference in growth of palms compared to trees displaying secondary growth, a fundamental similarity may be that all plants must grow. To paraphrase Bob Dylan, not growing means dying. Since fronds cannot live indefinitely, their replacement in arborescent species necessitates stem extension, and while it appears that frond turnover may slow down with palm age and size (De Steven et al. 1987; Lugo and Rivera Batlle 1987; Renninger and Phillips 2010), vertical growth is an unavoidable and irreversible condition in the life of an arborescent palm. On the other hand, vertical growth may cease in trees exhibiting secondary growth even as girth growth continues. In aging, tall statured forests, and in an era of environmental change which may under some conditions increase the growth of both palms and other trees while it simultaneously poses risks to them (e.g., windthrow), the costs and benefits of constrained girth growth versus vertical growth, a potentially larger respiration load, and limited options in terms of canopy form may make for a relatively more precarious future for palms. However, palms, in one form or another, have been in existence since the late Cretaceous (Berry 1914; Kvacek and Herman 2004; Harley 2006) experiencing both warmer and colder global climates than the present as well as a strong capacity to compete with dicot species in the tropics.

Conclusions

Based on the physiological constraints of palms compared with dicot species as well as studies of palm distributions and species richness patterns at various scales, water availability and hydrologic factors are of critical importance in explaining current patterns of palm abundance and richness. Therefore, alterations of precipitation patterns and soil moisture regimes with climate change have the potential to alter palm abundance and species richness in tropical and subtropical forests with subsequent changes in forest functioning. Palms play an important role in many tropical ecosystems as a food source for mammals (Wright et al. 2000; Silva and Tabarelli 2001; Svenning 2001; Fragoso et al. 2003; Galetti et al. 2006) with the most important seed dispersers in the Neotropics being ungulates, primates and rodents (Fig. 9, Andreazzi et al. 2009). Fruits from palms were found to be available throughout the year and during the dry season making them an important food source for frugivores (Terborgh 1986; De Steven et al. 1987; Peres 1994; Ratsirarson et al. 1996). In addition, permanently waterlogged tropical locations tend to be dominated by palms that provide a large resource of nesting habitat for macaws (*Ara sp.*) (Bonadie and Bacon 2000; Brightsmith and Bravo 2006). Palms may also be important for reforestation of disturbed, seasonally dry forests as their columnar structure prevents colonization by lianas and may facilitate the growth of late successional tree species (Salm et al. 2005). Therefore, changes in precipitation patterns and the associated changes in hydrology have the potential to alter palm distributions in the tropics and could affect the associated ecosystem services that palms provide. Likewise, increasing temperatures may expand the range of palms to

Fig. 9 Spider monkey
(*Ateles belzebuth* E. Geoffroy)
in the canopy of an *Iriartea
deltoidea* (Ruiz & Pav.) palm.
These primates are one of the
main seed dispersal agents of
this palm species (Henderson
1990)

higher elevations (Kessler 2000) and higher latitudes provided they are not limited
by dispersal and that their agents of seed dispersal (birds, mammals) are not affected
by human activities. What remains clear is that palms have survived for millions of
years and withstood significant global change providing evidence of their adaptability as a plant family and/or their ability to disperse to their preferred environmental conditions. Their unique anatomical and physiological attributes have
proved successful across the tropics and have allowed them to successfully compete
with woody dicot tree species.

Acknowledgments HJR acknowledges the National Science Foundation East Asia and Pacific
Summer Institute program for support during 2008 (NSF grant OISE – 0813242). NGP and HJR
acknowledge the National Science Foundation for research support (NSF grant IOB #0517521).

References

Alves LF, Martins FR, Santos FAM (2004) Allometry of a neotropical palm, *Euterpe edulis* Mart.
    Acta Botanica Brasil 18:369–374
Andreazzi CS, Pires AS, Fernandez FAS (2009) Mamíferos e palmeiras neotropicais: Interações
    em paisagens fragmentadas. Oecologia Brasiliensis 13:554–574
Archibald SB, Morse GE, Greenwood DR, Mathewes RW (2014) Fossil palm beetles refine
    upland winter temperatures in the early eocene climatic optimum. Proc Natl Acad Sci USA
    111:8095–8100

Ataroff M, Schwarzkopf T (1992) Leaf production, reproductive patterns, field germination and seedling survival in *Chamaedorea bartlingiana*, a dioecious understory palm. Oecologia 92:250–256

Avalos G, Fernández Otárola M (2010) Allometry and stilt root structure of the neotropical palm *Euterpe precatoria* (Arecaceae) across sites and successional stages. Am J Bot 97:388–394

Avalos G, Sylvester O (2010) Allometric estimation of total leaf area in the neotropical palm *Euterpe oleracea* at La Selva, Costa Rica. Trees 24:969–974

Avalos G, Salazar D, Araya AL (2005) Stilt root structure in the neotropical palms *Iriartea deltoidea* and *Socratea exorrhiza*. Biotropica 37:44–53

Baker WJ, Couvreur TLP (2012) Biogeography and distribution patterns of Southeast Asian palms. In: Gower D, Johnson K, Richardson J, Rosen B, Rüber L, Williams S (eds) Biotic evolution and environmental change in Southeast Asia. Cambridge University Press, Cambridge, pp 164–190

Baker WJ, Couvreur TLP (2013a) Global biogeography and diversification of palms sheds light on the evolution of tropical lineages. II. Diversification history and origin of regional assemblages. J Biogeogr 40:286–298

Baker WJ, Couvreur TLP (2013b) Global biogeography and diversification of palms sheds light on the evolution of tropical lineages I. Historical biogeography. J Biogeogr 40:274–285

Balslev H, Luteyn JL, Øllgaard B, Holm-Nielsen LB (1987) Composition and structure of adjacent unflooded and floodplain forest in Amazonian Ecuador. Opera Bot 92:37–57

Baslam M, Qaddoury A, Goicoechea N (2014) Role of native and exotic mycorrhizal symbiosis to develop morphological, physiological and biochemical responses coping with water drought of date palm, *Phoenix dactylifera*. Trees 28:161–172

Berry EW (1914) The upper cretaceous and eocene floras of South Carolina and Georgia No. 84. US Government Printing Office, Washington DC, USA

Bichet A, Wild M, Folini D, Schär C (2012) Causes for decadal variations of wind speed over land: Sensitivity studies with a global climate model. Geophys Res Lett 39. doi:10.1029/2012GL051685

Bjorholm S, Svenning JC, Skov F, Balslev H (2005) Environmental and spatial controls of palms (Arecaceae) species richness across the Americas. Global Ecol Biogeogr 14:423–429

Bjorholm S, Svenning JC, Baker WJ, Skov F, Balslev H (2006) Historical legacies in the geographical diversity patterns of New World palm (Arecaceae) subfamilies. Bot J Linn Soc 151:113–125

Blach-Overgaard A, Svenning JC, Dransfield J, Greve M, Balslev H (2010) Determinants of palm species distributions across Africa: the relative roles of climate, non-climatic environmental factors, and spatial constraints. Ecography 33:380–391

Bonadie WA, Bacon PR (2000) Year-round utilisation of fragmented palm swamp forest by Red-bellied macaws (*Ara manilata*) and Orange-winged parrots (*Amazona amazonica*) in the Nariva Swamp (Trinidad). Biol Conserv 95:1–5

Breure CJ (1988) The effect of palm age and planting density on the partitioning of assimilates in oil palm (*Elaeis guineensis*). Exp Agric 24:53–66

Brightsmith D, Bravo A (2006) Ecology and management of nesting blue-and-yellow macaws (*Ara ararauna*) in *Mauritia* palm swamps. Biodivers Conserv 15:4271–4287

Bullock SH, Heath D (2006) Growth rates and age of native palms in the Baja California desert. J Arid Environ 67:391–402

Cámara-Leret R, Paniagua-Zambrana N, Balslev H, Barfod A, Copete JC, Macía MJ (2014) Ecological community traits and traditional knowledge shape palm ecosystem services in northwestern South America. For Ecol Manage 334:28–42

Cavaleri M, Oberbauer SF, Ryan MG (2008) Foliar and ecosystem respiration in an old-growth tropical rain forest. Plant, Cell Environ 31:473–483

Cepeda-Cornejo V, Dirzo R (2010) Sex-related differences in reproductive allocation, growth, defense and herbivory in three dioecious neotropical palms. PLoS ONE 5. doi:10.1371/journal.pone.0009824

Chao CCT, Krueger RR (2007) The date palm (*Phoenix dactylifera* L.): overview of biology, uses, and cultivation. HortScience 42:1077–1082

Chazdon RL (1985) Leaf display, canopy structure, and light interception of two understory palm species. Am J Bot 72:1493–1502

Chazdon RL (1986) The costs of leaf support in understory palms: economy versus safety. Am Nat 127:9–30

Chown SL, Gaston KJ (2000) Areas, cradles and museums: the latitudinal gradient in species richness. Trends Ecol Evol 15:311–315

Clark DA, Clark DB, Sandoval RM, Vinicio Castro MC (1995) Edaphic and human effects on landscape-scale distributions of tropical rain forest palms. Ecology 76:2581–2594

Clement CR, Weber JC, van Leeuwen J, Astorga Domian C, Cole DM, Arévalo Lopez LA, Argüello H (2004) Why extensive research and development did not promote use of peach palm fruit in Latin America. Agrofor Syst 61:195–206

Clement CR, Santos RP, Desmouliere SJM, Ferreira EJL, Tomé J, Neto F (2009) Ecological adaptation of wild peach palm, its *in situ* conservation and deforestation-mediated extinction in Southern Brazilian Amazonia. PLoS ONE 4. doi:10.1371/journal.pone.0004564

Cochrane MA (2003) Fire science in rainforests. Nature 421:913–919

Cochrane MA (2009) Tropical fire ecology: Climate change, land use and ecosystem dynamics. Praxis Publishing Ltd, Chichester

Cohen Y, Alchanatis V, Prigojin A, Levi A, Soroker V, Cohen Y (2012) Use of aerial thermal imaging to estimate water status of palm trees. Precision Agric 13:123–140

Coley PD, Kursar TA (1996) Causes and consequences of epiphyll colonization. In: Mulkey S, Chazdon R, Smith A (eds) Tropical forest plant physiology. Chapman Hall, New York, pp 337–362

Corley RHV (1983) Photosynthesis and age of oil palm leaves. Photosynthetica 17:97–100

Cornett JW (1991) Population dynamics of the palm, *Washingtonia filifera*, and global warming. San Bernadino County Mus Assoc Q 39:46–47

Couvreur TLP (2014) Odd man out: why are there fewer plant species in African rain forests? Plant Syst Evol. doi:10.1007/s00606-014-1180-z

Couvreur TLP, Baker WJ (2013) Tropical rain forest evolution: palms as a model group. BMC Biol 11. doi:10.1186/1741-7007-11-48

Couvreur TLP, Forest F, Baker WJ (2011) Origin and global diversification patterns of tropical rain forests: inferences from a complete genus-level phylogeny of palms. BMC Biol 9. doi:10. 1186/1741-7007-9-44

Davis TA (1961) High root-pressure in palms. Nature 192:277–278

de Azevedo PV, de Sousa I, da Silva B, da Silva VPR (2006) Water-use efficiency of dwarf-green coconut (*Cocos nucifera* L.) orchards in northeast Brazil. Agric Water Manage 84:259–264

De Steven D, Windsor DM, Putz FE, de Leon B (1987) Vegetative and reproductive phenologies of a palm assemblage in Panama. Biotropica 19:342–356

Dransfield J, Uhl NW, Asmussen CB, Baker WJ, Harley MM, Lewis CE (2008) Genera Palmarum: the evolution and classification of palms. Kew Publishing, Kew

Dufrene E, Saugier B (1993) Gas exchange of oil palm in relation to light, vapour pressure deficit, temperature and leaf age. Funct Ecol 7:97–104

Dufrene E, Dubos B, Rey H, Quencez P, Saugier B (1992) Changes in evapotranspiration from an oil palm stand (*Elaeis guineensis* Jacq.) exposed to seasonal soil water deficits. Acta Oecol 13:299–314

Duryea ML, Blakeslee GM, Hubbard WG, Vasquez RA (1996) Wind and trees: a survey of homeowners after Hurricane Andrew. J Arboric 22:44–50

Eiserhardt WL, Bjorholm S, Svenning JC, Rangel TF, Balslev H (2011a) Testing the water-energy theory on American palms (Arecaceae) using geographically weighted regression. PLoS ONE 6. doi:10.1371/journal.pone.0027027

Eiserhardt WL, Svenning JC, Kissling WD, Balslev H (2011b) Geographical ecology of the palms (Arecaceae): determinants of diversity and distributions across spatial scales. Ann Bot 108:1391–1416

El-Juhany LI (2010) Degradation of date palm trees and date production in Arab countries: causes and potential rehabilitation. Aust J Basic Appl Sci 4:3998–4010

Eskafi FM, Basham HG, McCoy RE (1986) Decreased water transport in lethal yellowing-diseased coconut palms. Trop Agric (Trinidad) 63:225–228

FAOSTAT Statistics Database (2014) UN Food and Agricultural Organisation. http://faostat3.fao.org/home/E. Accessed 21 Oct 2014

Fisher JB, Burch JN, Noblick LR (1996) Stem structure of the Cuban belly palm (Gastrococos crispa). Principes 40:125–128

Fowler D, Nemitz E, Misztal P, Di Marco C, Skiba U, Ryder J, Helfter C, Cape JN, Owen S, Dorsey J, Gallagher MW, Coyle M, Phillips G, Davison B, Langford B, MacKenzie R, Muller J, Siong J, Dari-Salisburgo C, Di Carlo P, Aruffo E, Giammaria R, Pyle JA, Hewitt CN (2011) Effects of land use on surface-atmosphere exchanges of trace gases and energy in Borneo: comparing fluxes over oil palm plantations and a rainforest. Philos Trans R Soc Lond, Ser B: Biol Sci 366:3196–3209

Fragoso JMV, Silvius KM, Correa JA (2003) Long-distance seed dispersal by tapirs increases seed survival and aggregates tropical trees. Ecology 84:1998–2006

Galetti M, Donatti CI, Pires AS, Guimarães PR Jr, Jordano P (2006) Seed survival and dispersal of an endemic Atlantic forest palm: the combined effects of defaunation and forest fragmentation. Bot J Linn Soc 151:141–149

Gatti MG, Campanello PI, Montti LF, Goldstein G (2008) Frost resistance in the tropical palm Euterpe edulis and its pattern of distribution in the Atlantic Forest of Argentina. Forest Ecol Manage 256:633–640

Gatti MG, Campanello PI, Goldstein G (2011) Growth and leaf production in the tropical palm Euterpe edulis: Light conditions versus developmental constraints. Flora 206:742–748

Germer J, Sauerborn J (2007) Estimation of the impact of oil palm plantation establishment on greenhouse gas balance. Environ Dev Sustain 10:697–716

Gerritsma W, Soebagyo FX (1999) An analysis of the growth of leaf area of oil palms in Indonesia. Exp Agric 35:293–308

Goldammer JG, Price C (1998) Potential impacts of climate change on fire regimes in the tropics based on MAGICC and a GISS GCM-derived lightning model. In: Markham A (ed) Potential impacts of climate change on tropical forest ecosystems. Kluwer Academic Publishers, Dordrecht, pp 133–156

Gomes FP, Prado CHBA (2007) Ecophysiology of coconut palm under water stress. Braz J Plant Physiol 19:377–391

Gomes FP, Oliva MA, Mielke MS, de Almeida AAF, Leite HG, Aquino LA (2008) Photosynthetic limitations in leaves of young Brazilian Green Dwarf coconut (Cocos nucifera L. 'nana') palm under well-watered conditions or recovering from drought stress. Environ Exp Bot 62:195–204

Harley MM (2006) A summary of fossil records for Arecaceae. Bot J Linn Soc 151:39–67

Harries H (1978) The evolution, dissemination and classification of Cocos nucifera L. Bot Rev 44:265–319

Hartmann KL, Klein Tank AMG, Rusticucci M, Alexander LV, Brönnimann S, Charabi Y, Dentener FJ, Dlugokencky EJ, Easterling DR, Kaplan A, Soden BJ, Thorne PW, Wild M, Zhai PM (2013) Observations: atmosphere and surface. In: Stocker TF, Qin D, Plattner GK et al. (eds) Climate change 2013: the physical basis. Contribution of working group I to the fifth assessment report of the Intergovernmental Panel on Climate Change. Cambridge University Press, Cambridge

Henderson A (1990) Arecaceae Part I. Introduction and the Iriarteinae. Flora Neotropica. New York Botanical Garden, New York

Henderson A, Galeano G, Bernal R (1995) Field guide to the palms of the Americas. Princeton University Press, Princeton

Henson IE (2004) Estimating maintenance respiration of oil palm. Oil Palm Bulletin 48:1–10

Henson IE, Chang KC (2000) Oil palm productivity and its component processes. Advances in oil palm research, vol 1. Malaysian Palm Oil Board, Kajang, Malaysia, pp 97–145

Hewitt CN, MacKenzie AR, Di Carlo P, Di Marco CF, Dorsey JR, Evans M, Fowler D, Gallagher MW, Hopkins JR, Jones CE, Langford B, Lee JD, Lewis AC, Lim SF, McQuaid J, Misztal P, Moller SJ, Monks PS, Nemitz E, Oram DE, Owen SM, Phillips GJ, Pugh TAM, Pyle JA, Reeves CE, Ryder J, Siong J, Skiba U, Stewart DJ (2009) Nitrogen management is essential to prevent tropical oil palm plantations from causing ground-level ozone pollution. Proc Natl Acad Sci USA 106:18447–18452

Hogan KP (1988) Photosynthesis in two newtropical palm species. Funct Ecol 2:371–377

Holbrook NM, Sinclair TR (1992a) Water balance in the arborescent palm, *Sabal palmetto*. I. Stem structure, tissue water release properties and leaf epidermal conductance. Plant, Cell Environ 15:393–399

Holbrook NM, Sinclair TR (1992b) Water balance in the arborescent palm, *Sabal palmetto*. II. Transpiration and stem water storage. Plant, Cell Environ 15:401–409

Ibrahim MH, Jaafar HZE (2012) Impact of elevated carbon dioxide on primary, secondary metabolites and antioxidant responses of *Eleais guineensis* Jacq. (oil palm) seedlings. Molecules 17:5195–5211

James KR, Haritos N, Ades PK (2006) Mechanical stability of trees under dynamic loads. Am J Bot 93:1522–1530

Janssen T, Bremer K (2004) The age of major monocot groups inferred from 800 + rbcL sequences. Bot J Linn Soc 146:385–398

Kahn F (1988) Ecology of economically important palms in Peruvian Amazonia. Adv Econ Bot 6:42–49

Kahn F, Mejia K (1990) Palm communities in wetland forest ecosystems of Peruvian Amazonia. For Ecol Manage 33–34:169–179

Kahn F, Mejia K (1991) The palm communities of two 'terra firme' forests in Peruvian Amazonia. Principes 35:22–26

Kasturi Bai KV, Rajagopal V (2000) Osmotic adjustment as a mechanism for drought tolerance in coconut (*Cocos nucifera* L.). Indian J Plant Physiol 5:320–323

Kasturibai KV, Voleti SR, Rajagopal V (1988) Water relations of coconut palms as influenced by environmental variables. Agric Meteorol 43:193–199

Kessler M (2000) Upslope-directed mass effect in palms along an Andean elevational gradient: A cause for high diversity at mid-elevations? Biotropica 32:756–759

Killmann W (1983) Some physical properties of the coconut palm stem. Wood Sci Technol 17:167–185

Kissling WD, Baker WJ, Balslev H, Barfod AS, Borchsenius F, Dransfield J, Govaerts R, Svenning JC (2012) Quaternary and pre-Quaternary historical legacies in the global distribution of a major tropical plant lineage. Global Ecol Biogeogr 21:909–921

Koh LP, Miettinen J, Liew SC, Ghazoul J (2011) Remotely sensed evidence of tropical peatland conversion to oil palm. Proc Natl Acad Sci USA 108:5127–5132

Kreft H, Sommer JH, Barthlott W (2006) The significance of geographic range size for spatial diversity patterns in Neotropical palms. Ecography 29:21–30

Kristiansen T, Svenning JC, Grández C, Salo J, Balslev H (2009) Commonness of Amazonian palm (Arecaceae) species: cross-scale links and potential determinants. Acta Oecol 35:554–562

Kristiansen T, Svenning JC, Pedersen D, Eiserhardt WL, Grández C, Balslev H (2011) Local and regional palm (Arecaceae) species richness patterns and their cross-scale determinants in the western Amazon. J Ecol 99:1001–1015

Kurup VVGK, Voleti SR, Rajagopal V (1993) Influence of weather variables on the content and composition of leaf surface wax in coconut. J Plant Crops 21:71–80

Kvacek J, Herman AB (2004) Monocotyledons from the Early Campanian (Cretaceous) of Grünbach, Lower Austria. Rev Palaeobot Palynol 128:323–353

Legros S, Mailet-Serra I, Clement-Vidal A, Caliman JP, Siregar FA, Fabre D, Dingkuhn M (2009) Role of transitory carbon reserves during adjustment to climate variability and source-sink imbalances in oil palm (*Elaeis guineensis*). Tree Physiol 29:1199–1211

Leon R, Santamaria JM, Alpizar L, Escamilla JA, Oropeza C (1996) Physiological and biochemical changes in shoots of coconut palms affected by lethal yellowing. New Phytol 134:227–234

Lockett L (2004) The Sabal palm: restoring a species we didn't know we had (Texas). Ecol Restor 22:137–138

Lugo AE, Rivera Batlle CT (1987) Leaf production, growth rate, and age of the palm *Prestoea montana* in the Luquillo Experimental Forest, Puerto Rico. J Trop Ecol 3:151–161

Mäkelä A, Sievänen R (1992) Height growth strategies in open-grown trees. J Theor Biol 159:443–467

Malhi Y, Baldocchi DD, Jarvis PG (1999) The carbon balance of tropical, temperate and boreal forests. Plant, Cell Environ 22:715–740

Martinez S, Cordova I, Maust BE, Oropeza C, Santamaria JM (2000) Is abscisic acid responsible for abnormal stomatal closure in coconut palms showing lethal yellowing? J Plant Physiol 156:319–322

Martínez-Ramos M, Alvarez-Buylla E, Sarukhán J, Piñero D (1988) Treefall age determination and gap dynamics in a tropical forest. J Ecol 76:700–716

Maust BE, Espadas F, Talavera C, Aguilar M, Santamaria JM, Oropeza C (2003) Changes in carbohydrates metabolism in coconut palms infected with the lethal yellowing phytoplasma. Phytopathology 93:976–981

McDonough J, Zimmermann MH (1979) Effect of lethal yellowing on xylem pressure in coconut palms. Principes 23:132–137

Mialet-Serra I, Clément A, Sonderegger N, Roupsard O, Jourdan C, Labouisse JP, Dingkuhn M (2005) Assimilate storage in vegetative organs of coconut (*Cocos nucifera* L.). Exp Agric 41:1–14

Mialet-Serra I, Clement-Vidal A, Roupsard O, Jourdan C, Dingkuhn M (2008) Whole-plant adjustments in coconut (*Cocos nucifera*) in response to sink-source imbalance. Tree Physiol 28:1199–1209

Milburn JA, Davis TA (1973) Role of pressure in xylem transport of coconut and other palms. Physiol Plant 29:415–420

Milburn JA, Zimmermann MH (1977) Preliminary studies on sapflow in *Cocos nucifera* L II. Phloem transport. New Phytol 79:543–558

Misztal PK, Nemitz E, Langford B, Di Marco CF, Phillips GJ, Hewitt CN, MacKenzie AR, Owen SM, Fowler D, Heal MR, Cape JN (2011) Direct ecosystem fluxes of volatile organic compounds from oil palms in South-East Asia. Atmos Chem Phys 11:8995–9017

Morley RJ (2000) Origin and evolution of tropical rain forests. Wiley, Chichester

Navarro MNV, Jourdan C, Sileye T, Braconnier S, Mailet-Serra I, Saint-Andre L, Dauzat J, Nouvellon Y, Epron D, Bonnefond JM, Berbigier P, Rouziere A, Bouillet JP, Roupsard O (2008) Fruit development, not GPP, drives seasonal variation in NPP in a tropical palm plantation. Tree Physiol 28:1661–1674

Niklas KJ (1992) Plant biomechanics: an engineering approach to plant form and function. The University of Chicago Press, Chicago

Niklas KJ (1994) Plant allometry: the scaling of form and process. The University of Chicago Press, Chicago

Niklas KJ (1999) A mechanical perspective on foliage leaf form and function. New Phytol 143:19–31

Niklas KJ, Cobb E, Marler T (2006) A comparison between the record height-to-stem diameter allometries of pachycaulis and leptocaulis species. Ann Bot 97:79–83

Oyama K (1993) Are age and height correlated in *Chamaedorea tepejilote* (Palmae)? J Trop Ecol 9:381–385

Parthasarathy MV, Klotz LH (1976) Palm "wood" II. Ultrastructural apsects of sieve elements, tracheary elements and fibers. Wood Sci Technol 10:247–271

Parthasarathy MV, Tomlinson PB (1967) Anatomical features of metaphloem in stems of *Sabal, Cocos* and two other palms. Am J Bot 54:1143–1151

Peres CA (1994) Composition, density, and fruiting phenology of arborescent palms in an Amazonian terra firme forest. Biotropica 26:285–294

Peters HA, Pauw A, Silman MR, Terborgh JW (2004) Falling palm fronds structure Amazonian rainforest sapling communities. Proc R Soc London, Ser B: Biol Lett 271:S367–S369

Phillips N, Buckley TN, Tissue DT (2008) Capacity of old trees to respond to environmental change. J Integr Plant Biol 50:1355–1364

Pivovaroff AL, Sack L, Santiago LS (2014) Coordination of stem and leaf hydraulic conductance in southern California shrubs: a test of the hydraulic segmentation hypothesis. New Phytol 203:842–850

Prado CHBA, Passos EEM, De Moraes JAPV (2001) Photosynthesis and water relations of six tall genotypes of *Cocos nucifera* in wet and dry seasons. S Afr J Bot 67:169–176

Pritchard Miller R, Nair PKR (2006) Indigenous agroforestry systems in Amazonia: from prehistory to today. Agrofor Syst 66:151–164

Rao AS (1989) Water requirements of young coconut palms in a humid tropical climate. Irrig Sci 10:245–249

Ratsirarson J, Silander JA Jr, Richard AF (1996) Conservation and management of a threatened Madagascar palm species, *Neodypsis decaryi*, Jumelle. Conserv Biol 10:40–52

Rees AR (1961) Midday closure of stomata in the oil palm *Elaeis guineensis* Jacq. J Exp Bot 12:129–146

Renninger HJ, Phillips N (2010) Intrinsic and extrinsic hydraulic factors in varying sizes of two Amazonian palm species (*Iriartea deltoidea* and *Mauritia flexuosa*) differing in development and growing environment. Am J Bot 97:1926–1936

Renninger HJ, Phillips N (2011) Hydraulic properties of fronds from palms of varying height and habitat. Oecologia 167:925–935

Renninger HJ, Phillips N (2012) "Secondary stem lengthening" in the palm *Iriartea deltoidea* (Arecaceae) provides an efficient and novel method for height growth in a tree form. Am J Bot 99:607–613

Renninger HJ, Phillips N, Hodel DR (2009) Comparative hydraulic and anatomic properties in palm trees (*Washingtonia robusta*) of varying heights: implications for hydraulic limitation to increased height growth. Trees 23:911–921

Renninger HJ, Phillips N, Salvucci GD (2010) Wet- vs. dry-season transpiration in an Amazonian rain forest palm *Iriartea deltoidea*. Biotropica 42:470–478

Renninger HJ, McCulloh KA, Phillips N (2013) A comparison of the hydraulic efficiency of a palm species (*Iriartea deltoidea*) with other wood types. Tree Physiol 33:152–160

Repellin A, Laffray D, Daniel C, Braconnier S, Zuily-Fodil Y (1997) Water relations and gas exchange in young coconut palm (*Cocos nucifera* L.) as influenced by water deficit. Can J Bot 75:18–27

Rich PM (1986) Mechanical architecture of arborescent rain forest palms. Principes 30:117–131

Rich PM (1987) Developmental anatomy of the stem of *Welfia georgii*, *Iriartea gigantea* and other arborescent palms; implications for mechanical support. Am J Bot 34:792–802

Rich PM, Helenurm K, Kearns D, Morse SR, Palmer MW, Short L (1986) Height and stem diameter relationships for dicotyledonous trees and arborescent palms of Costa Rican tropical wet forest. Bull Torrey Bot Club 113:241–246

Rich PM, Holbrook NM, Luttinger N (1995) Leaf development and crown geometry of two Iriarteoid palms. Am J Bot 82:328–336

Roupsard O, Bonnefond JM, Irvine M, Berbigier P, Nouvellon Y, Dauzat J, Taga S, Hamel O, Jourdan C, Saint-André L, Mailet-Serra I, Labouisse JP, Epron D, Joffre R, Braconnier S, Rouzière A, Navarro M, Bouillet JP (2006) Partitioning energy and evapo-transpiration above and below a tropical palm canopy. Agric For Meteorol 139:252–268

Rüggeberg M, Speck T, Paris O, Lapierre C, Pollet B, Koch G, Burgert I (2008) Stiffness gradients in vascular bundles of the palm *Washingtonia robusta*. Proc R Soc Biol Sci Ser B 275:2221–2229

Rüggeberg M, Speck T, Burgert I (2009) Structure-function relationships of different vascular bundle types in the stem of the Mexican fanpalm (*Washingtonia robusta*). New Phytol 182:443–450

Ryan MG, Yoder BJ (1997) Hydraulic limits to tree height and tree growth. Bioscience 47:235–242

Ryan MG, Phillips N, Bond BJ (2006) The hydraulic limitation hypothesis revisited. Plant, Cell Environ 29:367–381

Salm R, Jalles-Filho E, Schuck-Paim C (2005) A model for the importance of large arborescent palms in the dynamics of seasonally-dry Amazonian forests. Biota Neotropica 5:1–6

Sayer J, Ghazoul J, Nelson P, Boedhihartono AK (2012) Oil palm expansion transforms tropical landscapes and livelihoods. Global Food Security 1:114–119

Schatz GE, Williamson GB, Cogswell CM, Stam AC (1985) Stilt roots and growth of arboreal palms. Biotropica 17:206–209

Shabani F, Kumar L, Taylor S (2012) Climate change impacts on the future distribution of date palms: A modeling exercise using CLIMEX. PLoS ONE 7. doi:10.1371/journal.pone.0048021

Shivashankar S, Kasturi Bai KV, Rajagopal V (1991) Leaf water potential, stomatal resistance and activities of enzymes during the development of moisture stress in the coconut palm. Trop Agric 68:106–110

Sillett SC, Van Pelt R, Koch GW, Ambrose AR, Carroll AL, Antoine ME, Mifsud BM (2010) Increasing wood production through old age in tall trees. For Ecol Manage 259:976–994

Silva MG, Tabarelli M (2001) Seed dispersal, plant recruitment and spatial distribution of *Bactris acanthocarpa* Martius (Arecaceae) in a remnant of Atlantic forest in northeast Brazil. Acta Oecol 22:259–268

Smith BG (1989) The effects of soil water and atmospheric vapour pressure deficit on stomatal behaviour and photosynthesis in the oil palm. J Exp Bot 40:647–651

Souza AF, Martins FR (2004) Population structure and dynamics of a neotropical palm in fire-impacted fragments of the Brazilian Atlantic Forest. Biodivers Conserv 13:1161–1632

Sperling O, Shapira O, Cohen S, Tripler E, Schwartz A, Lazarovitch N (2012) Estimating sap flux densities in date palm trees using the heat dissipation method and weighing lysimeters. Tree Physiol 32:1171–1178

Sperling O, Lazarovitch N, Schwartz A, Shapira O (2014a) Effects of high salinity irrigation on growth, gas-exchange, and photoprotection in date palms (*Phoenix dactylifera* L., cv. Medjool). Environ Exp Bot 99:100–109

Sperling O, Shapira O, Tripler E, Schwartz A, Lazarovitch N (2014b) A model for computing date palm water requirements: as affected by salinity. Irrig Sci 32:341–350

Sperry JS (1986) Relationship of xylem embolism to xylem pressure potential, stomatal closure, and shoot morphology in the palm *Rhapis excelsa*. Plant Physiol 80:110–116

Spicer R (2005) Senescence in secondary xylem: heartwood formation as an active developmental program. In: Holbrook NM, Zwieniecki MA (eds) Vascular transport in plants. Elsevier Academic Press, Burlington, pp 457–475

Spicer R, Holbrook NM (2007) Parenchyma cell respiration and survival in secondary xylem: does metabolic activity decline with cell age? Plant, Cell Environ 30:934–943

Stephenson NL, Das AJ, Condit R, Russo SE, Baker PJ, Beckman NG, Coomes DA, Lines ER, Morris WK, Rüger N, Álvarez E, Blundo C, Bunyavejchewin S, Chuyong G, Davies SJ, Duque Á, Ewango CN, Flores O, Franklin JF, Grau HR, Hao Z, Harmon ME, Hubbell SP, Kenfack D, Lin Y, Makana JR, Malizia A, Malizia LR, Pabst RJ, Pongpattananurak N, Su SH, Sun IF, Tan S, Thomas D, van Mantgem PJ, Wang X, Wiser SK, Zavala MA (2014) Rate of tree carbon accumulation increases continuously with tree size. Nature 507:90–93

Svenning JC (2001) On the role of microenvironmental heterogeneity in the ecology and diversification of neotropical rain-forest palms (Arecaceae). Bot Rev 67:2–53

Svenning JC, Borchsenius F, Bjorholm S, Balslev H (2008) High tropical net diversification drives the New World latitudinal gradient in palm (Arecaceae) species richness. J Biogeogr 35:394–406

Takenaka A, Takahashi K, Kohyama T (2001) Optimal leaf display and biomass partitioning for efficient light capture in an understorey palm, *Licuala arbuscula*. Funct Ecol 15:660–668

Terborgh J (1986) Keystone plant resources in the tropical forest. In: Soulé M (ed) Conservation biology: the science of scarcity and diversity. Sinauer, New York, pp 330–344

Thenkabail PS, Stucky N, Griscom BW, Ashton MS, Diels J, van der Meer B, Enclona E (2004) Biomass estimations and carbon stock calculations in the oil palm plantations of African derived savannas using IKONOS data. Int J Remote Sens 25:5447–5472

Thomas R, De Franceschi D (2013) Palm stem anatomy and computer-aided identification: The Coryphoideae (Arecaceae). Am J Bot 100:289–313

Ter Steege H, Pitman NCA, Sabatier D, Baraloto C, Salomão RP, Guevara JE, Phillips OL, Castilho CV, Magnusson WE, Molino JF, Monteagudo A, Núñez Vargas P, Montero JC, Feldpausch TR, Coronado ENH, Killeen TJ, Mostacedo B, Vasquez R, Assis RL, Terborgh J, Wittmann F, Andrade A, Laurance WF, Laurance SGW, Marimon BS, Marimon Jr BH, Guimarães Vieira IC, Amaral IL, Brienen R, Castellanos H, Cárdenas López D, Duivenvoorden JF, Mogollón HF, de Almeida Matos FD, Dávila N, García-Villacorta R, Stevenson Diaz PR, Costa F, Emilio T, Levis C, Schietti J, Souza P, Alonso A, Dallmeier F, Duque Montoya AJ, Fernandez Piedade MT, Araujo-Murakami A, Arroyo L, Gribel R, Fine PVA, Peres CA, Toledo M, Aymard C. GA, Baker TR, Cerón C, Engel J, Henkel TW, Maas P, Petronelli P, Stropp J, Zartman CE, Daly D, Neill D, Silveira M, Ríos Paredes M, Chave J, de Andrade Lima Filho D, Møller Jørgensen P, Fuentes A, Schöngart J, Cornejo Valverde F, Di Fiore A, Jimenez EM, Peñuela Mora MC, Phillips JF, Rivas G, van Andel TR, von Hildebrand P, Hoffman B, Zent EL, Malhi Y, Prieto A, Rudas A, Ruschell AR, N S, Vos V, Zent S, Oliveira AA, Cano Schutz A, Gonzales T, Nascimento MT, Ramirez-Angulo H, Sierra R, Tirado M, Umaña Medina MN, Van der Heijden G, Vela CIA, Torre EV, Vriesendorp C, Wang O, Young KR, Baider C, Balslev H, Ferreira C, Mesones I, Torres-Lezama A, Urrego Giraldo IH, Milliken W, Palacios Cuenca W, Pauletto D, Sandoval EV, Gamarra LV, Dexter KG, Feeley KJ, Lopez-Gonzalez G, Silman MR (2013) Hyperdominance in the Amazonian tree flora. Science 342. doi:10.1126/science.1243092

Tomlinson PB (1961) Anatomy of the Monocotyledons. II. Palmae. Clarendon Press, Oxford

Tomlinson PB (1962) The leaf base in palms its morphology and mechanical biology. J Arnold Arboretum 43:23–50

Tomlinson PB (1990) The structural biology of palms. Oxford University Press, Oxford

Tomlinson PB (2006) The uniqueness of palms. Bot J Linn Soc 151:5–14

Tomlinson PB, Huggett B (2012) Cell longevity and sustained primary growth in palm stems. Am J Bot 99:1891–1902

Tregear JW, Rival A, Pintaud JC (2011) A family portrait: unravelling the complexities of palms. Ann Bot 108:1387–1389

Tripler E, Ben-Gal A, Shani U (2007) Consequence of salinity and excess boron on growth, evapotranspiration and ion uptake in date palm (*Phoenix dactylifera* L., cv. Medjool). Plant Soil 297:147–155

Tripler E, Shani U, Mualem Y, Ben-Gal A (2011) Long term growth, water consumption and yield of date palm as a function of salinity. Agric Water Manage 99:128–134

Tyree MT, Ewers FW (1991) The hydraulic architecture of trees and other woody plants. New Phytol 119:345–360

Van Looy T, Carrero GO, Mathijs E, Tollens E (2008) Underutilized agroforestry food products in Amazonas (Venezuela): a market chain analysis. Agrofor Syst 74:127–141

Van Nieuwstadt MGL, Sheil D (2005) Drought, fire and tree survival in a Borneo rain forest, East Kalimantan, Indonesia. J Ecol 93:191–201

Vegas-Vilarrubia T, Baritto F, López P, Meleán G, Ponce ME, Mora L, Gómez O (2010) Tropical histosols of the lower Orinoco Delta, features and preliminary quantification of their carbon storage. Geoderma 155:280–288

Venceslau Aguiar A, Tabarelli M (2010) Edge effects and seedling bank depletion: The role played by the early successional palm *Attalea oleifera* (Arecaceae) in the Atlantic forest. Biotropica 42:158–166

Vormisto J, Svenning JC, Hall P, Balslev H (2004) Diversity and dominance in palm (Arecaceae) communities in *terra firme* forests in the western Amazon basin. J Ecol 92:577–588

Walther GR, Gritti ES, Berger S, Hickler T, Tang Z, Sykes MT (2007) Palms tracking climate change. Global Ecol Biogeogr 16:801–809

Waterhouse JT, Quinn FLS, Quinn CJ (1978) Growth patterns in the stem of the palm *Archontophoenix cunninghamiana*. Bot J Linn Soc 77:73–93

Way DA, Oren R (2010) Differential responses to changes in growth temperature between trees from different functional groups and biomes: a review and synthesis of data. Tree Physiol 30:669–688

Williams RJ, Cook GD, Gill AM, Moore PHR (1999) Fire regime, fire intensity and tree survival in a tropical savanna in northern Australia. Aust J Ecol 24:50–59

Winter DF (1993) On the stem curve of a tall palm in a strong wind. SIAM Rev 35:567–579

Woodruff DR, Bond BJ, Meinzer FC (2004) Does turgor limit growth in tall trees? Plant, Cell Environ 27:229–236

Wright SJ, Zeballos H, Domínguez I, Gallardo MM, Moreno MC, Ibáñez R (2000) Poachers alter mammal abundance, seed dispersal, and seed predation in a neotropical forest. Conserv Biol 14:227–239

Zimmerman JK, Everham EM III, Waide RB, Lodge DJ, Taylor CM, Brokaw VL (1994) Responses of tree species to hurricane winds in subtropical wet forest in Puerto Rico: implications for tropical tree life histories. J Ecol 82:911–922

Zimmermann MH (1973) The monocotyledons: their evolution and comparative biology IV. Transport problems in arborescent monocotyledons. Q Rev Biol 48:314–321

Zimmermann MH (1983) Xylem structure and the ascent of sap. Springer, New York

Zimmermann MH, Sperry JS (1983) Anatomy of the palm *Rhapis excelsa*, IX. Xylem structure of the leaf insertion. J Arnold Arboretum 64:599–609

Zimmermann MH, Tomlinson PB (1965) Anatomy of the palm *Rhapis excelsa* I. Mature vegetative axis. J Arnold Arboretum 46:160–181

Zimmermann MH, Tomlinson PB (1972) The vascular system of monocotyledonous stems. Bot Gaz 133:141–155

Carbon Allocation and Water Relations of Lianas Versus Trees

Paula I. Campanello, Eric Manzané, Mariana Villagra, Yong-Jiang Zhang, Adela M. Panizza, Débora di Francescantonio, Sabrina A. Rodriguez, Ya-Jun Chen, Louis S. Santiago and Guillermo Goldstein

Abstract Despite lianas being fundamental components of tropical and subtropical forest ecosystems throughout the world, the physiological characteristics of this growth form are not well known. Different behaviors at the seedling stage were until recently largely unnoticed. In one extreme of a continuum of adaptive traits, freestanding liana seedlings invest a large proportion of biomass in self-support tissue while on the other extreme support-seeker seedlings invest more resources in rapid elongation of slender stems with an efficient hydraulic conductive system. Adult lianas often have lower wood density and higher specific leaf area than trees and have most of their leaves deployed at the top of the canopy, experiencing high

P.I. Campanello (✉) · M. Villagra · A.M. Panizza · D. di Francescantonio · S.A. Rodriguez
Laboratorio de Ecología Forestal y Ecofisiología, Instituto de Biología Subtropical,
CONICET, FCF, Universidad Nacional de Misiones, Puerto Iguazú, Argentina
e-mail: pcampanello@gmail.com; pcampanello@conicet.gov.ar

M. Villagra
e-mail: marian.villagra@gmail.com

A.M. Panizza
e-mail: adelamariapanizza@gmail.com

D. di Francescantonio
e-mail: debodifra@gmail.com

S.A. Rodriguez
e-mail: sabrinarodriguez78@yahoo.com.ar

E. Manzané · L.S. Santiago
Smithsonian Tropical Research Institute, Balboa, Panama Apartado Postal 0843-03092,
Republic of Panama
e-mail: bioerman@gmail.com

L.S. Santiago
e-mail: santiago@ucr.edu

Y.-J. Zhang
Department of Organismic and Evolutionary Biology, Harvard University, Cambridge, MA
02138, USA
e-mail: yongjiangz@126.com

© Springer International Publishing Switzerland 2016   103
G. Goldstein and L.S. Santiago (eds.), *Tropical Tree Physiology*,
Tree Physiology 6, DOI 10.1007/978-3-319-27422-5_5

irradiance and transpirational demands, which requires effective regulation of water loss to avoid desiccation. Recent studies show that lianas have faster stomatal responses to increasing vapor pressure deficit (VPD) and exhibit stronger partial stomatal closure compared to trees. Strong stomatal control and efficient water transport help lianas maintain leaf water potential (Ψ_{leaf}) within a safe hydraulic range to avoid xylem dysfunction despite their low stem water storage capacity, which is achieved at a minimum cost in terms of carbon assimilation. Liana colonization of tree crowns can significantly reduce tree growth and transpiration with consequences for carbon and water economy at individual tree and ecosystem levels.

Keywords Capacitance · Freestanding liana seedlings · Support-seeker liana seedlings · Stomatal conductance · Transpiration · Long distance water transport

Introduction

Lianas or woody vines are fundamental components of tropical and subtropical forests. Their importance in forest ecology and functioning has been recently recognized with increasing studies all around the world, and particularly in tropical forests, where lianas can contribute up to 35 % of woody plant diversity and 40 % of stem density (Gerwing and Farias 2000; Chave et al. 2001; Schnitzer and Bongers 2002, 2011). For more than 30 years, research mainly focused on community and population-level processes and evaluation of the detrimental consequences that lianas have on trees, rendering a relatively good understanding of their effect on forest dynamics (Putz 1984a, b; Pinard and Putz 1994; Alvira et al. 2004). More recently, ecophysiological studies comparing lianas and trees have enhanced our knowledge of the physiological and morphological behavior of lianas

Y.-J. Zhang · Y.-J. Chen
Key Laboratory of Tropical Forest Ecology, Xishuangbana Tropical Botanical Garden, Chinese Academy of Sciences, Mengla 666303, Yunnan, China
e-mail: chenyj@xtbg.org.cn

L.S. Santiago
Department of Botany & Plant Sciences, University of California, Riverside, CA 92521, USA

G. Goldstein
Laboratorio de Ecología Funcional, Departamento de Ecología Genética y Evolución, Instituto IEGEBA (CONICET-UBA), Facultad de Ciencias Exactas y naturales, Universidad de Buenos Aires, Buenos Aires, Argentina
e-mail: gold@bio.miami.edu; goldstein@ege.fcen.uba.ar

G. Goldstein
Department of Biology, University of Miami, Coral Gables, FL 33146, USA

(Cai et al. 2008, 2009; Johnson et al. 2013; van der Sande et al. 2013; Santiago et al. 2015). However their water and carbon economy are still not well known.

Lianas tend to grow rapidly and climb suitable host trees covering most of the upper canopy where incoming solar radiation is high. Individuals reaching tree crowns also act as climbing supports, facilitating the access of other lianas to the upper canopy (Putz 1984a, b; Nabe-Nielsen 2001; Campanello et al. 2007). Lianas may colonize new gaps delaying gap-phase regeneration processes and the recovery of a tall mature forest (Schnitzer et al. 2000). They also have detrimental direct effects on trees, decreasing growth and fecundity and increasing mortality rates (Nabe-Nielsen et al. 2009; Hegarty and Caballe 1991; Grauel and Putz 2004; Kainer et al. 2006; Nabe-Nielsen et al. 2009; van der Heijden and Phillips 2009; Ingwell et al. 2010; Schnitzer and Carson 2010). This positive feedback loop between disturbance and liana abundance may partially explain increasing liana abundance and tree infestation in disturbed neotropical Forests (Fig. 1), with profound consequences for biogeochemical cycles (Phillips et al. 2002; Schnitzer and Bongers 2011; Wright et al. 2004a). In contrast to neotropical forests, liana abundance in African and Asian tropical forests is stable or even decreasing in some cases (Schnitzer and Bongers 2011). Global change effects, such as increasing evaporative demand and CO_2 concentration or nitrogen deposition have all been pointed out as responsible for this pattern (Schnitzer 2015). However, the causes of increasing liana biomass in neotropical forest remain unknown (Wright et al. 2015).

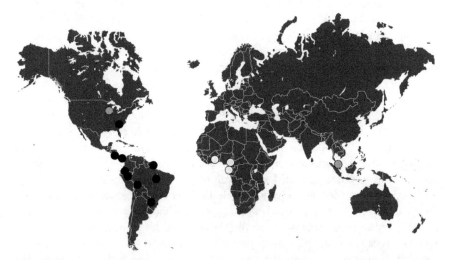

Fig. 1 Location of study sites around the world examining the abundance change of lianas through time. *Black circles* indicate sites where liana abundance or density increased, *white circles* indicate sites where liana abundance or density decreased and *grey circles* indicate no change over time. Data are from Phillips et al. (2002), Wright et al. (2004a, b, 2015), Wright and Calderón (2006), Ingwell et al. (2010), Schnitzer et al. (2012), Laurance et al. (2014), Yorke et al. (2013), Enquist and Enquist (2011), Chave et al. (2008), Caballé and Martin (2001), Bongers and Ewango (2015), Thomas et al. (2015), Allen et al. (2007), Londré and Schnitzer (2006) and Campanello et al. (2012)

Recognizing functional and morphological differences among lianas is essential for understanding spatial and temporal patterns of liana abundance. We need to move from a simplistic perspective in which all lianas conform to a unique "climbing habit" towards a more mechanistic view in which species are recognized as having different physiological behaviors and life history traits. Are all liana species increasing or decreasing in abundance in forests where temporal patterns of liana change have been documented? Do freestanding versus support-seeking strategies differ in their colonization ability? Are lianas able to regulate water loss to avoid desiccation at the top of the canopy where they experience high transpirational demands? To address these and other questions, in this chapter we explore how different liana species balance their carbon and water economies compared to trees, as well as the impacts of lianas on their host species.

Liana Seedlings in Tropical Forests

Plant seedlings in general, and liana seedlings in particular, due to their need for using hosts to climb and reach a reproductive age, are susceptible to high mortality rates. Liana seedlings vary in how they access host plants, with freestanding seedlings, which can grow without mechanical support and can remain as small woody plants for a relatively long period of time, at one end of the spectrum, and support-seeking seedlings, which reach for a host tree very early in their life cycle at the other end of the spectrum (Putz 1984a, b; Rowe and Speck 2005; Manzané 2012). Contrasting seedling forms have important implications for carbon allocation because freestanding liana seedlings may invest more in mechanical tissues and less in stem and root extension than support-seekers. This pattern of biomass allocation is comparable to life-history characteristics described for early—versus late-successional tree seedlings (Denslow 1980; Swaine and Whitmore 1988; Kitajima 1994). To assess for potential differences in life history traits of both types of liana seedlings, seeds of eight species were collected in lowland tropical forests in Panama and were grown in a shade house (55.70 μmol m^{-2} s^{-1} photon flux density, PFD) without any structural support for 12 months. The percentage of the seedlings that bent more than 45° from the upright position was recorded and used along with other characteristics to determine freestanding or support seeking behavior (Table 1). After 3 months, seedlings of three of the liana species started to bend naturally without the presence of any structural support, and after 12 months of growth, 30–60 % of the individuals of those three species deviated substantially from the vertical position (Fig. 2). None of the other seedlings bent after 12 months of growth. The support-seekers had smaller stem cross sectional area and total leaf area per individual than the freestanding seedlings, but all species conformed to the same functional relationship between leaf and stem cross sectional area (Fig. 3). Total leaf surface area per plant was 350 cm^2 for the four support-seekers and 850 cm^2 for the four freestanding species.

Table 1 Species used for the experiments, including species authority name, family and the hypothesized functional group

Species	Family	Functional group
Acacia hayesii Benth.	Fabaceae	SS
Amphilophium crucigerum (L.) L.G. Lohmann	Bignoniaceae	SS
Serjania atrolineata C. Wright	Sapindaceae	SS
Machaerium milleflorum Pittier	Fabaceae	SS
Petrea volubilis L.	Verbenaceae	F
Cnestidium rufescens Planch.	Connaraceae	F
Callichlamys latifolia (Rich.) K. Schum.	Bignoniaceae	F
Tontelea passiflora (Vell.) Lombardi	Celastraceae	F

SS = support-seekers, F = freestanding. Adapted from Manzané (2012)

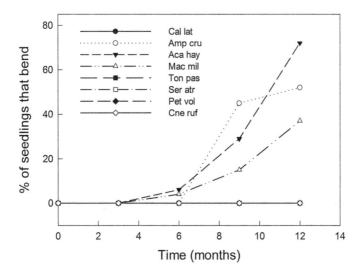

Fig. 2 Liana seedling species bending without the presence of any structural support during 12 months of growth in a greenhouse. Species abbreviations are: Cal lat, *Callichlamys latifolia*; Amp cru, *Amphilophium crucigerum*; Aca hay, *Acacia hayesii*; Mac mil, *Machaerium milleflorum*; Ton pas, *Tontelea passiflora*; Ser atr, *Serjania atrolineata*; Pet vol, *Petrea volubilis* and Cne ruf, *Cnestidium rufescens*. Each value corresponds to a species mean. Adapted from Manzané (2012)

The support-seekers had higher specific leaf area (leaf area divided by leaf mass; SLA), suggesting that these leaves had lower construction costs (Poorter et al. 2009), and lower maximum photosynthetic electron transport rate (ETR_{max}) per leaf area, than freestanding species (Fig. 4). These differences persist in a mass-based estimation of ETR_{max}, indicating that freestanding seedlings with a larger total leaf surface area and greater photosynthetic capacity per leaf area and mass should assimilate carbon and accumulate biomass faster than support-seekers.

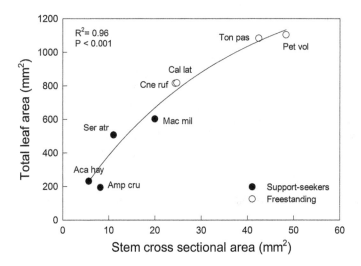

Fig. 3 Relationship between total leaf area and stem cross sectional area for support-seeker and freestanding liana saplings after 18 months of growing in greenhouse conditions. An exponential function was fitted to the data, y = 1475.5*(1−exp(−0.0302*x)). Species abbreviations are as in Fig. 2. Each value corresponds to a species mean. Adapted from Manzané (2012)

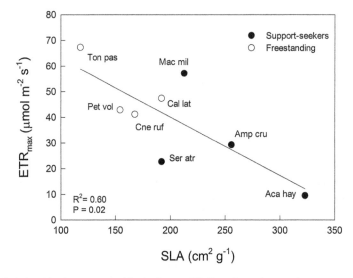

Fig. 4 Relationship between specific leaf area (SLA) and maximum electron transport rate (ETR$_{max}$), for support-seeker and freestanding liana saplings. A linear function was fitted to the data, y = 85.67−0.23x. Species abbreviations are as in Fig. 2. Each value corresponds to a species mean. Adapted from Manzané (2012)

There appear to be tradeoffs in resource allocation across both liana groups. Species with freestanding seedlings invest more in self-support tissue as shown by higher cross sectional stem area than support-seekers, whereas support-seekers invest more resources for rapid elongation of comparatively slender stems. Additionally, freestanding species have a less efficient water transport system than support-seekers at a comparable age (Manzané 2012). The differences in carbon allocation may mean that freestanding seedlings remain under forest canopy for long periods of time and perhaps are more shade tolerant than support-seekers. Support-seekers, in contrast, may rapidly seek higher light levels by finding suitable host plants and climbing earlier (Gianoli 2015), and may do so with a relatively low biomass cost of thin, high SLA leaves with low photosynthetic capacity. This ecological correlate of the physiological trait differences between support-seekers and freestanding species needs to be experimentally tested, particularly through studies exploring ontogenetic changes. For example, support-seekers might completely change their metabolism after finding a support plant and growing towards light (Gianoli 2015).

There is currently relatively little data available comparing the physiological traits of liana and trees seedlings. However, in one recent study, Pasquini et al. (2015) hypothesized that lianas should be more limited by soil resources than trees because of their fast growth rates, prevalence in regenerating forest, and affinity for high-nutrient sites. Using a fully replicated factorial nutrient addition experiment with nitrogen, phosphorus, and potassium, applied alone and in all combinations for more than a decade, the authors showed that phosphorus limits the photosynthetic performance of both trees and lianas in deeply shaded understory habitats, but that lianas always showed significantly greater ETR_{max}, photosynthetic quenching, and saturating light levels across all treatments (Pasquini et al. 2015). There was little evidence for nutrient-by-growth form interactions, indicating that lianas were not disproportionately favored in nutrient-rich habitats.

Patterns of Liana Growth and Crown Colonization

Studies on biomechanical properties, morpho-functional traits and biomass allocation could help to understand patterns of spatial distribution of lianas at a local and global scale as well as trends in liana abundance changes trough time. Some lianas can develop effective climbing behavior and specialized flexible stems without highly specialized organs of attachment. This is linked to a high degree of developmental plasticity in early stages of growth (Menard et al. 2009). The stiffest stems, those having the highest values of Young's modulus of elasticity, belong to saplings and leaning stems. Only when climbing stems are securely anchored to host trees does the liana develop highly flexible stem properties (low values of Young's modulus of elasticity). The change in stiffness during ontogeny is linked to the development of wood with numerous large vessels and thin-walled fibers (Menard et al. 2009). Lianas that climb via twining, tendrils or other specialized attachment organs may not necessarily require developmental plasticity during ontogeny in terms of biomechanics and stem anatomy.

Adult liana species differ in stem anatomical characteristics. For example, if freestanding liana seedlings conserve their physiological and anatomical characteristics after climbing the host tree, they should have higher wood density and narrower vessels than support-seeker lianas after reaching the canopy. The vascular systems of many lianas have specialized tissues that prevent stem breakage, dissipate torsional stress, increment parenchyma storage tissue, and maintain xylem and phloem transport during stem twisting (Putz and Holbrook 1991). Several neotropical lianas have stems with xylem cylinders distributed into several stems that form crowns in the canopy organized into multiple small crowns, while lianas with a unique xylem cylinder tend to occupy the canopy of the support tree with only one densely packed crown (Caballé 1993).

This different pattern of crown colonization of lianas associated with different stem anatomy and the predominance of either freestanding or support-seeking behavior may partially explain the increase or decrease in liana abundance in tropical forests around the world. For example, in the tropical forest of Barro Colorado Island, Panama, almost all of the support-seeking liana species are increasing in abundance, while freestanding seedlings tend to increase at slower rates with some species even decreasing in population size (Manzané 2012).

Biomass allocation to leaves and climbing stems should differ among tropical and subtropical lianas. In a temperate Japanese forest species, lianas that are abundant in forest edges, allocate a larger proportion of shoot biomass to climbing stems, whereas liana species growing inside the forest stands allocate relatively more biomass to leaves (Ichihashi et al. 2010; Ichihashi and Tateno 2011). Both groups of liana species have different impacts on tree growth. Studies of tropical lianas focusing on biomass allocation patterns and climbing behavior and the relationship with their physiological performance are urgently needed.

Carbon and Water Economy of Lianas and Trees

Lianas are considered to be structural parasites because they use the stems of other plants to reach the forest canopy (Schnitzer and Bongers 2002). As a consequence of limited allocation to stems, lianas are supposed to have a higher proportion of photosynthetic biomass per whole plant biomass compared to trees. This allocation pattern apparently enables them to obtain more resources at faster rates, but this hypothesis is based on limited data (Santiago et al. 2015). In a subtropical forest in Argentina, liana species were able to recolonize the canopy of tree species 10 years after after a cutting treatment was applied (Campanello et al. 2012). Yet, lianas can lose substantial biomass in reaching the canopy, in part because shoots may fail to find a host plant or due to the breakage of branches to which they are attached. Consistent with low mechanical demands, the low density wood in lianas could help to compensate for large biomass losses during climbing (Putz 1983), as well as providing a hydraulically efficient xylem. Wood density and saturated water content is shown in Table 2 for dominant tree species >10 cm diameter and abundant liana

Table 2 Traits of trees and lianas measured in a subtropical forest in northern Argentina

Species	Family	SLA	WD	WSWC
Trees				
Ceiba speciosa (A.St.-Hil., A. Juss. and Cambess.) Ravenna	Bombacaceae	150.7 ± 10.6	0.38 ± 0.02	2.12 ± 0.01
Cedrela fissilis Vell.	Meliaceae	121.7 ± 4.1	0.46 ± 0.01	1.42 ± 0.00
Cordia trichotoma (Vell.) Arráb. Ex Steud.	Borraginaceae	68.5 ± 5.3	0.54 ± 0.02	1.08 ± 0.01
Cabralea canjearana (Vell.) Mart.	Meliaceae	141.3 ± 6.9	0.56 ± 0.02	1.18 ± 0.00
Ocotea diospyrifolia (Meisn.) Mez.	Lauraceae	120.0 ± 5.8	0.57 ± 0.01	1.20 ± 0.00
Chrysophyllum gonocarpum (Mart. and Eichler)Engl.	Sapotaceae	124.4 ± 9.7	0.64 ± 0.02	0.89 ± 0.00
Balfourodendron riedelianum (Engl.) Engl.	Rutaceae	127.4 ± 8.0	0.71 ± 0.02	0.72 ± 0.01
Lonchocarpus muehlbergianus Hassl.	Fabaceae	137.6 ± 4.9	0.71 ± 0.02	0.78 ± 0.01
Parapiptadenia rigida (Benth.) Brenan	Fabaceae	126.6 ± 27.2	0.78 ± 0.01	0.62 ± 0.00
Holocalyx balansae Micheli	Fabaceae	134.4 ± 18.1	0.82 ± 0.01	0.62 ± 0.00
Lianas				
Adenocalymma marginatum (Cham.) DC.	Bignoniaceae	91.5 ± 2.1	0.44 ± 0.01	1.53 ± 0.07
Amphilophium paniculatum (L.) Kunth	Bignoniaceae	63.1 ± 4.8		
Aristolochia triangularis Cham.	Aristolochiaceae	44.5 ± 7.6	0.27 ± 0.00	2.70 ± 0.04
Schnella microstachya Raddi	Fabaceae	57.4 ± 1.2		
Condylocarpon isthmicum (Vell.) A. DC.	Apocynaceae	59.4 ± 6.4	0.44 ± 0.00	1.42 ± 0.01
Cratylia intermedia (Hassl.) L. P. Queiroz & R. Monteiro	Fabaceae	52.4 ± 5.3		
Dicella nucifera Chodat	Malpighiaceae	79.5 ± 8.8		
Dioclea violacea Mart. ex Benth.	Fabaceae	65.4 ± 1.4		
Forsteronia glabrescens Müll. Arg.	Apocynaceae	95.1 ± 7.0		
Fridericia chica (Bonpl.) L.G. Lohmann	Bignoniaceae	56.4 ± 6.3		
Fridericia florida (DC.) L.G. Lohmann	Bignoniaceae	84.3 ± 7.3		
Gouania sp.	Rhamnaceae	50.9 ± 4.2	0.44 ± 0.02	1.39 ± 0.10
Mansoa difficilis (Cham.) Bureau & K. Schum.	Bignoniaceae	81.1 ± 2.7	0.45 ± 0.02	1.56 ± 0.04

(continued)

Table 2 (continued)

Species	Family	SLA	WD	WSWC
Marsdenia macrophylla (Humb. & Bonpl. ex Schult.) E. Fourn.	Apocynaceae	58.4 ± 6.4	0.35 ± 0.01	2.09 ± 0.03
Mascagnia divaricata (Kunth) Nied.	Malpighiaceae	49.6 ± 8.5		
Peltastes peltatus (Vell.) Woods	Apocynaceae	47.6 ± 7.9		
Pereskia aculeata Mill.	Cactaceae	53.5 ± 1.0		
Pisonia aculeata L.	Nyctaginaceae	63.9 ± 4.4	0.43 ± 0.01	1.67 ± 0.10
Pristimera celastroides (Kunth) A.C. Sm.	Celastraceae	100.5 ± 5.6	0.46 ± 0.01	1.41 ± 0.04
Tanaecium mutabile (Bureau & K. Schum.) L.G. Lohmann	Bignoniaceae	70.8 ± 4.7	0.56 ± 0.01	1.11 ± 0.03
Tetracera oblongata DC.	Dilleniaceae	139.6	0.40 ± 0.01	1.21 ± 0.03
Tynanthus micranthus Corr. Méllo ex K. Schum.	Bignoniaceae	75.2 ± 21.3		

Species name, species authority, family, specific leaf area (SLA, cm^2 g^{-1}), sapwood density (WD, g cm^{-3}) and sapwood saturated water content (WSWC, %) are included. Values are means ± SE. Unpublished information from Panizza et al.

species with stems >2 cm diameter in a subtropical forest in Argentina. Lianas have on average lower wood density than trees and consequently higher saturated water content (Fig. 5), a measure of potential maximum amount of stem water storage,

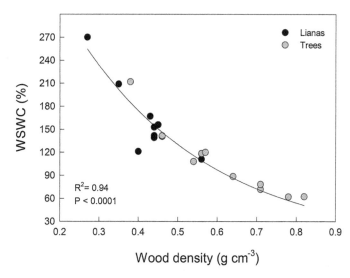

Fig. 5 Relationship between wood saturated water content (WSWC) and wood density for lianas and trees. An exponential function was fitted to the data, y = 571.89 * exp (−2.94*x). Each value corresponds to a species mean

which could contribute to maintain an adequate short-term water balance (Bucci et al. 2004). Both lianas and trees fit the same negative exponential relationship between wood saturated water content and wood density as shown in Fig. 5.

Studies of comparative maximum photosynthetic rates of trees and lianas across a variety of tropical climates have been accumulating throughout the early 2000s (Wright et al. 2004b; Santiago and Wright 2007; Cai et al. 2009; Zhu and Cao 2009). This data, combined with an expansive global dataset of liana and tree nutrient concentration and chemistry on more than 7000 tree and liana species from 48 sites now clearly show that lianas have higher foliar concentrations of key physiological elements such as nitrogen and phosphorus in sites with less than 2500 mm of annual precipitation, but that at greater precipitation or lower temperature this advantage is lost and trees and lianas largely converge in leaf physiology (Asner and Martin 2012). Thus when climate is taken into account, differences in maximum photosynthetic rates of lianas and trees at drier sites (Cai et al. 2009; Zhu and Cao 2009, 2010), and lack of differences at wetter sites (Santiago and Wright 2007), are entirely consistent with global patterns of liana and tree leaf traits (Santiago et al. 2015). Regardles of differences in net CO_2 assimilation between lianas and trees, the apparently higher proportion of photosynthetic biomass at the whole plant scale for lianas compared to trees, might enable lianas to grow relatively fast, which could be advantageous for reaching the upper forest canopy.

Results of studies on subtropical forests in Northern Argentina are beginning to show patterns similar to other studies on tropical forests (Wright et al. 2004b; Santiago and Wright 2007; Zhu and Cao 2009); liana species have higher SLA values than trees (Fig. 6a), consistent with the lower construction costs in lianas than in trees. Furthermore, ETR_{max} in trees is higher than lianas in subtropical forests in Argentina on a leaf area basis (Fig. 6b), but not on a mass basis (Fig. 6c), consistent with the finding that leaf scale photosynthetic advantages in lianas disappear as annual precipitation increases because mean annual precipitation in the Argentinean study site is 2000 mm per year with no dry season. Subtropical forests

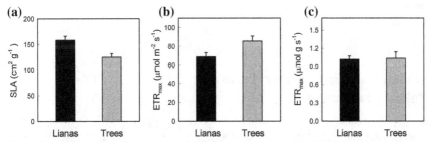

Fig. 6 Specific leaf area (SLA) (**a**), maximum electron transport rate (ETR_{max}) expressed on an area basis (**b**) and on a mass basis (**c**), for lianas and trees. Bars are means + SE. The means were compared between lianas and trees using a t-test SLA: t = 2.329, P = 0.027; ETR_{max} area: t = 2. P = 0.032; ETR_{max} mass: 0.162, P = 0.872. Unpublished results from Panizza et al.

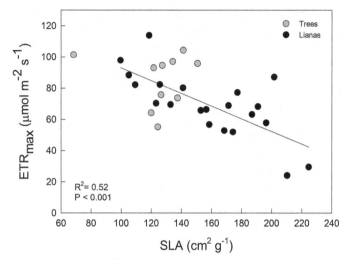

Fig. 7 Relationship between specific leaf area (SLA) and maximum electron transport rate (ETR$_{max}$) for trees and lianas. A regression line was only fitted to liana species, y = 133.64–0.41x. Each value corresponds to a species mean

in Argentina are geographically and taxonomically linked to seasonally dry forests (Pennington et al. 2009). In this ecosystem tree species from humid and seasonally dry forests co-occur, and a similar pattern was recently described for lianas (Malizia et al. 2015). Interestingly, ETR$_{max}$ on an area basis was negatively correlated with SLA for lianas, and for lianas and trees (Fig. 7), but no correlation was found for trees alone. This relationship for adult lianas in the subtropical forest in Argentina was also observed for saplings in Panama as described above (Fig. 4), and suggests that as a given mass of leaf tissue is spread over a smaller area, more photosynthetic components can be packed into the leaf. According to the leaf economics spectrum, photosynthetic capacity per unit leaf area is not necessarily expected to be directly related to SLA (Wright et al. 2004b). But for lianas, higher SLA and likely thinner leaves with lower construction costs are related to lower photosynthetic capacity per unit leaf area.

Stomatal Control, Water Transport, and Diurnal Water Balance in Lianas

Lianas, with most of their leaves on the top of the canopy, experience high tran-spirational demands at midday, which requires them to effectively regulate water loss to avoid desiccation. Leaf physiological processes can respond to changes in soil water availability and vapor pressure deficit (VPD). The temporal scale of changes in both environmental variables is different, with soil water availability changing over the course of several days or weeks while VPD may change strongly over the course of the day. A decrease in soil water and an increase in VPD may induce water deficits and constrain carbon assimilation. By regulating the stomatal aperture, lianas can limit water loss, keeping xylem tension within a safe range and thus reducing the risks of xylem dysfunction (Brodribb and Holbrook 2003; Brodribb et al. 2003; Johnson et al. 2011; Zhang et al. 2014). Variation in stomatal conductance (g_s) is mainly governed by stem (Ψ_{stem}) and leaf (Ψ_{leaf}) water potentials, VPD, and abscisic acid (Comstock and Mencuccini 1998; Salleo et al. 2000; Brodribb et al. 2003; Bunce 2006; Chapotin et al. 2006). Failure of stomata to respond quickly to the rapid increases in VPD could result in excessive dehydration and consequently irreversible hydraulic dysfunction in the vascular system. Stomatal regulation is thus an important safety valve, especially for canopy plants, which frequently experience high evaporative demand.

Trees tend to have larger internal stem water storage than lianas, and thus trees can use water from their stems to replace water loss through transpiration, whereas lianas rely mostly on soil water (Chen et al. 2015). Exposure to the stressful environment of the upper canopy does not limit lianas from colonizing the canopy. Instead, lianas readily thrive and may form a carpet-like leafy layer at the top of the forest in many tropical and subtropical forests. Lianas typically have wider and longer vessels than trees and are reported to be hydraulically efficient but vulnerable to drought induced cavitation (Isnard and Silk 2009; Zhu and Cao 2009; Johnson et al. 2013; Chen et al. 2015). How lianas balance hydraulic efficiency with safety and the role of stomatal regulation may be the key to understand the ability of lianas to aggressively colonize treefall gaps in wet and seasonally dry forests (DeWalt et al. 2000; Schnitzer 2005; Ledo and Schnitzer 2014).

Leaf and stem water potentials decline in both liana and tree species of a subtropical forest in China during the daytime, with seven of the eight studied species losing leaf turgor at around midday (Chen et al., unpublished information). Lianas show a trend of larger disequilibrium between leaf and stem water potential than tree species during the daytime (Table 3). The difference between minimum stem and minimum leaf water potential ($\Delta\Psi$) is 1.50 MPa for lianas and 0.69 MPa for trees. Clearly more data is needed to fully evaluate this, but if these differences between lianas and trees in the driving force for water transport in the terminal portion of the hydraulic continuum are robust when other species are measured, this would suggest that lianas have smaller internal stem water storage compared to trees and that the main source of water for transpiring leaves in lianas is the soil available water. Trees often have relatively large stem water storage or capacitance that can be used to replace transpiration water loss, particularly in the morning (Goldstein et al. 1998). Minimizing the distances between water sources through stem water storage appears to reduce differences between stem and leaf water potentials in trees during periods of high evaporative demand.

Table 3 Traits of trees and lianas measured in a subtropical forest in China

Species	Family	K_s	SE	$\Delta\Psi$	g_s	A
Trees						
Celtis tetrandra Roxb.	Cannabaceae	6.21	0.6	1.19	0.27	17.9
Ficus concinna (Miq.) Miq.	Moraceae	3.95	0.7	1.13	0.14	5.8
Harpullia ramiflora Radlk.	Sapindaceae	2.20	0.2	0.91	0.11	7.8
Michelia gioi (A. Chev.) Sima and H. Yu	Magnoliaceae	0.69	0.2	0.17	0.03	2.5
Streblus asper Lour.	Moraceae	1.86	0.4	0.51	0.16	9.6
Lianas						
Celastrus paniculatus Willd.	Celastraceae	8.60	1.6	1.39	0.04	4.5
Marsdenia sinensis Hemsl.	Apocynaceae	10.53	1.3	1.44	0.07	8.5
Ventilago calyculata Tul.	Rhamnaceae	4.80	0.6	1.11	0.15	9.5

Specific hydraulic conductivity ($K_S \pm$ SE, Kg m^{-1} s^{-1} MPa^{-1}), Maximum difference between leaf and stem water potentials during the day ($\Delta\Psi$, MPa), stomatal conductance at midday (g_s, mol H$_2$O m^{-2} s^{-1}), average daily net CO$_2$ assimilation rates (A, µmol CO$_2$ m^{-2} s^{-1}). Unpublished information from Chen et al.

Figure 8 shows the relationship between the maximum daytime differences of disequilibrium between stem and leaf water potential versus maximum specific hydraulic conductivity (Ks) for trees and lianas. The difference between stem and leaf water potentials can be considered as an indirect measure of stem capacitance. The difference in water potentials was multiplied by two, making the relative index to range from 0 to 2. If both stem and leaf water potential differences are small (both water potentials are well coupled) that means that the stem internal water storage is relatively small. Across tree and liana species, the indirect index of stem capacitance decreases exponentially with increasing Ks. In addition, it has been shown before that liana stems tend to be narrow, additional evidence of their relatively low stem water storage. Stomatal regulation is important for lianas not only because internal stem water storage is relatively small but also because they have a higher proportion of leaves exposed to the top of the canopy, where the VPD is high (Putz 1983). Lianas might have faster stomatal responses compared to trees and exhibit stronger partial stomatal closure than trees (Table 3). Lianas might also reach maximal g_s at relatively lower (<1 kPa) VPD than trees, helping them to maximize carbon assimilation in the morning. Strong stomatal control and efficient water transport would help lianas maintain Ψ_{leaf} within a safe hydraulic range to avoid xylem dysfunction despite their low stem water storage capacity, which is achieved at a minimum cost in terms of carbon assimilation. Average daily net CO$_2$ assimilation is 7.5 µmol m^{-2} s^{-1} for lianas and 8.7 µmol m^{-2} s^{-1} for trees (Table 3).

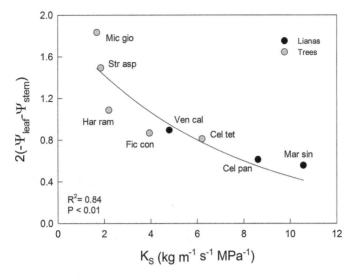

Fig. 8 A normalized index of water storage capacity $|(2(-\Psi_{leaf}-\Psi_{stem}))|$ in MPa as a function of specific hydraulic conductivity (Ks) for trees (*grey circles*) and lianas (*black circles*) from a tropical forest in SE China. An exponential function was fitted to the data, $y = 1.671\ e^{-0.114x}$. Each value corresponds to a species mean. Species abbreviations are: Cel pan, *Celastrus paniculatus*; Mar sin, *Marsdenia sinensis*; Ven cal, *Ventilago calyculata*; Cel tet, *Celtis tetrandra*; Fic con, *Ficus concinna*; Har ram, *Harpullia ramiflora*; Mic gio, *Michelia gioi* and Str asp, *Streblus asper*. Unpublished information from Chen et al.

Liana Impact on Tree Carbon and Water Economy

Liana colonization of tree crowns was identified early as a limitation for timber production in tropical forests. Removal of lianas before harvesting is traditionally used to decrease tree damage during logging operations (Putz and Holbrook 1991; Vidal et al. 1997; Alvira et al. 2004), and liana biomass in tree crowns reduces growth rates of canopy trees between harvesting cycles (Putz and Holbrook 1991). The effect of liana colonization on cumulative stem growth for nine tree species differing in wood density in a relatively undisturbed subtropical forest in northern Argentina is shown in Fig. 9. Trees with their canopy covered by more than 50 % with liana leaves tend to grow slower than trees with little liana infestation.

Liana colonization on tree crowns in a disturbed subtropical forest in Argentina increased from 69 to 83 % during a 10 years period (Campanello et al. 2012), and is one of the highest liana abundance increases reported for neotropical forests (van der Heijden et al. 2008; Ingwell et al. 2010). Taking into account the results shown in Fig. 9, the potential tree biomass increment lost due to liana colonization should be of considerable concern. Not all trees are equally affected by lianas. Trees of

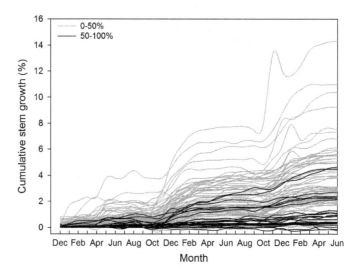

Fig. 9 Monthly changes of cumulative stem growth from December 2011 to June 2014 of nine tree species with more than 50 %, and less than 50 % of the crown covered by lianas. The species studied were: *Ceiba speciosa, Cedrela fissilis, Cordia trichotoma, Cabralea canjearana, Ocotea diospyrifolia, Chrysophyllum gonocarpum, Balfourodendron riedelianum, Parapiptadenia rigida* and *Holocalyx balansae.* Unpublished information from di Francescantonio et al.

slow growing species are normally severely infested, because older trees provide more opportunities and time for liana colonization. In addition, tree species with long branch-free boles and smooth bark are likely to be free of lianas while the opposite is expected for trees with branched trunks or rough bark (Campbell and Newbery 1993; Campanello et al. 2007). Liana colonization is a dynamic process with liana shedding occasionally occurring for some species (Ingwell et al. 2010; Campanello et al. 2012).

Lianas also have a substantial impact on tree water utilization. In Fig. 10, water consumption of trees with and without lianas infestation is shown for three species growing in a subtropical forest. The native forest studied has an average transpiration of 954 mm per year, mostly from trees and palms and only less than 1 % of this value was due to liana water consumption. Clearly, increasing liana colonization could considerably reduce transpiration with impacts not only on the water economy and nutrient absorption for individual trees but also with potential consequences at the ecosystem level.

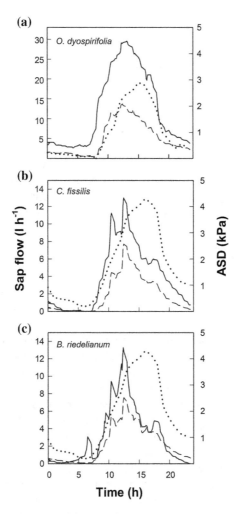

Fig. 10 Diurnal patterns of volumetric sap flow measured at the base of the main stem for three tree species with crowns free of lianas (*solid line*) or colonized by lianas (*dashed line*) and air saturation deficit (ASD)(*dotted line*). (**a**) *Ocotea diospyrifolia* is an evergreen species. Measurements were made in dominant trees without lianas or slightly infested (*Diameter* 69 and 62 cm respectively); (**b**) *Cedrela fissilis* is a deciduous species. Measurements were made in co-dominant trees without lianas or slightly infested (*Diameter* 36 and 31 cm respectively); (**c**) *Balfourodendron riedelianum* is a brevideciduous species. Measurements were made in dominant trees without lianas or severely infested (*Diameter* 42 cm). Adapted from Rodriguez (2015)

Concluding Remarks

For more than thirty years, research had mainly focused on the ecological conse-quences of liana colonization of trees and forest gaps and the subsequent impact on ecosystem processes. It was not until recently that the differences in physiological behavior between lianas and trees had been addressed. However, lianas are far from a homogeneous group of species. A comprehensive understanding of liana eco-physiology is fundamental for interpreting patterns of liana abundance and their consequences in a changing environment worldwide. In neotropical forests, lianas are increasing in abundance relative to trees. This increased species richness may reflect modifications of drivers of global change factors including increased temperature, atmospheric CO_2, habitat disturbance, and drought. However, questions remain as to the specific mechanisms facilitating the liana responses. We hope that this chapter provides information on specific physiological mechanisms that can help to explain aspects of the patterns of liana abundance increase as well as to identify key questions that need to be addressed before we can have a more comprehensive picture of functional interactions between environmental resource acquisition, life history traits, and physiology of lianas and host trees in forest ecosystems.

Liana seedlings fall along a continuum between two extremes represented by support-seekers and freestanding species, which may, at least partially, relate to the continuum in life history traits of shade-tolerance versus light requiring responses of trees. On one side, support-seeking seedlings invest relatively little biomass in thin leaves with low construction costs and high SLA, but stems grow fast towards suitable support. On the opposite side, freestanding lianas develop a wider and more resistant stem, and expand a larger leaf area of more expensive leaves with low SLA, that appears to enable growth in shaded environments. Both contrasting climbing strategies appear to characterize species with different growth patterns and with differential impacts on host trees. The more crown-invasive support-seekers may respond fast to disturbances with increased growth and canopy expansion.

Photosynthetic capacity and carbon assimilation rates in lianas are greater or similar compared to trees, depending on climate. However, lianas have stems and leaves with lower constructions costs than trees and exhibit a comparatively higher biomass turnover. Despite having low density wood, lianas have lower capacitance compared to trees because their stems are relatively thin. This limitation is partially offset by a more efficient water transport system and stronger stomatal control that allow lianas to cope with high evaporative demands at the top of the canopy. When colonizing tree crowns, lianas can negatively affect growth rates and transpiration of their tree hosts, with obvious consequences for the cycling of carbon and water at the ecosystem level.

References

Allen BP, Sharitz RR, Goebel PC (2007) Are lianas increasing in importance in temperate floodplain forests in the southeastern United States? For Ecol Manag 242:17–23

Alvira D, Putz FE, Fredericksen TS (2004) Liana loads and post-logging liana densities after liana cutting in a lowland forest in Bolivia. For Ecol Manage 190:73–86

Asner GP, Martin RE (2012) Contrasting leaf chemical traits in tropical lianas and trees: implications for future forest composition. Ecol Lett 15:1001–1007

Bongers F, Ewango CEN (2015) Dynamics of lianas in DR Congo. In: Schnitzer SA, Bongers F, Burnham RJ, Putz FE (eds) The ecology of lianas. Wiley-Blackwell, Oxford, pp 23–35

Brodribb TJ, Holbrook NM (2003) Stomatal closure during leaf dehydration, correlation with other leaf physiological traits. Plant Physiol 132:2166–2173

Brodribb TJ, Holbrook NM, Edwards EJ, Gutiérrez MV (2003) Relations between stomatal closure, leaf turgor and xylem vulnerability in eight tropical dry forest trees. Plant, Cell Environ 26:443–450

Bucci SJ, Goldstein G, Meinzer FC, Scholz FG, Franco AC, Bustamante M (2004) Functional convergence in hydraulic architecture and water relations of tropical savanna trees: from leaf to whole plant. Tree Physiol 24:891–899

Bunce JA (2006) How do leaf hydraulics limit stomatal conductance at high water vapour pressure deficits? Plant, Cell Environ 29:1644–1650

Caballé G (1993) Liana structure, function and selection: a comparative study of xylem cylinders of tropical rainforest species in Africa and America. Bot J Linn Soc 113:41–606

Caballé G, Martin A (2001) Thirteen years of change in trees and lianas in a Gabonese rainforest. Plant Ecol 152:167–173

Cai ZQ, Poorter L, Han Q, Bongers F (2008) Effects of light and nutrients on seedlings of tropical *Bauhinia* lianas and trees. Tree Physiol 28:1277–1285

Cai ZQ, Schnitzer SA, Bongers F (2009) Seasonal differences in leaf-level physiology give lianas a competitive advantage over trees in a tropical seasonal forest. Oecologia 161:25–33

Campanello PI, Garibaldi JF, Gatti MG, Goldstein G (2007) Lianas in a subtropical Atlantic Forest: host preference and tree growth. For Ecol Manage 242:250–259

Campanello PI, Villagra M, Garibaldi JF, Ritter LJ, Araujo JJ, Goldstein G (2012) Liana abundance, tree crown infestation, and tree regeneration ten years after liana cutting in a subtropical forest. For Ecol Manage 284:213–221

Campbell E, Newbery D (1993) Ecological relationships between lianas and trees in lowland rain forest in Sabah, East Malaysia. J Trop Ecol 9:469–490

Chapotin SM, Razanameharizaka JH, Holbrook NM (2006) Baobab trees (*Adansonia*) in Madagascar use stored water to flush new leaves but not to support stomatal opening before the rainy season. New Phytol 169:549–559

Chave J, Riéra B, Dubois MA (2001) Estimation of biomass in a neotropical forest of French Guiana: spatial and temporal variability. J Trop Ecol 17:79–96

Chave J, Olivier J, Bongers F et al (2008) Above-ground biomass and productivity in a rain forest of eastern South America. J Trop Ecol 24:355–366

Chen YJ, Cao KF, Schnitzer SA, Fan ZX, Zhang JL, Bongers F (2015) Water use advantage for lianas over trees in tropical seasonal forests. New Phytol 205:128–136

Comstock J, Mencuccini M (1998) Control of stomatal conductance by leaf water potential in *Hymenoclea salsola* (T. & G.), a desert subshrub. Plant, Cell Environ 21:1029–1038

Denslow JS (1980) Gap partitioning among tropical rainforest trees. Biotropica 12:47–55

DeWalt SJ, Schnitzer SA, Denslow JS (2000) Density and diversity of lianas along a chronosequence in central Panamanian lowland forest. J Trop Ecol 16:1–9

Enquist BJ, Enquist CA (2011) Long-term change within a Neotropical forest: assessing differential functional and floristic responses to disturbance and drought. Glob Change Biol 17:1408–1424

Gerwing JJ, Farias DL (2000) Integrating liana abundance and forest stature into an estimate of total aboveground biomass for an eastern Amazonian forest. J Trop Ecol16:327–335

Gianoli E (2015) The behavioural ecology of climbing plants. AoB PLANTS 7

Goldstein G, Andrade JL, Meinzer FC, Holbrook NM, Cavelier J, Jackson P, Celis A (1998) Stem water storage and diurnal patterns of water use in tropical forest canopy trees. Plant, Cell Environ 21:397–406

Grauel WT, Putz FE (2004) Effects of lianas on growth and regeneration of Prioria copaifera in Darien, Panama. For Ecol Manag 190:99–108

Hegarty EE, Caballé G (1991) Distribution and abundance of vines in forest communities. In: Putz FE, Mooney HA (eds) The biology of vines. Cambridge University Press, Cambridge, pp 313–335

Ichihashi R, Tateno M (2011) Strategies to balance between light acquisition and the risk of falls of four temperate liana species: to overtop host canopies or not? J Ecol 99:1071–1080

Ichihashi R, Nagashima H, Tateno M (2010) Biomass allocation between extension-and leaf display-oriented shoots in relation to habitat differentiation among five deciduous liana species in a Japanese cool-temperate forest. Plant Ecol 211:181–190

Ingwell LL, Wright J, Becklund KK, Hubbell SP, Schnitzer SA (2010) The impact of lianas on 10 years of tree growth and mortality on Barro Colorado Island, Panama. J Ecol 98:879–887

Isnard S, Silk WK (2009) Moving with climbing plants from Charles Darwin's time into the 21st century. Am J Bot 96:1205–1221

Johnson D, McCulloh K, Meinzer F, Woodruff D, Eissenstat D (2011) Hydraulic patterns and safety margins, from stem to stomata, in three eastern US tree species. Tree Physiology:tpr050

Johnson DM, Domec JC, Woodruff DR, McCulloh KA, Meinzer FC (2013) Contrasting hydraulic strategies in two tropical lianas and their host trees. Am J Bot 100:374–383

Kainer KA, Wadt LH, Gomes-Silva DA, Capanu M (2006) Liana loads and their association with Bertholletia excelsa fruit and nut production, diameter growth and crown attributes. J Trop Ecol 22:147–154

Kitajima K (1994) Relative importance of photosynthetic traits and allocation patterns as correlates of seedling shade tolerance of 13 tropical trees. Oecologia 98:419–428

Laurance WF, Andrade AS, Magrach A et al (2014) Long-term changes in liana abundance and forest dynamics in undisturbed Amazonian forests. Ecology 95:1604–1611

Ledo A, Schnitzer SA (2014) Disturbance and clonal reproduction determine liana distribution and abundance and maintain liana diversity in a tropical forest. Ecology 95:2169–2178

Londre RA, Schnitzer SA (2006) The distribution of lianas and their change in abundance in temperate forests over the past 45 years. Ecology 87:2973–2978

Malizia A, Campanello PI, Villagra M, Ceballos S (2015) Geographical, taxonomical and ecological aspects of lianas in subtropical forests of Argentina. In: Parthasarathy N (ed) Biodiversity of lianas. Springer International Publishing, Switzerland

Manzané EJ (2012) Freestanding and support-seeker liana seedlings: Spatial distribution, life history and physiological traits in tropical forests of central Panama. In: PhD Dissertation. University of Miami

Menard L, McKey D, Rowe NP (2009) Developmental plasticity and biomechanics of treelets and lianas in *Manihot* aff. *quinquepartita* (Euphorbiaceae): a branch-angle climber of French Guiana. Ann Bot 103:1249–1259

Nabe-Nielsen J (2001) Diversity and distribution of lianas in a neotropical rain forest, Yasuní National Park, Ecuador. J Trop Ecol 17:1–19

Nabe-Nielsen J, Kollmann J, Peña-Claros M (2009) Effects of liana load, tree diameter and distances between conspecifics on seed production in tropical timber trees. For Ecol Manage 257:987–993

Pasquini SC, Wright SJ, Santiago LS (2015) Lianas always outperform tree seedlings regardless of soil nutrients: results from a long-term fertilization experiment. Ecology 96:1866–1876

Pennington RT, Lavin M, Oliveira-Filho A (2009) Woody plant diversity, evolution, and ecology in the tropics: perspectives from seasonally dry tropical forests. Annu Rev Ecol Evol Syst 40:437–457

Phillips OL, Vásquez MR, Arroyo L et al (2002) Increasing dominance of large lianas in Amazonian forests. Nature 418:770–777

Pinard M, Putz F (1994) Vine infestation of large remnant trees in logged forest in Sabah, Malaysia: biomechanical facilitation in vine succession. J Trop Forest Sci 6:302–309

Poorter H, Niinemets Ü, Poorter L, Wright IJ, Villar R (2009) Causes and consequences of variation in leaf mass per area (LMA): a meta-analysis. New Phytol 182:565–588

Putz FE (1983) Liana biomass and leaf area of a tierra firme forest in the Rio Negro Basin, Venezuela. Biotropica 15:185–189

Putz FE (1984a) How trees avoid and shed lianas. Biotropica 16:19–23

Putz FE (1984b) The natural history of lianas on Barro Colorado Island, Panama. Ecology 65:1713–1724

Putz F, Holbrook N (1991) Biomechanical studies of vines. In: Putz FE, Mooney HA (eds). The biology of vines. Cambridge University Press: Cambridge:73–97

Rowe N, Speck T (2005) Plant growth forms: an ecological and evolutionary perspective. New Phytol 166:61–72

Salleo S, Nardini A, Pitt F, Gullo MAL (2000) Xylem cavitation and hydraulic control of stomatal conductance in laurel (*Laurus nobilis* L.). Plant, Cell Environ 23:71–79

Santiago LS, Wright SJ (2007) Leaf functional traits of tropical forest plants in relation to growth form. Funct Ecol 21:19–27

Santiago LS, Pasquini SC, De Guzman ME (2015) Physiological implications of the liana growth form. In: Schnitzer SA, Bongers F, Burnham R, Putz FE (eds) Ecology of lianas. Wiley, Oxford, UK, pp 288–298

Schnitzer SA (2005) A mechanistic explanation for global patterns of liana abundance and distribution. Am Nat 166:262–276

Schnitzer SA (2015) Increasing liana abundance in Neotropical forests: causes and consequences. In: Schnitzer SA, Bongers F, Burnham RJ, Putz FE (eds) The ecology of lianas. Wiley-Blackwell, Oxford, pp 451–464

Schnitzer SA, Bongers F (2002) The ecology of lianas and their role in forests. Trends Ecol Evol 17:223–230

Schnitzer SA, Bongers F (2011) Increasing liana abundance and biomass in tropical forests: emerging patterns and putative mechanisms. Ecol Lett 14:397–406

Schnitzer SA, Carson WP (2010) Lianas suppress tree regeneration and diversity in treefall gaps. Ecol Lett 13:849–857

Schnitzer SA, Dalling JW, Carson WP (2000) The impact of lianas on tree regeneration in tropical forest canopy gaps: evidence for an alternative pathway of gap-phase regeneration. J Ecol 88:655–666

Schnitzer SA, Mangan SA, Dalling JW et al (2012) Liana abundance, diversity, and distribution on Barro Colorado Island Panama. PloS One 7:e52114

Swaine MD, Whitmore TC (1988) On the definition of ecological species groups in tropical rain forests. Vegetatio 75:81–86

Thomas D, Burnham RJ, Chuyong G, Kenfack D, Sainge MN (2015) Liana abundance and diversity in Cameroon's Korup National Park. In: Schnitzer SA, Bongers F, Burnham RJ, Putz FE (eds) The ecology of lianas. Wiley-Blackwell, Oxford, pp 13–22

van der Heijden GMF, Phillips OL (2009) Environmental effects on Neotropical liana species richness. J Biogeogr 36:1561–1572

van der Heijden GM, Healey JR, Phillips OL (2008) Infestation of trees by lianas in a tropical forest in Amazonian Peru. J Veg Sci 19:747–756

van der Sande MT, Poorter L, Schnitzer SA, Markesteijn L (2013) Are lianas more drought-tolerant than trees? A test for the role of hydraulic architecture and other stem and leaf traits. Oecologia 172:961–972

Vidal E, Johns J, Gerwing JJ, Barreto P, Uhl C (1997) Vine management for reduced-impact logging in eastern Amazonia. For Ecol Manage 98:105–114

Wright SJ, Calderón O (2006) Seasonal, El Nino and longer term changes in flower and seed production in a moist tropical forest. Ecol Lett 9:35–44

Wright SJ, Calderón O, Hernandéz A, Paton S (2004a) Are lianas increasing in importance in tropical forests? A 16-year record from Barro Colorado Island, Panamá. Ecology 85:485–489

Wright IJ, Reich PB, Westoby M, Ackerly DD, Baruch Z, Bongers F, Cavender-Bares J, Chapin T, Cornelissen JHC, Diemer M, Flexas J, Garnier E, Groom PK, Gulias J, Hikosaka K, Lamont BB, Lee T, Lee W, Lusk C, Midgley JJ, Navas ML, Niinemets U, Oleksyn J, Osada N, Poorter H, Poot P, Prior L, Pyankov VI, Roumet C, Thomas SC, Tjoelker MG, Veneklaas EJ, Villar R (2004b) The worldwide leaf economics spectrum. Nature 428:821–827

Wright SJ, Sun IF, Pickering M, Fletcher CD, Chen YY (2015) Long-term changes in liana loads and tree dynamics in a Malaysian forest. Ecology 96:2748–2757

Yorke SR, Schnitzer SA, Mascaro J, Letcher SG, Carson WP (2013) Increasing liana abundance and basal area in a tropical forest: the contribution of long-distance clonal colonization. Biotropica 45:317–324

Zhang Y-J, Rockwell FE, Wheeler JK, Holbrook NM (2014) Reversible deformation of transfusion tracheids in *Taxus baccata* is associated with a reversible decrease in leaf hydraulic conductance. Plant Physiol 165:1557–1565

Zhu SD, Cao KF (2009) Hydraulic properties and photosynthetic rates in co-occurring lianas and trees in a seasonal tropical rainforest in southwestern China. Plant Ecol 204:295–304

Zhu SD, Cao KF (2010) Contrasting cost-benefit strategy between lianas and trees in a tropical seasonal rain forest in southwestern China. Oecologia 163:591–599

Part II
Adaptive Responses of Woody Plants to Particular Tropical Habitats

Flood Tolerant Trees in Seasonally Inundated Lowland Tropical Floodplains

Pia Parolin, Leandro V. Ferreira, Maria Teresa F. Piedade,
Cátia Nunes da Cunha, Florian Wittmann and Mauricio E. Arias

Abstract This chapter focuses on trees and their responses to flooding in large fresh water flood-pulsed ecosystems. The regularity and predictability of the flood pulse has allowed for the development of adaptations by which a large number of tree species are able to grow in ecosystems subjected to seasonally inundated sites. By comparing diversity and tree responses in four floodplain ecosystems on different continents, we attempt to improve our understanding of the factors influencing spatial distribution of plants, diversity of species and adaptations and thus contribute to our knowledge of tropical wetland ecology. In this way, we hope to assist in the successful restoration of degraded floodplains and promote the sustainable use and conservation of these highly valuable ecosystems.

Keywords Anoxic soils · Fresh water flooding · Growth rings · Hypoxia · Waterlogging

P. Parolin (✉)
Biocentre Klein Flottbek and Botanical Garden, University of Hamburg,
Mittelweg 177, 20148 Hamburg, Germany
e-mail: pparolin@botanik.uni-hamburg.de

P. Parolin
INRA French National Institute for Agricultural Research,
Univ. Nice Sophia Antipolis, CNRS, UMR 1355-7254
Institut Sophia Agrobiotech, Sophia Antipolis 06903, France

L.V. Ferreira
Museu Paraense Emílio Goeldi, Av. Perimetral 1901,
Bairro Terra Firme, Belém, Pará 66077-530, Brazil
e-mail: lvferreira@museu-goeldi.br

M.T.F. Piedade
INPA, CP 478, Manaus, Amazônia 69011, Brazil
e-mail: maitepp@internext.com.br

C.N. da Cunha
Departamento de Botânica e Ecologia, Universidade Federal de Mato Grosso,
Cuiabá, Mato Grosso, Brazil
e-mail: catianc@ufmt.br

© Springer International Publishing Switzerland 2016
G. Goldstein and L.S. Santiago (eds.), *Tropical Tree Physiology*,
Tree Physiology 6, DOI 10.1007/978-3-319-27422-5_6

Introduction

Tree physiology is subjected to a particular challenge when it comes to flooded tropical environments. When roots are waterlogged, or when whole plants are submerged, species can display a range of responses. Most tropical ecosystems where plants are subjected to flooding, however, are not well understood, and tree responses to environmental change can be difficult to comprehend and foresee. The following chapter focuses on responses and adaptations of tropical trees to different types of flooding, what we know, where our knowledge is lacking, and what we can expect regarding climatic and basin wide changes in floodplain ecosystems. Worldwide, the largest numbers of tree studies in tropical freshwater floodplains are performed in the Amazon Basin. Thus, we start off with the knowledge we have about tree physiology and morphology in the Amazon floodplains (Fig. 1), as this is the flooded ecosystem which to date is best understood. Then we outline information about trees in other large floodplains of the world, such as the Okavango Delta in Africa, the Tonle Sap floodplain of the Mekong River in Asia, and some other vast wetlands colonized by trees. We try to assess the number of tropical flood tolerant trees, and we end with some results, which may give insight into the responses of flood adapted trees to environmental change and increasing frequency and severity of flooding or drought events.

Responses and Adaptations to Flooding

The responses and adaptations to flooding in tropical environments depend largely on hydric conditions, including flood duration, depth, seasonality, periodicity and predictability. So far, a huge variety in responses and adaptations can be found in Amazonian floodplains, and in other tropical wetlands of the world. All trees inhabiting these areas are species whose survival and growth is not severely inhibited by flooding (Gill 1970). Morphological adaptations may be remnants of pre-adaptations from the non-flooded terra firme species where many floodplain trees originate (Kubitzki 1989; Wittmann et al. 2013).

F. Wittmann
Department of Biogeochemistry, Max Planck Institute for Limnology,
55020 P.O. Box 3060, Mainz, Germany
e-mail: florian.wittmann@mpic.de

M.E. Arias
Department of Organismic and Evolutionary Biology, 26 Oxford Street, Cambridge, MA
02138, USA
e-mail: mauricio_arias@hks.harvard.edu

Fig. 1 Várzea floodplain forest in Caxiuanã (Eastern Amazonia), Brazil (*Photos* Pia Parolin, June 2014)

Periods of limited growth last only a few weeks, are linked to high water, and also to periods of strong drought (Parolin et al. 2010). In most trees from Amazonian floodplains, new leaf flushes, flowering, fruiting and wood growth occur while they are flooded for several months (Worbes 1986, 1989, 1997; Parolin et al. 2002; Schöngart et al. 2002).

In the tropics, the aquatic phase occurs in a period in which temperature and light conditions are optimal for plant growth and development. In contrast to temperate floodplains where trees often persist in a dormant state, in tropical floodplains trees grow vigorously during most of the year, including the aquatic period (Parolin et al. 2004). The regularity of flooding may have enhanced the evolution of specific traits, which are known from floodplain trees in other tropical and in temperate regions. As a consequence of regular annual growth reductions, growth rings are formed in the wood of most flood tolerant tree species (Worbes 1986, 1989; Schöngart et al. 2011).

Constraints imposed on trees by waterlogging and by submergence are clearly different. Therefore, different types of flooding have different impacts on tree growth, and it is important to consider flood duration, depth, seasonality, periodicity, and predictability of the flooded periods if one wants to understand the ecology and regeneration of the plants inhabiting the ecosystem. Due to the immense changes of the annual high water levels, trees are seasonally submerged only in very few ecosystems, including the Amazon and the Tonle Sap floodplain in the Mekong, where regular differences between high and low water exceed 8 m every year. Furthermore, whether flooding occurs at periodical intervals as a predictable flood pulse, as a result of unpredictable rain events or man-made changes, is fundamental for the evolution of adaptations to withstand flooding (Junk et al. 1989). In the case of unpredictable events, the evolution of adaptations to cope with flooding stress is far less efficient.

Constraints Imposed by Waterlogging

Adult trees subjected to waterlogging in floodplains have to cope with a lack of oxygen in the rhizosphere (Crawford 1989). Inundated soils turn hypoxic or anoxic within a few hours as a result of oxygen consumption by respiring roots and microorganisms, and insufficient diffusion of oxygen through water (Crawford 1989). Oxygen depletion is accompanied by increased levels of CO_2, anaerobic decomposition of organic matter, increased solubility of mineral substances, and reduction of the soil redox potential (Joly and Crawford 1982; Kozlowski 1984), which is followed by accumulation of many potentially toxic compounds, caused by alterations in the composition of the soil micro flora (Ponnamperuma 1984).

Constraints Imposed by Submergence

Only seldom, will adult trees be fully submerged, but seedlings established on low levels in the flooding gradient are submerged in a predictable way in the Amazon floodplains (Parolin 2002). Establishing seedlings are not expected to have a fast height increase since they cannot escape from submergence of an 8 m water column. Thus, only species with a high tolerance of submergence can survive in seasonally inundated sites. They need other physiological and morphological adaptations than fast growth to resist several weeks to months of submergence, (Siebel and Blom 1998). Seedlings typically establishing on those sites have significantly lower shoot extension than on higher sites (Parolin 2000, 2002). Large and small seeds produce seedlings with high shoot elongation, enhancing the chances of non-submergence for the seedlings at high elevations. In igapó high seed mass compensates for the lack of nutrients of the environment. In nutrient-rich várzea, the environment supplies nutrients and seeds are smaller, but they may produce seedlings, which are just as tall as on low nutrient sites (Parolin 2002). It is assumed that the zonation observed in Amazonian floodplains is very closely linked to the submergence tolerance and establishment strategy of seedlings, and less to the tolerance of the adult trees.

Tree Adaptations to Flooding

Different adaptations are found at the level of structural, physiological, and phenological traits (Parolin et al. 2004). These are common to most floodplain trees, of temperate and tropical regions. Depending on their respective sets of adaptations and growth strategies, most species are restricted to specific areas with a determined pattern of flooding and sedimentation, leading to clear vegetation zonations along the flooding gradients (Wittmann et al. 2004).

Fig. 2 Root systems in várzea floodplain forest in Eastern Amazonia, some examples from *left* to *right*: lateral knee roots of *Symphonia globulifera*; lateral adventitious roots sprouting from the tree stem of *Virola surinamensis*; roots of the palm *Euterpe oleracea* (*Photos* Pia Parolin, June 2014)

Morphological and anatomical adaptations include the root systems of floodplain trees (Fig. 2), such as hypertrophy of lenticels, formation of adventitious roots, plank-buttressing and stilt rooting, development of aerenchyma, and the deposition of cell wall biopolymers such as suberin and lignin in the root peripheral cell layers (Schlüter and Furch 1992; Schlüter et al. 1993; Waldhoff et al. 1998; De Simone et al. 2002a, 2002b; Haase et al. 2003; Parolin et al. 2004). Different types of aboveground roots—stilt roots, plank buttressing (tabular roots)—are closely related to flooding duration and habitat dynamics (Wittmann and Parolin 2005) for Amazonian forests.

Stem-nodulation and nodulated adventitious roots were observed in various species despite periodical flooding and are assumed to be adaptations that allow legumes to fix N_2 in a flooded environment (James et al. 2001). Radial oxygen losses are avoided by developing strong suberin deposits in radial and tangential cell walls of the root hypodermis, starting immediately behind the root tip. This was found in various species of Amazonian floodplains (De Simone et al. 2003).

The leaves of Amazonian floodplain trees exhibit traits which are generally considered as xeromorphic (Waldhoff et al. 2002; Parolin et al. 2004). Physiological adaptations to flooding include changes of decreasing transpiration (Parolin et al. 2004), stomatal closure (Schlüter and Furch 1992), water potential (Scholander and Perez 1968), leaf chlorophyll (Furch 1984) and nitrogen content, as well as reductions of leaf size, growth and biomass production (Parolin et al. 2004).

Phenological adaptations in the Amazonian várzea are visible in that leaves are shed mainly in the flooded period and new leaves are flushed when water recedes. Flowering peaks at the beginning of the flooded period in igapó, while in várzea the peak is at highest water levels. The periodic behaviour of vegetative phenology is determined by the tree's water status, which is a function of the interaction between water potential of the environment and the structural and functional state of the tree (Reich and Borchert 1984; Schöngart et al. 2002).

Dispersal and seedling establishment are also linked to water. The main means of dispersal are hydro-and ichthyochory, emphasized by a close correlation between the timing of flooding and fruit maturation (Ziburski 1991; Gottsberger 1978;

Goulding 1980; Moegenburg 2002; Parolin et al. 2002; Piedade et al. 2006; Parolin et al. 2013). The diaspores show morphological adaptations for dispersal by water, which enhance floatation, like spongy tissues or large air-filled spaces (Kubitzki and Ziburski 1994). Seeds germinate very quickly upon emergence (Parolin and Junk 2002).

Responses and Adaptations in Other Large Tropical Flooded Environments

The patterns described so far for Amazonia may in part be encountered in other large wetlands of the world. Here, we focus on large tropical flood-pulsed floodplains and riverine forests with naturally occurring trees and a clearly predictable flood pulse (*sensu* Junk et al. 1989). Many environmental factors are common to all tropical floodplain forests with a flood pulse that represents a major influence on all organisms (Junk 1989; Junk et al. 1989) and a dominating stress, which requires a suite of adaptations for survival (Parolin et al. 2004).

However, there may be striking differences related to the presence or absence of salt and fire, height and duration of floods, and pressures from surrounding uplands. Those, which are dominated by grasslands like the African Okavango and Northern Australian floodplains are subjected to regular fire (Heinl et al. 2004, 2006, 2007; McGregor et al. 2010), whereas in the forest-dominated floodplains of Amazonia fire plays no significant role. In the Okavango, the high evapotranspiration causes increased salinity (Parolin and Wittmann 2010). Also, flooding amplitudes vary widely between the ecosystems, with about 2–4 m in the Pantanal, Okavango and Northern Australian, 8 m in the Mekong and 9 m in the Amazon floodplains (Table 1). This implies that complete submergence of saplings and trees occurs only in the Mekong and Amazon, posing different constraints for plant life than merely waterlogging of roots and stems (Parolin 2009).

Table 1 Characteristics of some big flood-pulsed wetlands with natural occurrence of forests

	Amazonia	Pantanal	Okavango delta	Tonle sap
Flood duration where trees grow	7 months	5 months	<1 months	5–10 months
Max flood height on tree stems	8 m	4 m	Root level	9 m
Influence of fire and salt	None	Fire!	Fire + salt!	Fire to some extent
Number of flood-tolerant tree species	>1000 (50 on lowest levels; many endemics)	60 (no endemics)	10	47 (very few endemics)
Density of human population (inhabitants per km^2)	3, 3–20	1–2	<6	115

Pantanal

Ecosystem and flooding pattern. The Pantanal is a huge wetland in the centre of South America, with an extension of 160,000 km^2. It has a pronounced monomodal predictable flood pulse with water levels of maximum 4 m in the sites flooded most heavily. Flood duration is about 8 months per year. The vegetation is mainly open grasslands and shrubs, but large forested areas also occur (Fig. 3). The trees are mostly semi-evergreen, such as *Vochysia divergens*, which sheds its leaves with the onset of the dry season or towards the end of the flooded period, and resprouts new leaves almost immediately thereafter. Only few species, such as those of the genus *Ceiba*, are deciduous and remain leafless for 2 months during the flooded phase. Many species growing on higher levels shed their leaves in the dry period. Flowering occurs mainly in the leafless period or with the onset of rain in December, before flooding. Linked to origins in surrounding uplands, the main dispersal vector for most tree species is wind, but hydrochory may occur. Fruits mature at high water, and some diaspores are adapted to dispersal by water with air filled spaces, or to dispersal by fish, with many fish species having a destructive impact (Galetti et al. 2008).

Diversity of species and functional groups. The Pantanal is highly diverse in flooded and non-flooded habitats. Different vegetation classifications have been developed mainly by association with height and period of seasonal inundations and physical and chemical soil properties (i.e., Prance and Schaller 1982; Ratter et al. 1988; Zeilhofer and Schessl 1994). Seasonal semi-deciduous forests and woodlands predominate on well-drained sites, while lower, periodically inundated floodplains are dominated by open savanna formations interspersed with evergreen forests (Prance and Schaller 1982).

Periodically inundated forests within the Pantanal are often composed of a few co-dominant species (Pott and Pott 1994; Oliveira et al. 2014). Pott and Pott (1994) listed 756 woody plant species in the Pantanal, but less than 20 % of them are considered tolerant to prolonged floods (Cunha and Junk 2001). On the other hand, more than 50 % of all tree species are restricted to non-flooded habitats.

Riparian forests, which accompany the main rivers and the secondary river-channels, are described as the most species-rich inundation forests within the Pantanal (Pott and Pott 1994; Damasceno-Junior et al. 2005; Wittmann et al. 2008;

Fig. 3 Floodplain forest in the Pantanal near Cuiabá (*Photos* Pia Parolin, March 2013)

Oliveira et al. 2014). Comparatively high species richness in riparian forests is most probably traced to high habitat diversity (Wittmann 2012).

There are no endemic floodplain tree species (Pott and Pott 1994; Montes and San José 1995; Veneklaas et al. 2005). Characteristic morpho-anatomical and/or physiological adaptations of tree species to flooding, such as they occur in many species of Amazonian floodplain forests are unknown within tree species of the Pantanal.

Flooding responses, adaptations and tree distribution. Responses to flooding are similar for many tropical floodplain species, with many species tolerating prolonged periods of waterlogging. Leaves are shed in the dry period and only very few species shed leaves in the high water period, such as species from the Bombacaceae. The fruit peak occurs at high and at low water, depending on the tree position in the flooding gradient. Flooding influences the physiology more than shading in the species of the Pantanal, and consistently higher photosynthesis and stomatal conductance have been measured under non-flooded conditions (Dalmolin et al. 2012). The dry period is thus the main growth period.

Okavango

Ecosystem and flooding pattern. The Okavango delta in northwestern Botswana is fed by the Okavango River which originates in Angola's western highlands and forms the world's largest inland delta. As described in Parolin and Wittmann (2010), the river discharges about 10 km^3 of water onto the delta fan each year, augmented by about 6 km^3 of rainfall, which sustains about 2500 km^2 of permanent wetland and up to 8000 km^2 of seasonal wetland. Interaction between this surface water and the groundwater strongly influences the structure and function of the wetland ecosystem. Soil pH and mineral content and ground water chemistry play major roles in the spatial distribution of plant communities. However, Bonyongo et al. (2000) state that timing and duration of seasonal flooding is the most important factor determining plant species composition.

Precipitation is low, with 10–320 mm of monthly rainfall. The Okavango delta experiences seasonal flooding starting toward the north between October and April and ending in May. In the dry season the floodwaters evaporate leaving salts on the ground, which leads to a strong influence of salt on the vegetation.

Fire plays a role in this ecosystem (Heinl et al. 2004, 2006, 2007). It is more frequent in floodplains than on drylands because of greater biomass and fuel load. Thus, the frequency of fires is determined by flooding frequency.

Diversity of species and functional groups. The floodplains consist mostly of grasslands with 1250 species (Fig. 4; Ellery and Tacheba 2003), mostly grasses interspersed with woody plants in the riverine forests. Impeded drainage and high precipitation in the rainy season lead to temporarily waterlogged soils. No woodlands can establish under these conditions, but humid grasslands dominated by

Fig. 4 Flooded grasslands in the Okavango delta (*Photos* Pia Parolin, January 2010)

dwarf-shrubs and small trees exist (Oldeland et al. 2013). Trees only grow in the Okavango on islands and savannahs that are seldom submerged (Murray-Hudson et al. 2014). Salt-rich islands are flooded less frequently and are dominated by acacias, mopane (*Colophospermum mopane*) and woody shrubs. 43 woody species occur in the dryland riverine woodland (Ellery and Tacheba 2003), none of which are endemic (Parolin and Wittmann 2010).

Flooding responses, adaptations and tree distribution. Although seldom floo-ded, the riparian woodland trees have their roots in the water table (Ellery and Tacheba 2003). Few woody species are tolerant of flooding and fire. Riparian woodlands are responsible for much of the water lost from the ecosystem and deplete ground water by transpiration (Ringrose 2003). Trees ensure that islands of vegetation function as 'kidneys' within the landscape (Ellery and Tacheba 2003).

The riparian trees remain green all year and partly sustain their growth as a result of groundwater uptake in dry periods. Renewal of leaf growth however is primarily related to rainfall, not to flood events in the distal Delta (Ringrose 2003). The Miombo forests are dominated by deciduous tree species which lose their leaves for a period of some months during the dry season, but deciduousness is not a response to flooding (Oldeland et al. 2013) Evergreen tree species with microphyllous compound leaves dominate extensive areas on sandy soils (Oldeland et al. 2013). Regenerative phenology is unknown, with a high percentage of even-aged stands of trees indicating that hydrological factors are important for tree regeneration because they provide spasmodically favorable circumstances for establishment (Hughes 1988). A study of leaf gas exchange of *Colophospermum mopane* in Northwest Botswana (Veenendaal et al. 2008) shows differences in physiological and mor-phological traits between tall and short forms of mopane (*Colophospermum mopane* (Kirk ex Benth.) Kirk ex J. Léonard) trees. These differences appeared attributable to differences in root depth and density between the physiognomic types (Veenendaal et al. 2008). So far, nothing is published about tree regeneration, dispersal syndromes, seed germination, seedling growth and physiology, seedling survival vs mortality, tree growth and productivity, or carbon stocks and seques-tration. There is clear vegetation zonation due to the hydrological regime characterized by depth, duration and timing of inundation, and soil and ground-water salinity (Ellery et al. 1993; Ellery and Tacheba 2003). Vegetation patterns are governed by prevailing flood regimes and depend on small differences of topography (Bonyongo et al. 2000; Oldeland et al. 2013). Trees are particularly important as they lower the water table beneath islands relative to the surrounding wetlands and cause a net inward flow of groundwater (McCarthy 2006). On these islands, broadleaf evergreen riparian trees dominate (Oldeland et al. 2013).

Changes and threats. Changes in the types of vegetation cover, due to both human and natural causes, have taken place since the first vegetation map was produced in 1971 (Ringrose et al. 2002). In the Southwest, shifts to thorn scrub vegetation prevail while in the eastern part of the country widespread bush encroachment is taking place. An increased human population density suggests that these are anthropogenic agrarian-degradation effects. Wherever broadleaved evergreen trees are cleared, widespread salinity occurs (Ellery et al. 2000). Projection of future vegetation changes to about 2050 indicates degeneration of the major vegetation types due to expected drying of the local climate (Ringrose et al. 2002).

Tonle Sap of the Mekong

Ecosystem and flooding pattern. The Mekong in is one of the world's longest rivers with a length of 4,800 km approximately (MRC 2005). Its lower basin includes extensive floodplains distributed mostly across Central Cambodia, Southern Vietnam and the Tonle Sap system. It is in this last one where most of the remaining flooded forests exist, whereas most of the other regions of the Mekong floodplains have been converted to agriculture. The Tonle Sap River forms a permanent lake of 2,600 km^2. During the dry season, the water in the Tonle Sap Lake is only 1–2 m deep. During the wet monsoonal season, it reaches a depth of 9 m with a water surface of 15,000 km^2. Average flood duration through the Tonle Sap floodplain is 5.4 ± 3.9 months per year (Arias et al. 2013). Fire plays a role in the Tonle Sap. Most grasslands, which cover about 15 % of the floodplain, have signs of contemporary fire. However, no quantifications of fire, or knowledge of whether causes are natural (lighting) or manmade are available.

Diversity of species and functional groups. The Tonle Sap floodplain is covered with a mosaic of natural, disturbed and agricultural habitats. Typically, most pristine habitats cover those areas in the lowest and middle portions of the floodplain, whereas rice paddies only occur in upper elevations that become flooded for short periods of time. Among these habitats, swamp shrub lands and forests cover 4,500 km^2 (Fig. 5; Arias et al. 2014a). Similar swamp forests are also present along floodplains of the Mekong and other major rivers in Cambodia. Swamp or gallery forests originally dominated the dry-season shoreline of Tonle Sap Lake in areas that are flooded about 9 months per year. The zone that is feasible for forests is

Fig. 5 Floodplain forest in the Tonle Sap/Mekong River floodplain (*Photos* Pia Parolin, July 2010)

about 657 km^2 (Arias et al. 2012, 2014b), but tall stands of forest currently cover a much smaller area.

It is estimated that up to 233 plant species occur in the Tonle Sap floodplain, with the greatest species density in the swamp shrubs and decreasing both towards lower and higher elevations (Arias et al. 2013; McDonald et al. 1997). Shrubs reach a maximum height of 4.4 ± 1.7 m and canopy cover of 58 ± 16 % (Arias et al. 2013). Swamp forests typically have less plant species density (4–8 per 100 m^2) than shrublands, but their canopy is much taller (8–11 m).

Flooding responses, adaptations and tree distribution. Little information is available on plant distribution and zonation along the Mekong river floodplains. In the Tonle Sap, the structure and composition of woody vegetation is largely a function of heterogeneity of soil moisture and seasonal flood dynamics (Arias et al. 2013). Human intervention also plays a key role as rice paddies have been established throughout the floodplain. The tallest trees grow closer to the permanent lake basin. McDonald et al. (1997) attributed the presence of shrub vegetation in this region to the short period optimal for growth that is restricted to the beginning and end of the dry season. Most woody species of the floodplain of Tonle Sap are deciduous, a probable adaptation to the periodic flood pulse. Instead of losing their leaves in the dry season, these species shed their leaves as a function of light limitation when submerged as the lake deepens. Flowering and fruit production in the floodplain trees and shrubs are delayed for several months after the flush of new leaves. Fruits reach maturity at the time of submergence suggesting that water movement and fauna may be important dispersal agents. No data on physiological responses to flooding and morphological adaptations were found.

Changes and threats. Most of the Mekong floodplains are heavily inhabited by fishermen and rice farmers. While the annual flood cycle of the Mekong provides resources for these people, it is a fragile balance. The Tonle Sap provides the largest remaining contiguous wetland in the region. Throughout the dry season burning is common, with fires used to clear land before ploughing or to facilitate access for cattle. Roads have permanently delimited the maximum extent of the continuous flooding. The greatest threat to the hydrological and ecological functioning of the Tonle Sap is the construction of large dams upstream along the Mekong and its tributaries (Cochrane et al. 2014). New dams bring hydrologic changes that could affect flooding regime, habitat distribution, and primary production of the Tonle Sap floodplain (Arias et al. 2014b).

Australian "Wet-Dry Tropics" of the Northern Territory

Ecosystem and flooding pattern. Floodplain wetlands are uncommon in the mostly arid continent of Australia. However, an important wetland, the Kakadu National Park in Northern Australia, extends over 99,000 km^2 with 2,900 km^2 of forested wetlands (Lowry and Finlayson 2004; Warfe et al. 2011; Ward et al. 2012). The annual rainfall is 600–1600 mm spread over 4–7 months. The wet season (November–December) lasts for 3–4 months (Taylor and Tulloch 1985). Water flows on a seasonal basis (Finlayson et al. 1990). Northern Australia wetlands are not contiguous and occur along several small streams. The forests are inundated by up to 1 m of water during the wet season but are dry at other times (Fig. 6).

Diversity of species and functional groups. Around 55 % of the terrestrial vegetation in the Kakadu Region is tropical tall grass savanna, composed of eucalypt-dominated open forest and woodland with a 1–2 m tall grassy understory (Finlayson 2005; Warfe et al. 2011; Ward et al. 2012). Closed canopy monsoon rainforests are restricted to floodplains. Gallery and floodplain forests in monsoonal northern Australia are mostly sclerophyllous and dominated by five closely related species of *Melaleuca* with a tree canopy cover of 10–70 % (Franklin et al. 2007).

Flooding responses, adaptations and tree distribution. Variability due to changes in the hydrological cycle has resulted in many specific adaptations that enable the plants to establish and grow (Finlayson et al. 1989; Finlayson 2005). Trees of the floodplains often have modified bark structures such as the corky bark of *Sesbania formosa* and *Barringtonia acutangula* and the distinctive papery bark of some melaleucas that possesses internal longitudinal gas-filled passages (Finlayson 2005). The majority of seed dispersal mechanisms involve water even though many

Fig. 6 Floodplain forest in the Kakadu National Park in Northern Australia (*Photos* Pia Parolin, July 2014)

parts of the floodplains are drier for a longer period than they are wet. Nothing seems to be published on physiological and morphological adaptations of trees of the Kakadu National Park. Using a heat-pulse method McJannet (2008) found a strong relationship between tree size and tree water use and showed that transpiration by *M. quinquenervia* was unaffected by inundation. This ability to transpire during flooding may be due to physiological adaptations of this species and to dynamic root systems that can quickly respond to rising and falling water tables and dense networks of fine ageotropic roots. Waterlogged *Melaleuca quinquenervia* also develops negatively gravitropic roots (Sena Gomes and Kozlowski 1980). A study by O'Grady et al. (2006) of *Corymbia bella* and *Melaleuca argentea* in riparian zones of Northern Territory, along the Daly River showed that throughout the dry season predawn leaf water potentials were above −0.5 MPa, indicating that neither species suffered significant unrelieved water deficit stress during the dry season. This study demonstrates strong gradients of tree water use within tropical riparian communities in Northern Australia.

The productivity of the floodplain vegetation changes with the annual cycle. This is indicated by periodic litterfall from *Melaleuca* trees. The distribution and density of trees on at least part of the floodplain was seen to change considerably between 1975 and 1990 (Finlayson 2005), indicating the dynamic nature of the wetland environment.

The hydrological cycle has been identified as important in shaping the pattern of the vegetation in the freshwater wetlands (Finlayson et al. 1989). The duration of flooding, depth of water and the velocity of water flow are major determinants of the vegetation composition of the floodplain (Finlayson et al. 1989). The changing pattern is a function of both the flooding and drying phases of the hydrological cycle (Finlayson et al. 1989, 1990).

Changes and threats. Fire and invasive plants and animal species have a significant impact on the extent and distribution of plant species and of the land cover (Finlayson et al. 1990). Damage to the natural levees that separate freshwater and saline wetland communities caused by climate change or by feral animals (especially water buffalo) may also change the vegetation. Notable responses by floodplain vegetation have already occurred following the removal of feral buffalo (Skeat et al. 1996).

Diversity of Flood Tolerant Plants Across Tropical Floodplains

In tropical floodplains, all plant growth forms are found. Trees, shrubs, vines and herbs dominate, but a large variety of life history traits is found, including hemi-epiphytic trees, stem succulent trees, palms and arborescent monocots, or arborescent tree forms near treeline. A wide range of shade tolerance from rapidly growing pioneer trees during gap phase regeneration to species that can survive by growing slowly in deep shade contribute to this diversity. In Amazonian flood-plains, all functional groups are present, from fast-growing short-lived pioneer trees to slow-growing millenarians (Schöngart et al. 2011), evergreen and deciduous species, those which produce large amounts of seeds every year such as *Salix* and *Cecropia* or those which produce some seeds only every four or more years, such as *Aldina latifolia*. The large number of adaptations for flooding allows for a partic-ularly high diversity of species, of growth forms, and of functional traits despite the strong pressure of flooding.

Despite the mostly unfavourable growth conditions for trees in floodplains, the number of angiosperm tree species is very high. With 30–150 species in Amazonian várzea (Balslev et al. 1987; Ayres 1993; Worbes 1997; Wittmann et al. 2002), species richness is clearly lower than in the adjacent Terra Firme, where 180–300 species ha^{-1} occur (Klinge and Rodrigues 1968). The overall diversity of species and life strategies in Pantanal and Amazonian floodplains is lower than in adjacent non-flooded uplands (Campbell et al. 1986; Balslev et al. 1987). On the other hand, diversity is still considerable with an estimated (Pott and Pott 1994) total number of about 756 woody plant species in the Pantanal. The number of species adapted to long-term flooding is small. A detailed analysis of the distribution of 85 tree species along the flood gradient in the northern part shows that 45 species are restricted to permanently dry areas and only 18 species show preference for habitats subject to extended annual flooding. Twenty-two species tolerate a very broad spectrum of flooded and dry conditions, a behaviour that is favoured by pluriannual extreme dry and wet periods (Nunes da Cunha and Junk 2001). If we consider this relationship representative for the entire Pantanal, 355 species (47 %) would show flood tol-erance to varying degrees.

Conclusions

What do we know? In the four floodplains we examined, the terrestrial non-flooded phase is when tree growth is most vigorous. However, many tree species retain actively flowing phloem sap even during flooding (Waldhoff et al. 2002; Visser et al. 2003) indicating that active sources and sinks for respirable substrates operate under these conditions. A set of metabolic adaptations are inevitably required to achieve this. We postulate that for most tree species of tropical floodplains, the primary morphological strategies in response to flooding are similar to those of temperate species (Jackson and Armstrong 1999) or in the well-analyzed tropical Amazonian floodplains (Parolin et al. 2004). In particular, there must be a development of gas-filled spaces in the roots and stems to allow diffusion of oxygen from the aerial portions of the plant into the roots. Morphological adaptations that favour this are hypertrophy of lenticels, formation of adventitious roots, plank-buttressing and stilt rooting, development of aerench-yma and the deposition of cell wall biopolymers such as suberin and lignin in the root peripheral cell layers. The formation of aerial roots may compensate for losses of respiration and function by roots affected by lack of oxygen in the soil. Under experimental conditions with stable water levels, most species show the potential to produce adventitious roots. But, in the field, they are seldom found, probably because their formation is hampered by a rapidly changing water level. Leaves of tropical forests in general and the Amazonian floodplain forests in particular commonly have xeromorphic structures (Waldhoff 2003). This attribute contributes to suppressing water loss at times of low water supply. This can apply to tree crowns during the aquatic phase and to periods of drought in the terrestrial phase.

The degree of flood tolerance depends, in large part, on the time taken to colonize the floodplains.

Morphological adaptations may be remnants of pre-adaptations from non-flooded upland tree species (Kubitzki 1989), which evolved further, leading over time to highly adapted species. Thus, the phylogenetic development of adaptations depends on the age of the ecosystem, and also on the dominant plant families, which colonized these ecosystems originally. In each of the four ecosystems, there are large differences in developmental age (Table 1). This implies different stages of adaptations among the organisms living there. Amazonian and African floodplains are extremely old, dating at least to the Pleistocene (>12 million years old), or even earlier. Such ancient landscapes have experienced several changes in climate and hydrology during the glacial and interglacial periods. In contrast, the Tonle Sap and the Australian floodplains are much younger and thought to be no older than 7500 and 4000 years, respectively (Junk et al. 2006).

Where are our major gaps? The data presented here for only a few huge wetlands is already very rudimentary. Other very large wetlands with forest patches were completely excluded from our overview, as we do not have any relevant data on them. For many of the world's largest wetlands, only basic data on hydrology and climatology are available with almost no information on plant distribution, tree adaptations and ecophysiology. It is important to raise attention to such poorly researched wetlands which are often inaccessible for social and political reasons but which are threatened by the ever-increasing human population and its need for water, food, waterways and hydroelectric power (Parolin and Wittmann 2010). The destruction is so fast that we may never learn of the adaptations underpinning the success of the tree species in these areas.

Throughout the world, wetland ecosystems are under increasing pressure from agriculture, urbanization of catchment areas, tourism and recreational activities, construction of impoundments and changes to hydrology and climate (Parolin and Wittmann 2010). By comparing diversity and tree responses in four floodplain ecosystems on different continents, we attempt to improve our understanding of the factors influencing spatial distribution of plants, diversity of species and adaptations, and thus contribute to our knowledge of tropical wetland ecology. In this way, we hope to assist in the successful restoration of degraded floodplains and promote the sustainable use and conservation of these highly valuable ecosystems.

What can we expect regarding climatic changes in floodplain ecosystems? Predictions suggest a transformation of the Amazon rainforest into cerrado vegetation (Oyama and Nobre 2003) or the shifting of the moist rainforests in the Amazon towards semi-deciduous forest types (Malhi et al. 2009). However, models and predictions often do not consider wetlands and the large percentage of flood-plain forests which are part of the Amazon basin and have a very important role in the water cycle of the whole Amazon basin. They form an efficient system for water retention, recharge the water table, decrease the amplitude of the creeks and rivers between floods and droughts, and contribute substantially to regional climate by the high rate of evapotranspiration. These environmental services are of utmost importance to mitigate the consequences of the changes in rainfall patterns of the future. All streams and small and large rivers are bordered by floodplain forests that will, in case the most alarming scenarios come true, form a dense network of flooded forests in a drier landscape. Floodplains, which by their environmentally drastic conditions are very resilient, are likely to become even greater hotspots for biodiversity and very significant biological corridors. Where possible, future studies should use methods that will allows comparisons to be made with confidence. Improving our understanding of floodplain functioning will underpin their preservation and effective future management.

References

Arias ME, Cochrane TA, Piman T, Kummu M, Caruso B, Killeen TJ (2012) Quantifying changes in flooding and habitats in the Tonle Sap Lake (Cambodia) caused by water infrastructure development and climate change in the Mekong Basin. J Environ Manage 112:53–66

Arias ME, Cochrane TA, Norton D, Killeen TJ, Khon P (2013) The flood pulse as the underlying driver of vegetation in the largest wetland and fishery of the Mekong Basin. Ambio 42(7):864–876

Arias ME, Cochrane TA, Elliott V (2014a) Modelling future changes of habitat and fauna in the Tonle Sap wetland of the Mekong. Environ Conserv 41:165–175

Arias ME, Cochrane TA, Kummu M, Lauri H, Koponen J, Holtgrieve GW, Piman T (2014b) Impacts of hydropower and climate change on drivers of ecological productivity of Southeast Asia's most important wetland. Ecol Model 272:252–263

Ayres JMC (1993) The várzea forests of Mamirauá. In: Estudos de Mamirauá. SociedadeCivilMamirauá (ed), Tefé, Brazil, vol I, pp 1–123.

Balslev H, Luteyn J, Oellgaard B, Holm-Nielsen LB (1987) Composition and structure of adjacent unflooded and floodplain forest in Amazonian Ecuador. Opera Botanica 92:37–57

Bonyongo CM, Veenendaal E, Bredenkamp G (2000) Floodplain vegetation in the Nxaraga Lagoon area, Okavango Delta, Botswana. S Afr J Bot 66:15–21

Campbell DG, Daly DC, Prance GT, Maciel UN (1986) Quantitative ecological inventory of terra firme and várzea tropical forest on the Rio Xingu, Brazilian Amazon. Brittonia 38:369–393

Cochrane TA, Arias ME, Piman T (2014) Historical impact of water infrastructure on water levels of the Mekong river and the tonle sap system. Hydrol Earth Sci Discuss 11:4403–4431

Crawford RMM (1989) The anaerobic retreat. In: Crawford RMM (ed) Studies in plant survival. Ecological case histories of plant adaptation to adversity. Studies in Ecology, vol 11. Blackwell Scientific Publications, pp 105–129

Cunha CN, Junk WJ (2001) Distribution of woody plant communities along the flood gradient in the Pantanal of Poconé, Mato Grosso, Brazil. Int J Ecol Environ Sci 27:63–70

Dalmolin AC, Dalmagro HJ, de Almeida Lobo F, Zortéa M, Junior A, Rogríguez Ortíz CE, Vourlitis GL (2012) Effects of flooding and shading on growth and gas exchange of *Vochysia divergens* Pohl (Vochysiaceae) of invasive species in the Brazilian Pantanal. Braz J Plant Physiol 24:75–84

Damasceno Júnior GA, Semir J, Santos FAM, Leitão Filho HF (2005) Structure, distribution of species, and inundation in a riparian forest of Rio Paraguai, Pantanal, Brazil. Flora 200:119–135

De Simone O, Müller E, Junk WJ, Schmidt W (2002a) Adaptations of Central Amazon tree species to prolonged flooding: root morphology and leaf longevity. Funct Plant Biol 29:1025–1035

De Simone O, Haase K, Müller E, Junk WJ, Gonsior GA, Schmidt W (2002b) Impact of root morphology on metabolism and oxygen distribution in roots and rhizosphere from two Central Amazon floodplain tree species. Funct Plant Biol 29:1025–1035

De Simone O, Haase K, Müller E, Junk WJ, Hartmann K, Schreiber L, Schmidt W (2003) Apoplasmic barriers and oxygen transport properties of hypodermal cell walls in roots from four Amazonian tree species. Plant Physiol (in press)

Ellery WN, Tacheba B (2003). Floristic diversity of the Okavango Delta, Botswana. In: Alonso LE, Nordin LA (eds) Chapter 5 in A rapid biological assessment of the aquatic ecosystems of the Okavango Delta, Botswana: High Water Survey. RAP Bulletin of Biological Assessment

Ellery WN, Ellery K, McCarthy TS (1993) Plant distribution in islands of the Okavango Delta, Botswana: determinants and feedback interactions. Afr J Ecol 31:118–134

Ellery WN, McCarthy TS, Dangerfield JM (2000) Floristic diversity in the Okavango Delta, Botswana as an endogenous product of biological activity. In: Gopal B., Junk W.J, Davis J.A. (eds.). Biodiversity in wetlands: assessment, function and conservation. Backhuis, Leiden. Vol. I

Finlayson CM (2005) Plant Ecology of Australia's Tropical Floodplain Wetlands: A Review. Ann Bot 96:541–555

Finlayson CM, Bailey BJ, Cowie ID (1989) Macrophytic Vegetation of the Magela Flood Plain, Northern Australia. Research Report No. 5, Office of the Supervising Scientist, Sydney, Australia, 38 pp

Finlayson CM, Cowie ID, Bailey BJ (1990) Characteristics of a seasonally flooded freshwater system in monsoonal Australia. In: Whigham DF, Good RE, Kvet J (eds) Wetland Ecology and Management: Case studies. Kluwer Aca-demic Publishers, Dordrecht, The Netherlands, pp 141–162

Franklin DC, Brocklehurst PS, Lynch D, Bowman DMJS (2007) Niche differentiation and regeneration in the seasonally flooded *Melaleuca* forests of northern Australia. J Trop Ecol 23:457–468

Furch B (1984) Untersuchungen zur Überschwemmungstoleranz von Bäumen der Várzea und des Igapó. Blattpigmente. Biogeographica 19:77–83

Galetti M et al (2008) Big fish are the best: seed dispersal of *Bactris glaucescens* by the Pacu fish (*Piaractus mesopotamicus*) in the Pantanal, Brazil. Biotropica 40:386–389

Gill CJ (1970) The flooding tolerance of woody species - a review. Forestry Abstracts 31:671–688

Gottsberger G (1978) Seed dispersal by fish in the inundated regions of Humaitá, Amazonia. Biotropica 10:170–183

Goulding M (1980) Interactions of fishes with fruits and seeds. In: Goulding M (ed) The fishes and the forest. Explorations in Amazonian natural history. University of California Press, pp 217–232

Haase K, De Simone O, Junk WJ, Schmidt W (2003) Internal oxygen transport in cuttings from flood-adapted varzea tree species. Tree Physiol 23(15):1069–1076

Heinl M, Sliva J, Tacheba B (2004) Vegetation changes after single fire-events in the Okavango Delta wetland, Botswana. S Afr J Bot 70:695–704

Heinl M, Neuenschwander A, Sliva J, Vanderpost C (2006) Interactions between fire and flooding in the Okavango Delta floodplains, Botswana. Landscape Ecol 21:699–709

Heinl M, Frost P, Vanderpost C, Sliva J (2007) Fire activity on drylands and floodplains in the southern Okavango Delta, Botswana. J Arid Environ 68:77–87

Hughes FMR (1988) The ecology of African floodplain forests in semi-arid and arid zones: a review. J Biogeogr 15:127–140

Jackson MB, Armstrong W (1999) Formation of aerenchyma and the processes of plant ventilation in relation to soil flooding and submergence. Plant Biol 1:274–287

James EK, Loureiro MD, Pott A, Pott VJ, Martins CM, Franco AA, Sprent JI (2001) Flooding-tolerant legume symbioses from the Brazilian Pantanal. New Phytol 150:723–738

Joly CA, Crawford RMM (1982) Variation in tolerance and metabolic responses to flooding in some tropical trees. J Exp Bot 33:799–809

Junk WJ (1989) Flood tolerance and tree distribution in Central Amazonian floodplains. In: Nielsen LB, Nielsen IC, Balslev H (eds) Tropical forests: Botanical dynamics, speciation and diversity. Academic Press, London, 47–64

Junk WJ, Bayley PB, Sparks RE (1989) The flood pulse concept in river-floodplain systems. In: Dodge DP (ed) Proceedings of the International Large River Symposium. Canadian Publications of Fisheries and Aquatic Sciences, vol 106. pp 110–127

Junk WJ, Brown MT, Campbell I, Finlayson M, Gopal B, Ramberg L, Warner BG (2006) The comparative biodiversity of seven globally important wetlands: a synthesis. Aquat Sci 68:400–414

Klinge H, Rodrigues W (1968) Litter production in an area of Amazonian terra firme forest. Part I. Litterfall, organic and total nitrogen contents of litter. Amazoniana 1:287–302

Kozlowski TT (1984) Responses of woody plants to flooding. In: Kozlowski TT (ed) Flooding and plant growth. Academic Press, Orlando, pp 129–163

Kubitzki K (1989) The ecogeographical differentiation of Amazonian inundation forests. Plant Syst Evol 162:285–304

Kubitzki K, Ziburski A (1994) Seed dispersal in flood plain forests of Amazonia. Biotropica 26:30–43

Lowry J, Finlayson CM (2004) A review of spatial data sets for wetland inventory in northern Australia. Darwin, Australia: Supervising Scientist Report 178, Supervising Scientist

Malhi Y, Aragao LEOC, Galbraith D, Huntingford C, Fisher R, Zelazowski P, Sitch S, McSweeney C, Meir P (2009) Exploring the likelihood and mechanism of a climate-change-induced dieback of the Amazon rainforest. PNAS 106:20610–20615

McCarthy TS (2006) Groundwater in the wetlands of the Okavango Delta, Botswana, and its contribution to the structure and function of the ecosystem. J Hydrol 320:264–282

McDonald A, Pech B, Phauk V, Leev B (1997) Plant communities of the Tonle Sap flood plain. Contribution to the nomination of the Tonle Sap as Biosphere Reserve for UNESCO's "Man in the Biosphere Program". UNESCO, Phnom Penh

McGregor S, Lawson V, Christophersen P, Kennett R, Boyden J, Bayliss P, Liedloff A, McKaige B, Andersen AN (2010) Indigenous wetland burning: conserving natural and cultural resources in Australia's World Heritage-listed Kakadu National Park. Human Ecol 38:721–730

McJannet D (2008) Water table and transpiration dynamics in a seasonally inundated Melaleuca quinquenervia forest, north Queensland, Australia. Hydrol Process 22:3079–3090

Moegenburg SM (2002) Spatial and temporal variation in hydrochory in Amazonian floodplain forest. Biotropica 34:606–612

Montes R, San José JJ (1995) Vegetation and soil analyses of topo-sequences in the Orinoco llanos. Flora 190:1–33

MRC (2005) Overview of the Hydrology of the Mekong Basin. Vientiane, Lao PDR

Murray-Hudson M, Wolski P, Murray-Hudson F, Brown M, Kashe K (2014) Disaggregating hydroperiod: components of the seasonal flood pulse as drivers of plant species distribution in floodplains of a tropical wetland. Wetlands

Nunes da Cunha C, Junk WJ (2001) Distribution of woody plant communities along the flood gradient in the Pantanal of Poconé, Mato Grosso, Brazil. Int J Ecol Env Sci 27:63–70

O'Grady AP, Eamus D, Cook PG, Lamontagne S (2006) Comparative water use by the riparian trees Melaleuca argentea and Corymbia bella in the wet-dry tropics of northern Australia. Tree Physiol 26:219–228

Oldeland J, Erb C, Finckh M, Jürgens N (2013) Environmental Assessments in the Okavango Region. Biodiversity & Ecology 5: 418. Biocentre Klein Flottbek and Botanical Garden, Hamburg

Oliveira MT, Damasceno-Junior GA, Pott A, Paranhos Filho AC, Suare YR, Parolin P (2014) Regeneration of riparian forests of the Brazilian Pantanal under flood and fire influence. For Ecol Manage 331:256–263

Oyama MD, Nobre CA (2003) A new climate-vegetation equilibrium state for Tropical South America. Geophysical Research Letters 30(23):2199–2203

Parolin P (2000) Phenology and CO_2-assimilation of trees in Central Amazonian floodplains. J Trop Ecol 16:465–473

Parolin P (2002) Submergence tolerance vs. escape from submergence: two strategies of seedling establishment in Amazonian floodplains. Environ Exp Bot 48:177–186

Parolin P (2009) Submerged in darkness: adaptations to prolonged submergence by woody species of the Amazonian Floodplains. Ann Bot 103:359–376

Parolin P, Junk WJ (2002) The effect of submergence on seed germination in trees from Amazonian floodplains. Boletim Museu Goeldi

Parolin P, Wittmann F (2010) Struggle in the flood – Tree responses to flooding stress in four tropical floodplain systems. AoB Plants

Parolin P, Armbrüster N, Wittmann F, Ferreira LV, Piedade MTF, Junk WJ (2002) A Review of tree phenology in central Amazonian floodplains. Pesquisas, Botânica 52:195–222

Parolin P, De Simone O, Haase K, Waldhoff D, Rottenberger S, Kuhn U, Kesselmeier J, Schmidt W, Piedade MTF, Junk WJ (2004) Central Amazon floodplain forests: tree survival in a pulsing system. Bot Rev 70:357–380

Parolin P, Lucas C, Piedade MTF, Wittmann F (2010) Drought responses of flood-tolerant trees in Amazonian floodplains. Ann Bot 105:129–139

Parolin P, Wittmann F, Ferreira L (2013) Fruit and seed dispersal in Amazonian floodplain trees – a review. Ecotropica 19:15–32

Piedade MTF, Parolin P, Junk WJ (2006) Phenology, fruit production and seed dispersal of *Astrocaryum jauari* (Arecaceae) in Amazonian black-water floodplains. Revista de Biología Tropical 54:1171–1178

Ponnamperuma FN (1984) Effects of flooding on soils. In: Kozlowski TT (ed) Flooding and plant growth. Academic Press, Orlando, pp 9–45

Pott A, Pott VJ (1994) Plantas do Pantanal. Empresa Brasileira de Pesquisa Agropecuária (EMBRAPA), Brasilia 320 pp

Prance GT, Schaller GB (1982) Preliminary study of some vegetation types of the Pantanal, Mato Grosso, Brazil. Brittonia 32:228–251

Ratter JA, Pott A, Pott VJ, Cunha CN, Haridasan M (1988) Observations on woody vegetation types in the Pantanal and at Corumbá. Brazil. Notes R Bot Gard Edinb 45:503–505

Reich PB, Borchert P (1984) Water stress and tree phenology in a tropical dry forest in the lowlands of Costa Rica. J Ecol 72:61–74

Ringrose S (2003) Characterisation of riparian woodlands and their potential water loss in the distal Okavango Delta, Botswana. Appl Geogr 23:281–302

Ringrose S, Chipanshi AC, Matheson W, Chanda R, Motoma L, Magole I, Jellema A (2002) Climate- and human-induced woody vegetation changes in Botswana and their implications for human adaptation. Environ Manage 30:98–109

Schlüter U-B, Furch B (1992) Morphologische, anatomische und physiologische Untersuchungen zur Überflutungstoleranz des Baumes *Macrolobium acaciaefolium*, charakteristisch für die Weiß- und Schwarzwasserüberschwemmungswälder bei Manaus, Amazonas. Amazoniana 12:51–69

Schlüter U-B, Furch B, Joly CA (1993) Physiological and anatomical adaptations by young *Astrocaryum jauari* Mart. (Arecaceae) in periodically inundated biotopes of Central Amazonia. Biotropica 25:384–396

Scholander PF, Perez MO (1968) Sap tension in flooded trees and bushes of the Amazon. Plant Physiol 43:1870–1873

Schöngart J, Piedade MFT, Ludwigshausen S, Horna V, Worbes M (2002) Phenology and stem-growth periodicity of tree species in Amazonian floodplain forests. J Trop Ecol 18:581–597

Schöngart J, Arieira J, Felfili Fortes C, Cezarine de Arruda E, Nunes da Cunha C (2011) Age-related and stand-wise estimates of carbon stocks and sequestration in the aboveground coarse wood biomass of wetland forests in the northern Pantanal, Brazil. Biogeosciences 8:3407–3421

Sena Gomes AR, Kozlowski TT (1980) Responses of *Melaleuca quinquenervia* seedlings to flooding. Physiol Plant 49:373–377

Siebel HN, Blom CWPM (1998) Effects of irregular flooding on the establishment of tree species. Acta Bot Neerl 47:231–240

Skeat AJ, East TJ, Corbett LK (1996) Impact of feral water buffalo. In: Finlayson CM, von Oertzen I (eds) Landscape and vegetation ecology of the Kakadu Region, Northern Australia. Kluwer Academic, Dordrecht. pp 155–177

Taylor JA, Tulloch D (1985) Rainfall in the wet-dry tropics: extreme events at Darwin and similarities between years during the period 1870–1983. Aust J Ecol 10:281–295

Veenendaal EM, Mantlana KB, Pammenter NW, Weber P, Huntsman-Mapila P, Lloyd J (2008) Growth form and seasonal variation in leaf gas exchange of *Colophospermum mopane* savanna trees in northwest Botswana. Tree Physiol 28:417–424

Veneklaas EJ, Fajardo A, Obregon S, Lozano J (2005) Gallery forest types and their environmental correlates in a Colombian Savanna landscape. Ecography 28:236–252

Visser EJW, Voesenek LACJ, Vartapetian BB, Jackson MB (2003) Flooding and plant growth. Ann Bot 91:107–109

Waldhoff D (2003) Leaf structure in trees of Central Amazonian floodplain forests (Brazil). Amazoniana 17:451–469

Waldhoff D, Junk WJ, Furch B (1998) Responses of three Central Amazonian tree species to drought and flooding under controlled conditions. Int J Ecol Environ 24:237–252

Waldhoff D, Junk WJ, Furch B (2002) Fluorescence parameters, chlorophyll concentration, and anatomical features as indicators for flood adaptation of an abundant tree species in Central Amazonia: *Symmeria paniculata*. Environ Exp Bot 48:225–235

Ward DP, Hamilton SK, Jardine TD, Pettit NE, Tews EK, Olley JM, Bunn SE, (2012) Assessing the seasonal dynamics of inundation, turbidity, and aquatic vegetation in the Australian wet–dry tropics using optical remote sensing. Ecohydrol

Warfe DM, Pettit NE, Davies PM, Pusey BJ, Hamilton SK, Kennard MJ, Townsend SA, Bayliss P, Ward DP, Douglas MM, Burford MA, Finn M, Bunn SE, Halliday IA (2011) The "wet–dry" in the wet–dry tropics drives river ecosystem structure and processes in northern Australia. Freshw Biol 56:2169–2195

Wittmann F (2012) Tree species composition and diversity in Brazilian freshwater floodplains. In: Pagano MC (ed) Mycorrhiza: occurrence in natural and restored environments. Nova Science Publ, New York, pp 223–263

Wittmann F, Parolin P (2005) Aboveground roots in Amazonian floodplain trees. Biotropica 37:609–619

Wittmann F, Anhuf D, Junk WJ (2002) Tree species distribution and community structure of Central Amazonian várzea forests by remote sensing techniques. J Trop Ecol 18:805–820

Wittmann F, Junk WJ, Piedade MTF (2004) The várzea forests in Amazonia: Flooding and the highly dynamic geomorphology interact with natural forest succession. For Ecol Manage 196:199–212

Wittmann F, Zorzi BT, Tizianel FAT, Urquiza MVS, Faria RR, Sousa NM, Módena ES, Gamarra RM, Rosa ALM (2008) Tree species composition, structure, and aboveground wood biomass of a riparian forest of the lower Miranda River, Southern Pantanal, Brazil. Folia Geobotanica 43:397–411

Wittmann F, Householder E, Piedade MTF, Assis RL, Schöngart J, Parolin P, Junk WJ (2013) Habitat specificity, endemism and the neotropical distribution of Amazonian white-water floodplain trees. Ecography 36:690–707

Worbes M (1986) Lebensbedingungen und Holzwachstum in zentralamazonischen Übersch wemmungswäldern. ScriptaGeobotanica, Erich Goltze, Göttingen 112

Worbes M (1989) Growth rings, increment and age of trees in inundation forests, savannas and a mountain forest in the neotropics. IAWA Bulletin 10:109–122

Worbes M (1997) The forest ecosystem of the floodplains. In: Junk WJ (ed) The Central Amazon floodplain: ecology of a pulsing system. Ecological Studies, vol 126. Springer, Heidelberg, pp 223–266

Zeilhofer P, Schessl M (1994) Relationship between vegetation and environmental conditions in the northern Pantanal of Mato Grosso, Brazil. J Biogeogr 27:159–168

Ziburski A (1991) Dissemination, Keimung und Etablierung einiger Baumarten der Überschwemmungswälder Amazoniens. In: Rauh W (ed) Tropische und subtropische Pflanzenwelt. Akademie der Wissenschaften und der Literatur 77:1–96

The Physiology of Mangrove Trees with Changing Climate

Catherine E. Lovelock, Ken W. Krauss, Michael J. Osland, Ruth Reef and Marilyn C. Ball

Abstract Mangrove forests grow on saline, permanently or periodically flooded soils of the tropical and subtropical coasts. The tree species that compose the mangrove are halophytes that have suites of traits that confer differing levels of tolerance of salinity, aridity, inundation and extremes of temperature. Here we review how climate change and elevated levels of atmospheric CO_2 will influence mangrove forests. Tolerance of salinity and inundation in mangroves is associated with the efficient use of water for photosynthetic carbon gain which underpins anticipated gains in productivity with increasing levels of CO_2. We review evidence of increases in productivity with increasing CO_2, finding that enhancements in growth appear to be similar to trees in non-mangrove habitats and that gains in productivity with elevated CO_2 are likely due to changes in biomass allocation. High levels of trait plasticity are observed in some mangrove species, which

C.E. Lovelock (✉) · R. Reef
School of Biological Sciences, The University of Queensland,
Brisbane St. Lucia, QLD 4072, Australia
e-mail: c.lovelock@uq.edu.au

R. Reef
e-mail: r.reef@uq.edu.au

K.W. Krauss · M.J. Osland
U.S. Geological Survey, Wetland and Aquatic Research Center,
Lafayette, LA 70506, USA
e-mail: kkrauss@usgs.gov

M.J. Osland
e-mail: mosland@usgs.gov

M.C. Ball
Research School of Biology, The Australian National University,
Canberra, ACT, Australia
e-mail: Marilyn.Ball@anu.edu.au

© Springer International Publishing Switzerland 2016
G. Goldstein and L.S. Santiago (eds.), *Tropical Tree Physiology*,
Tree Physiology 6, DOI 10.1007/978-3-319-27422-5_7

potentially facilitates their responses to climate change. Trait plasticity is associated with broad tolerance of salinity, aridity, low temperatures and nutrient availability. Because low temperatures and aridity place strong limits on mangrove growth at the edge of their current distribution, increasing temperatures over time and changing rainfall patterns are likely to have an important influence on the distribution of mangroves. We provide a global analysis based on plant traits and IPCC scenarios of changing temperature and aridity that indicates substantial global potential for mangrove expansion.

Keywords Elevated CO_2 · Flooding · Plasticity · Salinity · Water uptake

Introduction

The trees of mangrove forests have fascinated physiologists for decades. The highly saline, tidally flooded environments of mangrove forests seem unlikely to support tree growth, yet mangroves are some of the most productive forests on the planet (Alongi 2009). Both the number of families and individual species of plants that have evolved the necessary traits to grow in mangrove habitats has been relatively small: 70 species in over 40 million years (Ricklefs et al. 2006), reflecting the complex suite of traits that are required for growth in intertidal environments. The position of these forests in the landscape, on the ecotone between terrestrial and marine habitats, also brings high levels of variation in soil conditions that range over a hierarchy of timescales: daily (e.g., tidal inundation), monthly (e.g., tidal cycles), annual (e.g. seasonal precipitation) and tens to hundreds of years (e.g., sea level rise). Such rhythmic and dynamic conditions require the trees that grow in these intertidal habitats to have high levels of plasticity.

In the present review, we focus on recent insights into the ecophysiological processes that enable mangrove forests to maintain productivity under both saline and anoxic soil conditions, how their physiology is limited by temperature and how these physiological attributes may affect responses of mangrove forests to the complex environmental changes anticipated under future conditions. Enhancements in our understanding of the underlying physiological bases of salinity tolerance in mangroves is important to the development of both salinity tolerant crops and predictive models for management of the wide range of ecosystem services provided by mangrove forests under changing environmental and climatic scenarios (Barbier et al. 2011). However, global climate and atmospheric change do not affect salinity in isolation. Other key environmental factors, namely atmospheric CO_2 concentration, temperature and sea level are also changing with far reaching consequences for the structure, function and distribution of mangrove systems.

Water Uptake in Saline Soils

Mangrove tree species tolerate a wide range of soil salinity (Lugo and Snedaker 1974; Odum et al. 1982; Hutchings and Saenger 1987) (Table 1) and are highly adapted to salt concentrations in soils that exceed concentrations tolerated by most other plants (Ball 1988a). However, both low and high salinity can limit mangrove growth and productivity (Clough and Sim 1989; Lin and Sternberg 1992; Ball 2002). Saline habitats present physiological challenges for plants because their survival depends on the extraction of almost freshwater from highly saline soils. The low osmotic potentials of saline soil water make water acquisition and transport more difficult than in wet, non-saline soils, leading to high carbon costs of water uptake and transport. These costs are reflected in the typically high water-use efficiency of mangroves which tends to increase with increases in both the salt tolerance of the species and the salinity in which the plants are grown (Ball 1988a). However, such water use characteristics come at the expense of other functions. Table 2 summarizes the traits associated with salinity tolerance in mangroves and indicates some of the putative costs associated with salinity tolerance, including reduced survival in the shade (Ball 2002; Lopez Hoffman et al. 2007), reduced growth rates (Ball 1988a) or loss of mechanical strength (Santini et al. 2012).

Growth in saline environments necessitates adaptations to maintain the low tissue water potentials needed to extract water from highly saline soils, and to limit the loss of extracted water from leaves. To this end, mangrove species can exclude, accumulate, and excrete salts; none of these are salt tolerance strategies per se, although each can be related to water uptake and the requirement for water conservation (Ball 1988a). Mangroves as a broad group are halophytes with a wide range of salinity tolerance among species (Krauss and Ball 2013; Reef and Lovelock 2014) (Table 1). All mangrove species exclude the majority of salt ions during water absorption by the roots (up to 80–95 %; Scholander et al.1962; Scholander et al. 1968; Popp et al. 1993). Casparian bands and suberin lamellae provide barriers to apoplastic water flow through the root endodermis and are well developed close to the root tip (Lawton et al. 1980). Root traits vary among species. For example *Bruguiera* possesses a large root cap, high levels of phenolic deposits in cells and rapid development of vasculature to prevent salts from entering xylem vessels through this pathway (Lawton et al. 1981). In contrast *Avicennia marina* has a smaller root cap and vascular development is delayed, which may allow greater salt and water uptake (Fig. 1). Greater development of root apoplastic barriers among species reduces bypass flow, forcing water through the endodermis and enhancing efficient salt exclusion (Krishnamurthy et al. 2014). Indeed, concentration of salts within soils can pose a real dilemma for mangroves; recent stable isotope studies have shown that mangroves utilize less saline water sources when freshwater is available (Sternberg and Swart 1987; Ewe et al. 2007; Lambs et al. 2008; Wei et al. 2013). For example, in Florida *R. mangle* went from using 100 % shallow soil water in the wet season when that water was fresh to a mix of 55 % shallow soil water and 45 % deeper groundwater during the dry season when deeper

Table 1 Tolerance of some mangrove species to high salinity, high aridity, and low temperatures

Species	Relative tolerance				Occurrence in extreme climatic conditions				
	Salinity	Aridity	Low temperatures	High temperatures	Aridity index	Precipitation (mm/yr)	Extreme min temp (°C)	Monthly mean min temp (°C)	Monthly mean max temp (°C)
Acrostichum aureum	Mid	Mid	Mid	Low	Semi-arid	800–1200	0–5	8–12	34–37
Aegialitis annulato	High	High	Mid	Mid	Arid	<400	0–5	8–12	37–40
Aegiceras corniculatum	Mid	High	High	Mid	Arid	<400	−5 to 0	4–8	37–40
Avicennia germinons	High	High	High	Mid	Arid	<400	<−5	4–8	37–40
Avkennia marina	High	High	High	High	Arid	<400	<−5	4–8	>40
Bruguiera gymnorrtiiza	Mid	High	High	Mid	Arid	<400	−5 to 0	4–8	37–40
Bruguiera sexangula	Low	Mid	Low	Low	Semi-arid	>1200	0–5	12–16	34–37
Ceriops australis	High	High	Mid	Mid	Arid	<400	0–5	8–12	37–40
Ceriops decandra	Low	Low	Low	Mid	Dry sub-humid	800–1200	5–10	12–16	37–40
Ceriops tagal	Mid	High	Mid	Mid	Arid	<400	−5–0	8–12	37–40
Exccecona agallocha	Low	Mid	High	Mid	Semi-arid	400–800	−5 to 0	4–8	37–40
Heritiera littoralis	Mid	Mid	High	Mid	Semi-arid	400–800	−5 to 0	8–12	37–40

(continued)

Table 1 (continued)

Species	Relative tolerance				Occurrence in extreme climatic conditions				
	Salinity	Aridity	Low temperatures	High temperatures	Aridity index	Precipitation (mm/yr)	Extreme min temp (°C)	Monthly mean min temp (°C)	Monthly mean max temp (°C)
Kandelia candel	Mid	Low	Low	Low	Dry sub-humid	>1200	5–10	>16	34–37
Kandelia obovata	Low	Low	High	Low	Dry sub-humid	>1200	<–5	4–8	34–37
Laguncularia racemosa	Mid	High	Mid	Mid	Arid	<400	<–5	8–12	37–40
Lumnitzera littorea	High	Low	Low	Mid	Dry sub-humid	800–1200	0–5	12–16	37–40
Nypa fruticans	Low	Low	Low	Low	Dry sub-humid	>1200	0–5	12–16	34–37
Osbornia > a octodonata	Mid	Mid	Mid	Mid	Arid	<400	0–5	8–12	37–40
Rhizophora apiculata	Mid	Mid	Mid	Mid	Semi-arid	400–800	0–5	8–12	37–40
Rhizophora mangle	High	High	Mid	High	Arid	<400	<–5	8–12	37–40
Rhizophora mucronata	Low	High	Mid	Mid	Arid	<400	0–5	8–12	>40
Rhizophora stylosa	High	High	High	Mid	Arid	<400	–5 to 0	4–8	37–40

(continued)

Table 1 (continued)

Species	Relative tolerance				Occurrence in extreme climatic conditions				
	Salinity	Aridity	Low temperatures	High temperatures	Aridity index	Precipitation (mm/yr)	Extreme min temp (°C)	Monthly mean min temp (°C)	Monthly mean max temp (°C)
Sonneratia alba	Mid	Mid	Mid	Mid	Semi-arid	400–800	0–5	8–12	37–40
Sonneratia larceolata	Low	Low	Low	Low	Dry sub-humid	>1200	5–10	>16	34–37
Xylocarpus granatum	Low	Mid	Mid	Mid	Semi-arid	400–800	0–5	8–12	37–40
Xylocarpus moluccensis	Mid	Mid	Mid	Mid	Semi-arid	400–800	0–5	8–12	37–40

Species-specific relative tolerances categories were determined using information in Clough (1992) and Reef and Lovelock (2014) in combination with the climatic tolerance data presented here. The final four columns provide estimates of the extreme climatic conditions in which a species is present (highest aridity, lowest precipitation, lowest temperature). Species range data were obtained from Spalding et al. (2010). The aridity index data were obtained from Zomer et al. (2006). The precipitation and monthly mean, minimum and maximum temperature data were obtained from Hijmans et al. (2005). The extreme minimum temperature data were obtained from Maurer et al. (2009)

Table 2 Mangrove plant traits associated with salinity tolerance in mangrove tree species and the putative costs of the salt tolerance trait

Trait	Function	Putative costs	References
Suberized root cell walls, highly developed casparian strip	Ion exclusion	Reduced capacity for water uptake under fresh water conditions; salinization of soils	Lawton et al. (1981), Passioura et al. (1992)
High dependence on symplastic water uptake	Ion exclusion	Low rates of water uptake	Reef et al. (2012)
High salt concentrations in cell vacuoles	Maintenance of water potential in the vacuole	Metabolic/nutrient costs	Takemura et al. (2000)
High concentrations of osmotically compatible solutes	Maintenance of water potential and ion exclusion in the cytoplasm?	Metabolic/nutrient costs; reducing xylem water flow	Popp and Polanía (1989), Zimmermann et al. (1994), but see Becker et al. (1997)
Salt excretion	Ion balance; decreased VPD/reduced water loss	Metabolic	Reef and Lovelock (2014)
Low stomatal conductance	Reduced xylem tensions (reduced potential for cavitation)	Low rates of transpiration and photosynthetic carbon gain	Ball (1988a), Sobrado (2000), Clough and Sim (1989), Krauss and Allen (2003), Vandegehuchte et al. (2014)
Steep leaf angles	Thermal regulation, reduced water loss	Reduced light capture, reduced CO_2 diffusion (less mass transfer— less air flow)	Ball (1988a), Lovelock and Clough (1992)
Reductions in leaf size	Thermal regulation, reducing water loss	Reduced light capture, increased structural construction costs	Ball et al. (1988b)
Abaxial stomata	Reduced water loss	Reduced capacity for CO_2 fixation	Cheeseman (1994)
Leaf succulence	Thermal regulation and maintenance of ion balance	Increased resistance to CO_2 diffusion, increased structural construction costs	Saenger (1982), Camilleri and Ribi (1983), Wang et al. (2007)
Thickened leaf cuticles	Reduced water loss	Reduced light capture, increased structural construction costs	Saenger (1982), Wang et al. (2007), Naidoo et al. (2011)

(continued)

Table 2 (continued)

Trait	Function	Putative costs	References
Leaf pubescence (hairs)	Uptake of atmospheric water vapour, reduced water loss, thermal regulation	Metabolic, reduced light capture	Reef and Lovelock (2014)
CO_2 uptake of non-leaf tissues	Reduced whole plant water loss for carbon gain	Metabolic (nutrient demand imposed by chlorophyll and RUBISCO)	Reef and Lovelock (2014)
Small vessel size	Reduced potential for cavitation	Low rates of photosynthetic carbon gain	Ewers et al. (2004), Lovelock et al. (2006), Stuart et al. (2007)
Successive cambia (Xylem/Phloem/parenchyma bundles) in *Avicennia* and *Aegialitis*	Increased capacity to repair embolisms	Decreased mechanical strength; metabolic/nutrient costs	Carlquist (2007), Robert et al. (2011), Schmitz et al. (2008), Santini et al. (2012), Yáñez-Espinosa et al. (2004)
Root growth in patches of fresher water	Improved water balance	Biomass allocation below ground	Ewe et al. (2007), Wei et al. (2013), Sternberg and Swart (1987), Lambs et al. (2008)

groundwater had lower salinity than shallow soil water (Sternberg and Swart 1987; Ewe et al. 2007).

Reducing Water Loss Under Saline and Arid Conditions

Once water is transported to the leaves, mangroves are highly efficient in the use of water during photosynthesis (Farquhar et al. 1982; Sobrado 2000). Mangroves, which use a C_3 photosynthetic pathway, were as much as 35–56 % more efficient in water use than nearby tropical lowland tree species (Ball 1996) and can even surpass stand-level water use efficiency of co-occurring C_4 grasses in some settings (Krauss et al. 2014a). Photosynthetic water use efficiency (PWUE) is often reported as the ratio of leaf photosynthetic CO_2 assimilation rate to transpiration rate, while the intrinsic PWUE is calculated as the ratio of assimilation rate to stomatal conductance of water vapour. These values can be extremely high in mangroves, with intrinsic PWUE ranging up to 153–212 μmol CO_2/mol H_2O (Table 3) compared to 40–80 μmol CO_2/mol H_2O typical in tropical trees, and often increasing with

Bruguiera gymnorrhiza   *Avicennia marina*

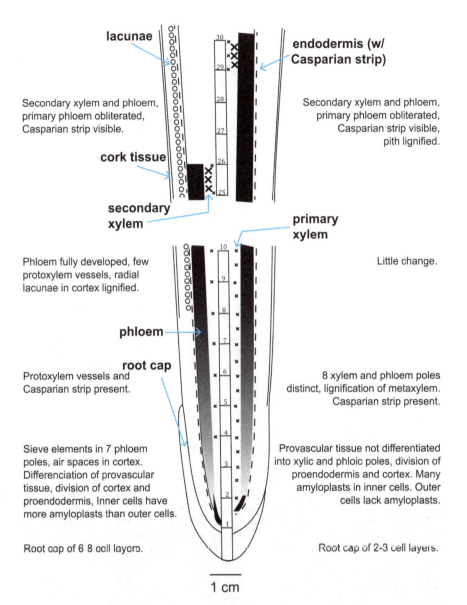

lacunae

endodermis (w/ Casparian strip)

Secondary xylem and phloem, primary phloem obliterated, Casparian strip visible.

Secondary xylem and phloem, primary phloem obliterated, Casparian strip visible, pith lignified.

cork tissue

secondary xylem

primary xylem

Phloem fully developed, few protoxylem vessels, radial lacunae in cortex lignified.

Little change.

phloem

root cap

Protoxylem vessels and Casparian strip present.

8 xylem and phloem poles distinct, lignification of metaxylem. Casparian strip present.

Sieve elements in 7 phloem poles, air spaces in cortex. Differenciation of provascular tissue, division of cortex and proendodermis, Inner cells have more amyloplasts than outer cells.

Provascular tissue not differentiated into xylic and phloic poles, division of proendodermis and cortex. Many amyloplasts in inner cells. Outer cells lack amyloplasts.

Root cap of 6 8 cell layers.

Root cap of 2-3 cell layers.

1 cm

Fig. 1 Comparative longitudinal sections of representative *Bruguiera gymnorrhiza* roots (*left*) and *Avicennia marina* roots (*right*) showing endodermal layers, vascular tissue, and root cap characteristics (after Lawton et al. 1981). Numbers along the pericycle represent approximate distance (cm) from root tip

Table 3 Intrinsic photosynthetic water use efficiencies (PWUE) measured in seedlings, saplings, and trees in mangroves globally (updated from Barr et al. 2009)

Species[a]	Life stage	PWUE, *intrinsic* (mmol CO_2/mol H_2O)	Source
RHST	Tree	80–135	Andrews and Muller (1985)
AVGE, COER	Tree	10–100	Smith et al. (1989)
AVGE, LARA, RHMA	Seedling	35–90	Pezeshki et al. (1990)
BRPA, BRGY	Tree	40–100	Cheeseman et al. (1991)
RHMA	Tree	40–55	Lin and Sternberg (1992)
RHMA, LARA	Tree	10–30	Martin and Loeschen (1993)
AVGE	Seedling	100–130	Naidoo and von Willert (1995)
AEAN, AECO	Seedling	60–100	Naidoo and von Willert (1995)
AVGE, LARA, RHMA, COER	Tree	35–40	Snedaker and Araujo (1998)
BRGY, AVMA	Tree	50–90	Naidoo et al. (1998)
AECO	Tree	43–120	Youssef and Saenger (1998b)
AVMA	Tree	60–80	Sobrado and Ball (1999)
RHMU, CETA	Tree	50–100	Theuri et al. (1999)
AVGE, LARA, RHMA	Tree	40–65	Sobrado (2000)
AVGE, LARA, RHMA	Seedling	69–153	Krauss et al. (2006)[b]
AVGE, LARA, RHMA	Sapling	51–95	Krauss et al. (2006)[b]
SOAP, SOCA, KACA, AVMA, EXAG	Tree	75–212	Chen et al. (2008)[b]
LARA, RHMA	Seedling	33–110	Cardona-Olarte et al. (2013)[b]

[a]RHST = *Rhizophora stylosa*, AVGE = *Avicennia germinans*, COER = *Conocarpus erectus*, LARA = *Laguncularia racemosa*, RHMA = *Rhizophora mangle*, BRPA = *Bruguiera parviflora*, BRGY = *Bruguiera gymnorrhiza*, AEAN = *Aegialitis annulata*, AECO = *Aegiceras corniculatum*, AVMA = *Avicennia marina*, RHMU = *Rhizophora mucronata*, CETA = *Ceriops tagal*, SOAP = *Sonneratia apetala*, SOCA = *Sonneratia caseolaris*, KACA = *Kandelia candel*, EXAG = *Excoecaria agallocha*
[b]Actual values obtained from author

incremental addition of salinity (Ball 1988a; Clough and Sim 1989; Smith et al. 1989; Krauss et al. 2008). In a broad survey of 19 different mangrove species in Australia and Papua New Guinea, Clough and Sim (1989) discovered that intrinsic PWUE did not drop below 49 μmol CO_2/mol H_2O for any species by site combination measured in the field, and ranged as high as 195 μmol CO_2/mol H_2O for

Avicennia marina where sites were highly saline. Changes in PWUE are also manifest at the stand level; eddy-flux-derived CO_2 uptake from a mangrove forest in south Florida decreased 5 % for each 10 parts per thousand (ppt) increment in salinity (Barr et al. 2013). High levels of photosynthetic efficiencies in water use of mangroves are a consequence of structurally imposed limitations on the rates of water supply to the leaves (Sobrado 2000; Lovelock et al. 2006; Hoa et al. 2009; Vandegehuchte et al. 2014), as well as tight regulation of water loss at the leaf level (Ball and Farquhar 1984; Clough and Sim 1989).

High levels of water use efficiency in mangrove tree species are associated with a range of traits, many of which lead to reductions in the leaf to air vapour pressure deficit (VPD). These traits include the presence of leaf pubescence (Reef and Lovelock 2014), the presence of salt on the leaf surface which occurs in salt secreting species and which may increase the humidity around the leaf (Reef and Lovelock 2014), and steep leaf orientations and small, thick leaves both of which affect the thermal balance of leaves (Ball 1988a). The characteristics that minimize VPD are likely to be particularly important in arid environments. Finally, photosynthetic CO_2 fixation in non-leaf tissues, which is acquired at lower water costs, could also be an important aspect of salinity tolerance. In *A. marina* re-fixation of respired CO_2 by corticular photosynthesis contributed up to 5 % of the CO_2fixation by the plant (Schmitz et al. 2012). CO_2 uptake by roots, although not yet studied in mangroves, has been shown to be significant in other submerged and wetland plants (Raven et al. 1988; Brix 1990; Rich et al. 2008).

Implications of Physiological and Structural Adaptations for Function of the Whole Forest

Much ecophysiological research on mangroves has been directed to leaf-level processes in seedlings, saplings, and occasionally, trees. Saenger (2002) reviews this literature and concludes that with the combination of ecophysiological strategies (e.g., high leaf-level PWUE) and adaptations for living in saline settings (e.g., salt exclusion at the roots, low stomatal conductance), mangrove forests are likely to be very conservative in water use. Ironically, quantifying the absolute rates of water use in mangroves has not been the central theme of many research programs; however, the available data on sap flux indicate that mangrove trees use water at rates over 3 times less than other forest trees per unit size (Fig. 2a). Individual tree water use ranged from 0.4–64.1 L H_2O/day in mangroves (Hirano et al. 1996; Muller et al. 2009; Krauss et al. 2007, 2014a; Lambs and Saenger 2011) compared with 116 ± 16 (SE) L H_2O/day from trees in other forest types (Wullschleger et al. 1998). Missing from this analysis are empirical data from large mangrove trees (>55 cm dbh).

Indeed, scaling ecophysiological processes from leaf to stand in mangroves provides insight into ecosystem CO_2 and H_2O fluxes. Lugo et al. (1975) found that two mangrove forests in Rookery Bay, Florida, USA took up 4.83 and 2.74 g

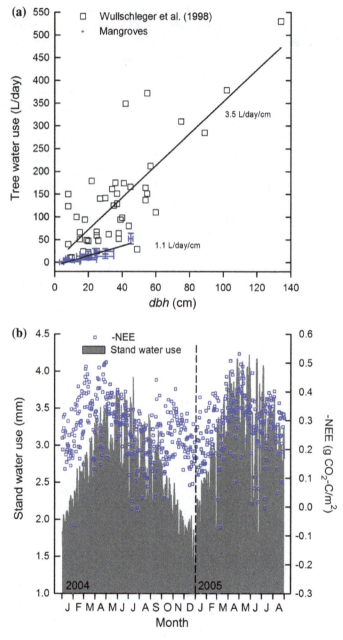

Fig. 2 **a** Water use (L/day) versus diameter at breast height (dbh) for individual trees from non-mangrove ecosystems (Wullschleger et al. 1998) versus data currently available from mangrove ecosystems. Bars presented for mangroves represent the absolute range of dbh versus water use values from specific studies (Krauss et al. 2007, 2014a, b; Muller et al. 2009; Lambs and Saenger 2011). **b** Stand water use (mm; Krauss et al. 2015) versus net ecosystem exchange (-NEE) of atmospheric CO_2 (g CO_2-C/m^2; Barr et al. 2010) from a mangrove forest along the Shark River, Everglades National Park, Florida, USA

C/m^2/day through net ecosystem exchange (-NEE) while using 2.57 and 1.57 mm H_2O/day, respectively. Of that transpiration, 95–97 % was associated with the canopy. Estimation of -NEE of carbon using eddy covariance along the Shark River in Everglades National Park, Florida was remarkably similar to Lugo et al.'s estimates, ranging from 2–5 g C/m^2/day (Barr et al. 2010). This was despite the fact that trees were nearly 8 m taller and forests had approximately 15 m^2/ha greater basal area along the Shark River than in Rookery Bay. During the same period of time, estimated water use of the dominant canopy ranged to 4.22 mm H_2O/day for Shark River mangroves, averaging 2.5 mm H_2O/day (Krauss et al. 2015; Fig. 2b). Thus, considering that the mangroves along the Shark River registered among the highest rates of carbon uptake among 49 forest types in North America (Amiro et al. 2010), it is truly remarkable that canopy water use was efficient enough in mangroves to represent only 63 % of regional rainfall and 66 % of ET (Krauss et al. 2015). Inherent to this are month-to-month fluctuations in annual rates of canopy-level PWUE that require additional study (Fig. 2b).

Elevated CO_2 Effects on Water Use

High levels of water use efficiency in mangroves and the increased PWUE with increasing salinity leads to the expectation that there could be considerable gains in productivity of mangroves with increasing levels of CO_2 in the atmosphere as stomatal limitations to CO_2 uptake are ameliorated. The effects of elevated CO_2 on plant performance have not been studied in mangroves as extensively as in other forest habitats. Only a handful of experimental studies have been conducted on the response of mangrove seedlings to elevated CO_2. Due to the difficulties posed by the intertidal habitat, Free-Air Concentration Enrichment of CO_2 (FACE) experiments are yet to be conducted in mangroves and thus we have no experimental data for the effects of elevated CO_2 on mature trees. Despite the scarcity of data, it is becoming apparent that while elevated CO_2 has a significant effect on PWUE in mangroves, the expected alleviation of salinity stress and subsequent improvement in performance at high salinity does not occur to the extent initially anticipated. The increase in mangrove seedling growth rates observed in response to elevated CO_2 ranged from a 12 to a 47 % increase in growth relative to that under ambient CO_2 concentrations (Table 4), which is overlapping with the mean and range recorded for well watered tropical tree seedlings of different species experiencing similar elevated CO_2 conditions (Cernusak et al. 2011; Krauss et al. 2014b). Studies that have incorporated a salinity treatment in elevated CO_2 experiments in mangroves conclude that at supra-optimal salinity conditions, elevated CO_2 does not significantly improve seedling growth, despite a significant improvement to PWUE (Ball et al. 1997; Reef et al. 2014) and that a fertilization effect is only observed within the low to mid salinity range. This is likely due to the fact that salinity constrains tree growth in a manner other than water stress (e.g. ion toxicity).

Table 4 The effects of elevated CO_2 on mangrove seedling growth, photosynthetic water use efficiency (PWUE) and specific leaf area (SLA) presented as the % change relative to values measured at ambient CO_2 concentrations

Species	Ambient (CO_2) ppm	Elevated $[CO_2]$ ppm	% increase in growth	% increase in WUE	% increase in SLA	Length of treatment (days)	Reference
Rhizophora mangle	350	700	47	1.72	NA	408	Farnsworth et al. (1996)
Rhizophora apiculata	340	700	31	15.2	−3	98	Ball et al. (1997)
Rhizophora stylosa	340	700	21	33.8	7.8	98	Ball (1997)
Avicennia germinans	365	720	17	NA	−10	540	McKee (2008)
Avicennia germinans	400	800	22	107	−1.1	132	Reef et al. (2014)
Avicennia germinans	280	800	12	218	−10	132	Reef et al. (2014)

In studies that included other factors (salinity, nutrient or humidity treatments) the CO_2 response was averaged over treatments

Mangroves respond to elevated CO_2 by reducing stomatal conductance and by producing leaves with lower stomatal densities (Farnsworth et al. 1996), but overall productivity of forests may increase. The observed reduction in transpiration rate on a leaf scale can have an effect on forest water use. However, elevated CO_2 can also lead to higher leaf area to total biomass ratios. In mangroves, increases in leaf area ratios with increasing $[CO_2]$ were observed in most studies (Farnsworth et al. 1996; Ball et al. 1997; Reef et al. 2014). Furthermore, there appears to be a lowering of the light compensation point (LCP) of photosynthesis. In *A. germinans* grown under elevated CO_2 the LCP reduced from 52.3 ± 1.36 μmol m^{-2} s^{-1} at (CO_2) of 280 ppm to 18.1 ± 9.6 μmol m^{-2} s^{-1} at $[CO_2]$ of 800 ppm, p = 0.03; R. Reef unpublished data) suggesting that a positive carbon balance in leaves can be maintained under shaded conditions at elevated CO_2. Declines in specific leaf area (SLA) were also observed in response to elevated CO_2 (Table 4) and in a time series study of herbarium specimens, SLA declined as CO_2 has increased over time in *A. marina* (but not for *R. stylosa*, Reef and Lovelock 2014). These responses to elevated CO_2 could result in increases in leaf level productivity and leaf area index (LAI) at the stand level with rising CO_2 levels. Increases in LAI at the stand level could offset the reduction in water use by individual leaves and result in enhanced production, but no overall change to transpiration on larger scales.

An important parameter that can influence the response of mangroves to elevated CO_2 in the field is nutrient availability. Under low nutrient conditions, similar to those measured in many scrub mangrove forests, seedling biomass did not increase in response to elevated CO_2 (McKee and Rooth 2008). However, when given higher nutrient concentrations, at levels similar to those measured in soils along creek banks, elevated CO_2 resulted in a significant 23 % increase in total biomass. Development of nutrient limitations to growth could also be the reason for a down-regulation of photosynthesis and reduced growth rates under elevated CO_2 conditions over time in *R. mangle* grown in pots (Farnsworth et al. 1996). Nutrient distributions are not uniform in mangrove forests due in part to their intertidal nature. Most nutrients are delivered by the tides, thus creating elevation driven gradients in nutrient availability. Furthermore, differences in nutrient inputs from rivers and other landward sources, soil salinity, inundation frequency and faunal activity affect the availability of nutrients (Reef et al. 2010, 2014). The response of mangroves to elevated CO_2 will thus likely be dependent on their position in the forest and local nutrient conditions.

Finally, in order to understand the effects of elevated CO_2 on forest structure and distribution, competitive interactions must be taken into account. Elevated CO_2 can improve mangrove salinity tolerance within the range of salinities already suitable for growth (Reef et al. 2014). This shift in the fundamental niche could translate into a range shift depending on competitive outcomes and could be a contributing factor in the observed mangrove encroachment into saltmarsh habitats observed at many locations (Saintilan et al. 2014). In the only experiment to investigate competitive interactions between mangroves and saltmarshes under elevated CO_2, McKee and Rooth (2008) found that elevated CO_2 does not improve competitive outcomes for the mangrove *Avicennia germinans* when grown in mixed culture with the saltmarsh species *Spartina alterniflora*. Although *S. alterniflora* is a very fast growing species and may not be representative of many saltmarsh species that have slower growth rates, this study highlights the difficulties in predicting the future of mangrove distributions in response to rising CO_2 concentrations.

Adaptations to Inundated Soils

Tolerance of periods of inundation by tidal water, flooding and storm surges are essential for mangrove tree species survival, and differences in species tolerances to flooding influences their distributions relative to changing hydroperiods with sea-level rise. Sea level has been relatively stable for the last five thousand years, but accelerating rates of sea level rise and associated geomorphological adjustments of the coast (Woodroffe 1990) are likely to result in vegetation transitions that are linked to inundation tolerance of species. A range of traits are linked to inundation tolerance (Table 5). Mangroves possess a number of elaborate, aerial root structures including prop roots (*Rhizophora*), pneumatophores (*Avicennia, Sonneratia*), knee roots (*Bruguiera*), and cable roots (*Xylocarpus, Heritiera*), prompting much early speculation into their role in aerating sub-soil roots and soils. Still, some mangrove species lack aerial roots (e.g., *Excoecaria*). Prominent on many of the aerial root structures and stems are lenticels, or gas exchange pores positioned above the soil surface. Air diffuses through these pores and via abundant aerenchyma tissues to belowground root structures facilitating aeration of roots embedded in oxygen-free

Table 5 Plant traits associated with tolerance of inundation in mangrove tree species and their putative costs

Trait	Function	Putative cost	Reference
Aerial roots with lenticels	Gas transport—air into below ground roots	Allocation of biomass to non-photosynthetic tissues; respiration	Scholander et al. (1955), Youssef and Saenger (1996), Skelton and Allaway (1996)
Rapid seedling root extention	Anchorage of seedlings	Reduced allocation to aboveground biomass	Delgado et al. (2001), Balke et al. (2011)
Aerenchyma	Gas transport and storage; oxidation of phytotoxic substances	Allocation of biomass to non-photosynthetic tissues; loss of mechanical strength	Scholander et al. (1955), Youssef and Saenger (1996, 1998a, b), Skelton and Allaway (1996), Purnobasuki and Suzuki (2004)
Suberized root cells	Limit oxygen leakage into soils; maintain availability of some essential elements close to the root surface	Enhance phytotoxic substances	Thibodeau and Nickerson (1986), Youssef and Saenger (1996), Reef et al. (2010)

soils (Scholander et al. 1955; Skelton and Allaway 1996; Allaway et al. 2001). The structure of mangrove roots facilitates gas exchange (McKee and Mendelssohn 1987; Youssef and Saenger 1996) which in addition to supporting respiration results in the oxidation of phytotoxic substances within roots (e.g., Fe^{2+}, H_2S) (Armstrong et al. 1992; Youssef and Saenger 1998a), although leakage of O_2 into the rhizosphere may affect the availability of some essential nutrients, particularly phosphorus which are more available under reduced conditions, and some micro-bial processes which are also favored under low oxygen concentrations (e.g. nitrogen fixation) (Reef et al. 2010) (Table 4). Early rapid root growth and investment in roots in seedlings are also important for the establishment of seedlings on exposed tidal flats and thus influences recruitment and forest expansion (Delgado et al. 2001; Balke et al. 2011).

There are differences among species in the capacity to withstand inundation. Experiments focused on gas exchange in root systems indicate that *A. germinans* and *Laguncularia racemosa* seedlings suffered a decrease in root oxygen concentrations when exposed to experimental hypoxia, while *Rhizophora mangle* did not (McKee 1996). McKee (1996) discovered that differences among species in response to anoxia were attributed to oxygenation of the roots through diffusive O_2 fluxes from the shoot, lower root respiration rates in *R. mangle* than *A. germinans* or *L. racemosa*, and less O_2 leakage from *R. mangle* roots to the surrounding soils. Oxygen tends to leak from *Avicennia* roots to a much greater degree than

Rhizophora roots (Thibodeau and Nickerson 1986). In a multi-species comparison, *Avicennia marina* and *Acanthus ilicifolius* had the highest concentrations of aerenchyma air space and the lowest diffusional resistance for O_2 to soil among eight mangrove species tested in Hong Kong (Pi et al. 2009) suggesting high levels of variation among Indo-Pacific species in their capacity to transport oxygen to and out of roots. Additionally, greater root porosity was found in pneumatophores than other root types of *Sonneratia alba* from Okinawa (Purnobasuki and Suzuki 2004) indicating variation in oxygen transport within root systems among wide-ranging species. Rates of oxygen leakage from roots not only vary among species but also with the strength of diffusion gradients between the root and soil (Sorrell and Armstrong 1994). O_2 leakage from the apical tips of *Kandelia candel* roots was higher than from the main root walls (Chiu and Chou 1993). Variation in oxygen leakage was attributed to the structure of the root surface, which is compacted and lignified in *Rhizophora* and *Aegiceras* and thinner with only 3–4 exodermal layers in *Avicennia* and *Bruguiera* (Youssef and Saenger 1996). The differences in mangrove species in the structure, growth and physiology of roots, including their ability to transport and retain O_2 within their roots, are likely to lead to differences in species responses to changing inundation regimes and associated hydrological change with sea level rise. Although many other factors are also likely to influence the distribution and composition of forests, including the space in the landscape for landward expansion, human and natural modifications of the coast (Doyle et al. 2010; Traill et al. 2011), underlying differences in species inundation tolerance are likely to be important.

Adaptations to Temperature Thresholds

Variations in temperature affect many processes in mangrove forests ranging from the fundamental metabolic processes of photosynthesis and respiration (e.g., Andrews and Muller 1985; Lovelock 2008) to carbon cycling (Alongi 2009) and reproductive success (Duke 1990). Temperature regimes greatly influence mangrove forest composition and structure with extreme temperature events playing an especially important role in some locations. Increasing global temperatures are likely to result in changes to growth and distribution patterns of mangrove forests on the edge of their ranges which are currently limited by low temperatures and in some locations aridity.

The effects of low temperature on mangrove physiology (Davis 1940; Stuart et al. 2007; Krauss et al. 2008; Ross et al. 2009) and distributions (e.g., West 1977; Sherrod and McMillan 1985; Woodroffe and Grindrod 1991; de Lange and de Lange 1994; Duke et al. 1998; Saenger 2002) have been widely studied, with recent studies focused on mangrove recruitment into warm-temperate salt marsh habitats (e.g., McKee et al. 2012; Osland et al. 2013; Cavanaugh et al. 2014; Saintilan et al. 2014). Mangrove species differ in sensitivity to low temperatures (Table 1), but none can survive the minimum temperatures that occur in cold-temperate climatic zones.

In general, mangrove forest biomass, structural development, and species richness are higher in wet tropical climatic zones. In colder climatic zones (e.g., subtropical or warm-temperate), low temperature stress typically produces mangrove trees that are short in stature with a shrub-like architecture (Woodroffe 1985; Osland et al. 2014a, b). Physiological stress due to low temperatures can be separated into chilling and freezing stress (Kozlowski and Pallardy 1997; Larcher 2003). Whereas chilling stress occurs at leaf temperatures above freezing (i.e., without ice formation), freezing stress occurs at leaf temperatures below freezing when intra- or extra-cellular ice formation occurs. The physiological effects and symptoms of freezing/chilling stress in mangrove trees include reduced metabolic rates, altered membrane structure and permeability (Markley et al. 1982), disrupted water and nutrient transport (Stuart et al. 2007), partial or complete loss of aboveground biomass (e.g., Osland et al. 2014a, b), reduced reproductive success (Duke 1990) and, in extreme cases, mortality (Ross et al. 2009).

Intra- and inter-specific differences in mangrove sensitivity to low temperature stress greatly influence the structure and composition of mangrove forests. There are many examples of differential species and life stage responses to low temperature stress (e.g., Lugo and Patterson-Zucca 1977; Lonard and Judd 1991; Olmsted et al. 1993; Ross et al. 2009; Chen et al. 2010; Pickens and Hester 2011). Climatic origin greatly influences intraspecific responses to chilling stress with mangrove individuals from colder climates typically being better adapted and more resistant to low temperatures (Markley et al. 1982; Sherrod and McMillan 1985). Species sensitivity to low temperature stress can be gauged using species distribution data in combination with multi-decadal climate data (e.g., Table 1; Quisthoudt et al. 2012; Osland et al. 2013). For example, whereas some mangrove species are highly sensitive to chilling stress and are only found in tropical climates (e.g., *Bruguiera sexangula, Sonneratia lanceolata*), other species have adaptations that enable them to be resistant to higher levels of chilling or freezing stress (e.g., *A. germinans, A. marina, Kandelia obovata, Aegiceras corniculatum*). Vulnerability to low-temperature induced xylem cavitation is especially high for mangrove trees due to the low xylem water potentials required for water transport in highly saline intertidal environments (Stuart et al. 2007). In a comparison of the most poleward mangrove species in Australia and the United States, Stuart et al. (2007) highlighted the effects of xylem embolism and show that species adapted to climates with colder mean annual minimum temperatures have smaller vessel diameters which enable them to better avoid embolism; however, narrow vessels also constrain water transport and productivity, possibly limiting mangrove forest structural development in these poleward locations. In addition to narrow vessels, variation in membrane properties also likely enables mangrove individuals and species from colder climates to maintain membrane fluidity during exposure to low temperatures (Markley et al. 1982).

On some continents, the frequency and intensity of extreme winter events greatly affect mangroves. For example, near the poleward mangrove limit in China and the southeastern United States, mean winter temperatures are not as cold as mean winter temperatures found near the poleward limit of mangroves in Australia or New Zealand. However, extreme minimum temperature events are more intense in

China and the southeastern United States which can result in sudden leaf loss, xylem embolism, branch and stem reductions, and, in the most extreme cases, tree mortality (Davis 1940; Lugo and Patterson-Zucca 1977; West 1977; Lonard and Judd 1991; Everitt et al. 1996). In these areas, the spatial extent of mangrove forests expands and contracts in response to the frequency and intensity of extreme winter events (West 1977; Sherrod and McMillan 1985; Stevens et al. 2006; Giri et al. 2011). In North America, *Avicennia germinans* is a species that is especially adapted to and resistant to extreme winter events. In parts of northern coastal Florida, Louisiana, and the northern coast of Texas, *A. germinans* individuals often lose a large portion of their aboveground biomass due to freeze events; however, *A. germinans* is capable of vigorous resprouting from the base of stems after freeze-damage due to the presence of epicormic buds (Lugo and Patterson-Zucca 1977; Tomlinson 1986; Osland et al. 2014a, b). Using data in Table 1 we graphically show the breadth of tolerance of mangrove species in the biogeographic provinces of the Atlantic and Eastern Pacific Ocean region (Fig. 3a) and the Indian and Western Pacific Ocean region (Fig. 3b) to temperature extremes and to aridity and salinity. Those species that are currently documented as expanding in their range (e.g., *A. germinans* and *A. marina*) have the broadest tolerance to low temperatures and to other environmental factors, while species that have restricted distributions (e.g., the palm *Nypa fruticans*) are less tolerant of low and high temperatures, high salinity and aridity.

In contrast, the effects of high temperatures on mangrove trees have not been as extensively considered. Early research indicates that photosynthesis in tropical species of Rhizophoraceae is depressed as leaf temperatures exceed 34 °C (e.g., Andrews and Muller 1985; Cheeseman et al. 1991). Most of this effect is associated with the strong stomatal closure required to minimize rates of water loss that can increase dramatically if leaf temperatures become higher than air temperatures or if high air temperatures are accompanied by low humidity (i.e. increasing leaf-to-air VPD). However, more research is required to understand how mangroves, particularly those in warm tropical climates that may already be close to their thermal limits, will respond to the projected increases in global temperature of at least 2 °C in the coming century. Clark (2004) suggested that productivity of tropical

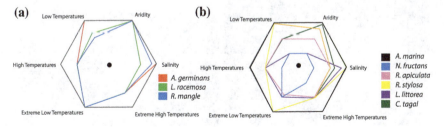

Fig. 3 Spider plots of the overlapping tolerances of mangrove species to high and low temperature and high and low temperature extremes, salinity and aridity (based on data from Table 1). **a** Common species from the Atlantic-East Pacific biogeographic region; **b** common species from the Indo-West Pacific region

rainforests could decline with increasing temperature due to increases in respiration and declining photosynthetic rates (Slot and Winter this volume). As yet there has been no assessment of this hypothesis in mangrove forests.

Plasticity of Traits Gives Rise to Different Capacity to Adjust to Climate Change

Plasticity of traits is a key feature of plants that allows them to acclimate to changing conditions, including climate (Jump and Penuelas 2005). Some mangrove species occur over broad ranges of environmental conditions and often assume different forms in different habitats, e.g., scrub and tall fringing forms of *R. mangle* in the Caribbean (Feller 1995; Medina et al. 2010) or for *A. marina* in New Zealand and southern Australia (Lovelock 2008; Martin et al. 2010) (Fig. 4), displaying high levels of both physiological and morphological plasticity (Lovelock et al. 2006). In fact, plasticity in PWUE among and within mangrove species is a primary reason why water use characteristics in mangroves have not been fully assessed; water use can depend strongly on site or experimental conditions limiting blanket assessments

Fig. 4 Examples of high levels of morphological plasticity within mangrove tree species. The *upper left* and *right* photos are of tall (15 m), forests fringing water ways and short (<2 m), scrub forests of *Rhizophora mangle*, respectively, from Belize. The *lower left* and *right* photos are of tall (5 m), fringing forests and short (<1 m), scrub forests, respectively, in New Zealand

Table 6 Variation in plasticity in growth (height, stem extension, biomass or relative growth rate, RGR) for mangrove species grown in comparative experiments

Species	Parameter	Treatment	Coefficient of variation	Reference
Ceriops australis	Stem height	Salinity and competition	0.21 ± 0.01 (14)a	Smith (1988)
Ceriops tagal			0.16 ± 0.02 (10)b	
Avicennia marina	Biomass	Salinity	0.20 (5)	Clough (1984)
Rhizophora stylosa			0.17 (5)	
Sonneratia alba	Stem height	Salinity	0.14 ± 0.03 (6)a	Ball and Pidsley (1995)
S. lanceolate			0.10 ± 0.02 (4)a	
R. stylosa	RGR	Salinity and humidity	0.14 ± 0.04 (6)b	Ball et al. (1998)
R. apiculate			0.13 ± 0.02 (8)b	
A. germinans	RGR	Salinity and nutrient availability	0.12 ± 0.02 (14)a	Ewe S., Lovelock C.E. unpublished data
Laguncularia racemosa			0.07 ± 0.01 (13)b	
R. mangle			0.24 ± 0.04 (15)c	
Aegiceras corniculatum	Stem height	Inundation	0.042 ± 0.002 (8)a	He et al. (2007)
A. marina			0.083 ± 0.007 (8)c	
Bruguiera gymnorrhiza			0.078 ± 0.005 (8) bc	
R. stylosa			0.067 ± 0.006 (8)b	
A. germinans	RGR	Oxygen around roots	0.11 (2)	McKee (1996)
L. racemose			0.19 (2)	
R. mangle			0.24 (2)	
A. germinans	Stem extension	Nutrient availability	0.35 ± 0.07 (3)a	Lovelock and Feller (2003)
L. racemose			0.36 ± 0.04 (3)a	
A. marina	Stem extension	Nutrient availability	0.22 ± 0.05 (3)a	Lovelock CE, Feller IC unpublished
C. australis			0.48 ± 0.10 (3)a	
R. stylosa	Stem extension	Nutrient availability	0.21 ± 0.03 (3)a	Lovelock CE unplublished
C. australis			0.50 ± 0.06 (3)b	

(continued)

Table 6 (continued)

Species	Parameter	Treatment	Coefficient of variation	Reference
Lumnitzera racemosa			0.19 ± 0.03 (3)a	
R. stylosa	Stem extension	Nutrient availability	0.15 ± 0.06 (3)a	Lovelock CE unpublished
C. australis			0.49 ± 0.13 (3)b	

Plasticity is represented by the coefficient of variation (based on a mean of means over treatments). The number of mean values used (N) is in parentheses. The most salt tolerant species is underlined. Where sufficient data were available differences among species were tested and significant differences at $P < 0.05$ are indicated with different letters after the mean

of water conservation in mangroves (see e.g., Becker et al. 1997 vs. Zimmermann et al. 1994; Krauss et al. 2015). In a comparison of the levels of plasticity in different traits over variation in fertility in *R. mangle*, whole plant architectural traits (e.g. leaf area index, shoot extension and hydraulic properties of stems) had much higher levels of plasticity than leaf level traits (e.g. photosynthesis, specific leaf area) (Lovelock et al. 2006). As the most plastic traits are often the ones that determine overall plant fitness (Agrawal 2001, Poorter and Lambers 1986; Callaway et al. 2003) this suggests that plasticity in growth rates, canopy development and hydraulic function are potentially the most important traits for successful dominance of mangrove habitats. A high level of plasticity in belowground growth is also likely, but as yet remains relatively unexplored (e.g., McKee 1996; Casteneda Moya et al. 2011; Lang'at et al. 2013). In Table 6 we contrast plasticity in growth (stem extension, relative growth rate or biomass expressed as a mean coefficient of variation) for a range of species where contrasts have been made over variation in treatments. Plasticity in growth varies significantly among mangrove species and tends to be greater in response to variation in nutrient availability than to other treatments. Species with the highest plasticity tend to be the most salt tolerant and widely distributed and may be favored with global climate change. The fitness cost of high levels of plasticity is difficult to determine, but could be associated with being inferior competitors; although with few competing species in mangrove forests (compared to tropical rainforests) there may be few disadvantages to high levels of plasticity, particularly in the biogeographic province of the Atlantic and East Pacific Ocean region where species diversity is particularly low.

Conclusions: Change in Distribution and Productivity of Mangrove Forests

As a group, mangrove species possess many physiological adaptations and life history characteristics that could enable them to adapt positively to future climate change. However, the complex interactions between climatic drivers are only just

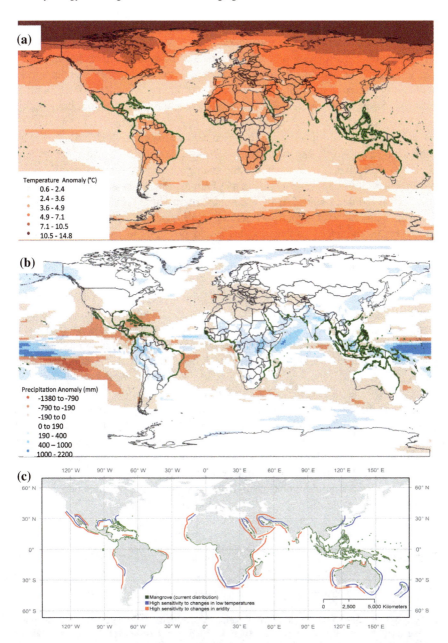

Fig. 5 The global mangrove distribution (*green*) overlain on: **a** predicted change in mean temperature (IPCC 2007, A1FI scenario); **b** predicted changes in annual precipitation (IPCC 2007, A1FI scenario); and **c** regions where mangrove distribution is sensitive to changes in aridity (*red*) or low temperatures (*blue*)

starting to become clear, and mangrove ecosystem responses to climate change will be highly context dependent (e.g., Saintilan et al. 2009; McKee et al. 2012). In the past, the global distribution and spatial extent of mangrove forests has expanded and contracted in response to changes in sea level, temperature and freshwater availability (Sherrod and McMillan 1985; Woodroffe and Grindrod 1991; Saintilan et al. 2014). In the future, climate change is expected to greatly alter the distribution, composition and ecological properties of mangroves forests and their adjacent ecosystems (i.e., salt marsh, salt flat, seagrass ecosystems) (Fig. 5). On some continents (e.g., North America, Northwest Asia, Australia), warmer winter temperatures will likely lead to poleward mangrove forest range expansion and development at the expense of salt marsh habitat (Osland et al. 2013; Cavanaugh et al. 2014; Saintilan et al. 2014). The southeastern United States is an area where the ecological effects of mangrove migration are expected to be especially large due to the large amount of salt marsh that could be replaced by mangrove forests (Osland et al. 2013). Moreover, increases in sea level, freshwater availability, elevated CO_2 and human influences including nutrient enrichment will also affect the productivity, composition and distribution of mangrove forests. In response to sea level rise, mangrove forests are expected to migrate landward where migration corridors exist (Doyle et al. 2010; Traill et al. 2011), although landward coastal wetland migration will be obstructed in some areas by natural and anthropogenic barriers (e.g., sea walls, hydrologic barriers). Species differences in inundation tolerance are likely to influence the composition of forests as sea level rise accelerates. In arid and semi-arid climatic zones, the abundance and composition of mangrove forests and other coastal wetland ecosystems will be greatly influenced by changes in freshwater availability (Smith and Duke 1987; Bucher and Saenger 1994; Saintilan et al. 2009; Semeniuk 2013; Osland et al. 2014a, b). Whereas increased aridity and/or reductions in freshwater input will likely result in reduced mangrove coverage and diversity, the converse is also true; in some areas, increased freshwater inputs, increases in humidity and increases in CO_2 could result in changes in communities, higher mangrove coverage, structural development and productivity (Reef and Lovelock 2014; Osland et al. 2014a, b).

References

Agrawal AA (2001) Phenotypic plasticity in the interactions and evolution of species. Science 294:321–326

Allaway WG, Curran M, Hollington LM, Ricketts MC, Skelton NJ (2001) Gas space and oxygen exchange in roots of *Avicennia marina* (Forssk.) Vierh. var. *australasica* (Walp.) Moldenke ex N. C. Duke, the grey mangrove. Wetlands Ecol Manage 9:211–218

Alongi DM (2009) The energetics of mangrove forests. Springer, Dordrecht

Amiro BD, Barr AG, Barr JG, Black TA, Bracho R, Brown M, Chen J, Clark KL, Davis KJ, Desai AR, Dore S, Engel V, Fuentes JD, Goldstein AH, Goulden ML, Kolb TE, Lavigne MB, Law BE, Margolis HA, Martin T, McCaughey JH, Misson L, Montes-Helu M, Noormets A, Randerson JT, Starr G, Xiao J (2010) Ecosystem carbon dioxide fluxes after disturbance in forests of North America. J Geophys Res 115:G00K02

Andrews TJ, Muller GJ (1985) Photosynthetic gas exchange of the mangrove, *Rhizophora stylosa* Griff. in its natural environment. Oecologia 65:449–455

Armstrong J, Armstrong W, Beckett PM (1992) *Phragmites australis*: veturi-and humidity-induced pressure flows enhance rhizome aeration and rhizosphere oxidation. New Phytol 120:197–207

Balke T, Bouma TJ, Horstman EM, Webb EL, Erftemeijer PLA, Herman PMJ (2011) Windows of opportunity: thresholds to mangrove seedling establishment on tidal flats. Mar Ecol Prog Ser 440:1–9

Ball MC (1988a) Ecophysiology of mangroves. Trees Struc Funct 2:129–142

Ball MC (1996) Comparative ecophysiology of mangrove forest and tropical lowland moist rainforest. In: Mulkey SS, Chazdon RL, Smith AP (eds) Tropical forest plant ecophysiology. Chapman and Hall, New York, pp 461–496

Ball MC (2002) Interactive effects of salinity and irradiance on growth: implications for mangrove forest structure along salinity gradients. Trees 16:126–139

Ball MC, Farquhar GD (1984) Photosynthetic and stomatal responses of two mangrove species, *Aegiceras corniculatum* and *Avicennia marina*, to long term salinity and humidity conditions. Plant Physiol 74:1–6

Ball MC, Pidsley SM (1995). Growth responses to salinity in relation to distribution of two mangrove species, *Sonneratia alba* and *S. lanceolata*, in northern Australia. Funct Ecol 77–85

Ball MC, Cochrane MJ, Rawson HM (1997) Growth and water use of the mangroves *Rhizophora apiculata* and *R. stylosa* in response to salinity and humidity under ambient and elevated concentrations of atmospheric CO_2. Plant Cell Environ 20:1158–1166

Ball MC, Cowan IR, Farquhar GD (1988b) Maintenance of leaf temperature and the optimisation of carbon gain in relation to water loss in a tropical mangrove forest. Australian Journal of Plant Physiology 15:263–276

Barbier EB, Hacker SD, Kennedy C, Koch EW, Stier AC, Silliman BR (2011) The value of estuarine and coastal ecosystem services. Ecol Monogr 81:169–193

Barr JG, Fuentes JD, Engel V, Zieman JC (2009) Physiological responses of red mangroves to the climate in the Florida Everglades. J Geophys Res 114:G02008

Barr JG, Engel V, Fuentes JD, Zieman JC, O'Halloran TL, Smith TJ III, Anderson GH (2010) Controls on mangrove forest-atmosphere carbon dioxide exchanges in western Everglades National Park. J Geophys Res 115:G02020

Barr JG, Engel V, Fuentes JD, Fuller DO, Kwon H (2013) Modeling light use efficiency in a subtropical mangrove forest equipped with CO_2 eddy covariance. Biogeoscience 10:2145–2158

Becker P, Asmat A, Mohamad J, Moksin M, Tyree MT (1997) Sap flow rates of mangrove trees are not unusually low. Trees Struc Funct 11:432–435

Brix H (1990) Uptake and photosynthetic utilization of sediment-derived carbon by *Phragmites australis* (Cav.) Trin. ex Steudel. Aquat Bot 38:377–389

Bucher D, Saenger P (1994) A classification of tropical and subtropical Australian estuaries. Aquat Cons: Mar Freshw Ecosyst 4:1–19

Callaway JC, Sullivan G, Zedler JB (2003) Species-rich plantings increase biomass and nitrogen accumulation in a wetland restoration experiment. Ecol App 13:1626–1639

Camilleri JC, Ribi G (1983) Leaf thickness of mangroves (*Rhizophora mangle*) growing in different salinities. Biotropica 15:139–141

Cardona-Olarte P, Krauss KW, Twilley RR (2013) Leaf gas exchange and nutrient use efficiency help explain the distribution of two Neotropical mangroves under contrasting flooding and salinity. Int J For Res 524–625

Carlquist S (2007) Successive cambia revisited: ontogeny, histology, diversity, and functional significance. The Journal of the Torrey Botanical Society 134:301–32

Castañeda-Moya E, Twilley RR, Rivera-Monroy VH, Marx BD, Coronado-Molina C, Ewe SM (2011) Patterns of root dynamics in mangrove forests along environmental gradients in the Florida Coastal Everglades, USA. Ecosystems 14:1178–1195

Cavanaugh KC, Kellner JR, Forde AJ, Gruner DS, Parker JD, Rodriguez W, Feller IC (2014) Poleward expansion of mangroves is a threshold response to decreased frequency of extreme cold events. Proc Nat Acad Sci (US) 111:723–727

Cernusak LA, Winter K, Turner BL (2011) Transpiration modulates phosphorus acquisition in tropical tree seedlings. Tree Physiol 31:878–885

Cheeseman JM (1994) Depressions of photosynthesis in mangrove canopies. In: Baker NR, Bowyer JR (eds) Photoinhibition of photosynthesis. Bios Scientific Publishers, Oxford, pp 377–389

Cheeseman JM, Clough BF, Carter DR, Lovelock CE, Eong OJ, Sim RG (1991) The analysis of photosynthetic performance in leaves under field conditions: a case study using *Bruguiera* mangroves. Photosyn Res 29:11–22

Chen L, Tam NFY, Huang J, Zeng X, Meng X, Zhong C, Wong Y, Lin G (2008) Comparison of ecophysiological characteristics between introduced and indigenous mangrove species in China. Estuar Coast Shelf Sci 79:644–652

Chen L, Wang W, Zhang Y, Huang L, Zhao C, Yang S, Yang Z, Chen Y, Xu H, Zhong C, Su B, Fang B, Chen N, Zeng C, Lin G (2010) Damage to mangroves from extreme cold in early 2008 in southern China. J Plant Ecol (Chinese Version) 34:186–194

Chiu C, Chou C (1993) Oxidation in the rhizosphere of mangrove *Kandelia candel* seedlings. Soil Sci Plant Nutr 39:725–731

Clark DA (2004) Tropical forests and global warming: slowing it down or speeding it up? Front Ecol Environ 2:73–80

Clough BF (1984) Growth and salt balance of the mangroves *Avicennia marina* (Forsk.) Vierh. and *Rhizophora stylosa* Griff. in relation to salinity. Funct Plant Biol 11:419–430

Clough BF (1992) Primary productivity and growth of mangrove forests. Tropical mangrove ecosystems. American Geophysical Union, Washington DC, pp 225–249

Clough BF, Sim RG (1989) Changes in gas exchange characteristics and water use efficiency of mangroves in response to salinity and vapour pressure deficit. Oecologia 79:38–44

Davis JH (1940) The ecology and geologic role of mangroves in Florida, vol 32. Carnegie Institue of Washington Publications. Papers Tortugas Lab, pp 303–412

de Lange WP, de Lange PJ (1994) An appraisal of factors controlling the latitudinal distribution of mangrove (*Avicennia marina* var. *resinifera*) in New Zealand. J Coast Res 10:539–548

Delgado P, Hensel PF, Jiménez JA, Day JW (2001) The importance of propagule establishment and physical factors in mangrove distributional patterns in a Costa Rican estuary. Aquat Bot 71:157–178

Doyle TW, Krauss KW, Conner WH, From AS (2010) Predicting the retreat and migration of tidal forests along the northern Gulf of Mexico under sea-level rise. For Ecol Manage 259:770–777

Duke NC (1990) Phenological trends with latitude in the mangrove tree *Avicennia marina*. J Ecol 78:113–133

Duke NC, Ball MC, Ellison JC (1998) Factors influencing biodiversity and distributional gradients in mangroves. Glob Ecol Biogeogr Lett 7:27–47

Everitt JH, Judd FW, Escobar DE, Davis MR (1996) Integration of remote sensing and spatial information technologies for mapping black mangrove on the Texas gulf coast. J Coast Res 12:64–69

Ewe SML, Sternberg LdSL, Childers DL (2007) Seasonal plant water uptake patterns in the saline Southeast Everglades ecotone. Oecologia 152:607–616

Ewers FW, Lopez-Portillo J, Angeles G, Fisher JB (2004) Hydraulic conductivity and embolism in the mangrove tree *Laguncularia racemosa*. Tree Physiol 24:1057–1062

Farnsworth EJ, Ellison AM, Gong WK (1996) Elevated CO_2 alters anatomy, physiology, growth, and reproduction of red mangrove (*Rhizophora mangle* L.). Oecologia 108:599–609

Farquhar GD, Ball MC, von Caemmerer S, Roksandic Z (1982) Effect of salinity and humidity on ^{13}C values of halophytes—evidence for diffusional isotopic fractionation determined by the ratio of intercellular/atmospheric CO_2 under different environmental conditions. Oecologia 52:121–137

Feller IC (1995) Effects of nutrient enrichment on growth and herbivory of dwarf red mangrove (Rhizophora mangle). Ecological monographs 65:477–505

Giri C, Long J, Tieszen L (2011) Mapping and monitoring Louisiana's mangroves in the aftermath of the 2010 Gulf of Mexico Oil Spill. J Coast Res 27:1059–1064

Hao G-Y, Jones TJ, Luton C, Zhang Y-J, Manzane E, Scholz FG, Bucci SJ, Cao KF, Goldstein G (2009) Hydraulic redistribution in dwarf *Rhizophora mangle* trees driven by interstitial soil water salinity gradients: impacts on hydraulic architecture and gas exchange. Tree Physiol 29:697–705

He B, Lai T, Fan H, Wang W, Zheng H (2007) Comparison of flooding-tolerance in four mangrove species in a diurnal tidal zone in the Beibu Gulf. Estuar Coast Shelf Sci 74:254–262

Hijmans RJ, Cameron SE, Parra JL, Jones PG, Jarvis A (2005) Very high resolution interpolated climate surfaces for global land areas. Int J Clim 25:1965–1978

Hirano T, Monji N, Hamotani K, Jintana V, Yabuki K (1996) Transpirational characteristics of mangrove species in southern Thailand. Environ Controls Biol 34:285–293

Hutchings P, Saenger P (1987) Ecology of mangroves. Queensland University Press

IPCC (2007) Intergovernmental panel on climate change. Climate change 2007: Synthesis report

Jump A, Penuelas J (2005) Running to stand still: adaptation and the response of plants to rapid climate change. Ecol Lett 8:1010–1020

Kozlowski TT, Pallardy SG (1997) Growth control in woody plants. Academic Press, San Diego

Krauss KW, Allen JA (2003) Influences of salinity and shade on seedling photosynthesis and growth of two mangrove species, *Rhizophora mangle* and *Bruguiera sexangula*, introduced to Hawaii. Aquat Bot 77:311–324

Krauss KW, Ball MC (2013) On the halophytic nature of mangroves. Trees Struc Funct 27:7–11

Krauss KW, Twilley RR, Doyle TW, Gardiner ES (2006) Leaf gas exchange characteristics of three Neotropical mangrove species in response to varying hydroperiod. Tree Physiol 26:959–968

Krauss KW, Young PJ, Chambers JL, Doyle TW, Twilley RR (2007) Sap flow characteristics of neotropical mangroves in flooded and drained soils. Tree Physiol 27:775–783

Krauss KW, Lovelock CE, McKee KL, López-Hoffman L, Ewe SML, Sousa WP (2008) Environmental drivers in mangrove establishment and early development: a review. Aquat Bot 89:105–127

Krauss KW, McKee KL, Hester MW (2014a) Water use characteristics of black mangrove (*Avicennia germinans*) communities along an ecotone with marsh at a northern geographical limit. Ecohydrol 7:354–365

Krauss KW, McKee KL, Lovelock CE, Cahoon DR, Saintilan N, Reef R, Chen L (2014b) How mangrove forests adjust to rising sea level. New Phytol 202:19–34

Krauss KW, Barr JG, Engel V, Fuentes JD, Wang H (2015) Approximations of stand water use versus evapotranspiration from three mangrove forests in southwest Florida, USA. Agric For Meteorol 213: 291–303

Krishnamurthy P, Jyothi-Prakash PA, Qin L, He J, Lin Q, Loh C, Kumar PP (2014) Role of root hydrophobic barriers in salt exclusion of a mangrove plant *Avicennia officinalis*. Plant Cell Environ 37:1656–1671

Lambs L, Saenger A (2011) Sap flow measurements of *Ceriops tagal* and *Rhizophora mucronata* mangrove trees by deuterium tracing and lysimetry. Rapid Commun Mass Spectrom 25:2741–2748

Lambs L, Muller E, Fromard F (2008) Mangrove trees growing in a very saline condition but not using seawater. Rapid Commun Mass Spectrom 22:2835–2843

Lang'at JKS, Kirui BK, Skov MW, Kairo JG, Mencuccini M, Huxham M (2013). Species mixing boosts root yield in mangrove trees. Oecologia 172: 271–278

Larcher W (2003) Physiological plant ecology: ecophysiology and stress physiology of functional groups. Springer, Berlin

Lawton JR, Todd A, Naidoo DK (1981) Preliminary investigations into the structure of the roots of the mangroves, *Avicennia marina* and *Bruguiera gymnorrhiza*, in relation to ion uptake. New Phytol 88:713–722

Lin G, Sternberg LDSL (1992) Comparative study of water uptake and photosynthetic gas exchange between scrub and fringe red mangroves, *Rhizophora mangle* L. Oecologia 90:399–403

Lonard RI, Judd FW (1991) Comparison of the effects of the severe freezes of 1983 and 1989 on native woody plants in the Lower Rio Grande Valley, Texas. Southwestern Nat 36:213–217

Lopez-Hoffman L, Anten NP, Martinez-Ramos M, Ackerly DD (2007) Salinity and light interactively affect neotropical mangrove seedlings at the leaf and whole plant levels. Oecologia 150:545–556

Lovelock CE (2008) Soil respiration and belowground carbon allocation in mangrove forests. Ecosystems 11:342–354

Lovelock CE, Feller IC (2003) Photosynthetic performance and resource utilization of two mangrove species coexisting in a hypersaline scrub forest. Oecologia 134:455–462

Lovelock CE, Ball MC, Choat B, Engelbrecht BM, Holbrook NM, Feller IC (2006) Linking physiological processes with mangrove forest structure: phosphorus deficiency limits canopy development, hydraulic conductivity and photosynthetic carbon gain in dwarf *Rhizophora mangle*. Plant Cell Environ 29:793–802

Lugo AE, Patterson-Zucca C (1977) The impact of low temperature stress on mangrove structure and growth. Trop Ecol 18:149–161

Lugo AE, Snedaker SC (1974) The ecology of mangroves. Annu Rev Ecol Syst 5:39–64

Lugo AE, Evink G, Brinson MM, Broce A, Snedaker SC (1975) Diurnal rates of photosynthesis, respiration, and transpiration in mangrove forests of south Florida. In: Golley FB, Medina E (eds) Tropical ecological systems. Springer, New York, pp 335–350

Markley JL, McMillan C, Thompson GA Jr (1982) Latitidinal differentiation in response to chilling temperatures among populations of three mangroves, *Avicennia germinans*, *Laguncularia racemosa*, and *Rhizophora mangle*, from the western tropical Atlantic and Pacific Panama. Can J Bot 60:2704–2715

Martin CE, Loeschen VS (1993) Photosynthesis in the mangrove species *Rhizophora mangle* L.: no evidence for CAM-cycling. Photosynthetica 28:391–400

Martin KC, Bruhn D, Lovelock CE, Feller IC, Evans JR, Ball MC (2010) Nitrogen fertilization enhances water-use efficiency in a saline environment. Plant Cell Environ 33:344–357

Maurer EP, Adam JC, Wood AW (2009) Climate model based consensus on the hydrologic impacts of climate change to the Rio Lempa basin of Central America. Hydrol Earth Syst Sci 13:183–194

McKee KL (1996) Growth and physiological responses of Neotropical mangrove seedlings to root zone hypoxia. Tree Physiol 16:883–889

McKee KL, Mendelssohn IA (1987) Root metabolism in the black mangrove (*Avicennia germinans* (L.)): response to hypoxia. Environ Exp Bot 27:147–156

McKee K, Rooth JE (2008) Where temperate meets tropical: multi-factorial effects of elevated CO_2, nitrogen enrichment, and competition on a mangrove-salt marsh community. Glob Change Biol 14:971–984

McKee K, Rogers K, Saintilan N (2012) Response of salt marsh and mangrove wetlands to changes in atmospheric CO_2, climate, and sea level. In: Middleton BA (ed) Global change and the function and distribution of wetlands: global change ecology and wetlands. Springer, Dordrecht, pp 63–96

Medina E, Cuevas E, Lugo AE (2010) Nutrient relations of dwarf *Rhizophora mangle* L. mangroves on peat in eastern Puerto Rico. Plant Ecol 207:13–24

Muller E, Lambs L, Fromard F (2009) Variations in water use by a mature mangrove of *Avicennia germinans*, French Guiana. Ann Sci For 66:803

Naidoo G, von Willert DJ (1995) Diurnal gas exchange characteristics and water use efficiency of three salt-secreting mangroves at low and high salinities. Hydrobiology 295:13–22

Naidoo G, Rogalla H, von Willert DJ (1998) Field measurements of gas exchange in *Avicennia marina* and *Bruguiera gymnorrhiza*. Mangroves Salt Marshes 2:99–107

Naidoo G, Hiralal O, Naidoo Y (2011) Hypersalinity effects on leaf ultrastructure and physiology in the mangrove *Avicennia marina*. Flora-Morphol, Distrib Funct Ecol Plants 206:814–820

Odum WE, McIvor CC, Smith III TJ (1982) The ecology of the mangroves of south Florida: a community profile. Virginia University Charlottsville Department of Environmental Sciences

Olmsted I, Dunevitz H, Platt WJ (1993) Effects of freezes on tropical trees in Everglades National Park Florida, USA. Trop Ecol 34:17–34

Osland MJ, Enwright N, Day RH, Doyle TW (2013) Winter climate change and coastal wetland foundation species: salt marshes versus mangrove forests in the southeastern US. Glob Change Biol 19:1482–1494

Osland MJ, Day RH, Larriviere JC, From AS (2014a) Aboveground allometric models for freeze-affected black mangroves (*Avicennia germinans*): equations for a climate sensitive mangrove-marsh ecotone. PLoS ONE 9(6):e99604

Osland MJ, Enwright N, Stagg CL (2014b) Freshwater availability and coastal wetland foundation species: ecological transitions along a rainfall gradient. Ecol 95:2789–2802

Passioura JB, Ball MC, Knight JH (1992) Mangroves may salinize the soil and in so doing limit their transpiration rate. Funct Ecol 6:476–481

Pezeshki SR, DeLaune RD, Patrick WH Jr (1990) Differential response of select mangroves to soil flooding and salinity: gas exchange and biomass partitioning. Can J For Res 20:869–874

Pi N, Tam NFY, Wong MH (2009) Root anatomy and spatial pattern or radial oxygen loss of eight true mangrove species. Aquat Bot 90:222–230

Pickens CN, Hester MW (2011) Temperature tolerance of early life history stages of black mangrove *Avicennia germinans*: implications for range expansion. Estuaries Coasts 34:824–830

Poorter H, Lambers H (1986) Growth and competitive ability of a highly plastic and a marginally plastic genotype of *Plantago major* in a fluctuating environment. Physiol Planta 67:217–222

Popp M, Polanía J (1989) Compatible solutes in different organs of mangrove trees. Ann Sci For 46:842s–844s

Popp M, Polanía J, Weiper M (1993) Physiological adaptations to different salinity levels in mangrove. In: Leith H, Al Masoom A (eds) Towards the rationale use of high salinity tolerant plants, vol 1. Kluwer Academic Publishers, Utrecht, The Netherlands, pp 217–224

Purnobasuki H, Suzuki M (2004) Aerenchyma formation and porosity in root of a mangrove plant, *Sonneratia alba* (Lythraceae). J Plant Res 117:465–472

Quisthoudt K, Schmitz N, Randin CF, Dahdouh-Guebas F, Robert EMR, Koedam N (2012) Temperature variation among mangrove latitudinal range limits worldwide. Trees Struct Funct 26:1919–1931

Raven JA, Handley LL, Macfarlane JJ (1988) The role of CO_2 uptake by roots and CAM in acquisition of inorganic C by plants of the isoetid life-form: a review, with new data on Eriocaulon decangulare L. New Phytol 108:125–148

Reef R, Lovelock CE (2014) Regulation of water balance in mangroves. Ann Bot. doi:10.1093/aob/mcu174

Reef R, Feller IC, Lovelock CE (2010) Nutrition of mangroves. Tree Physiol 30:1148–1160

Reef R, Schmitz N, Rogers BA, Ball MC, Lovelock CE (2012) Differential responses of the mangrove Avicennia marina to salinity and abscisic acid. Funct Plant Biol 39:1038–1046

Reef R, Winter K, Morales J, Adame MF, Reef DL, Lovelock CE (2014) The effect of atmospheric carbon dioxide concentrations on the performance of the mangrove *Avicennia germinans* over a range of salinities. Physiol Planta. doi:10.1111/ppl.12289

Rich SM, Ludwig M, Colmer TD (2008) Photosynthesis in aquatic adventitious roots of the halophytic stem-succulent *Tecticornia pergranulata* (formerly *Halosarcia pergranulata*). Plant Cell Environ 31:1007–1016

Ricklefs RE, Schwarzbach AE, Renner SS (2006) Rate of lineage origin explains the diversity anomaly in the world's mangrove vegetation. Am Nat 168:805–810

Robert EMR, Schmitz N, Boeren I, Driessens T, Herremans K, De Mey J, Van de Casteele E, Beeckman H, Koedam N (2011) Successive cambia: a developmental oddity or an adaptive structure? PloS one 6(1):e16558

Ross MS, Ruiz PL, Sah JP, Hanan EJ (2009) Chilling damage in a changing climate in coastal landscapes of the subtropical zone: a case study from south Florida. Glob Change Biol 15:1817–1832

Saenger P (1982) Morphological, anatomical and reproductive adaptations of Australian mangroves. In: Clough BF (ed) Mangrove ecosystems in Australia. Australian National University Press, Canberra, pp 153–191

Saenger P (2002) Mangrove ecology, silviculture and conservation. Kluwer Academic Publishers, Dordrecht

Saintilan N, Rogers K, McKee K (2009) Saltmarsh-Mangrove interactions in Australasia and the Americas. Chapter 31. In: Perillo GME, Wolanski E, Cahoon DR, Brinson MM (eds) Coastal wetlands; an integrated ecosystems approach. Elsevier, Atlanta, pp 855–883

Saintilan N, Wilson NC, Rogers K, Rajkaran A, Krauss KW (2014) Mangrove expansion and salt marsh decline at mangrove poleward limits. Glob Chan Biol 20:147–157

Santini N, Schmitz N, Lovelock C (2012) Variation in wood density and anatomy in a widespread mangrove species. Trees 26:1555–1563

Schmitz N, Robert EMR, Verheyden A, Kairo JG, Beeckman H, Koedam N (2008) A patchy growth via successive and simultaneous cambia: key to success of the most widespread mangrove species Avicennia marina? Ann Bot 101:49–58

Schmitz N, Egerton JJG, Lovelock CE, Ball MC (2012) Light-dependent maintenance of hydraulic function in mangrove branches: do xylary chloroplasts play a role in embolism repair? New Phytol 195:40–46

Scholander PF (1968) How mangroves desalinate seawater. Physiol Plant 21:251–261

Scholander PF, van Dam L, Scholander SI (1955) Gas exchange in the roots of mangroves. Am J Bot 42:92–98

Scholander PF, Hammel HT, Hemmingsen EA, Garey W (1962) Salt balance in mangroves. Plant Physiol 37:722–729

Semeniuk V (2013) Predicted response of coastal wetlands to climate changes: a Western Australian model. Hydrobiologia 708:23–43

Sherrod CL, McMillan C (1985) The distributional history and ecology of mangrove vegetation along the northern Gulf of Mexico coastal region. Contrib Mar Sci 28:129–140

Skelton NJ, Allaway WG (1996) Oxygen and pressure changes measured in situ during flooding in roots of the grey mangrove Avicennia marina (Forssk.) Vierh. Aquat Bot 54:165–175

Smith TJ (1988) Differential distribution between subspecies of the mangrove Ceriops tagal: competitive interactions along a salinity gradient. Aquat Bot 32:79–89

Smith TJ III, Duke NC (1987) Physical determinants of inter-estuary variation in mangrove species richness around the tropical coastline of Australia. J Biogeogr 14:9–19

Smith JAC, Popp M, Lüttge U, Cram WJ, Diaz M, Griffiths H, Lee HSJ, Medina E, Schäfer C, Stimmel K-H, Thonke B (1989) Ecophysiology of xerophytic and halophytic vegetation of a coastal alluvial plain in Northern Venezuela. VI. Water relations and gas exchange of mangroves. New Phytol 111:293–307

Snedaker SC, Araújo RJ (1998) Stomatal conductance and gas exchange in four species of Caribbean mangroves exposed to ambient and increased CO₂. Mar Freshw Res 49:325–327

Sobrado MA (2000) Relation of water transport to leaf gas exchange properties in three mangrove species. Trees Struct Funct 14:258–262

Sobrado MA, Ball MC (1999) Light use in relation to carbon gain in the mangrove, Avicennia marina, under hypersaline conditions. Aust J Plant Physiol 26:245–251

Sorrell BK, Armstrong W (1994) On the difficulties of measuring oxygen release by root systems in wetland plants. J Ecol 82:177–183

Spalding MD, Kainuma M, Collins L (2010) World Mangrove Atlas. Earthscan, with International Society for Mangrove Ecosystems, Food and Agriculture Organization of the United Nations, The Nature Conservancy, UNEP World Conservation Monitoring Centre, United Nations Scientific and Cultural Organisation, United Nations University, London. 319 pp

Sternberg LSL, Swart PK (1987) Utilization of freshwater and ocean water by coastal plants of southern Florida. Ecology 68:1898–1905

Stevens PW, Fox SL, Montague CL (2006) The interplay between mangroves and saltmarshes at the transition between temperate and subtropical climate in Florida. Wetland Ecol Manage 14:435–444

Stuart SA, Choat B, Martin KC, Holbrook NM, Ball MC (2007) The role of freezing in setting the latitudinal limits of mangrove forests. New Phytol 173:576–583

Takemura T, Hanagata N, Sugihara K, Baba S, Karube I, Dubinsky Z (2000) Physiological and biochemical responses to salt stress in the mangrove, *Bruguiera gymnorrhiza*. Aquat Bot 68:15–28

Theuri MM, Kinyamario JI, van Speybroeck D (1999) Photosynthesis and related physiological processes in two mangrove species, *Rhizophora mucronata* and *Ceriops tagal*, at Gazi Bay, Kenya. Afr J Ecol 37:180–193

Thibodeau FR, Nickerson NH (1986) Differential oxidation of mangrove substrate by *Avicennia germinans* and *Rhizophora mangle*. Am J Bot 73:512–516

Tomlinson PB (1986) The botany of mangroves. Cambridge University Press, New York

Traill LW, Perhans K, Lovelock CE, Prohaska A, McFallan S, Rhodes JR, Wilson KA (2011) Managing for change: wetland transitions under sea-level rise and outcomes for threatened species. Divers Distrib 17:1225–1233

Vandegehuchte MW, Guyot A, Hubau M, De Groote SRE, De Baerdemaeker NJF, Hayes M, Welti N, Lovelock CE, Lockington DA, Steppe K (2014) Long-term versus daily stem diameter variation in co-occurring mangrove species: environmental versus ecophysiological drivers. Agric For Meteor 192–193:51–58

Wang W, Xiao Y, Chen L, Lin P (2007) Leaf anatomical responses to periodical waterlogging in simulated semidiurnal tides in mangrove *Bruguiera gymnorrhiza* seedlings. Aquat Bot 86:223–228

Wei L, Lockington DA, Poh SC, Gasparon M, Lovelock CE (2013) Water use patterns of estuarine vegetation in a tidal creek system. Oecologia 172:485–494

West RC (1977) Tidal salt-marshes and mangal formations of Middle and South America. In: Chapman VJ (Ed) Ecosystems of the world 1. Wet coastal ecosystems, pp. 193–211. Elsevier Scientific Publishing Co., Amsterdam. 428 pp

Woodroffe CD (1985) Studies of a mangrove basin, Tuff Crater, New Zealand: I. Mangrove biomass and production of detritus. Estuarine, Coastal and Shelf Science 20:265-280

Woodroffe CD, Grindrod J (1991) Mangrove biogeography: the role of Quaternary environmental and sea-level change. J Biogeogr 5:479–492

Wullschleger SD, Meinzer FC, Vertessy RA (1998) A review of whole-plant water use studies in trees. Tree Physiol 18:499–512

Yáñez-Espinosa L, Terrazas T, López-Mata L, Valdez-Hernández J (2004) Wood variation in Laguncularia racemosa and its effect on fibre quality. Wood Sci Technol 38:217–226

Youssef T, Saenger P (1996) Anatomical adaptive strategies to flooding and rhizosphere oxidation in mangrove seedlings. Aust J Bot 44:297–313

Youssef T, Saenger P (1998a) Photosynthetic gas exchange and accumulation of phytotoxins in mangrove seedlings in response to soil physico-chemical characteristics associated with waterlogging. Tree Physiol 18:317–324

Youssef T, Saenger P (1998b) Photosynthetic gas exchange and water use in tropical and subtropical populations of the mangrove *Aegiceras corniculatum*. Mar Freshw Res 49:329–334

Zimmermann U, Zhu JJ, Meinzer FC, Goldstein G, Schneider H, Zimmermann G, Benkert R, Thürmer F, Melcher P, Webb D, Haase A (1994) High molecular weight osmotic compounds in the xylem sap of mangroves: implications for long-distance water transport. Botanica Acta 107:218–229

Zomer RJ, Trabucco A, van Straaten O, Bossio DA (2006) Carbon, land and water: hydrologic dimensions of climate change mitigation through afforestation and reforestation. IWMI Research Report 101. International Water Management Institute, Colombo

Functional Diversity in Tropical High Elevation Giant Rosettes

Fermín Rada

Abstract Strong daily temperature variations, seasonal soil water availability and high air evaporative demands play an essential role in adaptive responses of tropical high elevation mountain plants. Giant rosettes are a perfect example of successful adaptations to these conditions, representing an important life-form of high elevation tropical mountains in the Andes, Hawaii and Africa, a well-known case of convergent evolution. Adaptive radiation resulted in a substantially large number of giant rosette species in the 'paramos', a local name given for tropical alpine Andean vegetation. Plant functional responses: plant water relations, gas exchange characteristics and freezing resistance in giant rosettes are described in order to understand their responses to extreme environmental conditions characteristic of high elevation tropical habitats. Giant rosettes have a large capacitance (water-storage pith) and strong stomatal control to cope with periods of water deficit, resulting in the maintenance of high leaf water potentials on a daily and seasonal basis. Maximum net CO_2 assimilation rates are variable among species (3–10 μmol m^{-2} s^{-1}), all showing photosynthetic decreases from wet to dry seasons. Giant rosettes rely on permanent supercooling of the leaves together with insulating structures protecting stems and apical buds to cope with freezing damage. Even though the general aspect and plant morphology of giant rosettes is similar across all high elevation tropical regions, responses to similar selective pressures resulted in different physiological characteristics in freezing resistance mechanisms, e.g. tolerance versus avoidance, and thermal balance of the rosette. Giant rosette responses under changing global environments are also discussed. The emphasis in the description of physiological and morphological characteristics will be on South American giant rosettes due to the large number of studies and the large number of species occurring in this region.

Keywords Gas exchange · Leaf pubescence · Stem capacitance · Supercooling · Water relations

F. Rada (✉)
Instituto de Ciencias Ambientales y Ecológicas de los Andes Tropicales (ICAE)
Facultad de Ciencias, Universidad de Los Andes, Mérida, Venezuela
e-mail: frada@ula.ve

© Springer International Publishing Switzerland 2016
G. Goldstein and L.S. Santiago (eds.), *Tropical Tree Physiology*,
Tree Physiology 6, DOI 10.1007/978-3-319-27422-5_8

181

Introduction

Tropical high mountains are characterized by strong daily temperature variations. Freezing temperatures frequently occur on any night of the year, while high incoming radiation during the day results in relatively high daytime temperatures at plant and soil surfaces. Hedberg (1964) characterized these environments as "summer" during the day and "winter" at night. Daily temperature variations far exceed seasonal ones (Troll 1968; Sarmiento 1986; Rundel 1994).

The páramos are tropical high Andean ecosystems dominated by tussock grasses, sclerophyllous shrubs and giant rosettes. In general, this vegetation type is found above the natural limit of the continuous forest line of the cloud forest (Monasterio 1979; Smith and Young 1987) from approximately 3000 m up to the permanent snow line at about 4600 m. This ecosystem occurs in higher altitudinal regions of Venezuela, Colombia and Ecuador, with minor extensions towards Panamá and Costa Rica to the north and in Northern Perú to the south between latitudes 11° N and 8° S (Monasterio 1980; Luteyn 1992). The páramo has been considered the most floristically diverse and holds the largest number of endemic species of any mountain ecosystem worldwide (Luteyn 1999).

Intense incoming solar radiation has an impact on the plant and/or soil surface thermal balance, increasing risks of rapid heating after nighttime frosts. Precipitation regimes have a wide range of patterns depending on elevation and slope orientation. In the particular case of the Venezuelan Andes, marked seasonal variations with a dry season between December and March occurs. Annual precipitation ranges between 650 mm for the driest to 1800 mm in the more humid Páramos (Sarmiento 1986). In summary, the combination of different environmental conditions of the high tropical Andes determine thermal and water stresses to which plants are constantly exposed (Fig. 1). Higher daytime incoming radiation and night reirradiation during the dry season result in a greater minimum-maximum temperature range generating significant day and nighttime temperature stresses. Increased cloud cover and fog presence during the wet season dampen day-night temperature extremes reducing thermal stresses; however infrequent clear skies may produce similar dry season temperature effects on some occasions. Wet season night temperatures remain relatively high compared to the dry season. In spite of different seasonal temperature patterns, tropical alpine environments present small oscillations in seasonal mean temperatures. In terms of water characteristics, increased cloud cover, fog and precipitation determine favorable soil water conditions and the low evaporative demand during the wet season drastically contrasts with the dry season. Seasonal water stress may be a key factor in plant adaptations to tropical alpine climates.

Even though the physiognomy of the vegetation of the tropical Andean mountains may vary between regions, dominant plant species belong to one of the following growth forms: acaulescent rosettes, cushion plants, tussock grasses, sclerophyllous shrubs or giant caulescent rosettes (Smith and Young 1987; Azócar and Rada 2006). The giant rosettes may be considered the result of particular

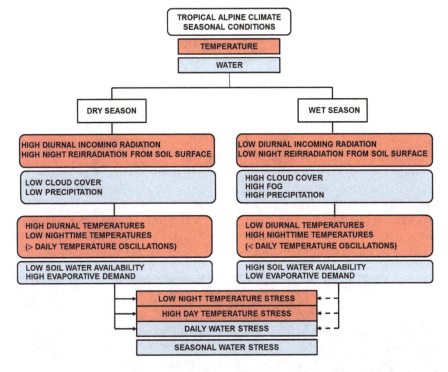

Fig. 1 Environmental conditions of the high tropical Andes, which determine persistent thermal and water stresses for plants. Temperature and water stresses constantly affect plants during the dry season (→) and occasionally occur during the wet season (◀--) in the presence of unusually clear skies. See text for a detailed description

adaptations to cold tropical climates. Different genera, consisting of a single stem supporting a rosette have been described for many tropical high mountain regions: *Coespeletia* and *Espeletia* in the Northern Andes, *Senecio*, *Dendrosenecio* and *Lobelia* for Equatorial Africa, *Argyroxiphium* in Hawaii, among others (Cuatrecasas et al. 1986; Monasterio and Vuillemier 1986; Smith and Young 1987; Bussmann 2006). Giant rosettes have been considered a classical example of convergent evolution which independently evolved in the different mountain ranges. However, Antonelli (2009) suggested that giant rosette 'Lobelioids' from Eastern Africa and the Hawaiian Islands are closely related and derived from a single woody African ancestor. Consequently, parallel evolution among tropical alpine environments should be further tested.

The *Espeletias* are considered a typical example of rapid diversification and adaptive radiation in a relatively recent ecosystem (Monasterio and Sarmiento 1991), a common event in 'island' environments (Carlquist 1974) such as the paramos: high mountain peaks surrounded by lower elevation forests. Berry et al. (1988) suggest a major role of interspecific hybridization in the adaptive radiation of these dominant giant rosettes. The Venezuelan Andes display the greatest diversity and number of giant rossette species and is the most probable site of origin of this functional group (Cuatrecasas et al. 1986). There are approximately 130 species within the former genus *Espeletia* of which 63 are endemic to the Merida Cordillera in Venezuela. Cuatrecasas (1976), using morphological attributes, suggested a classification of the genus *Espeletia* into seven genera included within a new subtribe, *Espeletiinae*: *Carramboa*, *Coespeletia*, *Espeletia*, *Espeletiopsis*, *Libanothamnus*, *Ruilopezia* and *Tamania*. However, Rauscher (2002) through phylogenetic studies showed that the *Espeletiinae* fall within the subtribe *Melampodiinae*, and coin instead the term 'Espeletia complex' to group these genera within the larger subtribe *Melampodiinae*.

The most representative genera of the Northern Andes, *Coespeletia* and *Espeletia*, are characterized by a single stem, which varies in size across species from a few cm to up to 3 or 4 m, and is covered by a compact, thick layer of dead attached leaves, termed marcescent leaves, which insulate the stem and water storage pith from subzero temperatures (Fig. 2a). This stem supports a rosette

Fig. 2 **a** *Coespeletia timotensis* at Páramo de Piedras Blancas (4200 m), **b** *Espeletia schultzii* rosette at Páramo de Mucubají (3550 m) and **c** *Ruilopezia atropurpurea* rosette at Páramo de San José (3150 m)

composed of a dense mass of very pubescent leaves and an apical bud surrounded by a large number of unexpanded pubescent young leaves (Fig. 2b). The central pith, made up of parenchymatous tissue, has a large capacity to store water (Goldstein et al. 1984). Other giant rosettes, for example the *Ruilopezias* (Fig. 2c), consist of large rosettes with very large leaves and reduced stems. Leaves are glabrous on its abaxial surface and pubescent on the adaxial surface.

In this chapter some relevant plant functional traits influencing plant water relations (how plants cope with water deficits and high evaporative demand conditions), gas exchange (how water availability and temperature conditions affect stomatal responses, transpirational losses and photosynthetic processes) and freezing resistance (how plants survive nocturnal subzero temperatures all year round) in giant rosettes of the Venezuelan Andes is reviewed. A comprehensive analysis of these functional characteristics will permit us to identify the diversity of responses and the adaptive significance of the responses in this dominant life-form, compare these functional characteristics to other Andean plant life-forms and aid in understanding how rosettes may be affected by global climate change. Information on giant rosettes from other high tropical mountains is also included in this analysis.

Stem Capacitance and Pith Water Volume Impact on Water Economy

As a general rule, plant height decreases with increasing altitude (Körner 2003). However, exceptions exist; Smith (1980) described the paradox of the giant rosettes of the tropical Andes. These rosettes exhibit different degrees of caulescence, from sessile plants up to rosettes with stems several meters in height. The tallest giant rosettes, mainly within the genus *Coespeletia*, are restricted to the higher paramos (>3800 m). *Espeletia schultzii*, species with the widest altitudinal range (2600–4400 m) in the Andes, increases stem size with altitude (Smith 1980; Meinzer et al. 1985). Meinzer et al. (1985) describe a ten-fold increase in pith volume for *E. schultzii* from 2600 to 4200 m. Additionally, leaf area decreases by 50 % at higher altitudes so that the relative water capacitance, defined by Goldstein et al. (1984) as the pith volume/leaf area ratio, exponentially increases with elevation. These authors describe differences along elevational gradients, not only with respect to pith volume, but in the relationship between transpiring leaf surface and water pith volume (Fig. 3). This water storage capacity facilitates gas exchange during early morning hours when low soil temperatures result in high root resistance for water uptake, and during drought periods. *C. timotensis*, a species with a large capacitance (Goldstein et al. 1985a), maintains relatively high and constant leaf water potentials even during the dry season (Goldstein et al. 1984). These authors suggest that the hydraulic connections between the pith and the xylem tissue are very efficient allowing this species to maintain a favorable water status.

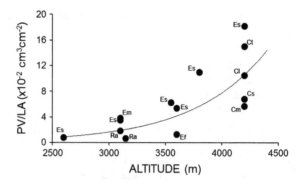

Fig. 3 Relationship between pith volume/leaf area (PV/LA) and elevation for different Andean giant rosettes (PV/LA = $0.015e^{0.0016Altitude}$, R^2 = 0.64). *Ruilopezia atropurpurea* (Ra), *Espeletia schultzii* (Es), *E. marcana* (Em), *E. flocossa* (Ef), *Coespeletia timotensis* (Ct), *C. lutescens* (Cl), *C. spicata* (Cs), *C. moritziana* (Cm). Data from Goldstein et al. (1984, 1985a), Meinzer et al. (1985), Navarro (2013)

The role of the insulating marcescent leaves on the water balance of *Coespeletia timotensis* was described by Goldstein and Meinzer (1983). Removal of the dead leaf layer around the stem changed the patterns of stem temperatures, which affected the plant's water balance (Smith 1979; Goldstein and Meinzer 1983). These effects were attributed to impeded water transport through the xylem when the marsescent leaves were experimentally removed due to embolisms or frost injury of the exposed pith tissue which occur at night. The role of the marcescent leaves and their effects on stem and pith temperature will be further discussed. Even though nutrient relations will not be treated in this chapter, it is necessary to emphasize the importance of this layer of marcescent leaves on nutrient cycling in the páramos. A large percentage of the rosette's biomass remains as standing necromass in these layers of dead leaves around the stem (Monasterio and Sarmiento 1991). In a nutrient limited environment such as the paramo, this slowly decomposing material that surrounds the stem may be considered a key nutrient reservoir for giant rosettes. With fine roots developing inside the marcescent leaves, and evidence of the presence of many decomposers, Garay (1981) indicated that this nutrient pathway which does not include the organic soil is a closed above-ground nutrient recycling circuit.

The Role of Leaf Pubescence

Different functions have been attributed to leaf pubescence. For hot and arid environments, this layer reduces light absorption by the leaf lamina resulting in lower leaf temperatures which in turn lower water losses by transpiration (Ehleringer 1984). Baruch and Smith (1979) suggest that pubescence in leaves of Andean giant rosettes may play a similar role as in hot-arid environments.

Meinzer and Goldstein (1985), on the other hand, find that leaf pubescence affects the thermal balance of *Coespeletia timotensis* by increasing the boundary layer thickness and reducing convective and latent heat transfer, while the effects on absorption of solar radiation are minor. The resulting effect of the pubescent boundary layer on energy transfer is a substantial leaf temperature increase during the daytime. This ability to increase leaf temperature is of adaptive importance in environments where air temperature is low all year round. The evolution of leaf pubescence in hot-arid habitats and tropical mountains has occurred in response to dissimilar selective pressures, even though both environments have high incoming solar radiation, the daily and seasonal patterns of air temperatures are very different in hot desserts and in high tropical environments (Meinzer et al. 1994).

Leaf Water Potentials and Seasonal Adjustments

Leaf water potentials of giant rosettes are relatively high, even during the dry season, suggesting that they do not experience water deficits. Different patterns are found between the high elevation *Coespeletias* and *Espeletia schultzii* and the lower altitude *Ruilopezia atropurpurea* and *Espeletia marcana* (Fig. 4). Minimum leaf water potentials (Ψ_L^{min}) were always above leaf water potentials at turgor loss (Ψ_L^{tlp}) which means that none of the species loses turgor even under drought conditions. *C. timotensis*, the species having the largest capacitance measured within the Andean giant rosettes (Meinzer and Goldstein 1986), maintains positive and relatively constant leaf water potentials all year round (Goldstein et al. 1984). Rada et al. (2012) observed lower Ψ_L^{min} and Ψ_L^{tlp} for *Coespeletia moritziana* compared to other *Coespeletia* species. Giant rosettes at lower altitudes, *R. atropurpurea* or *E. marcana*, exhibit similar Ψ_L^{min} and Ψ_L^{tlp} compared to the higher elevation

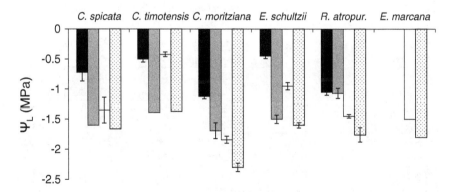

Fig. 4 Mean minimum leaf water potential (Ψ_L^{min}, wet (■) and dry (□) seasons) and leaf water potential at the turgor loss point (Ψ_L^{tlp}, wet (▨) and dry (▨) seasons) for different Andean giant rosette species (Goldstein et al. 1984, 1985a, b; Estrada et al. 1991; Rada et al. 1998, 2012; Navarro 2013). The *Coespeletias* and *Espeletia schultzii* (4200 m), *Ruilopezia atropurpurea* (3150 m) and *Espeletia marcana* (3100 m)

Coespeletias and *Espeletias*. *C. moritziana* (Rada et al. 2012) and *R. atropurpurea* (Navarro 2013) are the only species which are able to osmotically adjust from wet to dry seasons allowing these two species to maintain turgor under the more stressful dry season conditions. *C. moritziana* is associated with poorly developed soils with many fractured rocks and limited water availability. *R. atropurpurea* shows the smallest pith volume and relative capacitance of all studied giant rosettes (Goldstein et al. 1985a; Meinzer and Goldstein 1986).

Seasonal Variations in Gas Exchange Characteristics

The magnitude of leaf stomatal opening and closure, estimated by leaf stomatal conductance (g_s), is regulated by different environmental factors. Among these environmental drivers of the plant's water status are soil water availability and evaporative demand of the air surrounding the rosette leaves. Of all giant rosettes studied at 4200 m, *E. schultzii* shows the largest reduction in g_s from one season to the other (67 %) (Table 1). This species, with a smaller capacitance, relies on stomatal control to maintain positive leaf water potentials during the dry season. *C. timotensis*, the species with the largest stem water storage capacity, shows a smaller seasonal stomatal decrease. *C. moritziana* exhibits a slight decrease (15 %) in seasonal stomatal closure from wet to dry season, suggesting that this species, with the lowest wet season g_s, maintains stomata partially opened all year round. This is probably due to soil conditions and low relative water capacitance described previously. *R. atropurpurea*, growing at a lowest elevation, has the highest stomatal conductances during the wet season, but a large decrease in the dry season (70 %). In general, low g_s observed in Andean giant rosettes explain their relatively low

Table 1 Gas exchange characteristics (leaf stomatal conductance (g_s), CO_2 assimilation rate (A)) for different giant rosettes of the Venezuelan páramos in wet (WS) and dry (DS) seasons

Species	Altitude (m)	g_s (mmol m^{-2} s^{-1})		A_{max} (µmol m^{-2} s^{-1})	
		WS	DS	WS	DS
Coespeletia spicata[a]	4200	100.3 ± 11.0	60.1 ± 5.3	8.2 ± 0.3	6.5 ± 0.3
Coespeletia timotensis[a]	4200	89.9 ± 13.2	70.7 ± 6.0	4.2 ± 0.3	2.8 ± 0.7
Coespeletia moritziana[b]	4200	61.9 ± 7.2	52.3 ± 3.9	8.0 ± 0.3	4.5 ± 0.4
Espeletia schultzii[c]	4200	122.3 ± 22.4	40.4 ± 7.6	2.6 ± 0.2	1.8 ± 0.3
Espeletia schultzii[c]	3550	202.1 ± 26.6	46.8 ± 11.3	3.1 ± 0.2	2.2 ± 0.2
Espeletia floccosa[d]	3550	87 ± 9.1	37 ± 4.3	6.7 ± 0.2	2.5 ± 0.4
Ruilopezia atropurpurea[e]	3150	231.1 ± 56.7	69.9 ± 12.0	9.8 ± 0.6	6.0 ± 0.1

Values are mean maximum ± one standard error
[a]Goldstein et al. (1989)
[b]Rada et al. (2012)
[c]Rada et al. (1998)
[d]Unpublished data
[e]Navarro (2013)

transpirational losses (maximum transpirational rates of 3 mmol m^{-2} s^{-1}) (Goldstein et al. 1984, 1989; Rada et al. 1998, 2012). The highest transpiration rates described for Andean giant rosettes correspond to *R. atropurpurea*, 4 mmol m^{-2} s^{-1}, during the wet season at a lower elevation (3150 m) (Navarro 2013).

Together with stomatal conductance, the water vapor pressure gradient between leaf and surrounding air (VPD) determine leaf transpirational losses. Different authors have described a very close relationship between stomatal conductance and VPD for Andean rosettes (Goldstein et al. 1989; Rada et al. 1998). The integration of these adaptive characteristics: large water reservoir in well-developed pith tissues and a high stomatal control lead to the maintenance of a favorable plant water balance throughout the year for Andean giant rosettes.

Mean maximum species-specific CO_2 assimilation rates for giant rosettes range from 3–10 μmol m^{-2} s^{-1} with an approximate mean of 6 μmol m^{-2} s^{-1} (Table 1). These low rates are comparable to other woody species of the high Andes: *Polylepis sericea*, a tree species, exhibits maximum rates of 7 μmol m^{-2} s^{-1} and *Hypericum laricifolium*, a shrub, exhibits maximum rates of 5 μmol m^{-2} s^{-1} (Azócar and Rada 2006). The highest rates were observed in the low elevation species *R. atropurpurea*. More importantly, the seasonal decreases in maximum assimilation rates are consistent with decreasing stomatal conductance.

Optimum temperature for CO_2 assimilation decreases with increasing elevation on altitudinal gradients. *E. schultzii* has an optimum temperature of 11.5 °C at 2950 m, which decreases to 7.9 °C at 4200 m (Rada et al. 1992). *C. moritziana*'s optimum temperature is 10.2 °C at 4200 m (Rada et al. 2012). With respect to temperature ranges in which CO_2 assimilation rates remain positive, *E. schultzii*'s low temperature compensation point occurs below 0 °C, while for *C. moritziana* it is approximately 3 °C. High temperature compensation points reach 25 °C for both species at 4200 m. In neither of the two species does air temperature seem to be a limitation to photosynthesis. Mean leaf temperatures for *E. schultzii* and *C. moritziana* were always within the 80 and 90 % of the optimum temperature for photosynthesis, respectively.

Plant Water Relations and Gas Exchange in Juvenile Rosettes

Due to the relatively small pith water volumes, stem capacitance does not play a significant role in the maintenance of a favorable water status in juveniles. Additionally, environmental conditions at ground level are far more extreme than those where adult plants inhabit. How do juvenile rosettes survive these harsher drought conditions near the ground level of high mountain environments? There are few studies on water relations and gas exchange in juveniles of Andean giant rosette plants. Low survival rates of seedlings and juveniles indicate that this stage in the life cycle is the most vulnerable due to the extreme environmental conditions at ground level (Smith 1981; Goldstein et al. 1985a; Estrada and Monasterio 1988; Guariguata and Azócar 1988). This high mortality is attributed to the inability of these plants to physiologically respond to drought stress. García-Varela (2000) indicated that the nurse effect allow juveniles of *C. spicata* and *C. timotensis* to survive associated with adult plants or rocks, both considered favorable microsites.

Orozco (1986) described the water relations of juveniles of three different *Coespeletia* species occupying contrasting habitats in the Desert Páramo. *C. spicata* occurs in valley bottoms where moisture conditions are more favorable. *C. timotensis* survives on slopes where soil water availability is less favorable. Even though there are no gas exchange studies for this species, high leaf water potentials in juveniles must be a consequence of strong stomatal closure during daily and/or seasonal drought periods. *C. moritziana* juveniles tolerate lower leaf water potentials (−2.3 MPa) and show an important osmotic adjustment from wet to dry season (Orozco 1986; Rada et al. 2012).

There is only one study which considers CO_2 assimilation rates for juveniles of Andean giant rosettes (Rada et al. 2012). *C. moritziana* juveniles have mean rates of 5.3 and 3.5 $\mu mol\ m^{-2}\ s^{-1}$ for wet and dry seasons, respectively. These rates are comparable to other *Coespeletia* species, suggesting that juveniles surviving the early stages of the life cycle respond in a similar manner to adult rosettes.

Freezing Resistance

Low temperature, and in particular subzero temperatures, is an important factor that influences a plant's physiology and performance, and determines, its distribution. Tropical alpine plants avoid or tolerate ice formation in their tissues at temperatures below the equilibrium freezing temperatures. Avoidance may occur through the insulation of organs from surrounding low temperatures. Avoidance may also be obtained through supercooling, the maintenance of water in a liquid state even though temperatures may be well below their equilibrium freezing temperatures. In the case of frost tolerance, ice formation occurs in the extracellular spaces without damaging their intracellular contents.

Plants from environments where minimum temperatures are not extreme and only last for a few continuous hours, i.e. tropical mountain plants, rely mainly on avoidance mechanisms, while those subjected to severe temperatures and longer duration rely on tolerance (Sakai and Larcher 1987). All Andean giant rosettes studied to date rely on avoidance mechanisms to survive the conditions of the tropical Andes.

Thermal Insulation and Nyctinastic Leaf Folding

Giant rosettes are continuously forming new pubescent leaves from an inner apical bud (Fig. 2b). At night adult leaves fold in, known as nyctinastic movements, to form a dense, compact layering protecting the apical bud. Hedberg (1964) described it as the 'night bud' in African giant rosettes. This structure delays heat loss from the bud towards the surrounding air avoiding freezing of internal water. Smith (1974) described these leaf nyctinastic movements in the giant Andean rosette *Espeletia schultzii*. Rada et al. (1985a) measured differences of up to 6 °C between nighttime apical bud and air temperatures in *Coespeletia timotensis* and *C. spicata* in the high Andes. Additionally, Ice nucleation and injury temperatures (−5 to −6 °C) for *C. timotensis* and *C. spicata* apical buds, were approximately 8 °C below their minimum measured temperature. Beck et al. (1982) report similar results for two Afroalpine giant rosettes, *Lobelia telekii* and *Senecio keniodendron*.

Stems of most giant rosettes of the genera *Espeletia* and *Coespeletia* are covered with a very dense layer of marcescent leaves (Fig. 2b). The leaves after senescence remain attached to the stem. This layer acts as a thermal insulator, maintaining internal stem tissue (e.g. phloem, xylem and pith) temperatures above 0 °C (Smith 1979; Goldstein and Meinzer 1983; Rada et al. 1985a). Minimum stem temperatures remained above 2 °C while air temperature was −3 °C (Rada et al. 1985a). Freezing of stem tissue never occurred in naturally growing giant rosettes.

Frost Injury and Supercooling

Supercooling functions in environments where minimum freezing temperatures are not extremely low and only last for a few hours (Beck 1994), as in the high tropical Andes. This short term supercooling is a risky strategy as ice nucleation experimentally induced in supercooled tissues will immediately result in damage due to abrupt freezing. This strategy has been described for different life forms of the Venezuelan high Andes, trees (Rada et al. 1985b; Dulhoste 2010), shrubs (Squeo et al. 1991) and giant rosettes. Leaves of all giant rosettes studied in the tropical Andes rely on permanent supercooling (sensu Larcher 1975) in order to survive. Larcher (1975) and Larcher and Wagner (1976) observed ice nucleation temperatures of −12 °C in leaves of *Espeletia semiglobulata*. Subsequently, Rada et al. (1985a) described permanent supercooling capacities of −16 and −14 °C for *Coespeletia spicata* and *C. timotensis*, respectively. Additional research in other giant rosette species confirmed that temperatures at which ice nucleation began coincided with that at which injury occurred, therefore indicating that leaves were not able to resist ice formation in their tissues (Goldstein et al. 1985b; Rada et al. 1987; Cavieres et al. 2000; Dulhoste 2010).

The supercooling capacity found in Andean giant rosettes may be explained by: (a) Sessile leaves which reduce wind effects and buffer ice nucleation, (b) presence of a hydrophobic pubescent layer on leaves as a barrier for external nucleators, (c) Small cell sizes and intercellular spaces to reduce the chance of ice crystal formation in these apoplastic spaces (Goldstein et al. 1985b; Rada et al. 1987).

The water status of a plant is known to affect its supercooling capacity (Burke et al. 1976; Levitt 1980). Goldstein et al. (1985b) and Rada et al. (1987) explained how supercooling capacity of Andean giant rosettes increases as Ψ_L becomes more negative, with plants from higher altitudes being more sensitive to these changes. This pattern was observed for a single species, *E. schultzii*, along a 2600–4200 m altitudinal gradient, and for a group of species from different altitudes (Rada et al. 1987; Goldstein et al. 1985b). Although this increased supercooling capacity with decreasing Ψ_L should not be considered a response, it is of great adaptive value since minimum air and leaf temperatures occur during the dry season, when plants exhibit their minimum leaf water potentials.

Supercooling capacity and injury temperatures of giant rosettes are closely related to elevation (Fig. 5). Species from lower elevations show ice nucleation and injury temperatures between −5 and −8 °C, with higher elevation species below −10 °C. Goldstein et al. (1985b) observed decreasing ice nucleation temperatures with increasing elevation in ten species of the genera *Ruilopezia*, *Espeletia* and *Coespeletia*. Rada et al. (1987) found similar results for different *E. schultzii* populations growing from 2600 to 4200 m in elevation. The decreases in apoplastic water content and cell size account for the changes in supercooling capacity, with lower apoplastic water content observed in higher elevation giant rosettes with higher supercooling capacity.

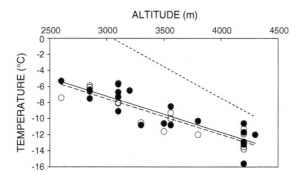

Fig. 5 Relationship between ice nucleation (○) and injury (●) temperatures and altitude for the different giant rosette species along an elevational gradient. For ice nucleation temperature: INT = −0.0045(altitude) + 6.26, $r^2 = 0.78$ and for injury temperature: IT = −0.0045(altitude) + 5.85, $r^2 = 0.83$. Absolute minimum temperature line (——)

Freezing Avoidance in Juvenile Rosettes

Night re-irradiation from the soil surface determines the low temperatures at ground level. Plants growing closer to the ground are subjected to more extreme subzero temperatures compared to taller rosettes which occur further above the ground level. Adult giant rosettes avoid freezing temperatures through isolating structures and supercooling. But how do juveniles, growing under a more stressful environment near the ground, survive freezing temperatures? García-Varela and Rada (2003) report avoidance mechanisms through permanent supercooling capacity in juveniles of *Coespeletia timotensis* and *C. spicata* (ice nucleation and injury temperatures of -15 °C). A high mortality rate during establishment may also be a consequence of extreme low nighttime temperatures (Estrada and Monasterio 1988). Established seedlings and juveniles are associated with favorable microhabitats such as near adult giant rosettes, rocks or cushion plants (García-Varela 2000; Smith 1981; Pérez 1984, 1989).

Functional Responses of Giant Rosettes Versus Other Andean Plant Life-Forms

The ability of giant rosettes to maintain turgor is comparable to other tropical Andean plant life-forms. Trees and shrubs exhibit mean Ψ_L^{tlp} around -1.8 MPa, while herbaceous life forms have values between -2.2 and -2.5 MPa (Fig. 6a). If we consider Ψ_L^{min} as a measure of tolerance to drought conditions, the giant rosettes have the most positive values of all life-forms, significantly above the turgor loss point. General morphological characteristics discussed above and positive mean leaf water potentials and minimum leaf water potentials which remain above the turgor loss point indicate that Andean giant rosettes are extremely drought avoidant species.

Injury in giant rosettes is comparable to trees, shrubs and forbs, while acaulescent rosettes, grasses and cushion having lower injury temperatures (Fig. 6b). Trees and giant rosettes rely exclusively on avoidance mechanisms (Rada et al. 1985a, b; Goldstein et al. 1985b), mainly through supercooling, while grasses (Márquez et al. 2006), acaulescent rosettes and cushion plants (Azócar et al. 1988; Squeo et al. 1991) tolerate ice formation, and thus cell dehydration, in their leaf tissues. The other two life-forms, shrubs and forbs, exhibited both avoidance and tolerance mechanisms (Squeo et al. 1991; Azócar 2006).

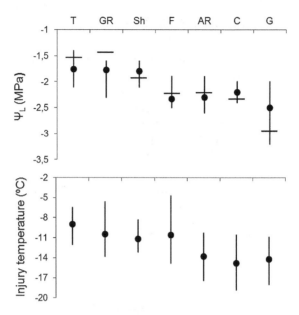

Fig. 6 **a** Mean leaf water potential (Ψ_L) at turgor loss (●) and mean minimum leaf water potential (—) for different plant life-forms of the Andean tropical mountain. *Vertical bars* extend up and down to absolute maximum and minimum leaf water potentials at turgor loss for each of the life-forms. **b** Mean injury temperature (●) for different plant life-forms of the Andean tropical mountain. *Vertical bars* extend up and down to absolute maximum and minimum injury temperatures for each of the life-forms. Trees (T), giant rosettes (GR), shrubs (Sh), forbs (F), acaulescent rosettes (AR), cushions (C), grasses (G)

Functional Responses of Andean Versus Other Tropical Alpine Giant Rosettes

Important differences are found when functional characteristics of Andean and other tropical giant rosettes are compared (Table 2). Although injury temperatures are similar for both Andean and Afroalpine rosettes growing above 3500 m, the former avoid frost while the latter tolerate extracellular ice formation in their leaf tissues. African giant rosettes (studied at 4200 m) have a high ice nucleation temperature while injury occurs at a much lower temperature, typical of plants that tolerate ice seeding in the extracellular spaces and thus cell dehydration. This difference in mechanisms highlights two important aspects of giant rosette plant function. As stated previously, plants rely on avoidance in less extreme environments compared to tolerant species (Sakai and Larcher 1987) which assumes that Afroalpine rosettes are adapted to harsher conditions than Andean rosettes. In terms of water relations, afroalpine rosettes being tolerant to freezing are also tolerant to drought. Even though daytime mean minimum leaf water potentials of Afroalpine giant rosettes (Schulze et al. 1985) may be comparable to the lowest minimum leaf

Table 2 Comparison between functional characteristics of Andean[a] and Afroalpine giant rosettes

Functional characteristics	Andean	Afroalpine
INT (°C)	−12.8	−5.3
IT(°C)	−12.7	−12.3
Ψ_L^{min} (MPa)	−1.5	−2.1 (−5.2)[b]
g_s (mmol m^{-2} s^{-1})	106/44[c]	−/130
A_{max} (µmol m^{-2} s^{-1})	7.3/4.3	−/9.5

INT ice nucleation temperature, IT injury temperature, Ψ_L^{min} Minimum leaf water potential, g_s mean leaf stomatal conductance, A_{max}: maximum CO_2 assimilation rate

[a]Data correspond to Andean giant rosettes at 4200 m so that they are comparable to the results presented for afroalpine rosettes

[b]Leaf water potential of frozen leaves (−4.4 °C) at night (Beck et al. 1984)

[c]Values for g_s and A_{max} correspond to wet/dry seasons

water potentials reported for *Coespeletia moritziana* (Rada et al. 2012), Beck et al. (1984) have observed survival of Afroalpine rosettes down to an absolute minimum leaf water potential of −5 MPa. This ability to withstand extremely low leaf water potentials allows Afroalpine rosettes to maintain stomates open during drought (Table 2). Mean stomatal conductance of Afroalpine giant rosettes during the dry season is three times higher than Andean giant rosettes during the dry season and is slightly higher than the mean stomatal conductance of Andean giant rosettes during the wet season. Consequently, Afroalpine giant rosettes maintain higher CO_2 assimilation rates (Schulze et al. 1985; Lüttge et al. 2001) which in turn results in higher growth rates.

Leaf pubescence affects the thermal balance of Andean giant rosettes by increasing the boundary layer thickness while light reflectance seems to have a minor role. This effect on the boundary layer results in an uncoupling of leaves and air temperatures, with leaves at several degrees above air temperature. Unlike Andean giant rosettes, the Hawaiian giant rosette *Argyroxiphium sandwicense*, with a relatively thin leaf pubescence, has similar leaf and air temperatures under high incoming solar radiation (Melcher et al. 1994). The leaf pubescence of this species is highly reflective. The temperature of the apical bud, which is located in the center of the parabolic rosette, is usually 25 °C higher than air temperatures at midday, because the incoming solar radiation is focused towards the apical bud. These authors suggest that this heating effect enhances the rates of physiological processes and growth rates of developing leaves.

Lipp et al. (1994) and Goldstein et al. (1996) found an interesting case in terms of freezing resistance adaptations in *Argyroxiphium sandwicense*. Adult *A. sandwicense* rosettes tolerate freezing, similar to the African counterpart; while seedlings avoid freezing of their leaf tissues. Ice nucleation and injury temperatures for these Hawaiian juvenile rosettes are similar to those of Andean rosette species (Goldstein et al. 1985b; Rada et al. 1985a; García-Varela and Rada 2003).

Giant Rosettes and Climate Change

Mountain ecosystems represent an important source of water and energy, maintain high endemism due to high altitude isolation and display a high biological diversity (Beniston 2003). Temperature and precipitation variations along elevational gradients are large, resulting in drastic changes in vegetation types along the slopes of tropical mountains. This close relationship between climate and environment in relatively short transitional zones makes them sensitive to any climatic change (Gosling and Bunting 2007). The combination of a fragile ecosystem and an amplified climate change illustrates the dramatic effect global change may have on mountain areas (Buytaert et al. 2010), important hotspots of biodiversity (Myers et al. 2000).

Global climate change will have a strong impact on water resources. Higher temperatures are likely to increase evapotranspiration, saturated atmospheric water vapor content and may result in changes in precipitation patterns and frequency of extreme events (Bates et al. 2008), mainly in mountain climates under the influence of continental or Mediterranean regimes (Beniston 2003). There is little agreement on the effects of climate change on precipitation patterns for mountain ranges (IPCC 2007). The Andes stand out as a region of high projection uncertainty, casting doubts on the reliability of future predictions. Large errors in the simulation of precipitation patterns over the Andes result when the model is compared to observational data (Buytaert et al. 2010).

A first approximation in relation to the response of mountain vegetation to climate change is that species will migrate upwards. However this hypothesis may be too simplistic, Walther et al. (2005) recognize that shifts are species-specific and do not occur as a vegetation 'front'. Moreover, projections that assume that plant migration occurs along isotherms overestimate change at the level of vegetation belts while the effect on single species may be underestimated (Körner 2003). These differences in species migration rates may be due to their different tolerance ranges, longevities, survival rates, competition abilities among other varying limiting factors (Beniston 2003). Migration evidence has not been documented for the Andes although they have been described for temperate and other tropical mountains (Chen et al. 2011; Pauli et al. 2012). The predicted upward shifting of plant species was evidenced as an effect of global warming through an increase in species richness at mountain summits (Grabherr et al. 1994, 2010; Pauli et al. 1996). These latter authors also state that the migration of vascular plants was far below the hypothetical rates of upward movement of temperature isolines indicating an important time lag between temperature increase and ecological responses. Climate induced differential migration rates could lead to the formation of new plant assemblages and result in changes in ecosystem functioning (Root et al. 2003).

Together with migration, plants may respond in two other ways to changes in climate conditions: through adaptation or local extinction (Thuiller et al. 2008). How plants will respond depends on their physiological and ecological characteristics. Studies on plant communities of the tropical Andes suggest that species with restricted ranges and those at higher elevations will show a large contraction in their habitat distribution and perhaps many species will be driven towards extinction (Cuesta and Becerra 2012). Furthermore, an elevational shift always implies a loss in habitat size as a consequence of diminishing and more fragmented surface area with increasing elevation. Therefore high mountain community assemblages are expected to be modified by increased habitat fragmentation and increased competition compared to species from lower elevations (Walther et al. 2005).

The paramo may be considered one of the most recently assembled ecosystems and its presence as a well differentiated and unique ecosystem may be due to: Altitudinal conditions above 3000 m, geographical location and incoming solar radiation, specific humidity and precipitation conditions in both horizontal and vertical gradients, and the central position of the northern Andes along the Alaska-Patagonia biological corridor permitting an important genetic exchange between North and South American species. These conditions determine certain important characteristics of the vegetation: a wide range of adaptability to extreme conditions of day-night cycles rather than seasonal ones, the fragility to climatic variations due to the very specialized physiology of most plants and the highest endemism index compared to any other ecosystem (Castaño-Uribe 2002).

Uncertainty in future projections is the rule for the tropical high Andes. Temperature is expected to rise over the next century due to anthropogenic activity (IPCC 2007). Existing projections on temperature shifts are rather consistent and suggest a 3 ± 1.5 °C rise for the Andes in the next century (Buytaert et al. 2011). For the last 30 years a 0.34 °C/decade increase has been observed for the northern Andes (Vuille et al. 2003). Again there are no clear projections in relation to precipitation patterns. Different reports suggest an approximate 300 mm year^{-1} increase in a century for Ecuador and most of Colombia, while Northern Colombia and Venezuela, dominated by trade winds, will show a decreasing rainfall trend (Buytaert et al. 2011).

Characteristic long life cycles of alpine vegetation represent a greater vulnerability to disturbance in comparison to low elevation plants (Körner 2003). Pioneer species, with rapid growth rates and production of large amounts of easily dispersed seeds, may take advantage of climate change by facilitating rapid upward movement (Salick et al. 2009). The extent to which physiological characteristics change via plasticity or evolution are essential in the projection of the effects of climate change. If plastic responses and/or evolutionary adjustments do not occur or are slower than environmental change, plant species will go extinct (Chown et al. 2010).

Resistance to freezing temperatures is a prerequisite for plant survival and determine to great extent plant distribution at high elevations. Increasing temperatures result in an upward displacement of the frost line giving the opportunity to plants from lower elevations to ascend and occupy sites of established alpine plants. Under this condition, competition with lower elevation vegetation would seem to play a more determining role in giant rosette distribution at the pre-páramo forest/páramo transition zone.

Increasing daytime temperatures would not seem to have an important effect on photosynthetic capacity of giant rosettes. CO_2 assimilation in giant rosettes is maintained at high levels in a wide temperature range. However, higher temperatures will determine higher ambient evaporative demands and lower soil water availability. Together with projected decreases in precipitation for the Northern Andes, they may have an effect on water relations and gas exchange of giant rosettes. Compared to other páramo life-forms, the giant rosettes may be labeled as extreme drought avoiders. In order to maintain high leaf water potentials a stronger stomatal control under unfavorable conditions will take place limiting plant physiological activities such as CO_2 assimilation and consequently affecting growth.

It is reasonable to assume that species that show similarities in essential functional characteristics will respond in a similar manner to changes in climatic conditions (Diaz and Cabido 1997). As a dominant functional group, all studied giant rosettes show a narrow range of responses to water stress, suggesting that they may be susceptible to the changing environmental conditions which in turn result in a low resilience and a fragile functional stability of the high elevation ecosystems dominated by giant rosette plants (Azócar et al. 2000).

Acknowledgements Financial support of the author's research by the Consejo de Desarrollo Científico, Humanístico, Tecnológico y de las Artes (CDCHTA) of the Universidad de Los Andes, the Fondo Nacional de Ciencia, Tecnología e Innovación (FONACIT) and the Inter-American Institute for Global Change Research (IAI) is gratefully acknowledged.

References

Antonelli A (2009) Have giant lobelias evolved several times independently? Life form shits and historical biogeography of the cosmopolitan and highly diverse subfamily Lobelioideae (Campanulaceae). BMC Biol 7:82

Azócar C (2006) Relación entre anatomía foliar, forma de vida y mecanismos de Resistencia a temperaturas congelantes en diferentes especies en el Páramo de Piedras Blancas. Masters Thesis. Universidad de Los Andes, Mérida, Venezuela

Azócar A, Rada F (2006) Ecofisiología de Plantas de Páramo. Publicaciones ICAE, Mérida, Venezuela, 182 pp

Azócar A, Rada F, Goldstein G (1988) Freezing tolerance in *Draba chionophylla*, a 'miniature' caulescent rosette species. Oecologia 75:156–160

Azócar A, Rada F, García-Núñez C (2000) Aspectos ecofisiológicos para la conservación de ecosistemas tropicales contrastantes. Boletín de la Sociedad Mexicana de Botánica 65:89–94

Baruch Z, Smith AP (1979) Morphological and physiological correlates of niche breadth in two species of Espeletia (Compositae), in the Venezuelan Andes. Oecologia 38:71–82

Bates BC, Kundzewics ZW, Wu S, Palutikof J (2008) Climate change and water. Technical paper of the intergovernmental panel on climate change, IPCC Secretariat, Geneva

Beck E (1994) Cold tolerance in tropical alpine plants. In: Rundel PW, Meinzer FC, Smith AP (eds) Tropical alpine environments: plant form and function. Cambridge University Press, Cambridge, pp 77–110

Beck E, Senser M, Scheibe R, Steiger H, Pongratz P (1982) Frost avoidance and freezing tolerance in afroalpine "giant rosette" plants. Plant Cell Environ 5:215–222

Beck E, Schulze ED, Senser M, Scheibe R (1984) Equilibrium freezing of leaf water and extracellular ice formation in afroalpine "giant rosette" plants. Planta 162:276–282

Beniston M (2003) Climatic change in mountain regions: A review of possible impacts. Clim Change 59:5–31

Berry PE, Beaujon S, Calvo R (1988) Hybridization in the evolution of the frailejones (Espeletia, Asteraceae). Ecotropicos 1:11–24

Burke MJ, Gusta LV, Quamme HA, Weiser CJ, Li PH (1976) Freezing and injury in plants. Annu Rev Plant Physiol 27:507–528

Bussmann RW (2006) Vegetation zonation and nomenclature of African mountains—an overview. Lyonia 11:41–66

Buytaert W, Vuille M, Dewulf A, Urrutia R, Karmalkar A, Celleri R (2010) Uncertainties in climate change projections and regional downscaling in the tropical Andes: implications for water resources management. Hydrol Earth Syst Sci 14:1247–1258

Buytaert W, Cuesta-Camacho F, Tobón C (2011) Potential impacts of climate change on the environmental services of humid tropical alpine regions. Glob Ecol Biogeogr 20:19–33

Carlquist S (1974) Island biology. Columbia University Press, New York

Castaño-Uribe C (2002) Colombia alto andina y la significancia ambiental del bioma páramo en el contexto de los Andes tropicales: Una aproximación a los efectos futuros por el cambio climático global (Global climatic tensor). En: C. Castaño-Uribe (ed), Páramos y Ecosistemas Alto Andinos de Colombia en Condición Hotspot & Global Climatic Tensor. IDEAM, pp 24–70

Cavieres L, Rada F, Azócar A, García-Núñez C, Cabrera HM (2000). Gas exchange and low temperature resistance in two tropical high mountain tree species from the Venezuelan Andes. Acta Oecol 21:203–211

Chen IC, Hill JK, Ohlemüller R, Roy DB (2011) Rapid range shifts of species associated with high levels of climate warming. Science 333:1024–1026

Chown SL, Hoffmann AA, Kristensen TN, Angilletta MJ Jr, Stenseth NC, Pertoldi C (2010) Adapting to climate change: a perspective from evolutionary physiology. Clim Res 43:3–15

Cuatrecasas J (1976) A new subtribe in the Heliantheae (Compositae) Espeletiinae. Phytologia 35:43–61

Cuatrecasas J, Vuilleumier F, Monasterio M (1986) Speciation and radiation of the Espeletiinae in the Andes. In: Vuilleumier F, Monasterio M (eds) High altitude tropical biogeography. Oxford University Press, Oxford, pp 267–303

Cuesta F, Becerra MT (2012) Biodiversidad y cambio climático en los Andes: Importancia del monitoreo y el trabajo regional. Revista virtual REDESMA 6:19–27

Díaz S, Cabido M (1997) Plant functional types and ecosystem function in relation to global change. J Veg Sci 8:463–474

Dulhoste R (2010) Estrés hídrico y térmico en especies leñosas de la zona de transición selva húmeda-páramo. Doctor's thesis, Universidad de Los Andes, Mérida, Venezuela

Ehleringer J (1984) Ecology and ecophysiology of leaf pubescence in North American desert plants. In: Rodríguez E, Healey P, Mehta I (eds) Biology and chemistry of plant trichomes. Plenum Press, New York, pp 113–132

Estrada C, Goldstein G, Monasterio M (1991) Leaf dynamics and water relations of *Espeletia spicata* and *E. timotensis*, two giant rosettes of the Desert Paramoin the tropical Andes. Acta Oecologica 12:603–616

Estrada C, Monasterio M (1988) Ecología poblacional de una roseta gigante, *Espeletia spicata* Sch Bip (Compositae) del páramo desértico. Ecotropicos 1:25–39

Garay I (1981) Le peuplement de microarthropodes dans la litière sur pied de *Espeletia timotensis* et *E. lutescens*. Revue Ecologie et Biologie du Sol 18:209–219

García-Varela S (2000) Mecanismos de resistencia a temperaturas congelantes en plantas jóvenes de *Espeletia spicata* y *Espeletia timotensis*. Undergraduate thesis, Universidad de Los Andes, Mérida, Venezuela

García-Varela S, Rada F (2003) Freezing avoidance mechanisms in juveniles of giant rosette plants of the genus Espeletia. Acta Oecol 24:165–167

Goldstein G, Meinzer FC (1983) Influence of insulating dead leaves and low temperatures on water balance in an Andean giant rosette plant. Plant Cell Environ 6:649–656

Goldstein G, Meinzer FC, Monasterio M (1984) The role of capacitance in the water balance of Andean giant rosette species. Plant Cell Environ 7:179–186

Goldstein G, Meinzer FC, Monasterio M (1985a) Physiological and mechanical factors in relation to size-dependent mortality in Andean giant rosette species. Acta Oecol Oecol Plant 6:263–275

Goldstein G, Rada F, Azócar A (1985b) Cold hardiness and supercooling along an altitudinal gradient in andean giant rosette species. Oecologia (Berlin) 68:147–152

Goldstein G, Rada F, Canales MO, Zabala O (1989) Leaf gas exchange of two giant caulescente rosete species. Oecologia Plant 10:359–370

Goldstein G, Drake DR, Melcher P, Giambelluca TW, Heraux J (1996) Photosynthetic gas exchange and temperature-induced damage in seedlings of the tropical alpine species *Argyroxiphium sandwicense*. Oecologia 106:298–307

Gosling WD, Bunting MJ (2007) A role for palaeoecology in anticipating future change in mountain regions? Palaeogeogr Palaeoclimatol Palaeoecol 259:1–5

Grabherr G, Gottfried M, Pauli H (1994) Climate effects on mountain plants. Nature 369:448–448

Grabherr G, Pauli H, Gottfried M (2010) A worldwide observation of effects of climate change on mountain ecosystems. In: Borsdorf A, Grabherr G, Heinrich K, Scott B, Stötter J (eds) Challenges for mountain regions-tackling complexity. Böhlau Verlag, Vienna

Guariguata MR, Azócar A (1988) Seed bank dynamics and germination ecology in *Espeletia timotensis* (Compositae), an Andean giant rosette. Biotropica 20:54–59

Hedberg O (1964) Features of afroalpine plant ecology. Acta Phytogeographica Suecica 49:1–44

IPCC (2007) Climate change 2007—impacts, adaptation and vulnerability. Cambridge University Press, Cambridge

Körner Ch (2003) Alpine plant life, functional plant ecology of high mountain ecosystems. Springer, Berlin, 344 pp

Larcher W (1975) Pflanzenokologische Beobachtungen in der paramostufe der Venezolanischen Anden. Anz Math-Naturw Kl Oest Akad Wissensch 112:194–213

Larcher W, Wagner J (1976) Temperaturgrenzen der CO_2-aufnahme und temperaturresistenz der blätter von gebirgspflanzen in vegetationsaktiven Zustand. Oecola Plant 11:361–374

Levitt J (1980) Responses of plants to environmental stresses, vol 1. Chilling, freezing and high temperature stresses, 2nd edn. Academic Press, New York

Lipp CC, Goldstein G, Meinzer FC, Niemezura W (1994) Freezing tolerance and avoidance in high elevation Hawaiian plants. Plant Cell Environ 17:1035–1044

Luteyn JL (1992) Páramos: why study them? In: Balslev H, Luteyn JL (eds) Páramo, an Andean ecosystem under human influence. Academic Press, pp 1–14

Luteyn JL (1999) Páramos: a checklist of plant diversity, geographical distribution and botanical literature. Memoirs of the New York Botanical Garden, vol 84

Lüttge U, Fetene M, Liebig M, Rascher U, Beck E (2001) Ecophysiology of niche occupation by two giant rosette plants, *Lobelia gibberoa* Hemsl and *Solanecio gigas* (Vatke) C. Jeffrey, in an afromontane forest valley. Ann Bot 88:267–278

Márquez EJ, Rada F, Fariñas MR (2006) Freezing tolerance in grasses along an altitudinal gradient in the Venezuelan Andes. Oecologia 150:393–397

Meinzer FC, Goldstein G (1985) Some consequences of leaf pubescence in the Andean giant rosette plant *Espeletia timotensis*. Ecology 66:512–520

Meinzer FC, Goldstein G (1986) Adaptations for water and termal balance in Andean giant rosette plants. In: Givnish TH (ed) On the economy of plant form and function. Cambridge University Press, Cambridge, pp 381–411

Meinzer FC, Goldstein G, Rundel PH (1985) Morphological changes along an altitude gradient and their consequences for an Andean giant rosette. Oecologia 65:278–283

Meinzer FC, Goldstein G, Rada F (1994) Paramo microclimate and leaf termal balance of Andean giant rosette plants. In: Rundel PW, Smith AP, Meinzer FC (eds) Tropical alpine environments: plant form and function. Cambridge University Press, pp 45–59

Melcher PJ, Goldstein G, Meinzer FC, Minyard B, Giambelluca TW, Loope LL (1994) Determinants of termal balance in the Hawaiian giant rosette plant, Argyroxiphium sandwicense. Oecologia 98:412–418

Monasterio M (1979) El páramo desértico em el altiandino de Venezuela. In: Salgado-Labouriau ML (ed) El Medio Ambiente Páramo. UNESCO-IVIC, Caracas, Venezuela, pp 150–159

Monasterio M (1980) Las formaciones vegetales de los páramos de Venezuela. In: Monasterio M (ed) Estudios Ecologicos en los Páramos Andinos. Universidad de Los Andes, Mérida, Venezuela, pp 93–158

Monasterio M, Sarmiento L (1991) Adaptive radiation of Espeletia in the cold Andean tropics. Trends Ecol Evol 6:387–391

Monasterio M, Vuilleumier F (1986) Introduction: high tropical mountain biota of the world. In: Vuilleumier F, Monasterio M (eds) High altitude tropical biogeography. Oxford University Press, Oxford, pp 3–7

Myers N, Mittermeier RA, Mittermeier CG, da Fonseca GAB, Kent J (2000) Biodiversity hotspots for conservation priorities. Nature 403:853–858

Navarro A (2013) Relaciones hídricas en *Ruilopezia atropurpurea* (A.C. Sm.) Cuatrec. a diferentes condiciones microambientales en el Páramo de San José, Estado Mérida. Master's thesis, Universidad de Los Andes, Mérida, Venezuela

Orozco A (1986) Economía hídrica en rosetas juveniles de Espeletia en el páramo desértico. Masters thesis, Universidad de Los Andes, Mérida, Venezuela

Pauli H, Gottfried M, Grabherr G (1996) Effects of climate change on mountain ecosystems—upward shifting of alpine plants. World Resour Rev 8:382–390

Pauli H, Gottfried M, Dullinger S, Abdaladze O, Akhalkatsi M, Alonso JLB, Coldea G, Dick J, Erschbamer B, Fernández Calzado R, Ghosn D, Holte JI, Kanka R, Kazakis G, Kollár J, Larsson P, Moiseev P, Moiseev D, Molau U, Molero Mesa J, Nagy L, Pelino G, Puscas M, Rossi G, Stanisci A, Syverhuset AO, Theurillat JP, Tomaselli M, Unterluggauer P, Villar L, Vittoz P, Grabherr G (2012) Recent plant diversity changes on Europe's mountain summits. Science 336:353–355

Pérez FL (1984) Striated soil in an Andean páramo of Venezuela: its origin and orientation. Arct Alp Res 16:277–289

Pérez FL (1989) Some effects of giant Andean stem-rosettes on ground microclimate, and their ecological significance. Int J Biometeorol 33:131–135

Rada F, Azócar A, Rojas-Altuve A (2012) Water relations and gas exchange in *Coespeletia moritziana* (Sch. Bip) Cuatrec., a giant rosette species of the high tropical Andes. Photosynthetica 50:429–436

Rada F, Azócar A, Briceño B, González J (1998) Leaf gas Exchange in *Espeletia schultzii* Wedd, a giant caulescent rosette along an altitudinal gradient in the Venezuelan Andes. Acta Oecol 19:73–79

Rada F, Goldstein G, Azócar A, Meinzer F (1985a) Freezing avoidance in andean giant rosette plants. Plant Cell Environ 8:501–507, Leicester, Gran Bretaña

Rada F, Goldstein G, Azócar A, Meinzer F (1985b) Daily and seasonal osmotic changes in a tropical treeline species. J. Exp. Botany 36(167):989–1000

Rada F, Goldstein G, Azócar A, Torres F (1987) Supercooling along an altitudinal gradient in Espeletia schulzii a caulescent giant rosette species. J Exp Bot 38:491–497

Rada F, González J, Briceño B, Azócar A, Jaimez R (1992) Net photosynthesis-leaf temperature relations in plant species with different height along an altitudinal gradient. Oecol Plant 13:535–542

Rauscher JT (2002) Molecular phylogenetics of the Espeletia complex (Asteraceae): evidence from mrDNA ITS sequences on the closest relatives of an Andean adaptive radiation. Am J Bot 89:1074–1084

Root TL, Price JT, Hall KR, Schneider SH, Rosenzweig C, Pounds A (2003) Fingerprints of global warming on wild animals and plants. Nature 421:57–60

Rundel PW (1994) Tropical alpine climates. In: Rundel PW, Smith AP, Meinzer FC (eds) Tropical alpine environments. Cambridge University Press, Cambridge, pp 21–44

Sakai A, Larcher W (1987) Frost survival of plants: responses and adaptations to freezing stress. Springer, Berlin, 321 pp

Salick J, Zhendong F, Byg A (2009) Eastern Himalayan alpine plant ecology, Tibetan ethnobotany, and climate change. Glob Environ Change 19:147–155

Sarmiento G (1986) Ecologically crucial features of climate in high tropical mountains. In: Vuilleumier F, Monasterio M (eds) High altitude tropical biogeography. Oxford University Press, Oxford, pp 11–45

Schulze ED, Beck E, Scheibe R, Ziegler P (1985) Carbon dioxide assimilation and stomatal response of afroalpine giant rosette plants. Oecologia 65:207–213

Smith AP (1974) Bud temperature in relation to nyctinastic leaf movement in an Andean giant rosette plant. Biotropica 6:263–266

Smith AP (1979) The function of dead leaves in *Espeletia schultzii* (Compositae) an Andean giant rosette species. Biotropica 11:43–47

Smith AP (1980) The paradox of plant height in an Andean giant rosette species. J Ecol 68:63–68

Smith AP (1981) Growth and population dynamics of *Espeletia* (Compositae) of the Venezuelan Andes. Smithsonian Contribut Botany 48:1–45

Smith AP, Young TP (1987) Tropical alpine plant ecology. Annu Rev Ecol Syst 18:137–158

Squeo F, Rada F, Azócar A, Goldstein G (1991) Freezing tolerance and avoidance in high tropical Andean plants: is it equally represented in species with different plant height? Oecologia 86:378–382

Thuiller W, Albert C, Araújo MB, Berry PM, Cabeza M, Guisan A, Hickler T, Midgley GF, Paterson J, Schurr FM, Sykes MT, Zimmermann NE (2008) Predicting global change impacts on plants' species distribution: future challenges. Perspect Plant Ecol Evol Systemat 9:137–152

Troll C (1968) The cordilleras of the tropical Americas. In: Troll C (ed) Geoecology of the mountainous regions of the tropical Americas. Dumbler, Bonn, pp 15–55

Vuille M, Bradley RS, Werner M, Keimig F (2003) 20th century climate change in the tropical Andes: observations and model results. Clim Change 59:75–99

Walther G, Beißner S, Pott R (2005) Climate change and high mountain vegetation shifts. Mountain ecosystems, pp 77–96

Physiological Significance of Hydraulic Segmentation, Nocturnal Transpiration and Capacitance in Tropical Trees: Paradigms Revisited

Sandra J. Bucci, Guillermo Goldstein, Fabian G. Scholz and Frederick C. Meinzer

Abstract Results from water relations and hydraulic architecture studies of trees from tropical savannas and humid tropical and subtropical forests were reanalyzed in view of paradigms related to the (i) physiological significance of hydraulic segmentation across trees with different life history traits and habitats, (ii) determinants of massive tree mortality, (iii) nocturnal transpiration, and (iv) the role of internal stem water storage. Stems and leaves of tropical and subtropical deciduous tree species are equally vulnerable to cavitation, whereas leaves of evergreen species are substantially more vulnerable than stems. Tree species from tropical ecosystems that do not experience seasonal droughts have stems and leaves with similar vulnerability to cavitation while trees from tropical ecosystems that experience seasonal droughts have leaves that are more vulnerable to drought induced cavitation compared to stems. Strong segmentation (whether hydraulic or vulnerability) during severe droughts may have an indirect negative impact on tree

S.J. Bucci (✉) · F.G. Scholz
Grupo de Estudios Biofísicos y Ecofisiológicos, Facultad de Ciencias Naturales, Universidad Nacional de la Patagonia San Juan Bosco, Comodoro Rivadavia, Chubut, Argentina
e-mail: sj_bucci@yahoo.com

F.G. Scholz
e-mail: fgscholz@yahoo.com

S.J. Bucci · F.G. Scholz
Consejo Nacional de Investigaciones Científicas y Técnicas (CONICET),
Buenos Aires, Argentina

G. Goldstein
Laboratorio de Ecología Funcional, Departamento de Ecología Genética y Evolución, Instituto IEGEBA (CONICET-UBA), Facultad de Ciencias Exactas y naturales, Universidad de Buenos Aires, Buenos Aires, Argentina
e-mail: goldstein@ege.fcen.uba.ar; gold@bio.miami.edu

G. Goldstein
Department of Biology, University of Miami, Coral Gables, FL 33146, USA

F.C. Meinzer
USDA Forest Service, Forestry Sciences Laboratory, 3200 SW Jefferson Way, Corvallis, OR 97331, USA
e-mail: rick.meinzer@oregonstate.edu

© Springer International Publishing Switzerland 2016
G. Goldstein and L.S. Santiago (eds.), *Tropical Tree Physiology*,
Tree Physiology 6, DOI 10.1007/978-3-319-27422-5_9

carbon balance. For example for *Sclerolobium paniculatum*, a widespread tree species in neotropical savannas and seasonally dry forests, the decrease in total leaf surface area per plant (which impact hydraulic architecture) during droughts help to maintain an adequate water balance but has large physiological costs: trees receive a lower return in carbon gain from their investment in stem and leaf biomass. Leaf hydraulic failure and carbon starvation may contribute to the massive, size-dependent mortality observed in this species. The functional significance of the widespread phenomenon of nocturnal transpiration in tropical trees is discussed. One of the most likely functions of nocturnal sap flow in savanna trees growing in nutrient poor soils appears to be enhanced nutrient acquisition from oligotrophic soils. Large capacitance plays a central role in the rapid growth patterns of tropical deciduous tree species facilitating rapid canopy access as these species are less shade tolerant than evergreen species. Higher growth rates in species with high capacitance could be achieved by keeping the stomata open for longer periods of time.

Keywords Carbon starvation · Mortality · Safety margin · Transpiration · Vulnerability to cavitation

Introduction

Physiological traits of tropical trees have been extensively studied for many years. Although characterization of tree hydraulic architecture was initially conducted on temperate species (e.g. Zimmermann 1983), it did not take too long for this approach to be applied for investigating aspects of water relations of woody plants in tropical ecosystems. The first forest canopy tower crane was erected in Panama in the early 1990s. Until then, direct observations of transpiration, stomatal behavior and other physiological attributes in the upper canopy of mature tropical forests were hampered by difficulty of canopy access. This problem had been partially overcome by a variety of canopy access systems. Towers have been used to make observations in the upper canopy of Amazonian forests (e.g. Roberts et al. 1990). Other access systems include a large canopy-supported raft used in Cameroon (Koch et al. 1994), but the real breakthrough for assessing the integration of whole-tree physiological behavior was the use of canopy cranes (e.g. Meinzer et al. 1993). Previously, studies on aspects of the hydraulic architecture of high-elevation woody giant rosettes in the tropics, and in particular on the role of internal water storage in the water economy of these woody plants, were done with state of the art instrumentation by researchers at the Universidad de Los Andes in Venezuela (e.g.

Goldstein and Meinzer 1983). Although these arborescent giant rosettes can be considered functionally as trees, they rarely attain heights greater than 4 m. Similarly, savanna trees, particularly in South America, are relatively short, and thus access to their upper crowns does not require tower cranes. The hydraulic architecture of this type of trees has been extensively investigated in Venezuela and Brazil for many years, resulting in an important body of knowledge for understanding how plant hydraulics, such as stem and leaf hydraulic resistances, stem capacitance, embolism formation and repair, hydraulic segmentation, and root morphological patterns in relation to hydraulic lift, influence tree water relations worldwide (e.g. Bucci et al. 2003, 2004, 2005, 2006, 2008; Goldstein et al. 1984, 1989; Hao et al. 2008; Sarmiento et al. 1985; Meinzer et al. 1999, Scholz et al. 2002, 2006, 2008a, b; Zhang et al. 2009).

In this chapter we focus on some aspects of the water economy and water relations of tropical trees that are perhaps more controversial. We explore the validity of some paradigms that are currently being debated. There are two general questions related to the hydraulic architecture of trees i.e., the assemblage of structural and functional traits that determine the dynamic patterns of water flow from roots to leaves that will be discussed. One question is related to hydraulic segmentation and the other is related to the potential role of hydraulic architecture traits on the massive mortality of trees. It is assumed that hydraulic segmentation (differences in the resistance to water flow in roots, stems and leaves according to Zimmermann 1983) is widespread across tree species. We will assess the extent to which differences in the degree of hydraulic segmentation among tropical trees are related to differences in life history traits (e.g. deciduous vs. evergreen) versus environmental characteristics where the trees occur (e.g. with and without a pronounced dry season). Increasing tree mortality rates as a consequence of more frequent droughts or an increase in the severity of the droughts has been recently observed worldwide (Allen et al. 2010). The mechanisms of drought-induced mortality are not well known. In this chapter we provide evidences of two potential mechanisms triggering dieback (hydraulic dysfunction and carbon starvation) mediated by strong hydraulic segmentation for a tropical tree species.

Nocturnal sap flow has been observed predominantly in tropical ecosystems (Forster 2014), however its functional significance has been little studied. In this chapter we provide strong evidence that enhanced nutrient uptake by nocturnal transpiration for plants growing in oligotrophic (nutrient poor) soils are likely to have an important role in the nutrient balance of some tropical tree species. Finally, we discuss the key role that internal stem water storage plays in many aspects of tropical tree functioning, and in particular for helping to maintain high growth rates in low wood density species (mostly deciduous). The large sapwood capacitance in deciduous species may also help to avoid catastrophic xylem embolism, and extend the period which stomata are able to remain open for carbon assimilation diurnally.

Hydraulic Segmentation and Vulnerability Segmentation

A widespread paradigm in studies of the water relations of vascular plants is that hydraulic segmentation prevents cavitation from propagating into less expendable portions of the water transport system. A key feature of hydraulic segmentation is that during periods of high evaporative demand over the course of a day or during a drought period, cavitation is mostly confined to more expendable organs (i.e. leaves and small branches) in favor of larger organs such as trunks that represent years of cumulative growth and carbohydrate investments. The range of variation in the extent and patterns of hydraulic segmentation among tropical trees is not well known. Another concept related to hydraulic segmentation is vulnerability segmentation. Hydraulic segmentation refers to the differential resistance to water flow in different parts of woody plant along the root-to-leaf continuum, whereas vulnerability segmentation refers to the differential vulnerability to embolism depending on the organ along the continuum. Implicit in the vulnerability segmentation concept and its relationship to hydraulic segmentation is that leaves are more vulnerable than branches and trunks. The hydraulic architecture of leaves and their vulnerability to embolism have only recently been studied intensively (e.g. Nardini et al. 2001; Broddrib et al. 2003; Sack and Holbrook 2006; Blackman et al. 2010; Johnson et al. 2011; Scoffoni et al. 2014), whereas the vulnerability of stems and roots to embolism has been studied for several decades (e.g. Tyree and Sperry 1989; Kavanagh et al. 1999; Maherali et al. 2004; Domec et al. 2006). Although leaves and stems represent consecutive points along a hydraulic continuum, there are relatively few studies of their hydraulic properties in the same individual (e.g. Hao et al. 2008; Bucci et al. 2012, 2013).

The extent of hydraulic segmentation is different depending on leaf phenology (i.e. deciduous versus evergreen species) and depending on seasonal water availability. A review of studies having simultaneous measurements of vulnerability to cavitation in stems and leaves suggests that stems and leaves of tropical and subtropical deciduous tree species are equally vulnerable to cavitation as assessed from Ψ_{50}, species-specific values of leaf or stem water potential inducing 50 % loss of the maximum leaf or stem hydraulic conductivity (Fig. 1a). Leaves of evergreen species appear to be substantially more vulnerable than stems, which on average were 1 MPa less vulnerable compared to leaves (Fig. 1a). Tropical tree species from ecosystems that do not experience pronounced seasonal droughts have stems and leaves with similar vulnerability to cavitation, whereas trees from ecosystems that experience seasonal droughts have leaves that are more vulnerable to drought-induced cavitation than stems (Fig. 1b). In this case, hydraulic segmentation may help to prevent hydraulic dysfunction of the trunks and stems during mild droughts. It appears that tree species from aseasonal climates have not been subjected to selective pressures resulting in a highly segmented hydraulic architecture. Implicit in this assessment of relationships between selective pressures and evolutionary responses is the assumption that there is a cost associated with building

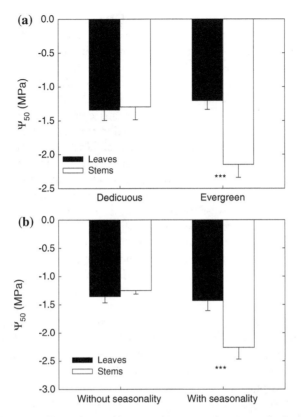

Fig. 1 Leaf water potential at 50 % loss of maximum leaf hydraulic conductance (Ψ_{50}, *open columns*) and stem water potential at 50 % loss of maximum stem hydraulic conductance (Ψ_{50}, *filled columns*) of 26 **a** deciduous and evergreen tree species from tropical and subtropical ecosystems and **b** tree species from ecosystems without rainfall seasonality and with rainfall seasonality. Significant differences between columns are indicated as ***($p < 0.0001$). Information obtained from Bucci et al. (2008), Hao et al. (2008), Chen et al. (2009), McCulloh et al. (2012), Johnson et al. (2012), Villagra et al. (2013)

leaves with high resistance to water flow. Across species, regardless if they are deciduous or evergreen or from seasonal or non-seasonal environments, mean leaf Ψ_{50} is similar with values close to -1.25 MPa.

Hydraulic Failure and Carbon Starvation as Determinants of Mortality

The precise mechanisms and plant traits underlying species-specific vulnerability to drought-induced mortality are poorly understood but are currently the subject of a debate and ongoing research (McDowell et al. 2011; Allen et al. 2012; Choat et al.

2012; Sala et al. 2012; Hartmann et al. 2013; Plaut et al. 2013). Two non-mutually exclusive mechanisms have been suggested to explain drought-induced mortality in trees (McDowell et al. 2008). Species with tight stomatal control to avoid hydraulic failure (isohydric) have reduced CO_2 assimilation during prolonged drought with negative consequences for their carbon economy, which may eventually lead to carbon starvation (Sala et al. 2010; Adams et al. 2013). At the other extreme, anisohydric species, which exhibit loose stomatal control allowing large fluctuations in leaf water potential may be predisposed to hydraulic failure if they operate with narrow hydraulic safety margins (Bucci et al. 2013). Knowledge of hydraulic safety margins may be more informative than only documenting xylem water potentials or xylem vulnerability to embolism because safety margins indicate how closely a plant operates to the loss of its hydraulic capacity. The safety margin is the difference between the lowest water potential experienced by plant organs such leaves and stems, and a reference point at which a given loss of xylem hydraulic function occurs. The Ψ_{50} is the most commonly used index of xylem resistance to embolism, however, this threshold value is not always the best reference point to compare the susceptibility to irreversible hydraulic failure across species (Scholz et al. 2014). Gymnosperms generally exhibit wider safety margins than angiosperms (Meinzer et al. 2009; Choat et al. 2012; Johnson et al. 2012).

A frequently cited paradigm is that the safety margin between leaves and stems is large enough to avoid any hydraulic dysfunction in the stems during droughts, assuming that the leaves are more vulnerable to cavitation than stems (vulnerability segmentation). However, strong segmentation (whether hydraulic or vulnerability) during severe droughts may have indirectly a negative impact on the tree carbon balance. An interesting case of dieback has been documented in a tree species having rapid growth and a relatively shallow root system that occurs in neotropical savannas and seasonal dry forests in Brazil (Zhang et al. 2009). Massive mortality in taller trees of *Sclerolobium paniculatum* appears to be the result of leaf hydraulic conductance loss due to strong hydraulic segmentation leading to leaf senescence during periods with low water availability (Zhang et al. 2009). This species is an evergreen tree species that usually renew the entire leaf crop during the dry season (the leaf life span is about 14 months). Even though its stems have a wide safety margin (2.2 MPa or larger depending on tree height) and are well protected by vulnerable leaves (Fig. 2a), the leaves in the top of the canopy of taller trees have lower leaf hydraulic conductance (K_{leaf}) compared to leaves of smaller trees (Fig. 2b). The threshold of water potential leading to hydraulic dysfunction can be reached frequently, particularly in leaves of taller trees resulting in leaf senescence and leaf drop, which decreases the photosynthetic surface for carbon assimilation. Furthermore, net CO_2 assimilation decreases substantially with increasing tree height (Fig. 2b). Lower K_{leaf} may result in more negative leaf water potentials at a given transpiration rate, lower stomatal conductance and lower rates of CO_2 diffusion to the sites of carboxylation in the leaf mesophyll. A decline in total leaf surface area per plant in taller individuals would temporarily help to maintain adequate water balance, but a substantial decrease in total leaf surface area and relatively low carbon uptake per unit leaf surface area have a large physiological

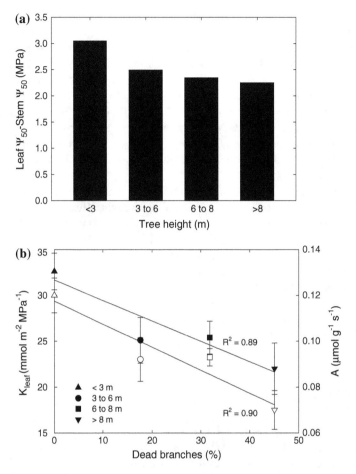

Fig. 2 a Safety margin between leaf Ψ_{50} and stem Ψ_{50} (leaf Ψ_{50}–stem Ψ_{50}) for *Sclerolobium paniculatum* trees with different stature and **b** Leaf hydraulic conductance (*black symbols*) and net assimilation rate (*open symbols*) versus the percentage of *S. paniculatum* dead branches for trees with different height. Values are mean ±1SE (n = 4–6 trees for K_{leaf} and n = 6–14 for **a**). Information obtained from Zhang et al. (2009)

cost: larger *S. paniculatum* trees receive a lower return in carbon gain from their investment in stem and leaf biomass compared to small trees. Two mechanisms were apparently responsible for massive mortality in tall *S. paniculatum* trees: leaf hydraulic failure and carbon starvation. Leaf hydraulic failure can trigger leaf senescence and leaf drop, and thus result in a substantial decrease in carbon uptake. New leaf production would be limited if growth will depend more on the depleted internal carbohydrate pool stored in permanent tree parts such as roots and/or stems than on exogenous carbon assimilation. Ultimately, carbon starvation will drive the taller trees to mortality. No apparent symptoms of herbivore damage was observed, thus biotic attacks may not play a role in the mortality of *S. paniculatum*.

A similar mass mortality phenomenon associated with strong vulnerability segmentation has been recently studied (Scholz et al. 2014). Dieback was observed in South America temperate forests during one of the most severe droughts of the 20th century (1998–1999). During this drought period *Nothofagus dombeyi* experienced symptoms of water stress, such as leaf wilting and abscission, before tree die-back occurred. Even though *N. dombeyi* stems had a threshold of -6.7 MPa before reaching nearly complete hydraulic failure (Ψ_{88}), leaf Ψ_{88} occurred at a leaf water potential of only -2 MPa, which probably is reached during extreme episodic droughts. Massive mortality in *N. dombeyi* appears to be the result of the total loss of leaf hydraulic conductance leading to leaf dehydration and leaf drop and ultimately to carbon starvation.

Another proposed mechanism, in addition to hydraulic failure and carbon starvation, is cellular metabolism limitation during drought. This hypothesis suggests that low tissue water potentials during drought may constrain cell metabolism (Würth et al. 2005; Ryan et al. 2006; Sala and Hoch 2009) thereby preventing the production and translocation of carbohydrates and secondary metabolites necessary for plant defense against biotic attacks and herbivory. Of course in addition to physiological mechanisms, there are other common modes of mortality in tropical forests that are related to biomechanical factors, and not necessarily to drought, such as uprooting, trunk snapping, and defoliation due to high winds (Putz et al. 1983; Kraft et al. 2010).

Nocturnal Transpiration

Nocturnal transpiration has been found to occur across many plant species, seasons and biomes (e.g. Caird et al. 2007; Dawson et al. 2007; Forster 2014). Nocturnal sap flow in trunks, in general, is the consequence of nocturnal leaf water loss through open stomata or the recharge of water storage. There is no general understanding as to how much nocturnal flow (Qn) occurs and whether it is a significant contribution to total daily sap flow (Q). A recent literature review assessing the magnitude of Qn as a proportion of Q across seasons, biomes, and phylogenetic groups found that among 98 species Qn was 12.03 % on average, but ranging up to 69 % with equatorial and tropical biomes having significantly higher percentages of Qn than warm temperate biomes (Forster 2014).

There is no consensus concerning the functional significance that nocturnal water flow may have for trees and ecosystems. Of course, some species may be incapable of complete stomatal closure at night, which presumably would not confer an adaptive advantage. Some of the suggested roles of Qn are (1) recharge of internal water storage, (2) embolism repair, (3) improving nutrient acquisition by driving bulk flow of soil water to the root surfaces, (4) sustaining carbohydrate export and other processes driven by dark respiration, particularly in fast growing-light requiring tree species, (5) and improving oxygen supply to the sapwood at night which may be necessary to sustain respiration in xylem parenchyma (Bucci et al. 2004; Scholz et al. 2006; Snyder et al. 2008). Regarding the role of Qn for improving nutrient acquisition it is necessary to mention that for many ions, it is their mobility and solubility in the soil, rather than the maximal inflow rate, that determines the rate at which roots can acquire them from the rhizosphere (Clarkson 1981).

The highest percentage of species with nocturnal water loss has been observed in tropical regions probably as a consequence of the widespread presence of oligotrophic and very nutrient poor soils (Scholz et al. 2006; Forster 2014). Nitrogen or phosphorus limitations on growth are common in tropical regions, and are particularly severe in the neotropical savannas of central Brazil (Cerrado) characterized by well-drained, old oxisols. These soils are not only highly nutrient deficient, and their pH and cation exchange capacities low, but their aluminum saturation levels are high, which may affect phosphorylation and reduce availability of calcium and phosphorus (Furley and Ratter 1988; Haridasan 2000). Moreover, seasonal drought, high irradiances and high atmospheric evaporative demand are characteristic of neotropical savannas. Although savanna trees can achieve high transpiration rates and high stomatal conductances, which could potentially increase nutrient uptake, hydraulic limitations and the high evaporative demand of the atmosphere during the daytime, particularly during the dry season, impose strong stomatal control of transpiration via partial stomatal closure (Meinzer et al. 1999; Bucci et al. 2005, 2008). Thus, imbalances in water uptake and loss during the daytime may substantially constrain plant capacity to optimize nutrient uptake.

There is evidence that the transpiration stream is largely responsible for supplying mobile soil nutrients (e.g., nitrate) to plant roots. When the nutrients arrives at the root surface, its transport across the plasma membrane may occur either by diffusion down an electrochemical potential gradient or mainly by active transport against an electrochemical potential gradient. For many ions, it is their mobility in the soil, rather than the maximum inflow rate, that determines the rate at which roots can acquire them from the rhizosphere (Clarkson 1981). Maintenance of water flux through the soil–plant–atmosphere continuum via nighttime transpiration could enhance nutrient supply to plant roots and hence improve the nutrient balance of savanna trees growing in oligotrophic soils. If this assumption is correct, then lower rates of nocturnal water loss should be observed in plants growing in nutrient-deficient soils where nutrient limitations have been relieved. Scholz et al. (2006) made use of a long-term fertilization experiment in a savanna ecosystem of Central Brazil in which N, P and N + P nutrients had been added twice a year from 1998 to 2006. *Qualea grandiflora*, a common savanna tree growing in non-fertilized plots exhibited relatively high rates of nocturnal sap flow and stomatal conductance compared with trees in fertilized plots (Fig. 3a, b). The percent of nocturnal water flow and stomatal conductance compared to total daily transpiration and stomatal conductance in *Q. grandiflora* in the non-fertilized plots were higher than in those in N and P fertilized plots suggesting that long term fertilization had decreased or completely removed the nutrient limitations of the oligotophic savanna soils, and that nocturnal transpiration has no longer an adaptive value.

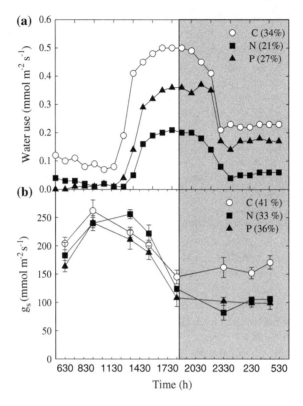

Fig. 3 Mean daily courses of **a** water loss per unit of leaf surface area in terminal branches and **b** stomatal conductance, during the dry season, for two trees of *Qualea grandiflora* in three treatments: Control (*C*), Nitrogen (*N*) added, and Phosphorus (*P*) added. The *gray area* indicates the dark period. Values within parentheses indicate the percent of nocturnal transpiration (**a**) and stomatal conductance (**b**) compared to total daily transpiration and stomatal conductance, respectively, calculated using the area under the *curve*. Information obtained from Scholz et al. (2006)

Total foliar nitrogen content per plant across five savanna tree species was negatively correlated with nocturnal water use calculated as the percent of total daily water use (Fig. 4). Total foliar N increased in trees fertilized with N during 5 years, and their nocturnal water use was substantially lower compared to non-fertilized trees. This functional relationship between N content and nocturnal water use is also consistent with a putative role of nocturnal transpiration in improving nutrient acquisition by savanna trees growing in nutrient poor soils. Other studies have not found evidence of nocturnal transpiration promoting nutrient uptake in plants (e.g. Howard and Donovan 2007, 2010; Matimati et al. 2014). However, some of these studies were conducted with short lived species growing in greenhouses, which constrain the analysis of the adaptive significance of their results. There is apparently a genetic basis for nighttime transpiration. A study with

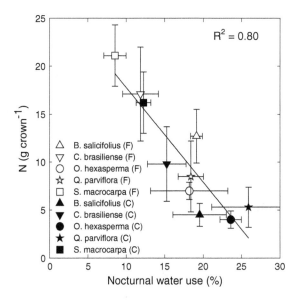

Fig. 4 Total amount of nitrogen in foliage per tree (N) in relation to nocturnal water use (percent of daily total water loss) for five savanna woody species. Each point represents the mean value ± SE of 3–4 trees per species. The line is linear regression fitted to the data: (**a**) y = 27.7–0.97x, $p < 0.001$. *Filled symbols* are plants in control plots and *open symbols* are plants in fertilized plots. Information from Bucci et al. (2006), Scholz et al. (2006) and unpublished data

different accessions of *Arabidopsis thaliana* indicated that nighttime stomatal conductance for different accessions of this species were well correlated with their native habitat vapor pressure deficit (VPD), suggesting lower nighttime stomatal conductance is favored by natural selection in drier habitats (Christman et al. 2009). It is likely that in neotropical trees adapted to extremely low soil nutrient availability, natural selection may have favored maintenance of relatively high nocturnal sap flow in their native habitats, and that upon removal of nutrient limitation by long-term fertilization; they can acclimate to their new nutrient status by decreasing the rate of nocturnal water loss. Furthermore, many neotropical savanna tree species have deep root systems where abundant water is available, which allow plants to maintain relatively constant daily minimum leaf water potential regardless of the duration of the dry season

Recharge of internal water storage at night may occur if water loss by nocturnal transpiration does not compete strongly with nocturnal recharge. In this case, the larger fraction of nocturnal water uptake should be directed toward refilling of internal storage compartments rather than used by transpiring leaves. In the Brazilian savanna tree species *Styrax ferrugineus*, sap flow measured near the base of the trunk continued during the night at rates up to 12 cm^{-3} h, but substantially decreased or became negligible when trees were covered to minimize transpiration, suggesting that a large part of Qn consisted of transpiration rather than recharge of storage compartments (Fig. 5). Diurnal changes in stem diameter can be used as a

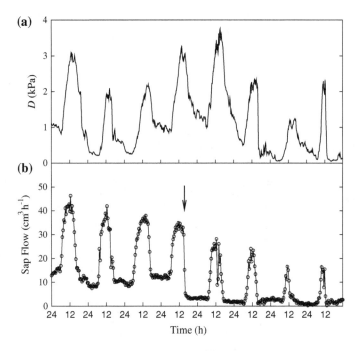

Fig. 5 Daily courses of air saturation deficit (*D*) and sap flow measured near leaves of one individual of *Styrax ferrugineus*. The *arrow* indicates the moment when the plant was covered with opaque plastic bags to restrict transpiration. Information from Bucci et al. (2004)

proxy for the dynamics of water movement during utilization and recharge of stem water (Scholz et al. 2008a). When water is withdrawn from the stem storage, stem diameter decreases, and when the stem storage is recharged, mainly at night, stem diameter increases. Based on continuous measurements of stem diameter, stem water storage tends to recharge faster when nocturnal transpiration is prevented (Fig. 6), suggesting that nocturnal transpiration and recharge of water storage compartments act as competing sinks for water taken up by roots. Another competing sink for water taken up by deep roots is hydraulic redistribution (Scholz et al. 2008b), the movement of water from deeper and wetter soil layers to upper and drier regions of the soil profile via plant roots (Scholz et al. 2002). During hydraulic redistribution, water taken up by deep roots moves out of shallow roots and into the upper soil layers instead of moving into the transpiration stream inside the tree. This can potentially occur when nocturnal transpiration is relatively low and the dry soil is the stronger sink (Scholz et al. 2008b).

Nocturnal water flow can also contribute to embolism repair. Embolism repair is unlikely to occur as a consequence of night time water flow alone unless there is an active mechanism in place for carbohydrate movement into empty conduits (Johnson et al. 2012). For trees species that cannot avoid embolism formation (see below), an alternative adaptive response includes the loss and recovery of hydraulic

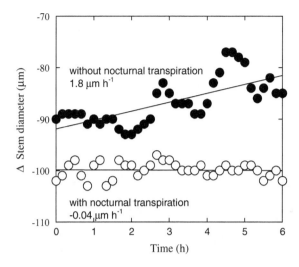

Fig. 6 Stem diameter changes from midnight to 6 h for exposed (with nocturnal transpiration, *open symbols*) and covered (without nocturnal transpiration, *filled symbols*) *Styrax ferrugineus* tree of Fig. 4. The mean rates of diameter increase or decrease are also shown. Information from Bucci et al. (2004)

conductivity on a daily basis. This may be an especially common phenomenon in tropical tree species with high wood density (negligible stem water reservoirs) (Meinzer et al. 2008a). Apparently for embolism repair to occur, sugars must be transported into the embolized conduits with water movement following along a gradient in osmotic potential (Bucci et al. 2003; Johnson et al. 2012). This gradient in osmotic potential is more likely to occur at night when tension driven gradients inside the trunks and stems are relatively low.

Capacitance

Wood capacitance seems to be a key functional trait that strongly affects many aspects of woody plant functioning. There is a strong correlation between sapwood capacitance and measures of short-term water balance in large trees, including the operating minimum stem and leaf water potentials and the safety margin (minimum water potential—water potentials causing catastrophic run-away embolisms) (Scholz et al. 2007; Oliva Carrasco et al. 2014). These correlations are likely mediated by interactions between sapwood capacitance, percentage of daily water use drawn from internal water storages and stomatal regulation of transpiration. The transient buffering effect of high sapwood capacitance plays a particularly important role in maintaining tree water balance in species with relatively low wood density and high growth rates (e.g. Meinzer et al. 2003; Scholz et al. 2007).

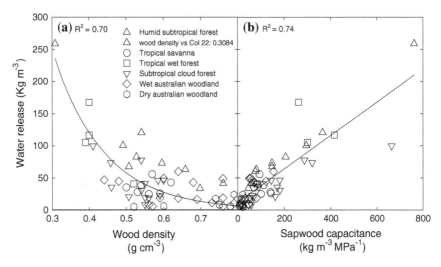

Fig. 7 Water released across the stem water potential operating range as a function of **a** wood density and **b** sapwood capacitance for 71 tropical and subtropical tree species. The *line* in **a** is the exponential decay function fitted to the data (y = 2268*exp (−7.3x), $p < 0.0001$), and in **b** is the linear regression fitted to the data (y = 12 + 0.26x, $p < 0.0001$). Information obtained from Meinzer et al. (2003, 2004, 2008a, b), Scholz et al. (2007), Zhang et al. (2009), Richards et al. (2014), Oliva Carrasco et al. (2014)

Species-specific stem water storage and stem capacitance are strongly correlated with wood density in tropical and subtropical tree species, with lower sapwood density resulting in larger capacitance (Bucci et al. 2004; Meinzer et al. 2003; Scholz et al. 2007; Richards et al. 2014; Oliva Carrasco et al. 2014). Figure 7 depicts the relationship between the amounts of water released to the transpiration stream by the stem storages, across the stem water potential operating range, as a function of sapwood density. The amount of water released increases exponentially with decreasing sapwood density (Fig. 7a) across several species from different tropical and subtropical ecosystems. Similarly, the amount of water released increased linearly, instead of exponentially, with increasing sapwood capacitance, which is generally defined as the ratio of change in tissue water volume to changes in its water potential, indicating the ability of the sapwood to provide water to the transpiring leaves (Fig. 7b). These relationships reflect species-specific anatomic differences of the sapwood, perhaps in the fraction lumen:wall of the fibers, which is one of the determinants of wood density (Chave et al. 2009) and appears to have an important influence on release of stored water (Richards et al. 2014).

Recently Wolfe and Kursar (2015) determined that water released from water storages increases during flushing leaves in deciduous species from seasonally dry tropical forests. Consistent with this finding, sapwood capacitance is significantly higher in deciduous compared to evergreen tropical tree species (Fig. 8a), which appears also to play a central role in their rapid growth patterns (Fig. 8b) facilitating rapid canopy access of the forests as these species are less shade tolerant than

Fig. 8 Sapwood capacitance (**a**) and cumulative growth rate (**b**) of deciduous and evergreen species. Values are mean ±SE from 62 evergreen and 9 deciduous species for sapwood capacitance and 13 evergreen and 5 deciduous species for cumulative growth. Information obtained from Meinzer et al. (2003), Scholz et al. (2007), Zhang et al. (2009), Richards et al. (2014), Oliva Carrasco et al. (2014)

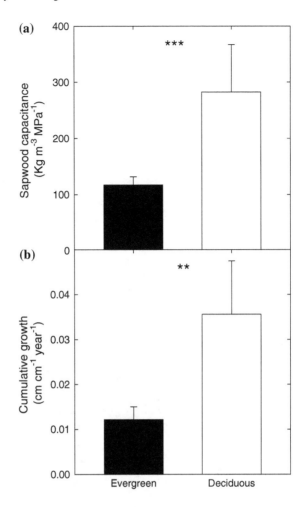

evergreen species. Higher growth rates in species with high capacitance could be achieved by keeping the stomata open for longer periods of time. For several tropical and subtropical species, the ratio of crown sap flow in the early afternoon and the maximum crown sap flow attained during the day (E_{1300}/E_{max}) increased asymptotically with the amounts of stored water use determined by comparing the daily sap flow patterns in the stem base and in the crown of the tree (Fig. 9). The E_{1300}/E_{max} ratio reflects the degree of stomatal control of transpirational losses with higher ratios implying less stringent stomatal control at midday. The asymptotic relationship suggests that in trees with larger amounts of water in the reservoirs carbon assimilation can remain relatively high throughout the day.

Several studies have shown that species with high capacitance are more vulnerable to vessel cavitation (less negative Ψ_{50}; Meinzer et al. 2008b; Sperry et al. 2008), presumably associated with low wood density. Consistent with this pattern,

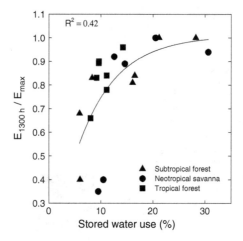

Fig. 9 The ratio of branch sap flow measured at 1300 h (E 1300 h) and maximum branch sap flow (E_{max}) as a function of stored water use (%) derived from sap flow measured at the stem base and in the crown of the tree for tree species from subtropical and tropical forests and tropical savannas. The line is the exponential rise to maximum function fitted to the data ($y = 1-\exp(-0.13x)$, $p < 0.05$). Information obtained from Goldstein et al. (1998), Meinzer et al. (2003, 2004), Scholz et al. (2008a, b), Oliva Carrasco et al. (2014)

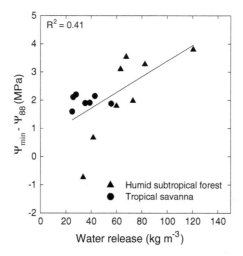

Fig. 10 Safety margin between minimum stem water potential (Ψ_{min}) and stem water potential at 88 % loss of maximum stem hydraulic conductance (stem Ψ_{88}) (Ψ_{min}–stem Ψ_{88}) as a function of water release for tropical savanna and subtropical tree species. The line is the linear regression fitted to the data ($y = 0.61 + 0.063x < p = 0.01$). Information obtained from Bucci et al. (2008), Hao et al. (2008), Villagra et al. (2013) and Oliva Carrasco et al. (2014)

the safety margin (Ψ_{min}–Ψ_{88}) of tropical and subtropical species increases linearly with water released into the sapwood (at the operating minimum stem water potential) (Fig. 10) suggesting a role for stem water storage in preventing embolism formation by reducing fluctuations in water potential due to water deficits. Although high wood density and high capacitance are at the end of a trait spectrum representing a trade-off between avoiding mechanical damage (by biotic or abiotic agents) and enhanced hydraulic efficiency, high capacity to release water to the transpiration steam may be a low cost adaptation to avoid water deficits during peak water use at midday and under seasonal or transient drought periods such as it occurs in wet subtropical forests.

Conclusions

Most trees species have a relatively short life span of 80–200 years (but see Russo and Kitajima this volume) in the wet tropics compared to more than a 1000 years in some temperate zone trees. The results of the unique selective pressures in the tropics have resulted in a large array of life history traits and in particular a large array of functional traits related to the water relations of trees. It is not well known if there are large species specific differences in physiological traits plasticity to environmental changes. Fast-growing tropical tree species in different light regimes, for example, are capable of plastic changes in hydraulic architecture, mainly intrinsic hydraulic conductivity, and increased water-transport efficiency in response to increases in light availability (Campanello et al. 2008). Slower growing tree species under similar environments however do not exhibit plastic responses. More information on the degree of plastic responses and the genetic variability within species will help to better predict changes in species composition in tropical ecosystems under different scenarios of global change.

The aspects of tree hydraulic architecture and water transport discussed in this chapter are relevant for understanding potential tree responses to global climate change and to the rapid land use changes, in particular deforestation. Multiple selective pressures may result in species-specific physiological traits that can have adaptive value under stable environmental regimes. However, under the current rapid environmental change scenario, a particular trait may become non-adaptive. The degree of hydraulic segmentation in tropical and subtropical trees varies depending on their life history characteristics and habitat type. Strong segmentation, whether it be hydraulic or vulnerability, may help to maintain an adequate water balance but may also have indirect negative impacts on tree carbon balance during severe droughts. Two mechanisms can trigger dieback in tree species with strong vulnerability segmentation: leaf hydraulic failure and carbon starvation. More research is necessary for assessing the generality of dieback mechanisms in tropical trees experiencing prolonged drought. Nocturnal water loss and/or nocturnal sap flow is widespread among many tropical and subtropical tree species. One of the most likely functions of nocturnal sap flow in savanna or forest trees growing in nutrient poor soils appears to be enhanced nutrient acquisition. Nocturnal water flow can also contribute to embolism repair. High capacitance appears to play a central role in the rapid growth of tropical deciduous tree species facilitating rapid canopy access by keeping the stomata open for longer periods of time, and thus facilitating carbon assimilation.

References

Adams HD, Germino MJ, Breshears DD, Barron-Gafford GA, Guardiola-Claramonte M, Zou CB, Huxman TE (2013) Nonstructural leaf carbohydrate dynamics of *Pinus edulis* during drought-induced tree mortality reveal role for carbon metabolism in mortality mechanism. New Phytol 197:1142–1151

Allen C, Alison K, Chenchouni H, Bachelet D, McDowell N, Vennetier M, Kitzberger T, Rigling A, Breshears D, Hogg E, Gonzalez P, Fensham R, Zhang Z, Castro J, Demidova N, Lim J, Allard G, Running S, Semerci A, Cobb B (2010) A global overview of drought and heat-induced tree mortality reveals emerging change risks for forests. Forest Ecol Manage 259:660–684

Allen CD, Macalady AK, Chenchouni H, Bachelet D, McDowell N, Vennetier M, Kitzberger T, Rigling A, Anderegg WRL, Berry JA, Smith DD, Sperry JS, Anderegg LDL, Field CB (2012) The roles of hydraulic and carbon stress in a widespread climateinduced forest die-off. Proc Natl Acad Sci USA 109:233–237

Blackman CJ, Brodribb TJ, Jordan GJ (2010) Leaf hydraulic vulnerability is related to conduit dimensions and drought resistance across a diverse range of woody angiosperms. New Phytol 188:1113–1123

Brodribb TJ, and Holbrook NM (2003) Stomatal closure during leaf dehydration, correlation with other leaf physiological traits. Plant Physiol. 132:2166–2173

Bucci SJ, Scholz FG, Goldstein G, Meinzer FC (2003) Dynamic diurnal changes in hydraulic specific conductivity in petiole of two savanna tree species: identifying factors and mechanisms contributing to the refilling of embolized vessels. Plant Cell Environ 26:1633–1645

Bucci SJ, Goldstein G, Meinzer FC Scholz FG, Franco AC, Bustamante M (2004) Functional convergence in hydraulic architecture and water relations of tropical savanna trees: from leaf to whole plant. Tree Physiol 24:891–899

Bucci SJ, Goldstein G, Meinzer FC, Franco AC, Campanello P, Scholz FG (2005) Mechanisms contributing to seasonal homeostasis of minimum leaf water potential and predawn disequilibrium between soil and plants in Neotropical savanna trees. Trees 19:296–304

Bucci SJ, Scholz FG, Goldstein G, Meinzer FC, Franco AC, Campanello PI, Villalobos-Vega R, Bustamante M, Miralles-Wilhelm (2006) Nutrient availability constrains the hydraulic architecture and water relations of savanna trees. Plant Cell Environ 29:2153–2167

Bucci SJ, Scholz FG, Goldstein G, Meinzer FC, Franco AC, Zhang Y, Hao GY (2008) Water relations and hydraulic architecture in Cerrado trees: adjustments to seasonal changes in water availability and evaporative demand. Braz J Plant Physiol 20:233–245

Bucci SJ, Scholz FG, Campanello PI, Montti L, Jimenez M, La Manna L, Rockwell A, Guerra P, Lopez-Bernal P, Troncoso O, Holbrook MN, Goldsein G (2012) Hydraulic differences along the water transport system of South American Nothofagus species: do leaves protect the stem functionality? Tree Physiol 32:880–892

Bucci SJ, Scholz FG, Peschiutta ML, Arias N, Meinzer FC, Goldstein G (2013) The stem xylem of Patagonian shrubs operates far from the point of catastrophic dysfunction and is additionally protected from drought- induced embolism by leaves and roots. Plant Cell Environ 36:2163–2174

Caird MA, Richards JH, Donovan LA (2007) Nighttime stomatal conductance and transpiration in C-3 and C-4 plants. Plant Physiol 143:4–10

Campanello PI, Gatti MG, Goldstein G (2008) Coordination between water- transport efficiency and photosynthetic capacity in canopy tree species at different growth irradiances. Tree Physiol 28:85–94

Chave J, Coomes D, Jansen S, Lewis SL, Swenson NG, Zanne AE (2009) Towards a worldwide wood economics spectrum. Ecol Lett 12:351–366

Chen J-W, Q Zhang, L X-S. Li and K-F. Cao (2009) Independence of stem and leaf hydraulic traits in six Euphorbiaceae tree species with contrasting leaf phenology. Planta 230:459–468

Choat B, Jansen S, Brodribb TJ, Cochard H, Delzon S, Bhaskar R, Bucci SJ, Feild TS, Gleason SM, Jacobsen AL, Lens F, Maherali H, Martinez-Vilalta J, Mayr S, Mencuccini M, Mitchell PJ, Nardini A, Pittermann J, Pratt RB, Sperry JS, Westoby M, Wright IJ, Zanne A (2012) Global convergence in the vulnerability of forests to drought. Nature 491:752–756

Christman MA, Donovan LA, Richards JH (2009) Magnitude of night-time transpiration does not affect plant growth or nutrition in well-watered *Arabidopsis*. Physiol Plant 136:264–273

Clarkson DT (1981) Nutrient interception and transport by root systems. In: Johnson CB (ed) Physiological factors limiting plant productivity. Butterworths, London, pp 307–314

Dawson TE, Burgess SSO, Tu KP, Oliveira RS, Santiago LS, Fisher JB, Simonin KA, Ambrose AR (2007) Nighttime transpiration in woody plants from contrasting ecosytems. Tree Physiol 27:561–576

Domec J-C, Scholz FG, Bucci SJ, Meinzer FC, Goldstein G, Villalobos-Vega R (2006) Diurnal and seasonal variation in root xylem embolism in neotropical savanna woody species: impact on stomatal control of plant water status. Plant Cell Environ 29:26–35

Forster M (2014) How significant is nocturnal sap flow? Tree Physiol 34:757–765

Furley PA, Ratter JA (1988) Soil resources and plant communities of the central Brazilian Cerrado and their development. J Biography 15:97–108

Goldstein G, Meinzer FC (1983) Influence of insulation of dead leaves and low temperatures on water balance in an Andean giant rosette plant. Plant Cell Environ 6:649–656

Goldstein G, Meinzer FC, Monasterio M (1984) The role of capacitance in the water balance of Andean giant rosette species. Plant, Cell Environ 21:397–406

Goldstein G, Rada F, Rundel P, Azocar A, Orozco A (1989) Gas exchange and water relations of evergreen and deciduous tropical savanna trees. Ann Sci Forestieres 46:448–453

Goldstein G, Andrade JL, Meinzer FC, Holbrook NM, Cavelier J, Jackson P, Celis A (1998) Stem water storage and diurnal patterns of water use in tropical forest canopy trees. Plant, Cell Environ 21:397–406

Hao G, Hoffmann WA, Scholz FG, Bucci SJ, Meinzer FC, Franco AC, Cao K-F, Goldstein G (2008) Stem and leaf hydraulics of congeneric tree species from adjacent tropical savanna and forest ecosystems. Oecologia 155:405–415

Haridasan M (2000) Nutrição mineral das plantas nativas do Cerrado. Rev Bras Fisiol Veg 12:54–64

Hartmann H, Ziegler W, Kolle O, Trumbore S (2013) Thirst beats hunger declining hydration during drought prevents carbon starvation in Norway spruce saplings. New Phytol 200:340–349

Howard AR, Donovan LA (2007) *Helianthus* nighttime conductance and transpiration respond to soil water but not nutrient availability. Plant Physiol 143:145–155

Howard AR, Donovan LA (2010) Soil nitrogen limitation does not impact night time water loss in Populus. Tree Physiol 30:23–31

Johnson DM, McCulloh KA, Meinzer FC, Woodruff DR, Eissenstat DM (2011) Hydraulic patterns and safety margins, from stem to stomata, in three eastern US tree species. Tree Physiol 31:659–668

Johnson DM, McCulloh KA, Woodruff DR, Meinzer FC (2012) Hydraulic safety margins and embolism reversal in stems and leaves: why are conifers and angiosperms so different? Plant Sci 195:48–53

Kavanagh KL, Bond BJ, Aitken SN, Gartner BL, Knowe S (1999) Shoot and root vulnerability to xylem cavitation in four populations of Douglas-fir seedlings. Tree Physiol 19:31–37

Koch GW, Amthor JS, Goulden M (1994) Diurnal patterns of leaf photosynthesis, conductance and water potential at the top of a lowland rain forest canopy in cameroon. measurements from the radeau des cimes. Tree Physiol 14:347–360

Kraft NJB, Metz MR, Condit RS, Chave J (2010) The relationship between wood density and mortality in a global tropical forest data set. New Phytol 188:1124–1136

Maherali H, Pockman WT, Jackson RB (2004) Adaptive variation in the vulnerability of woody plants to xylem cavitation. Ecology 85:2184–2199

Matimati I, Verboom GA and Cramer MD (2014). Do hydraulic redistribution and nocturnal transpiration facilitate nutrient acquisition in Aspalathus linearis? Oecologia 175:1129–1142

McCulloh KA, Johnson DM, Meinzer FC, Voelker SL, Lachenbruch B and Domec J-C (2012). Hydraulic architecture of two species differing in wood density: opposing strategies in co-occurring tropical pioneer trees, Plant Cell Environ 35:116–125

McDowell N, Pockman WT, Allen CD, Breshears DD, Cobb N, Kolb T, Plaut J, Sperry J, West A, Williams DG et al (2008) Mechanisms of plant survival and mortality during drought: why do some plants survive while others succumb to drought? New Phytol 178:719–739

McDowell NG, Beerling DJ, Breshears DD, Fisher RA, Raffa KF, Stitt M (2011) The interdependence of mechanisms underlying climate driven vegetation mortality. Trends Ecol Evol 26:523–532

Meinzer F, Goldstein G, Holbrook NM, Jackson P, Cavelier J (1993) Stomatal and environmental control of transpiration in a lowland tropical forest. Plant, Cell Environ 16:429–436

Meinzer FC, Goldstein G, Franco AC, Bustamante M, Igler E, Jackson P, Caldas L. Rundel PW (1999) Atmospheric and hydraulic limitations on transpiration in Brazilian cerrado woody species. Funct Ecol 13:273–282

Meinzer FC, James SA, Goldstein G, Woodruff D (2003) Whole-tree water transport scales with sapwood capacitance in tropical forest canopy trees. Plant, Cell Environ 26:1147–1155

Meinzer FC, James SA, Goldstein G (2004) Dynamics of transpiration, sap flow and use of stored water in tropical forest canopy trees. Tree Physiol 24:901–909

Meinzer FC, Campanello PI, Domec J-C, Gatti MG, Goldstein G, Villalobos-Vega R, Woodruff DR, (2008a) Constraints on physiological function associated with branch architecture and wood density in tropical forest trees. Tree Physiol 28:1609–1617

Meinzer FC, Woodruff DR, Domec J-C, Goldstein G, Campanello PI, Gatti G, Villalobos-Vega R (2008b) Coordination of leaf and stem water transport properties in tropical forest trees. Oecologia 156:31–41

Meinzer FC, Johnson DM, Lachenbruch B, McCulloh KA, Woodruff DR (2009) Xylem hydraulic safety margins in woody plants: coordination of stomatal control of xylem tension with hydraulic capacitance. Funct Ecol 23:922–930

Nardini A, Tyree MT, Salleo S (2001) Xylem cavitation in the leaf of *Prunus laurocerasus* and its impact on leaf hydraulics. Plant Physiol 125:1700–1709

Oliva Carrasco L, Bucci SJ, Di Francescantonio D, Lezcano OA, Campanello PI, Scholz FG, Rodríguez S, Madanes N, Cristiano PM, Hao G-Y, Holbrook NM, Goldstein G (2014) Water storage dynamics in the main stem of subtropical tree species differing in wood density, growth rate, and life history traits. Tree Physiol 31:1–12

Plaut JA, Wadsworth WD, Pangle R, Yepez EA, McDowell NG, Pockman WT (2013) Reduced transpiration response to precipitation pulses precedes mortality in a piñon-juniper woodland subject to prolonged drought. New Phytol 200:375–387

Putz FE, Coley PD, Lu K, Montalvo A, Aiello A (1983) Uprooting and snapping of trees: structural determinants and ecological consequences. Can J Res 13:1011–1020

Richards AE, Wright IJ, Lenz TI, Zanne AE (2014) Sapwood capacitance is greater in evergreen sclerophyll species growing in high compared to low-rainfall environments. Funct Ecol 28:734–744

Roberts J, Cabral OMR, De Aguiar LF (1990) Stomatal and boundary-layer conductances in an Amazonian terra firme rain forest. J Appl Ecol 27:336–353

Ryan M, Phillips N, Bond B (2006) The hydraulic limitation hypothesis revisited. Plant, Cell Environ 29:367–381

Sack L, Holbrook NM (2006) Leaf hydraulics. Ann Rev Plant Biol 57:361–381

Sala A, Hoch G (2009) Height-related growth declines in ponderosa pine are not due to 183 carbon limitation. Plant, Cell Environ 32:22–30

Sala A, Piper F, Hoch G (2010) Physiological mechanisms of drought-induced tree mortality are far from being resolved. New Phytol 186:274–281

Sala A, Woodruff DR, Meinzer FC (2012) Carbon dynamics in trees: feast or famine? Tree Physiol 32:764–775

Sarmiento G, Goldstein G, Meinzer FC (1985) Adaptive strategies of woody species in neotropical savannas. Biol Rev 60:315–355

Scholz FG, Bucci SJ, Goldstein G, Meinzer FC, Franco AC (2002) Hydraulic redistribution of soil water by neotropical savanna trees. Tree Physiol 22:603–612

Scholz FG, Bucci SJ, Goldstein G, Meinzer FC, Franco AC, Miralles-Welheim F (2006) Removal of nutrient limitations by long-term fertilization decreases nocturnal water loss in savanna trees. Tree Physiol 27:551–559

Scholz FG, Bucci SJ, Goldstein G, Meinzer FC, Franco AC, Miralles-Wilhelm F (2007) Biophysical properties and functional significance of stem water storage tissues in neo-tropical savanna trees. Plant, Cell Environ 30:236–248

Scholz FG, Bucci SJ, Goldstein G, Meinzer FC, Franco AC, Miralles-Wilhelm F (2008a) Temporal dynamics of stem expansion and contraction in savanna trees: withdrawal and recharge of stored water. Tree Physiol 28:469–480

Scholz FG, Bucci SJ, Goldstein G, Moreira MZ, Meinzer FC, Domec J_C, Villalobos R, Franco AC, Miralles-Wilhelm F (2008b) Biophysical and life history determinants of hydraulic lift in Neotropical savanna trees. Func Ecol 22:773–786

Scholz FG, Bucci SJ, Goldstein G (2014) Strong hydraulic segmentation and leaf senescence may trigger die-back in *Nothofagus dombeyi* under severe droughts: a comparison with the co-occurring *Austrocedrus chilensis*. Trees 28:1425–1487

Scoffoni C, Vuong C, Diep S, Cochard H, Sack L (2014) Leaf shrinkage with dehydration: coordination with hydraulic vulnerability and drought tolerance. Plant Physiol 164(4):1772–1788

Snyder KA, Jaes JJ, Richards JH, Donovan LA (2008) Does hydraulic lift or nighttime transpiration facilitate nitrogen acquisition? Plant Soil 306:159–166

Sperry JS, Meinzer FC, McCulloh KA (2008) Safety and efficiency conflicts in hydraulic architecture: scaling from tissues to trees. Plant, Cell Environ 31:632–645

Tyree MT, Sperry JS (1989) Vulnerability of xylem to cavitation and embolism. Ann Rev Plant Physiol Plant Mol Biol 40:19–38

Villagra M, Campanello PI, Bucci SJ, Goldstein G (2013) Functional relationships between leaf hydraulics and leaf economics in response to nutrient addition in subtropical tree species. Tree Physiol 33:1308–1318

Wolfe BT, Kursar TA (2015) Diverse patterns of stored water use among saplings in seasonally dry tropical forests. Oecologia. doi:10.1007/s00442-015-3329-z

Würth MKR, Peláez-Riedl S, Wright SJ, Körner C (2005) Non-structural carbohydrate pools in a tropical forest. Oecologia 143:11–24

Zhang YJ, Meinzer FC, Hao GY, Scholz FG, Bucci SJ, Franco AC, Villalobos Vega R, Giraldo JP, Cao KF, Hoffmann WA, Goldstein G (2009) Size-dependent 1 mortality in a Neotropical savanna tree: the role of height-related adjustments in hydraulic architecture and carbon allocation. Plant, Cell Environ 32:1456–1466

Zimmermann MH (1983) Xylem structure and the ascent of sap. Springer, New York

Maintenance of Root Function in Tropical Woody Species During Droughts: Hydraulic Redistribution, Refilling of Embolized Vessels, and Facilitation Between Plants

F.G. Scholz, S.J. Bucci, F.C. Meinzer and G. Goldstein

Abstract Most woody dominated tropical ecosystems are subjected to drought periods of different lengths, from few days or weeks in wet forests to several months in drought deciduous forests and savanna ecosystems. The roots during these low soil water availability periods may experience hydraulic and metabolic dysfunctions resulting not only in a substantial decrease in root growth rates, but also in reduced water and nutrient uptake activity. We discuss three groups of processes: (1) hydraulic redistribution, (2) refilling of embolized vessels, and (3) ecological facilitation, which can contribute together or in isolation to continuous root activities despite low soil water availability. Furthermore they can help avoid physiological uncoupling between above and belowground plant parts and thus help to preserve the integration of the root, stem and leaf hydraulic continuum within trees. The two mechanisms for refilling embolized vessels that help repair the functionality of the water transport system discussed are the transient pressure imbalance between the xylem and surrounding tissues, and positive root pressures.

F.G. Scholz (✉) · S.J. Bucci
Grupo de Estudios Biofísicos y Ecofisiológicos, Facultad de Ciencias Naturales,
Universidad Nacional de la Patagonia San Juan Bosco, 9000 Comodoro Rivadavia,
Chubut, Argentina
e-mail: fgscholz@unpata.edu.ar; fgscholz@yahoo.com

F.G. Scholz · S.J. Bucci
Consejo Nacional de Investigaciones Científicas y Técnicas (CONICET),
Buenos Aires, Argentina

F.C. Meinzer
USDA Forest Service, Forestry Sciences Laboratory, 3200 SW Jefferson Way,
Corvallis, OR 97331, USA

G. Goldstein
Laboratorio de Ecología Funcional, Departamento de Ecología Genética y Evolución,
Instituto IEGEBA (CONICET-UBA), Facultad de Ciencias Exactas y naturales,
Universidad de Buenos Aires, Buenos Aires, Argentina

G. Goldstein
Department of Biology, University of Miami, Coral Gables, FL 33146, USA

© Springer International Publishing Switzerland 2016
G. Goldstein and L.S. Santiago (eds.), *Tropical Tree Physiology*,
Tree Physiology 6, DOI 10.1007/978-3-319-27422-5_10

227

Tree root systems not only obtain water from the soil to compensate for the water lost through transpiration, but they can also move water from moist to dry layers within the soil profile along gradients in soil water potential. This process known as hydraulic lift or hydraulic redistribution provides water for reabsorption by roots of the same plant or by neighboring individuals with active roots in the soil where water is released through the process of ecological facilitation.

Keywords Belowground processes · Savanna · Soil · Water uptake · Water transport

Introduction

Roots not only anchor plants to a substrate but are also the main pathway for water and nutrient uptake. Knowledge of root distribution and hydraulic determinants of uptake of available soil resources is critical for understanding the physiology of trees. However, we know relatively little about roots compared to the aboveground parts of plants, which reflects in part the inherent difficulties of studying roots in their natural environment. In some ecosystems, the amount of biomass in tree roots is higher than 50 % of the total tree biomass (Jackson et al. 1996; Castro and Kauffman 1998). This is particularly true for some tropical savanna trees that are exposed to 4–5 months of drought and which sustain a high rate of physiological and metabolic activity during this period (Sarmiento et al. 1985; Goldstein et al. 1986, 2008; Franco 1998; Franco et al. 2005; Bucci et al. 2006). In addition to root mass and surface area, access to adequate soil moisture during dry periods depends on the presence of roots deep enough to tap more abundant water in deeper soil layers (Scholz et al. 2008). Thus, the metabolic cost involved in maintaining adequate root performance is high, but it is essential for maintaining homeostasis of whole-tree carbon, nutrient and water economy.

Maximum average rooting depth for tropical deciduous forests (seasonally dry forests) is 7.3 ± 2.8 m and for tropical evergreen forests is 15.0 ± 5.4 m (Canadell et al. 1996). This range of rooting depth for tropical evergreen forest is larger than usually assumed suggesting that trees in wet tropical forests with little or no seasonality in precipitation allocate a substantial fraction of their total biomass to roots. Length may not translate necessarily to biomass because a long root does not mean that it is a thick one. However it is likely that due to allometric constraints longer roots require more energy and biomass investment than shorter roots. It is possible that the wide range of rooting depth reflects adaptations of some tree species to climatic anomalies resulting in prolonged droughts. The relatively small amount of carbon allocated to roots in seasonally dry forests (root:shoot ratio = 0.34; Jackson et al. 1996) is associated with most tree species in this forest type being drought-deciduous (Murphy and Lugo 1986). Losing leaves during the dry season substantially reduces the rate of water loss and prevents water deficits. The maintenance of leaves during the dry period would be costly because it requires a large allocation of carbon to underground biomass. Root systems in savannas are known to be quite deep, permitting water to be tapped from deep soil layers where water availability remains high during the dry season (Jackson et al. 1999; Scholz et al. 2008). Although many of the tree species in this type of ecosystem have access to deep, stable water sources, root architecture differs across species with deciduous and brevideciduous species having dimorphic root systems with active roots exploring several soil layers, and evergreen species having monomorphic roots that tap water from the same soil layer (Scholz et al. 2008). In this chapter we focus on root hydraulic processes and their consequences for carbon allocation at the whole-tree level, which contribute to continuous functioning of tropical trees subjected to diurnal and/or seasonal water deficits.

Xylem Dysfunction and Refilling of Embolized Conduits in Roots

Properties of the root system are often the primary limitation on water movement along the soil-to-leaf continuum (Nobel and Cui 1992). Typically, 50 % or more of the total resistance to water flow occurs below-ground despite the tendency for xylem conduit diameters to be largest in roots (Passioura 1984). Transpirational water loss generates tension, which is transmitted through continuous water columns down to the vascular system in the roots, making this portion of the water transport pathway susceptible to embolism. In roots, loss of functional xylem due to embolisms reduces whole-plant hydraulic conductance and may partially prevent water uptake (Linton and Nobel 1999), until the conduits are refilled with water (Ewers et al. 1997; Johnson et al. 2012). Roots of trees are potentially more vulnerable to drought-induced embolism than stems (e.g. Sperry and Saliendra 1994; Alder et al. 1996). The vessels tend to be larger in roots compared to stems and leaves (Zimmermann and Potter 1982; Gartner 1995; Pate et al. 1995; Ewers et al. 1997), which may partially explain their greater vulnerability to embolism.

   One of the hypothesized mechanisms for refilling embolized conduits in roots that has been assessed experimentally involves transient pressure imbalance between the xylem and surrounding tissues. According to this repair mechanism, embolized conduits are refilled on a daily basis while the surrounding xylem is still under tension (McCully et al. 1998; Melcher et al. 1998). Bucci and coauthors (2003) hypothesized that the refilling process in leaf petioles is driven by an increase in the amount of osmotically active sugars due to starch hydrolysis in cells surrounding the vascular bundles. This increase in osmotically active solutes drives water uptake by parenchyma cells resulting in a transient pressure imbalance because water uptake is mechanically constrained by the petiole cortex. The development of internal pressure was hypothesized to induce radial movement of water into embolized vessels. Along the same line of reasoning, Schmitz and coworkers (2012) showed that leaf-specific hydraulic conductivity of photosynthetic stems in mangroves was lower in covered than in uncovered branches,

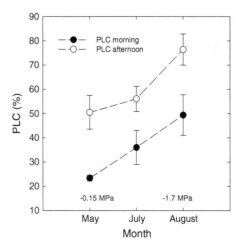

Fig. 1 Mean diurnal changes in percentage loss of root conductivity (PLC) in four savanna woody species (*Blepharocalyx salicifolius, Byrsonima crassa, Kielmeyera coriaceae* and *Qualea parviflora*). *Closed* and *open symbols* indicate morning predawn and midday measurements respectively. Values above the x-axis correspond to the maximum (May) and the minimum (August) soil water potentials at the depth where roots were sampled. Data from Domec et al. (2006)

implicating stem photosynthesis in the maintenance of hydraulic function. Given the proximity of the sites of exogenous CO_2 uptake to the xylem vessels in the stems, the results of this manipulative experiment suggest that xylary chloroplasts may play a role in light-dependent repair of embolized xylem vessels.

There is little information on the prevalence of repair of embolized root xylem conduits in tropical woody species. To our knowledge, daily formation and reversal of root embolism has only been documented in four Brazilian savanna tree species (Domec et al. 2006) and may be a more general phenomenon in roots of many tropical tree species. These diurnal changes in root hydraulic conductivity are consistent with the diurnal changes in hydraulic conductivity observed in stems and petioles of temperate trees (Salleo et al. 1996; Zwieniecki and Holbrook 1998). During the dry season, roots of Brazilian savanna species lose more than 30 % of their hydraulic conductivity from morning to afternoon (Fig. 1). The results of field manipulations in roots of the savanna species *Blepharocalyx salicifolius* were consistent with the operation of a transient pressure imbalance mechanism proposed for petioles of savanna species (Bucci et al. 2003), in which the diurnal refilling of embolized vessels was prevented by incisions in the cortex (Fig. 2). Consequently, the percent loss of hydraulic conductivity (PLC) in roots with incisions did not decrease overnight and was significantly higher than PLC observed the previous morning prior to the experimental treatment (Fig. 2). In contrast, hydraulic function of control roots was restored overnight. Moreover, Domec et al. (2006) observed a common relationship between maximum daily stomatal conductance and afternoon root conductivity over the course of the dry season across the four savanna woody species studied. Maximum stomatal conductance declined linearly with increasing

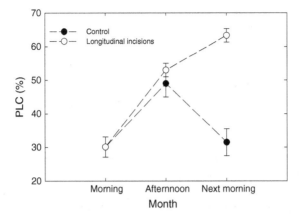

Fig. 2 Diurnal changes in percentage loss of root conductivity (PLC) for control roots (*closed symbols*) and for roots with longitudinal phloem incisions (*open symbols*) in *Blepharocalyx salicifolius* during the dry season in a neotropical savanna. Modified from Domec et al. (2006)

afternoon loss of root conductivity, suggesting that daily cycles of embolism and refilling in roots may be an inherent component of a hydraulic signaling mechanism that coordinates stomatal behavior and transpiration with fluctuations in liquid water transport capacity.

Another hypothesized mechanism for refilling embolized conduits in roots, which has also been assessed experimentally, is root pressure. This mechanism of embolism reversal appears to be common in bamboos. A survey of root pressure in 53 Asian species of tropical and subtropical bamboo (Wang et al. 2011a) revealed that all possessed root pressure and in some species root pressure was high enough to cause guttation of liquid from leaves overnight. Xylem exudates and guttation fluid have an osmotic pressure greater than root pressure, which can be explained by the low reflection coefficient of the root endodermis to solutes influx (Wang et al. 2011a). Root pressure is caused by active extrusion of mineral nutrient ions into the root xylem, which accumulate and lower the xylem water potential when transpiration is very low, commonly at night, which drives osmotic uptake of water from the soil into the root xylem. Root pressure in bamboos thus generates a force, which refills embolized vessels by pushing water up the stem to the highest leaves. This is of crucial importance for bamboos because they do not have lateral meristematic tissues and thus lack the ability to produce new xylem tissue laterally. Results of a recent study of daily water use strategies of *Sinarundinaria nitida*, an abundant subtropical bamboo species, showed that stem hydraulic conductivity did not decrease during the day whereas the leaf relative water content, leaf hydraulic and stomatal conductance exhibited a distinct decrease at midday (Yang et al. 2012). Diurnal down-regulation in K(leaf) and stomatal conductance slow down potential water loss in stems protecting the stem hydraulic pathway from cavitation. Since K(leaf) did not recover during late afternoon, refilling of bamboo leaf embolisms probably fully depends on nocturnal root pressure. The embolism refilling mechanism by root pressure could be helpful for the growth and persistence of this giant monocot species (Yang et al. 2012).

Hydraulic Redistribution in Tropical Savanna and Forest Trees

The root systems of trees not only obtain water from the soil to compensate for the water lost through transpiration, but also can move water upward from moist to drier layers within the soil profile along gradients in soil water potentials (Caldwell et at. 1998). This process, which was initially termed "hydraulic lift", is also referred to as hydraulic redistribution because roots can also transport water downward (e.g. Burgess et al. 1998, 2001) and laterally (e.g. Bauerle et al. 2008). Reverse sap flow along roots (from the plant to the soil) is an important part of the hydraulic redistribution process. Hydraulically redistributed water is available for reabsorption by roots of the same plant or by neighboring plants of the same or other species that have active roots in the soil layer where water is released. The magnitude and consequences of hydraulic redistribution are governed by multiple factors including water potential gradients between various points within the soil-plant-system, the hydraulic resistances to water flow in the soil, and the spatial distribution of the roots (Scholz et al. 2008). During the dry season, the upper soil layers typically have lower water content and thus more negative water potential compared to deeper soil layers. Hydraulic redistribution driven by differences in water potential within the soil profile is observed in tropical ecosystems with seasonal changes in precipitation such as savannas and at the edges of Amazonian forests. Evergreen trees in the Cerrado of Brazil do not exhibit hydraulic lift because they have only deep fine roots, however deciduous trees with both shallow and deep fine roots (dymorphic root systems) can carry out hydraulic lift (Scholz et al. 2008). *Byrsonima crassa* is a dominant deciduous savanna species that drops and expands new leaves during the dry season. Typical of species that can carry out hydraulic lift during the middle and end of the dry season, the shallow roots of this species exhibit reverse flow mainly at night (negative values of sap flow in Fig. 3b, c). At the beginning of the dry season, there are no differences in water potential between deep and shallow soil layers and thus hydraulic redistribution does not occur due to a lack of a driving force for passive water movement between roots in different soil layers (Fig. 3a). However, near the end of the dry season, this species is nearly leafless and thus the strength of the transpirational sink is relatively small, allowing reverse sap flow to occur throughout the entire 24-h cycle (Fig. 3c).

Hydraulic redistribution can also be driven by changes in salinity within the soil as observed in dwarf mangrove ecosystems along the Atlantic coastline (Hao et al. 2009). Tall *Rhizophora mangle* trees growing along the coastline of Biscayne National Park in Florida, USA, are continuously flooded and drained by the advance

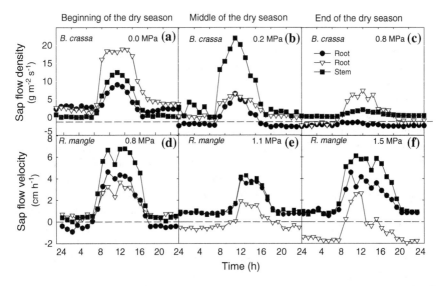

Fig. 3 **a–c** Sap flow density of *Byrsonima crassa* trees in a savanna ecosystem of Central Brazil on representative days at the beginning of the dry season, at the middle of the dry season and at the end of dry season, and **d–f** sap flow velocity of dwarf *Rhizophora mangle* trees in the inland areas of the Biscayne National Park, USA. *Squares* represent the main stem and circles and *triangles* represent lateral roots. Negative values of sap flow indicate reverse sap flow when there are differences in water potential at two soil depths. Values on each panel are (**a–c**) soil water potential gradient between 30 and 100 cm depth, and (**d–f**) soil osmotic potential gradient between 10 and 100 cm depth. Modified from Scholz et al. (2008) and Hao et al. (2009)

and retreat of tides, whereas dwarf *R. mangle* trees in adjacent inland areas are only flooded by sea water during occasional high tides. Surface soils in the inland zones are inundated with seawater during the wet season and gradually dry out during the dry season, while the source of deeper soil water is fresh water coming from small rivers or canals. Particularly during the dry season, a large water potential gradient develops between shallow and deep soil layers, which drives hydraulic redistribution in dwarf *R. mangle* at the inland sites. At the end of the wet season (December), the difference in the osmotic potential of the interstitial soil water (Ψ_o) between 10 and 100 cm depth is 0.8 MPa (Fig. 3d). During the dry season, the Ψ_o of the surface soil becomes more negative due to evaporation but Ψ_o remains high in deeper soil layers because of the continuous influx of underground fresh water. At the peak of the dry season (May) the largest difference in Ψ_o (c.a. −1.5 MPa) is observed (Fig. 3f). Reverse sap flow is consistently recorded in prop roots of dwarf *R. mangle* from the beginning to the end of the dry season (Fig. 3e, f) and its magnitude increases as Ψ_o increases. Reverse sap flow in dwarf trees was found on 93 % of the measurement days. Most of the reverse sap flow in dwarf *R. mangle* occurs during the night.

Approximately half of the evergreen forests of the Amazon Basin are subjected to a 2–3 month dry season with less than 0.5 mm d^{-1} of precipitation. Hydraulic redistribution by some evergreen eastern Amazonian tree species helps to avoid water deficits during the short dry season (Oliveira et al. 2005). Although hydraulic redistribution increases transpiration and plant growth during normal dry seasons, under extreme droughts during El Niño years, utilization of water sources by deep roots can eventually reduce dry season transpiration and net primary productivity during the end of the extended dry season. If soil water storage is not able to sustain forest transpiration through a long dry season, hydraulic redistribution may enhance moisture depletion early in the dry season thereby reducing water availability during the latter portion of the dry period (Wang et al. 2011b). It may appear at a first sight counterintuitive, but deeply rooted species having fine shallow and deep roots may be at a disadvantage under a global change scenario with an extended dry season because of an accelerated depletion of deep water sources. This implies that hydraulic lift may favor survival of certain tree species at the expense of others, depending on specific morphological and physiological traits. The shift of the wet evergreen tropical forests to a different type of forest will depend on the combined effects of prolonged drought and hydraulic redistribution. However, the nature of the substrate should also be included in models predicting the future trajectory of forest composition. At the periphery of the Amazonian forest many species are drought deciduous not only because of prolonged drought, but also due to high soil nutrient availability. Under similar climatic conditions, both evergreen forests and drought deciduous forests can occur depending on soil type. Evergreen-dominated forests can be found on nutrient poor soils while nutrient rich soils mostly contain deciduous trees. Consequently, soil nutrient availability has the potential to greatly improve analyses of plant processes above and beyond climate data (Maire et al. 2015).

Hydraulic redistribution requires seasonal changes in the amount of precipitation and thus a change in the magnitude of the soil to root water potential gradients from very small to large gradients from wet to dry seasons, respectively. Alternatively, besides seasonal changes of precipitation, perpetually arid ecosystems can develop a vertical soil water potential gradient if there is a shallow water table, with a capillary fringe above the water table, such as in Southern Arizona and California along the Colorado River floodplain. Neotropical savannas are characterized by 4- to 5-month dry seasons. Reverse sap flow from the roots to the soil increases as the difference in water potential between roots and soil rises at the beginning of the dry season. The percent loss of hydraulic conductivity in shallow lateral roots from wet to dry season is species-specific and depends on rates of reverse sap flow (Fig. 4). Species with lower rates of reverse sap flow exhibit larger hydraulic conductivity drop during the peak of the dry season. This implies that for each species the functionality of the roots growing in the upper dry soil layers during the dry season is a function of the amounts of water discharge around the rhizosphere by hydraulic redistribution. These results are consistent with the initial suggestion that a reduction of soil drying rates through hydraulic redistribution would extend fine root survival and growth (Caldwell et al. 1998). Bauerle et al. (2008) tested this

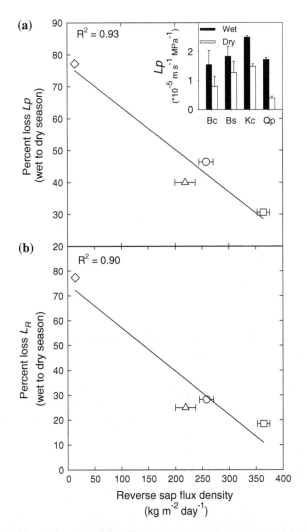

Fig. 4 a Percent loss of total root hydraulic conductance (LP) and **b** percent loss of radial conductance (LR) from the wet to the dry season in relation to total daily reverse sap flux in lateral roots of savanna woody species at the peak of the dry season. Values of reverse sap flux are means (±SE) of three to six roots in different trees. Values of LP and LR are means of three to six different roots measured during the wet seasons (January 2004) and the dry (August 2004) in different trees. A linear regression was fitted to each relationship **a** y = 77 − 0.13x, P = 0.035; **b** y = 74.5 − 0.17x, P = 0.05. Symbols are: (○) *Byrsonima crassa*, (△) *Kielmeyera coriacea*, (□) *Blepharocalyx salicifolius* and (◇) *Qualea parviflora*. The *inset* in panel **a** is *Lp* in m s^{-1} MPa^{-1} measured during the wet (*filled bars*) and dry (*open bars*) for the same species. Modified from Bucci et al. (2008) and Scholz et al. (2008)

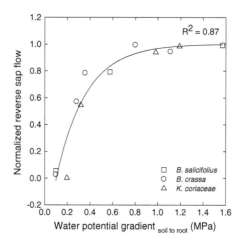

Fig. 5 Normalized total daily reverse sap flow in lateral roots of three savanna woody species in relation to the water potential gradient between soil and roots for different days from the beginning to the end of the dry season. Reverse sap flow was normalized respect to the maximum value reached for a particular root during the dry season. Modified from Scholz et al. (2008)

hypothesis using two *Vitis sp.* cultivars. Nighttime illumination of plants irrigated on only one side of the row strongly suppressed hydraulic redistribution. Root water potentials were twice as negative in plants where hydraulic redistribution was suppressed than in plants where hydraulic redistribution occurred. A simulation model developed by Scholz et al. (2010) for neotropical savannas, and validated with field data, predicted that without hydraulic redistribution, water potential in the upper soil layers would drop to very low values during the dry season, resulting in negative effects on tree functioning, dieback of herbaceous plants, and disruption of the biogeochemical cycles leading to a substantial decrease in rates of soil organic matter decomposition.

Near the end of the dry season, reverse sap flow in lateral roots of woody savanna species tends to initially increase sharply but then remain nearly constant as the root-to-soil water potential gradient increases even further (Fig. 5). A seasonal decrease in the axial and radial hydraulic conductivity of lateral roots as the soil dry out could (inset Fig. 5) constitute a compensatory mechanism to limit losses of water to shallow soil layers via reverse sap flow. The term "rectification" has been used to describe this process resulting from both physiological and morphological changes in root characteristics (Nobel and Sanderson 1984). In addition to rectification, the asymptotic relationship shown in Fig. 5 may be the result of nocturnal transpiration and recharge of stem internal water storage, which may compete for internal water resources during the dry season (Scholz et al. 2008).

Facilitation Between Hydraulic Redistributors and Neighboring Trees

Hydraulically redistributed water may be available for uptake by neighboring individuals with active roots in the layer to which water is distributed and represents an example of ecological facilitation (Moreira et al. 2003; Ludwig et al. 2004; Hawkins et al. 2009; Pang et al. 2012; Sardans and Peñuelas 2013). The effect that species engaged in hydraulic redistribution have on their neighbors is still unclear. There are reports of positive, neutral or even negative effects depending on ecosystem type, plant life form or whether donor and receiver species shared common ecto- and endomycorrhizal networks (Prieto et al. 2011). A water isotope labeling experiment conducted with deuterium (^{2}H, D) on tropical savanna trees from central Brazil was used to answer the question of whether hydraulically lifted water is utilized by plants neighboring labeled individuals (Moreira et al. 2003). Uptake of hydraulically redistributed water by neighboring trees was confirmed by stable isotopic measurements of deuterium (D) in their stem water. A solution containing 75 % D_2O was fed to intact tap roots and stem samples of D-labeled plants were obtained during the next three consecutive days. All treated plants had at least two neighbors with stem water showing deuterium abundance above background, suggesting that roots of those trees were able to obtain water released into the soil by reverse flow through roots of the treated trees (Fig. 6). However, there may be other pathways for water transfer between neighboring trees such as root grafting or mycorrhizal connections. The mycorrhizal pathway seems to be the most efficient, as twice as much water was transported between donor and receiver Ponderosa pine trees (*Pinus ponderosa*) directly via mycorrhizal connections compared to a non mycorrhizal pathway (Warren et al. 2008). In this study, the rate of water transport between large trees and seedlings was high (0.16–0.63 m d^{-1}), which may be an indication that large amounts of water could potentially be transferred through mycorrhizal networks. Thus, neighbors associated with plants engaged in hydraulic redistribution could benefit from the extra moisture through increased survival rates, which could potentially increase their fitness as they may establish in the community and live long enough to grow and reproduce in subsequent seasons (Schöb 2014).

Belowground competition between East African *Acacia* trees and grasses may overwhelm the facilitative effects of hydraulic lift. Prevention of tree-grass interactions through root trenching led to increased soil water content, indicating that trees took up more water from the upper soil layer than they released through hydraulic lift (Ludwig et al. 2004). Grasses used hydraulically lifted water provided by trees, or alternatively they were able to tap deep soil water directly by growing deep roots when competition with trees occurred. Competition for water resources may therefore modulate the facilitation effect of hydraulic lift on neighboring species. Field manipulation experiments, such as those done by Ludwig et al. (2004) are needed to disentangle facilitation versus competition effects.

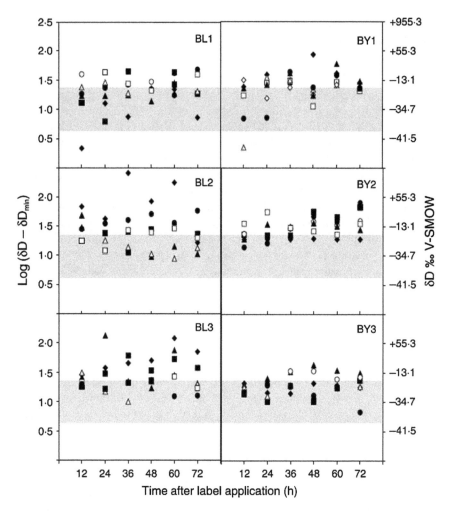

Fig. 6 Log differences between δD of stem water from plants neighboring treated plants fed with deuterated water and those of control plants in a Brazilian Neotropical savanna. Three individuals of two Cerrado tree species were used: *Byrsonima crassa* (BY) and *Blepharocalyx salicifolius* (BL): Background abundance of D in stem samples was collected in individuals of the same species more than 20 m away from the labeled individuals. The delta (δ) notation is used to quantify stable isotope as relative ratios. Actual δD values equivalent to those reported on the log scale are shown on the right axis. Values shown outside the grey area are significantly different from control values at $P < 0.05$. From Moreira et al. (2003)

Conclusions

Most woody dominated tropical ecosystems are subjected to drought periods of different lengths, from few days or weeks to several months. Roots during these low soil water availability periods may experience hydraulic and metabolic dysfunctions resulting not only in a substantial decrease in root growth rates, but also in reduced water and nutrient uptake activity. Maintenance of root functioning is thus key for the overall tree growth and reproductive performance. Roots constitute one of the plant segments that are most vulnerable to drought-induced embolism. The vessels tend to be larger in roots compared to stems and leaves, which may partially explain their greater vulnerability to embolism. We discussed two of the hypothesized mechanisms for refilling embolized conduits in roots that have been assessed experimentally: transient pressure imbalance between the xylem and surrounding tissues, and positive root pressures. The root systems of trees not only obtain water from the soil to compensate for the water lost through transpiration, but also can move water upward from moist to drier layers within the soil profile along gradients in soil water potentials. This process known as hydraulic lift or hydraulic redistribution can also transport water downward or laterally. Hydraulically redistributed water is available for reabsorption by roots of the same plant or by neighboring plants of the same or other species that have active roots in the soil layer where water is released. The effect that species engaged in hydraulic redistribution have on their neighbors apparently depends on ecosystem type, plant life form, whether donor and receiver species share common ecto- and endomycorrhizal networks, and by the intensity of belowground competition which may overwhelm the facilitative effects of hydraulic lift.

References

Alder NN, Sperry JS, Pockman WT (1996) Root and stem xylem embolism, stomatal conductance, and leaf turgor in *Acer grandidentatum* populations along a soil moisture gradient. Oecologia 105:293–301

Bauerle TL, Richards JH, Smart DR, Eissenstat DM (2008) Importance of internal hydraulic redistribution for prolonging the lifespan of roots in dry soil. Plant Cell Environ 31:177–186

Bucci SJ, Scholz FG, Goldstein G, Meinzer FC (2003) Dynamic diurnal changes in hydraulic specific conductivity in petiole of two savanna tree species: identifying factors and mechanisms contributing to the refilling of embolized vessels. Plant Cell Environ 26:1633–1645

Bucci SJ, Scholz FG, Goldstein G, Meinzer FC, Franco AC, Campanello PI, Villalobos-Vega R, Bustamante M, Miralles-Wilhelm F (2006) Nutrient availability constrains the hydraulic architecture and water relations of savanna trees. Plant Cell Environ 29:2153–2167

Bucci SJ, Scholz FG, Goldstein G, Meinzer FC, Franco AC, Zhang Y, Hao G-Y (2008) Water relations and hydraulic architecture in Cerrado trees: adjustments to seasonal changes in water availability and evaporative demand. Braz J Plant Physiol 20:233–245

Burgess SSO, Adams MA, Turner NC, Ong CK (1998) The redistribution of soil water by tree root systems. Oecologia 115:306–311

Burgess SSO, Adams MA, Turner NC, White DA, Ong CK (2001) Tree roots: conduits for deep recharge of soil water. Oecologia 126:158–165

Caldwell MM, Dawson TE, Richards JH (1998) Hydraulic lift—consequences of water efflux from the root of plants. Oecologia 113:151–161

Canadell J, Jackson RB, Ehleringer JR, Mooney HA, Sala OE, Schulze E-D (1996) Maximum rooting depth of vegetation types at the global scale. Oecologia 108:583–595

Castro LHR, Kauffman JB (1998) Ecosystem structure in the Brazilian Cerrado: a vegetation gradient of aboveground biomass, root mass and consumption by fire. J Trop Ecol 14:263–283

Domec JC, Scholz FG, Bucci SJ, Meinzer FC, Goldstein G, Villalobos-Vega R (2006) Diurnal and seasonal variation in root xylem embolism in neotropical savanna woody species: impact on stomatal control of plant water status. Plant Cell Environ 29:26–35

Ewers FW, Carlton MR, Fisher JB, Kolb KJ, Tyree MT (1997) Vessel diameters in roots versus stems of tropical lianas and other growth forms. IAWA Journal 18:261–279

Franco AC (1998) Seasonal patterns of gas exchange, water relations and growth of *Roupala montana*, an evergreen species. Plant Ecol 136:69–76

Franco AC, Bustamante M, Caldas LS, Goldstein G, Meinzer FC, Kozovits AR, Rundel P, Coradin VTR (2005) Leaf functional traits of neotropical savanna trees in relation to seasonal water deficit. Trees 19:326–335

Gartner BL (1995) Patterns of xylem variation within a tree and their hydraulic and mechanical consequences. In: Gartner BL (ed) Plant stems: physiology and functional morphology. Academic Press, San Diego, pp 125–149

Goldstein G, Sarmiento G, Meinzer FC (1986) Patrones diarios y estacionales en las relaciones hídricas de árboles siempreverdes de la sabana tropical. Acta Oecol (Oecol Plant) 7:107–119

Goldstein G, Meinzer FC, Bucci SJ, Scholz FG, Franco AC, Hoffmann WA (2008) Water economy of Neotropical savanna trees: some paradigms revisited. Tree Physiol 28:395–404

Hao GY, Jones TJ, Luton C, Zhang YJ, Manzane E, Scholz F, Bucci S, Cao KF, Goldstein G (2009) Hydraulic redistribution in dwarf *Rhizophora mangle* trees driven by interstitial soil water salinity gradients: impacts on hydraulic architecture and gas exchange. Tree Physiol 29:697–705

Hawkins H, Hettasch H, West AG, Cramer MD (2009) Hydraulic redistribution by Protea 'Sylvia' (Proteaceae) facilitates soil water replenishment and water acquisition by an understorey grass and shrub. Funct Plant Biol 36:752–760

Jackson RB, Canadell J, Ehleringer JR, Mooney HA, Sala OE, Schulze ED (1996) A global analysis of root distributions for terrestrial biomes. Oecologia 108:389–411

Jackson PC, Meinzer FC, Bustamante M, Goldstein G, Franco AC, Rundel PW, Caldas L, Igler E, Causin F (1999) Partitioning of soil water among tree species in a Brazilian Cerrado ecosystem. Tree Physiol 19:717–724

Johnson DM, McCulloh KA, Woodruff DR, Meinzer FC (2012) Hydraulic safety margins and embolism reversal in stems and leaves: why are conifers and angiosperms so different? Plant Sci 195:48–53

Linton MJ, Nobel PS (1999) Loss of water transport capacity due to xylemcavitation in roots of two CAM succulents. Am J Bot 86(11):1538–1543

Ludwig F, Dawson TE, Prins HHT, Berendse F, Kroon H (2004) Below-ground competition between trees and grasses may overwhelm the facilitative effects of hydraulic lift. Ecol Lett 7:623–631

Maire V, Wright IJ, Prentice C, Batjes NH, Bhaskar R, Bodegom Pv, Cornwell WK, Ellsworth DS, Niinemets Ü, Ordoñez J, Reich PB, Santiago LS (2015) Global soil and climate effects on leaf photosynthetic traits and rates. Glob Ecol Biogeogr 6:706–717

McCully M, Huang CX, Ling LEC (1998) Daily embolism and refilling of xylem vessels in the roots of field-grown maize. New Phytol 138:327–342

Melcher PJ, Meinzer FC, Yount DE, Goldstein G, Zimmermann U (1998) Comparative measurements of xylem pressure in transpiring and non-transpiring leaves by means of the pressure chamber and the xylem pressure probe. J Exp Bot 49:1757–1760

Moreira MZ, Scholz FG, Bucci SJ, Sternberg LS, Goldstein G, Meinzer FC, Franco AC (2003) Hydraulic lift in a Neotropical savanna. Funct Ecol 17:573–581

Murphy PG, Lugo AE (1986) Ecology of tropical dry forest. Annu Rev Ecol Syst 17:67–88

Nobel P, Cui M (1992) Hydraulic conductances of the soil, the root-soil air gap, and the root: changes for desert succulents in drying soil. J Exp Bot 43:319–326

Nobel PS, Sanderson J (1984) Rectifier-like activities of roots of two desert succulents. J Exp Bot 35:727–727

Oliveira RS, Dawson TE, Burgess SO, Nepstad DC (2005) Hydraulic redistribution in three Amazonian trees. Oecologia 145:354–363

Pang Y, Wang Y, Lambers H, Tibbett M, Siddiquec KHM, Ryan MH (2012) Commensalism in an agroecosystem: hydraulic redistribution by deep-rooted legumes improves survival of a droughted shallow-rooted legume companion. Physiol Plant 149:79–90

Passioura JB (1984) Hydraulic resistance of plants. I. Constant or variable? Aust J Plant Physiol 11:333–339

Pate JS, Jeschke WD, Aylward MJ (1995) Hydraulic architecture and xylem structure of the dimorphic root systems of South-West Australian species of Proteaceae. J Exp Bot 46:907–915

Prieto I, Padilla FM, Armas C, Pugnaire F (2011) The role of hydraulic lift on seedling establishment under a nurse plant species in a semi-arid environment. Perspect Plant Ecol 13:181–187

Salleo S, Lo Gullo MA, de Paoli D, Zippo M (1996) Xylem recovery from cavitation-induced embolism in young plants of *Laurus nobilis*: a possible mechanism. New Phytol 132:47–56

Sardans J, Penuelas J (2013) Hydraulic redistribution by plants and nutrient stoichiometry: shifts under global change. Ecohydrology. doi:10.1002/eco.1459

Sarmiento G, Goldstein G, Meinzer FC (1985) Adaptative strategies of woody species in neotropical savannas. Biol Rev 60:315–355

Schmitz N, Egerton JJG, Lovelock CE, Ball MC (2012) Light-dependent maintenance of hydraulic function in mangrove branches: do xylary chloroplasts play a role in embolism repair? New Phytol 195:40–46

Schöb C et al (2014) Consequences of facilitation: one plant's benefit is another plant's cost. Funct Ecol 28(2):500–508

Scholz FG, Bucci SJ, Goldstein G, Moreira MZ, Meinzer FC, Domec J-C, Villalobos-Vega R, Franco AC, Miralles-Wilhelm F (2008) Biophysical and life history determinants of hydraulic lift in Neotropical savanna trees. Funct Ecol 22:773–786

Scholz FG, Bucci SJ, Hoffmann WA, Meinzer FC, Goldstein G (2010) Hydraulic lift in a neotropical savanna: experimental manipulation and model simulation. Agr Forest Meteorol 150:629–639

Sperry J, Saliendra NZ (1994) Intra- and inter-plant variation in xylem cavitation in *Betula occidentalis*. Plant Cell Environ 17:1233–1241

Wang F, Tian X, Ding Y, Wan X, Tyree MT (2011a) A survey of root pressure in 53 Asian species of bamboo. Ann For Sci 68:783–791

Wang G, Alo C, Mei R, Sun S (2011b) Droughts, hydraulic redistribution, and their impact on vegetation composition in the Amazon forest. Plant Ecol 212:663–673

Warren JM, Brooks JR, Meinzer FC, Eberhart JL (2008) Hydraulic redistribution of water from *Pinus ponderosa* trees to seedlings: evidence for an ectomycorrhizal pathway. New Phytol 178:382–394

Yang S-J, Zhang Y-J, Sun M, Goldstein G, Cao K-F (2012) Recovery of diurnal depression of leaf hydraulic conductance in a subtropical woody bamboo species: embolism refilling by nocturnal root pressure. Tree Physiol 32:414–422

Zimmermann MH, Potter D (1982) Vessel-length distribution in branches, stems, and roots of *Acer rubrum*. International Association of Wood Anatomists (IAWA). Bulletin 3:103–109

Zwieniecki MA, Holbrook NM (1998) Diurnal variation in xylem hydraulic conductivity in white ash (*Fraxinus americana* L.) red maple (*Acer rubrum* L.) and red spruce (*Picea rubens* Sarg.) Plant Cell Environ 21:1173–1180

Drought Survival Strategies of Tropical Trees

Louis S. Santiago, Damien Bonal, Mark E. De Guzman
and Eleinis Ávila-Lovera

Abstract Climate change is predicted to increase the occurrence of extreme droughts, which are associated with elevated mortality rates in tropical trees. Drought-induced mortality is thought to occur by two main mechanisms: hydraulic failure or carbon starvation. This chapter focuses on the strategies that plants use to survive these two drought-induced mortality mechanisms and how these mechanisms are distributed among the immense diversity of tropical tree species. The traits that tropical trees may use to survive drought include (1) xylem that is resistant to drought-induced cavitation, (2) high sapwood capacitance that protects xylem from critically low water potentials, (3) drought deciduousness, (4) photosynthetic stems that have the potential to assimilate carbon at greater water-use efficiency than leaves, (5) deep roots, (6) regulation of gas exchange to reduce leaf water loss or to maintain photosynthesis at low leaf water potential and (7) when all else fails, low cuticular conductance from exposed tissues during extended drought. To date, most research has focused on deciduousness, resistant xylem, soil water, gas exchange behavior and sapwood capacitance, whereas little is known about the role of photosynthetic stems or cuticular conductance during extreme extended drought, making these processes a high priority for a complete understanding of tropical tree physiology during drought.

L.S. Santiago (✉) · M.E. De Guzman · E. Ávila-Lovera
Department of Botany & Plant Sciences, University of California,
2150 Batchelor Hall, Riverside, CA 92521, USA
e-mail: santiago@ucr.edu

M.E. De Guzman
e-mail: mark.deguzman@email.ucr.edu

E. Ávila-Lovera
e-mail: eleinis.avilalovera@email.ucr.edu

L.S. Santiago
Smithsonian Tropical Research Institute, Apartado Postal 0843-03092 Balboa
Panama, Republic of Panama

D. Bonal
INRA, UMR EEF—Université de Lorraine/INRA, 54280 Champenoux, France
e-mail: bonal@nancy.inra.fr

© Springer International Publishing Switzerland 2016   243
G. Goldstein and L.S. Santiago (eds.), *Tropical Tree Physiology*,
Tree Physiology 6, DOI 10.1007/978-3-319-27422-5_11

Introduction

Drought-induced tree mortality has now been documented in forests worldwide (Allen et al. 2010; Anderegg et al. 2012). With global change, pronounced shifts of rainfall patterns, as well as increased frequency and severity of extreme droughts are expected (Field 2014). Even in wet tropical forests, which are some of the world's most humid ecosystems, elevated mortality of seedlings and mature trees associated with drought occurs (Condit et al. 1995; Slik 2004; Engelbrecht et al. 2005, 2006; Comita and Engelbrecht 2009; Phillips et al. 2010; Feeley et al. 2011). Predicting how tropical forests will respond to increased drought depends on understanding which species are most susceptible to mortality by drought and distinguishing these from species that are likely to survive drought and contribute to future tropical forest composition. This is a grand challenge for ecophysiologists because tropical forests may contain more than 300 species per hectare, creating a large amount of data needed to develop predictions of future drought survival and mortality. Such predictions are critical for understanding how potential mortality of tropical trees affect carbon cycling and subsequent feedbacks to the climate system because tropical forests harbor the greatest biodiversity, productivity and terrestrial carbon stocks of any biome and thus play a disproportionately large role in the global climate system (Beer et al. 2010; Saatchi et al. 2011). In fact, the largest uncertainties in modeling how tropical forest carbon pools might alter in response to climate change are associated with plant physiological responses (Huntingford et al. 2013).

Much of the study of woody plant susceptibility to water deficit has focused on the vulnerability of xylem to drought-induced cavitation (Choat et al. 2012; Craine et al. 2013). Indeed xylem cavitation is a critical step in drought-induced mortality. However, plants also posses a diversity of features that can buffer xylem from reaching critical water potentials and prolong survival during drought conditions (Pivovaroff et al. 2016). The traits that tropical trees use to survive drought include: (1) xylem that is resistant to drought-induced cavitation, (2) high sapwood capacitance that protects xylem from critically low water potentials, (3) drought decid-uousness, (4) photosynthetic stems that have the potential to assimilate carbon at greater water-use efficiency than leaves, (5) deep roots, (6) regulation of gas exchange to reduce leaf water loss or to maintain photosynthesis at low leaf water potential and (7) when all else fails, low cuticular conductance from exposed tissues during extended drought (Sperry 2000; Ackerly 2004; Meinzer et al. 2008; Choat et al. 2012; Ávila et al. 2014). Similar to other ecological strategies, drought survival strategies thus consist of the presence or absence of drought survival traits within species and are reflected by particular trait combinations that occur repeatedly in unrelated taxa under similar environments (Grime 1974; Chapin 1980; Westoby et al. 2002). In this chapter, we examine what is known about these traits in tropical tree species as a framework for understanding drought survival strategies and predicting which tropical trees may be most susceptible to climate change-type drought.

During drought, lack of soil water is not the only problem faced by tropical trees. High temperatures that accompany drought have consequences for photosynthesis, and respiration, as well as stomatal regulation through the effects of temperature on atmospheric vapor pressure deficit. Thus, the limitation on carbon assimilation imposed by drought is also a critical component of drought-induced mortality mechanisms in woody species. Two physiological mechanisms of woody plant mortality involving water and carbon have been proposed and provide a useful conceptual framework (McDowell et al. 2008). The first is the hydraulic failure hypothesis, which suggests that drought reduces plant water potential such that the xylem water column cavitates and cessation of water supply to leaves leads to plant death under conditions of intense, short-term drought (McDowell et al. 2008). The second is the carbon starvation hypothesis, which suggests that drought causes stomatal closure and eventually leads to exhaustion of non-structural carbohydrate reserves and thus leads to plant death under conditions of extended, low-intensity droughts (McDowell et al. 2008; McDowell and Sevanto 2010). Carbohydrate exhaustion may furthermore increase vulnerability to pathogens and herbivores, which may be the ultimate cause of mortality (McDowell et al. 2008). Isohydric species close stomata early during drought and maintain plant water status at the expense of declining carbon intake, making them more likely to die of carbon starvation. In contrast, anisohydric species maintain stomata open further into drought and maintain carbon intake at the expense of declining plant water status, making them more likely to die of hydraulic failure. Drought effects on water and carbon supply also interact in complex ways when limited water supply reduces carbon uptake, transport, and utilization, or when limited carbon supply reduces xylem refilling (Salleo et al. 2004; McDowell 2011; McDowell et al. 2013). Therefore, it is now recognized that both mechanisms are involved and do not act separately (Hartmann et al. 2015). For example, both hydraulic architecture and carbon relations have been shown to be important for mortality of tropical tree seedlings under drought (Kursar et al. 2009; O'Brien et al. 2015).

Most of our understanding of tropical tree mortality during drought is based on long-term forest census plots that are measured every 3–5 years (Condit et al. 1995; Condit 1998; Phillips et al. 2010) and on large throughfall reduction experiments (Nepstad et al. 2002; Fisher et al. 2007). These studies demonstrate that light-wooded trees tend to suffer higher mortality than dense-wooded trees (Condit et al. 1995, 1996; Phillips et al. 2010), consistent with the finding that low wood density is correlated with high vulnerability to xylem cavitation (Hacke et al. 2001). These studies also demonstrate that small seedlings and large-stature trees are highly susceptible to mortality by drought (Condit et al. 1995, 1996; Engelbrecht et al. 2006; Phillips et al. 2010). High seedling susceptibility is consistent with the relatively small root systems, and thus limited access to water by seedlings. Large trees may suffer elevated mortality during drought because of exposure to high temperatures and radiation (Phillips et al. 2010), or due to altered water and carbon relations in tall trees (Ryan and Yoder 1997; Burgess and Dawson 2007). Yet within any size class, fine physiological differences and the presence of drought survival traits should determine mortality during drought. Therefore, this chapter addresses the seven drought survival traits described above in the context of their occurrence in tropical trees and the particular aspects of physiological regulation that stave off mortality. We also explore mechanisms of drought recovery as related to drought survival traits. Although recovery is not strictly a drought survival trait, there is evidence that the necessity to continue growing and the degree of resource expenditure during drought largely determine the ability to recover in subsequent years (Doughty et al. 2015).

Xylem Resistance to Drought-Induced Cavitation

Quantifying the tension under which xylem cavitates and loses the ability to transport water is a principal approach to characterizing drought resistance in woody plant species (Tyree and Sperry 1989; Choat et al. 2012). The process of cavitation during declining water potential is caused by air entry into the xylem lumen when negative xylem pressure exceeds the surface tension of water at the border of gas-filled pit membranes (Sperry et al. 1988). Smaller pores in pit membranes have the benefit of minimizing cavitation risk, but are associated with smaller xylem vessels and carry the cost of reduced water transport. Vessel size is also related to wood density, with dense-wooded species generally having smaller vessels, greater resistance to drought-induced xylem cavitation and lower rates of water transport (Hacke et al. 2001). Each species has a particular value of tension under which air begins to enter into xylem vessels and restrict hydraulic conductivity and thus the ability to transport water to leaves (Choat et al. 2012). Among species, the tension at which the xylem experiences 50 % loss of conductivity (PLC) is usually used as a benchmark for comparison (P_{50}; MPa). The P_{50} parameter is determined by plotting the PLC as a function of plant water potential. Figure 1 illustrates the trajectory of declining hydraulic conductivity of terminal stems of two species from Amazonian forest in Paracou, French Guiana, as stem water potential declines during a dry down under laboratory conditions. The more vulnerable species, *Symphonia globulifera*, has among the lowest wood density of canopy tree species in the Paracou plot, whereas the more resistant species, *Licania alba*, is one of the species with the highest wood density found in the Paracou plot (Fig. 1). Thus the vulnerability curve space between these two end-member species likely represents the entire range of xylem vulnerability for canopy trees at Paracou.

Data on vulnerability to xylem-induced cavitation on woody plant species from around the globe have now been assembled into a single analysis (Choat et al. 2012). This compendium shows that arid systems contain both vulnerable and resistant species, and that humid systems, such as wet tropical forest, only contain species that are relatively vulnerable to cavitation. Thus tropical forests represent one of the world biomes that should be most vulnerable to drought-induced mortality during extreme drought associated with climate change or El Niño years. However, analysis of vulnerability curves and comparative P_{50} does not tell us when the xylem of a particular species will reach the point of catastrophic hydraulic failure, but rather

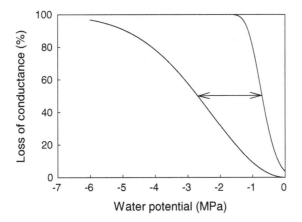

Fig. 1 Vulnerability curves of upper canopy branches of two tree species from Paracou, French Guiana. The more resistant species, *Licania alba*, begins losing water transport capacity at a more negative water potential than the more vulnerable species, *Symphonia globulifera*. The *arrow* marks the water potential at 50 % loss of hydraulic conductivity (P_{50}) and illustrates the vulnerability curve space between these two end-member species representing the range of xylem vulnerability for canopy trees at Paracou (Santiago et al. unpublished data)

what happens when xylem tissue arrives at such dangerous water potentials. For example, xylem of riparian trees is often very vulnerable to drought-induced cavitation, but it is not likely that these trees will reach critically low water potentials due to high water availability (Pockman and Sperry 2000; Ackerly 2004). Therefore, the use of vulnerability curve analysis to predict survival or mortality during drought is most useful when combined with other drought survival traits.

Sapwood Capacitance

Sapwood capacitance, the stored water in stems that can temporarily supply the water for transpiration and protect xylem from precipitous drops in water potential (Bucci, this volume), has recently emerged as one of the most important hydraulic factors for understanding the actual risk of cavitation (Meinzer et al. 2008; Barnard et al. 2011). Indeed tropical tree species with high capacitance have been shown to survive drought even without high cavitation resistance because they are buffered by stored water and are at lower risk of reaching the xylem tensions that cause hydraulic failure (Meinzer and Goldstein 1996; Meinzer et al. 2008, 2009; Sperry et al. 2008). Capacitance is normally determined with a sapwood water release curve plotted as cumulative water released as a function of sapwood water potential (Fig. 2). Thus sapwood capacitance reflects the amount of water that can be mobilized per unit water potential, and together with vulnerability curve analysis provides keen insight to the functional limits of xylem performance under conditions of declining water potential.

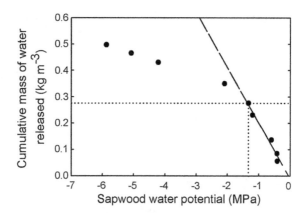

Fig. 2 Sapwood water release curve of one upper canopy branch of *Hiraea reclinata* from Parque Metropolitano, Panama, showing relationship between sapwood water potential and water released. *Dotted lines* indicate value of water potential and water released corresponding to turgor loss point. *Dashed line* is a linear regression fit to the initial, nearly linear portion of the curve, the slope of which is sapwood capacitance in units of kg m^{-3} MPa^{-1} (De Guzman, unpublished data)

Drought Deciduousness

Changes in deciduousness are known to occur along the great precipitation gradients of the tropics (Schimper 1903; Givnish 2002; Santiago et al. 2004). Forests on the dry side of precipitation gradients commonly have a higher proportion of dry-season deciduous canopy species that tend to lose their leaves during the dry season when low soil water availability may limit physiological activity. Dry-season deciduousness is therefore considered to be a characteristic that reduces whole-plant transpiration and respiration, providing an advantage over evergreen phenology in tropical dry forest, savanna and caatinga vegetation during annual seasonal drought. However, dry-season deciduousness may also impart an advantage during supra-annual extreme droughts by providing an ability to maintain dormancy while minimizing resource expenditure. The cost of this potential drought survival trait would be suspended carbon income, which could make dry-season deciduous species more vulnerable to mortality by carbon starvation.

Within the context of dry-season deciduousness, it is important to distinguish seasonally deciduous species from normally evergreen species that begin to shed leaves during extreme drought. Whereas many evergreen species will thin their canopy throughout a normal dry season (Santiago and Mulkey 2005), with this behavior continuing into extreme drought situations that extend beyond normal dry seasons, some evergreen species do not posses the ability to shed leaves and can experience mortality of entire limbs with crisp, brown leaves still attached to the plant (Santiago, personal observation). Currently, there is little data to classify species in terms of leaf shedding beyond the deciduous versus evergreen classification scheme based on year-to-year phenology during normal annual dry seasons.

However, the advantage of minimizing water loss and carbon costs of respiration during drought is consistent with the dominance of dry-season deciduous species in the most arid regions of the tropics.

Stem Photosynthesis

A close look at the bark of species in arid ecosystems quickly reveals the existence of green stems (Fig. 3, Ávila-Lovera and Ezcurra, this volume). In many species that remain leafless for most or part of the year, CO_2 fixation by stems represents an important contribution to whole plant photosynthesis (Pfanz 2008; Saveyn et al. 2010; Steppe et al. 2015). Stem photosynthesis can be classified into two types based on physiological processes and anatomical characteristics (Ávila et al. 2014): (1) Stem net photosynthesis (SNP), which includes net CO_2 fixation by stems with stomata in the epidermis and net cortical CO_2 fixation in suberized stems, and (2) Stem recycling photosynthesis (SRP), defined as assimilation of respired CO_2 in suberized stems. Under drought stress, when stomatal closure or leaf shedding (see above) limit leaf photosynthetic carbon income, stem photosynthesis is expected to become an increasingly large proportion of total carbon income because it is less reduced than leaf photosynthesis (Nilsen 1995). It is also thought that stem photosynthesis can occur at greater water-use efficiency (WUE; carbon gain per unit water loss) than leaf photosynthesis because of lower rates of water loss during stem photosynthesis. To evaluate this hypothesis, we assembled the limited data

Fig. 3 Leaves and photosynthetic stems of *Parkinsonia praecox* from Margarita Island, Venezuela. Photo by Wilmer Tezara

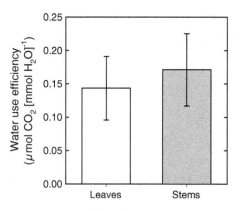

Fig. 4 Photosynthetic water use efficiency (mean ± standard error) of leaves and stems of eight species taken from the literature (Osmond et al. 1987; Comstock and Ehleringer 1988; Tinoco-Ojanguren 2008; Wittmann and Pfanz 2008). Comparison of means indicates no statistically significant difference ($t = 0.4177$; $p = 0.6825$)

available for WUE of stem and leaf photosynthesis in the same species (Fig. 4). The results from eight cases (Osmond et al. 1987; Comstock and Ehleringer 1988; Tinoco-Ojanguren 2008; Wittmann and Pfanz 2008) indicate that whereas WUE trended higher in stems than leaves, the difference was not statistically significant ($t = 0.4177$; $p = 0.6825$). However, it is important to note that all of the available data in this limited dataset comes from temperate species, raising the question of what we will find as similar measurements are undertaken on tropical trees.

Soil Water Access

Belowground processes are notoriously difficult to measure, but access to soil water as determined by rooting depth or rooting in moist microsites should be a major determinant of how tropical trees will fare during extreme drought. Matching the stable isotopic composition of hydrogen (^{2}H, deuterium, D) and oxygen (^{18}O) in plant sap with water in the soil profile is useful for assessing water sources used by plants (Allison 1982; Ehleringer and Dawson 1992). In fact, stable isotopic analysis has revealed patterns that are difficult to observe with any other technique and the use of labelling approaches provides further tools to estimate soil water extraction depth (Stahl et al. 2013). For example, in seasonal tropical forest on Barro Colorado Island in Panama, evergreen tree species take water from deeper soil layers than deciduous species, suggesting that deep roots allow evergreen species to maintain their leaves and continue physiological activity during the dry season (Jackson et al. 1995). In some instances, smaller trees have been shown to preferentially tap deeper water sources relative to larger trees, suggesting that the presence of deep roots in smaller trees could substantially reduce the risk of mortality during the dry season

(Meinzer et al. 1999). Yet in other instances, smaller trees extract water from 0–2 m depth and larger trees extract water mainly from deeper layers (Stahl et al. 2013). In tropical dry forest in Yucatan, Mexico, evergreen trees were also shown to use deeper water sources compared to deciduous trees, but this difference was only observed in the earliest successional stage (>10 years), and disappeared in later successional stages (Hasselquist et al. 2010). Furthermore, Hasselquist et al. (2010) showed that among deciduous species, larger individuals tapped deeper water sources, but there was no relationship between tree size and the depth of water uptake for evergreen species. These results highlight the importance of rapidly developing deep roots as a strategy to overcome seasonal water limitations for evergreen tree species.

Although the presence of roots at a particular depth does not necessarily signify water uptake from that depth, deeper root profiles should promote survival during drought. In tropical dry forest of Yucatan, Mexico, live fine roots have been observed near the water table at 9 m (Estrada-Medina et al. 2013). Furthermore, live fine roots extend to 18 m depth in eastern Amazonian forest (Nepstad et al. 1994; Trumbore et al. 1995). Therefore, different tropical ecosystems may have very different sensitivity to drought based on absolute rooting depth of the system relative to the subterranean water table.

Regulation of Gas Exchange

With the onset of drought, plants are faced with the dilemma of either closing stomata to minimize water loss at the expense of reduced carbon intake (isohydric behavior), or opening stomata to maintain carbon intake at the expense of declining water potential (anisohydric behavior) (McDowell et al. 2008). These two extremes of stomatal behavior during drought are thought to lead to mortality by carbon starvation when stomata close, or to hydraulic failure when stomata open. Yet, these extreme examples may not capture the nuanced behavior of the diversity of tree species in tropical forest. The available data suggest that there is a wide variability in the degree of isohydry among tropical trees (Bonal and Guehl 2001). Modeling exercises based on Bornean tropical forests indicate that under moist conditions, anisohydric species would have higher productivity than isohydric species, but as moisture availability decreases, the mortality of anisohydric plants drastically increases whereas that of isohydric plants remains relatively constant and low (Kumagai and Porporato 2012). Furthermore, based on a throughfall reduction experiment in eastern Amazonia, the reduction in modeled sap flow during the dry season was four times greater than what one would predict based purely on reduced soil water potential, consistent with isohydric control of leaf water potential (Fisher et al. 2006). Thus the degree of isohydry measured when water is available may starkly differ from the degree of isohydry during drought. Additionally, the phenomenon of midday depression, when tropical canopy trees reduce gas exchange during the heat of midday (Franco and Lüttge 2002), is composed of stomatal and non-stomatal limitations to photosynthesis (Pons and Welschen 2003), highlighting the fact that the sequence of physiological regulation of gas exchange during drought necessitates further study.

Cuticular Conductance

When all else fails, restricting water from leaves before they are senesced should be an important determinant of the trajectory of a tropical tree during drought. Diffusion of water across the cuticle involves water dissolving in the lipophilic medium of the cuticle and desorption from the outer surface of the cuticle membrane into the atmosphere (Kerstiens 1996). The principle diffusion barrier is located in a narrow band near the outer surface of the cuticle and has been called the limiting skin (Schreiber and Riederer 1996). Thus the overall thickness of the cuticule is not correlated with water permeability. The term cuticular conductance is used to describe loss of water vapor under conditions of maximum stomatal closure, yet when there is uncertainty about the contribution of stomatal conductance to overall water loss, the term 'minimum conductance' (g_{min}) is often used. Previous compendiums of g_{min} clearly demonstrate that there is large variation in this parameter among species (Kerstiens 1996; Schreiber and Riederer 1996). Figure 5 shows patterns of water loss from upper canopy sun leaves of 13 species of trees from Amazonian forest in French Guiana (Santiago et al. unpublished data). The slope of the linear portion of the curve is used to calculate cuticular conductance. These data indicate that there is large variation among leaves from a single canopy layer in tropical forest, demonstrating that a greater knowledge of variation in cuticle conductance among tropical tree species should allow physiologists and modelers to better simulate drought responses.

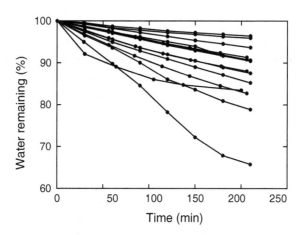

Fig. 5 Evaporative water loss as a function of time for upper canopy excised leaves of 13 Amazonian tree species from Paracou, French Guiana. Measurements were conducted in low (>20 μmol m^{-2} s^{-1} photon flux density) light to maximize stomatal closure

Recovery Mechanisms

Upon alleviation of extended dry periods, the degree of resource expenditure on growth, defenses, and reproduction during drought are likely to determine post-drought survival in subsequent years. There is evidence that Amazonian trees prioritize growth over maintenance and defense during and after drought because any decrease in growth would give neighboring trees an advantage in terms of resource acquisition and put that particular individual at a competitive disadvantage through loss of resources (Franklin et al. 2012; Doughty et al. 2015). Although there is little empirical data to predict which species expend more on maintenance and defense during drought, which could improve survival following drought, it is likely that species normally allocating more to maintenance and defense, late successional species, are less likely to "overspend" during drought than early successional species. Beyond carbon allocation, recovery of trees suffering partial mortality of branches during drought may depend on the ability to re-sprout. This process has not been explicitly linked to drought recovery in tropical forest, but databases on sprouting ability exist (Paciorek et al. 2000), and may help develop predictions of drought survival. Finally, reproductive output during or after drought has the potential to contribute genes to future generations and could be particularly important in monocarpic species (Kitajima and Augspurger 1989), or in species that increase allocation to reproduction following drought due to the increased light energy absorbed during cloudless drought periods (Wright and van Schaik 1994).

Future Directions

Overall, the framework of drought survival traits presented offers several pathways forward for developing forecasts of species-specific drought sensitivity for tropical trees. First, characterizing more tropical tree species in terms of the presence or absence of drought survival traits will allow us to identify which particular trait combinations occur under which environmental circumstances, rather than approaching each species in a completely idiosyncratic manner. Second, where physiologists can characterize drought survival traits in association with large-scale studies, such as throughfall exclusion experiments, or watershed studies, progress can be made linking physiological and ecosystem scales. Third, where experiments are not available, variation of physiological processes as a function of water availability can be obtained along natural environmental gradients, thus utilizing spatial variation as a proxy for temporal climatic variation (Santiago et al. 2004). Finally, many of the physiological measurements needed to characterize drought survival strategies among species are exactly the types of data inputs needed for physiological models that simulate when trees die based on carbon and water relations (Sperry et al. 1998; Höltta et al. 2006).

Acknowledgments The authors thank Christopher Baraloto, Bettina Engelbrecht, Gregory Goldsmith, Omar Lopez, Yadvinder Malhi, Patrick Meir, John Sperry, Klaus Winter and Joe Wright for enlightening discussions on the topics presented; Leonor Álvarez-Cansino, Benoît Burban, Jean-Yves Goret for work together in the field and lab, and the Botany & Plant Sciences Department of the University of California, Smithsonian Tropical Research Institute, Labex CEBA (Centre d'Etude de la Biodiversité Amazonienne; Investissement d'Avenir grant from the ANR, ANR-10-LABX-0025), and the USDA National Institute of Food and Agriculture.

References

Ackerly D (2004) Functional strategies of chaparral shrubs in relation to seasonal water deficit and disturbance. Ecol Monogr 74:25–44

Allen CD, Macalady AK, Chenchouni H, Bachelet D, McDowell N, Vennetier M, Kitzberger T, Rigling A, Breshears DD, Hogg EH, Gonzalez P, Fensham R, Zhang Z, Castro J, Demidova N, Lim JH, Allard G, Running SW, Semerci A, Cobb N (2010) A global overview of drought and heat-induced tree mortality reveals emerging climate change risks for forests. For Ecol Manage 259:660–684

Allison GB (1982) The relationship between O-18 and deuterium in water in sand columns undergoing evaporation. J Hydrol 55:163–169

Anderegg WRL, Berry JA, Smith DD, Sperry JS, Anderegg LDL, Field CB (2012) The roles of hydraulic and carbon stress in a widespread climate-induced forest die-off. Proc Natl Acad Sci USA 109:233–237

Ávila E, Herrera A, Tezara W (2014) Contribution of stem CO_2 fixation to whole-plant carbon balance in nonsucculent species. Photosynthetica 52:3–15

Barnard DM, Meinzer FC, Lachenbruch B, McCulloh KA, Johnson DM, Woodruff DR (2011) Climate-related trends in sapwood biophysical properties in two conifers: avoidance of hydraulic dysfunction through coordinated adjustments in xylem efficiency, safety and capacitance. Plant Cell Environ 34:643–654

Beer C, Reichstein M, Tomelleri E, Ciais P, Jung M, Carvalhais N, Rodenbeck C, Arain MA, Baldocchi D, Bonan GB, Bondeau A, Cescatti A, Lasslop G, Lindroth A, Lomas M, Luyssaert S, Margolis H, Oleson KW, Roupsard O, Veenendaal E, Viovy N, Williams C, Woodward FI, Papale D (2010) Terrestrial gross carbon dioxide uptake: global distribution and covariation with climate. Science 329:834–838

Bonal D, Guehl JM (2001) Contrasting patterns of leaf water potential and gas exchange responses to drought in seedlings of tropical rainforest species. Funct Ecol 15:490–496

Burgess SSO, Dawson TE (2007) Predicting the limits to tree height using statistical regressions of leaf traits. New Phytol 174:626–636

Chapin FS III (1980) The mineral nutrition of wild plants. Annu Rev Ecol Syst 11:233–260

Choat B, Jansen S, Brodribb TJ, Cochard H, Delzon S, Bhaskar R, Bucci SJ, Feild TS, Gleason SM, Hacke UG, Jacobsen AL, Lens F, Maherali H, Martinez-Vilalta J, Mayr S, Mencuccini M, Mitchell PJ, Nardini A, Pittermann J, Pratt RB, Sperry JS, Westoby M, Wright IJ, Zanne AE (2012a) Global convergence in the vulnerability of forests to drought. Nature 491:752–755

Choat B, Jansen S, Brodribb TJ, Cochard H, Delzon S, Bhaskar R, Bucci SJ, Feild TS, Gleason SM, Hacke UG, Jacobsen AL, Lens F, Maherali H, Martinez-Vilalta J, Mayr S, Mencuccini M, Mitchell PJ, Nardini A, Pittermann J, Pratt RB, Sperry JS, Westoby M, Wright IJ, Zanne AE (2012b) Global convergence in the vulnerability of forests to drought. Nature 491:752

Comita LS, Engelbrecht BMJ (2009) Seasonal and spatial variation in water availability drive habitat associations in a tropical forest. Ecology 90:2755–2765

Comstock JP, Ehleringer JR (1988) Contrasting photosynthetic behavior in leaves and twigs of *Hymenoclea salsola*, a green-twigged warm desert shrub. Am J Bot 75:1360–1370

Condit R (1998) Ecological implications of changes in drought patterns: Shifts in forest composition in Panama. Clim Change 39:413–427

Condit R, Hubbell SP, Foster RB (1995) Mortality rates of 205 Neotropical tree and shrub species and the impact of a severe drought. Ecol Monogr 65:419–439

Condit R, Hubbell SP, Foster RB (1996) Assessing the response of plant functional types to climatic change in tropical forests. J Veg Sci 7:405–416

Craine JM, Ocheltree TW, Nippert JB, Towne EG, Skibbe AM, Kembel SW, Fargione JE (2013) Global diversity of drought tolerance and grassland climate-change resilience. Nat Clim Change 3:63–67

Doughty CE, Metcalfe DB, Girardin CAJ, Farfan Amezquita F, Galiano Cabrera D, Huaraca Huasco W, Silva-Espejo JE, Araujo-Murakami A, da Costa MC, Rocha W, Feldpausch TR, Mendoza ALM, da Costa ACL, Meir P, Phillips OL, Malhi Y (2015) Drought impact on forest carbon dynamics and fluxes in Amazonia. Nature 519:78–82

Ehleringer JR, Dawson TE (1992) Water uptake by plants: perspectives from stable isotope composition. Plant Cell Environ 15:1073–1082

Engelbrecht BMJ, Kursar TA, Tyree MT (2005) Drought effects on seedling survival in a tropical moist forest. Trees Struct Funct 19:312–321

Engelbrecht BMJ, Dalling JW, Pearson TRH, Wolf RL, Galvez DA, Koehler T, Tyree MT, Kursar TA (2006) Short dry spells in the wet season increase mortality of tropical pioneer seedlings. Oecologia 148:258–269

Estrada-Medina H, Santiago LS, Graham RC, Allen MF, Jiménez-Osornio JJ (2013) Source water, phenology and growth of two tropical dry forest tree species growing on shallow karst soils. Trees Struct Funct 27:1297–1307

Feeley KJ, Davies SJ, Perez R, Hubbell SP, Foster RB (2011) Directional changes in the species composition of a tropical forest. Ecology 92:871–882

Field CB (2014) Climate change 2014: impacts, adaptation, and vulnerability. Contribution of working group ii to the fifth assessment report of the intergovernmental panel on climate change

Fisher RA, Williams M, Da Costa AL, Malhi Y, Da Costa RF, Almeida S, Meir P (2007) The response of an Eastern Amazonian rain forest to drought stress: results and modelling analyses from a throughfall exclusion experiment. Glob Change Biol 13:2361–2378

Fisher RA, Williams M, Do Vale RL, Da Costa AL, Meir P (2006) Evidence from Amazonian forests is consistent with isohydric control of leaf water potential. Plant Cell Environ 29:151–165

Franco AC, Lüttge U (2002) Midday depression in savanna trees: coordinated adjustments in photochemical efficiency, photorespiration, CO_2 assimilation and water use efficiency. Oecologia 131:356–365

Franklin O, Johansson J, Dewar RC, Dieckmann U, McMurtrie RE, Brannstrom A, Dybzinski R (2012) Modeling carbon allocation in trees: a search for principles. Tree Physiol 32:648–666

Givnish TJ (2002) Adaptive significance of evergreen vs. deciduous leaves: solving the triple paradox. Silva Fennica 36:703–743

Grime JP (1974) Vegetation classification by reference to strategies. Nature 250:26–31

Hacke UG, Sperry JS, Pockman WT, Davis SD, McCulloh KA (2001) Trends in wood density and structure are linked to prevention of xylem implosion by negative pressure. Oecologia 126:457–461

Hartmann H, Adams HD, Anderegg WR, Jansen S, Zeppel MJ (2015) Research frontiers in drought-induced tree mortality: crossing scales and disciplines. New Phytol 205:965–969

Hasselquist NJ, Allen MF, Santiago LS (2010) Water relations of evergreen and drought-deciduous trees along a seasonally dry tropical forest chronosequence. Oecologia 164:881–890

Höltta T, Vesala T, Sevanto S, Peramaki M, Nikinmaa E (2006) Modeling xylem and phloem water flows in trees according to cohesion theory and Munch hypothesis. Trees Struct Funct 20:67–78

Huntingford C, Zelazowski P, Galbraith D, Mercado LM, Sitch S, Fisher R, Lomas M, Walker AP, Jones CD, Booth BBB, Malhi Y, Hemming D, Kay G, Good P, Lewis SL, Phillips OL, Atkin OK, Lloyd J, Gloor E, Zaragoza-Castells J, Meir P, Betts R, Harris PP, Nobre C, Marengo J, Cox PM (2013) Simulated resilience of tropical rainforests to CO_2-induced climate change. Nat Geosci 6:268–273

Jackson PC, Cavelier J, Goldstein G, Meinzer FC, Holbrook NM (1995) Partitioning of water resources among plants of a lowland tropical forest. Oecologia 101:197–203

Kerstiens G (1996) Cuticular water permeability and its physiological significance. J Exp Bot 47:1813–1832

Kitajima K, Augspurger CK (1989) Seed and seedling ecology of a monocarpic tropical tree, *Tachigalia versicolor*. Ecology 70:1102–1114

Kumagai T, Porporato A (2012) Strategies of a Bornean tropical rainforest water use as a function of rainfall regime: isohydric or anisohydric? Plant Cell Environ 35:61–71

Kursar TA, Engelbrecht BMJ, Burke A, Tyree MT, El Omari B, Giraldo JP (2009) Tolerance to low leaf water status of tropical tree seedlings is related to drought performance and distribution. Funct Ecol 23:93–102

McDowell NG (2011) Mechanisms linking drought, hydraulics, carbon metabolism, and vegetation mortality. Plant Physiol 155:1051–1059

McDowell NG, Sevanto S (2010) The mechanisms of carbon starvation: how, when, or does it even occur at all? New Phytol 186:264–266

McDowell N, Pockman WT, Allen CD, Breshears DD, Cobb N, Kolb T, Plaut J, Sperry J, West A, Williams DG, Yepez EA (2008) Mechanisms of plant survival and mortality during drought: why do some plants survive while others succumb to drought? New Phytol 178:719–739

McDowell NG, Fisher RA, Xu CG, Domec JC, Höltta T, Mackay DS, Sperry JS, Boutz A, Dickman L, Gehres N, Limousin JM, Macalady A, Martínez-Vilalta J, Mencuccini M, Plaut JA, Ogée J, Pangle RE, Rasse DP, Ryan MG, Sevanto S, Waring RH, Williams AP, Yepez EA, Pockman WT (2013) Evaluating theories of drought-induced vegetation mortality using a multimodel-experiment framework. New Phytol 200:304–321

Meinzer FC, Goldstein G (1996) Scaling up from leaves to whole plants and canopies for photosynthetic gas exchange. In: Mulkey SS, Chazdon RL, Smith AP (eds) Tropical forest plant ecophysiology. Chapman & Hall, New York, pp 114–138

Meinzer FC, Andrade JL, Goldstein G, Holbrook NM, Cavelier J, Wright SJ (1999) Partitioning of soil water among canopy trees in a seasonally dry tropical forest. Oecologia 121:293–301

Meinzer FC, Woodruff DR, Domec JC, Goldstein G, Campanello PI, Gatti MG, Villalobos-Vega R (2008) Coordination of leaf and stem water transport properties in tropical forest trees. Oecologia 156:31–41

Meinzer FC, Johnson DM, Lachenbruch B, McCulloh KA, Woodruff DR (2009) Xylem hydraulic safety margins in woody plants: coordination of stomatal control of xylem tension with hydraulic capacitance. Funct Ecol 23:922–930

Nepstad DC, Decarvalho CR, Davidson EA, Jipp PH, Lefebvre PA, Negreiros GH, Dasilva ED, Stone TA, Trumbore SE, Vieira S (1994) The role of deep roots in the hydrological and carbon cycles of Amazonian forests and pastures. Nature 372:666–669

Nepstad DC, Moutinho P, Dias MB, Davidson E, Cardinot G, Markewitz D, Figueiredo R, Vianna N, Chambers J, Ray D, Guerreiros JB, Lefebvre P, Sternberg L, Moreira M, Barros L, Ishida FY, Tohlver I, Belk E, Kalif K, Schwalbe K (2002) The effects of partial through fall exclusion on canopy processes, aboveground production, and biogeochemistry of an Amazon forest. J Geophys Res Atmos 107

Nilsen ET (1995) Stem photosynthesis: extent, patterns and role in plant carbon economy. In: Gartner B (ed) Plant stems: physiology and functional morphology. Academic Press, San Diego, pp 223–240

O'Brien MJ, Burslem DFRP, Caduff A, Tay J, Hector A (2015) Contrasting nonstructural carbohydrate dynamics of tropical tree seedlings under water deficit and variability. New Phytol 205:1083–1094

Osmond CB, Smith SD, Guiying B, Sharkey TD (1987) Stem photosynthesis in a desert ephemeral, *Eriogonum inflatum*. Characterization of leaf and stem CO_2 fixation and H_2O vapor exchange under controlled conditions. Oecologia 72:542–549

Paciorek CJ, Condit R, Hubbell SP, Foster RB (2000) The demographics of resprouting in tree and shrub species of a moist tropical forest. J Ecol 88:765–777

Pfanz H (2008) Bark photosynthesis. Trees Struct Funct 22:137–138

Phillips OL, van der Heijden G, Lewis SL, Lopez-Gonzalez G, Aragao L, Lloyd J, Malhi Y, Monteagudo A, Almeida S, Davila EA, Amaral I, Andelman S, Andrade A, Arroyo L, Aymard G, Baker TR, Blanc L, Bonal D, de Oliveira ACA, Chao KJ, Cardozo ND, da Costa L, Feldpausch TR, Fisher JB, Fyllas NM, Freitas MA, Galbraith D, Gloor E, Higuchi N, Honorio E, Jimenez E, Keeling H, Killeen TJ, Lovett JC, Meir P, Mendoza C, Morel A, Vargas PN, Patino S, Peh KSH, Cruz AP, Prieto A, Quesada CA, Ramirez F, Ramirez H, Rudas A, Salamao R, Schwarz M, Silva J, Silveira M, Slik JWF, Sonke B, Thomas AS, Stropp J, Taplin JRD, Vasquez R, Vilanova E (2010) Drought-mortality relationships for tropical forests. New Phytol 187:631–646

Pivovaroff AL, Pasquini SC, De Guzman ME, Alstad KP, Stemke J, Santiago LS (2016) Multiple strategies for drought survival among woody plant species. Funct Ecol. doi:10.1111/1365-2435.12518

Pockman WT, Sperry JS (2000) Vulnerability to xylem cavitation and the distribution of Sonoran desert vegetation. Am J Bot 87:1287–1299

Pons TL, Welschen RAM (2003) Midday depression of net photosynthesis in the tropical rainforest tree *Eperua grandiflora*: contributions of stomatal and internal conductances, respiration and Rubisco functioning. Tree Physiol 23:937–947

Ryan MG, Yoder BJ (1997) Hydraulic limits to tree height and tree growth. Bioscience 47:235–242

Saatchi SS, Harris NL, Brown S, Lefsky M, Mitchard ETA, Salas W, Zutta BR, Buermann W, Lewis SL, Hagen S, Petrova S, White L, Silman M, Morel A (2011) Benchmark map of forest carbon stocks in tropical regions across three continents. Proc Natl Acad Sci USA 108:9899–9904

Salleo S, Lo Gullo MA, Trifilo P, Nardini A (2004) New evidence for a role of vessel-associated cells and phloem in the rapid xylem refilling of cavitated stems of Laurus nobilis L. Plant Cell Environ 27:1065–1076

Santiago LS, Mulkey SS (2005) Leaf productivity along a precipitation gradient in lowland Panama: patterns from leaf to ecosystem. Trees 19:349–356

Santiago LS, Kitajima K, Wright SJ, Mulkey SS (2004) Coordinated changes in photosynthesis, water relations and leaf nutritional traits of canopy trees along a precipitation gradient in lowland tropical forest. Oecologia 139:495–502

Saveyn A, Steppe K, Ubierna N, Dawson TE (2010) Woody tissue photosynthesis and its contribution to trunk growth and bud development in young plants. Plant Cell Environ 33:1949–1958

Schimper AFW (1903) Plant geography upon a physiological basis. Clarendon, Oxford

Schreiber L, Riederer M (1996a) Determination of diffusion coefficients of octadecanoic acid in isolated cuticular waxes and their relationship to cuticular water permeabilities. Plant Cell Environ 19:1075–1082

Schreiber L, Riederer M (1996b) Ecophysiology of cuticular transpiration: comparative investigation of cuticular water permeability of plant species from different habitats. Oecologia 107:426–432

Slik JWF (2004) El Niño droughts and their effects on tree species composition and diversity in tropical rain forests. Oecologia 141:114–120

Sperry JS (2000) Hydraulic constraints on plant gas exchange. Agric For Meteorol 104:13–23

Sperry JS, Donnelly JR, Tyree MT (1988) A method for measuring hydraulic conductivity and embolism in xylem. Plant Cell Environ 11:35–40

Sperry JS, Adler FR, Campbell GS, Comstock JP (1998) Limitation of plant water use by rhizosphere and xylem conductance: results from model. Plant Cell Environ 21:347–359

Sperry JS, Meinzer FC, McCulloh KA (2008) Safety and efficiency conflicts in hydraulic architecture: scaling from tissues to trees. Plant Cell Environ 31:632–645

Stahl C, Hérault B, Rossi V, Burban B, Bréchet C, Bonal D (2013) Depth of soil water uptake by tropical rainforest trees during dry periods: does tree dimension matter? Oecologia 173:1191–1201

Steppe K, Sterck F, Deslauriers A (2015) Diel growth dynamics in tree stems: linking anatomy and ecophysiology. Trends Plant Sci 20:335–343

Tinoco-Ojanguren C (2008) Diurnal and seasonal patterns of gas exchange and carbon gain contribution of leaves and stems of Justicia californica in the Sonoran desert. J Arid Environ 72:127–140

Trumbore SE, Davidson EA, Decamargo PB, Nepstad DC, Martinelli LA (1995) Belowground cycling of carbon in forests and pastures of eastern Amazonia. Global Biogeochem Cycles 9:515–528

Tyree MT, Sperry JS (1989) Vulnerability of xylem to cavitation and embolism. Ann Rev Plant Physiol Plant Mol Biol 40:19–38

Westoby M, Falster DS, Moles AT, Vesk PA, Wright IJ (2002) Plant ecological strategies: some leading dimensions of variation between species. Annu Rev Ecol Syst 33:125–159

Wittmann C, Pfanz H (2008) Antitranspirant functions of stem periderms and their influence on corticular photosynthesis under drought stress. Trees Struct Funct 22:187–196

Wright SJ, van Schaik CP (1994) Light and the phenology of tropical trees. Am Nat 143:192–199

Part IV
Nutrient Limitation of Ecophysiological Processes in Tropical Trees

Nutrient Availability in Tropical Rain Forests: The Paradigm of Phosphorus Limitation

James W. Dalling, Katherine Heineman, Omar R. Lopez, S. Joseph Wright and Benjamin L. Turner

Abstract A long-standing paradigm in tropical ecology is that phosphorus (P) availability limits the productivity of most lowland forests, with the largest pool of plant-available P resident in biomass. Evidence that P limits components of productivity is particularly strong for sites in Panama and the Amazon basin. Analyses of forest communities in Panama also show that tree species distributions are strongly affected by P availability at the regional scale, but that their local distributions in a single site on Barro Colorado Island (BCI) are as frequently correlated with base cations as with P. Traits associated with species sensitivity to P availability require more detailed exploration, but appear to show little similarity with those associated with N limitation in temperate forests. Recent research indicates that a large fraction of P in tropical forests exists as organic and microbial

J.W. Dalling (✉)
Department of Plant Biology, University of Illinois at Urbana-Champaign,
505 S. Goodwin Avenue, Urbana, IL 61801, USA
e-mail: dalling@illinois.edu

J.W. Dalling · O.R. Lopez
Smithsonian Tropical Research Institute, Apartado Postal,
0843-03092 Panama, Republic of Panama
e-mail: olopez@indicasat.org.pa

K. Heineman
Program in Ecology, Evolution and Conservation Biology,
University of Illinois at Urbana-Champaign, 505 S. Goodwin Avenue,
Urbana, IL 61801 USA
e-mail: kheineman@life.illinois.edu

O.R. Lopez
Instituto de Investigaciones Científicas y Servicios de Alta Tecnología,
Apartado Postal, 0843-01103 Ciudad de Saber, Panama, Republic of Panama

S.J. Wright · B.L. Turner
Smithsonian Tropical Research Institute, Apartado Postal,
0843-03092 Balboa, Panama, Republic of Panama
e-mail: WRIGHTJ@si.edu

B.L. Turner
e-mail: TurnerBL@si.edu

© Springer International Publishing Switzerland 2016
G. Goldstein and L.S. Santiago (eds.), *Tropical Tree Physiology*,
Tree Physiology 6, DOI 10.1007/978-3-319-27422-5_12

P in the soil; plant adaptations to access organic P, including the synthesis of phosphatase enzymes, likely represent critical adaptations to low P environments. Plants also cope with low P availability through increases in P use-efficiency resulting from increased retention time of P in biomass and decreased tissue P concentration. Although foliar P responds strongly to P addition, we show here that foliar P and N:P are highly variable within communities, and at BCI correlate with regional species distributional affinity for P. An improved understanding of P limitation, and in particular the plasticity of responses to P availability, will be critical to predicting community and ecosystem responses of tropical forests to climate change.

Keywords Ecosystem nutrient budget · Fertilization experiments · Nutrient uptake · Physiological traits · Soil nutrients

Introduction

Humid low elevation tropical forests support among the highest above-ground biomass and net primary productivity of any ecosystem (Scurlock and Olson 2002; Houghton 2005). Yet, many tropical forests that have been cleared for agriculture fail to support crop yields for more than a few years without large fertilizer inputs (Nye and Greenland 1960; Sanchez et al. 1983; Lal 1986). This apparent paradox suggested that much of the nutrient reserve of tropical forests is stored in plant tissue, resulting in critical nutrient limitation once biomass is removed. A synthesis of datasets on nutrient concentrations in leaves and senescent tissue returning to the ecosystem as litterfall highlighted phosphorus (P) as the nutrient that most clearly distinguishes the stoichiometry of tropical from temperate forests, leading to the paradigm that P availability constrains the productivity of most lowland tropical forests (Vitousek 1984). Here we review the evidence for this paradigm, highlighting the insights gained over the past three decades on the distribution of P in tropical ecosystems, its availability to plants, adaptations of plants for P acquisition and use, and evidence that plant growth and forest diversity reflects not only P supply, but also co-limitation by nitrogen (N) and base cations.

Nutrient Limitation in Tropical Forests

Many tropical soils are strongly weathered, reflecting warm, moist conditions acting on stable land surfaces that were not directly affected by Quaternary glaciations (Baillie 1996). These older, well-weathered 'Oxisol' (Soil Taxonomy system; Soil

Survey Staff 1999) or 'Ferralsol' soils (IUSS Working Group WRB 2014) are characterized by a moderately organically enriched surface horizon, and deep, relatively uniform yellow-red subsoils, reflecting weathering of clay minerals into ferric sesquioxides (Baillie 1996). Chemically, these soils are acidic, highly leached, with a low cation exchange capacity and high aluminum saturation. Beyond this broad generalization, however, lies a diversity of soil physical properties (notably influencing drainage patterns and water-holding capacity) and chemical properties reflecting heterogeneous parent materials, substrate age, and topographic effects on soil development. Nonetheless, a feature of many soil orders associated with tropical forests is limited P availability and low total P, resulting from long periods of leaching and strong adsorption or occlusion of P with iron and aluminium oxides (summarized in Table 10.12 in Baillie 1996).

The central importance of P limitation to the productivity of lowland forests reached prominence after comparisons of litterfall in temperate and tropical forests (Vitousek 1984). If P is a key limiting nutrient, then plants might be expected either to function with less of it (reflected in lower tissue P concentrations), or to cycle it more efficiently (reflected in part in higher resorption efficiency of P as tissues senesce). Analysis of nutrient concentrations and nutrient ratios in leaf litter is consistent with both mechanisms. Litterfall from lowland tropical forests, particularly those on older, more weathered soils of the Amazon basin, had low P concentrations, whereas tissue N and Ca concentrations were mostly comparable to temperate forests (Vitousek 1984). This study also highlighted differences in nutrient limitation between lowland and montane tropical forests. In montane forests, P availability can be low, and foliar P generally declines with elevation (Tanner et al. 1998; Benner et al. 2010). However, N limitation appears to be much more important in the mountains than in the lowlands, probably due to temperature and moisture effects on rates of mineralization (e.g. Grubb 1977; Vitousek et al. 1994; reviewed in Benner et al. 2010).

Fertilization experiments have generally supported the view of P limitation in lowland forests, and N or N and P limitation in montane forests (reviewed by Sayer and Banin, this volume, Dalling et al. 2015). However, insights from fertilization experiments are incomplete with regard to the scale and generality of nutrient limitation. This is because factorial fertilization experiments are expensive to establish and therefore restricted in the number of nutrient addition treatments that can be included, and may take many years to yield clear treatment effects (Sullivan et al. 2014). Fertilization experiments may therefore miss co-limitation scenarios that could arise either because productivity in different plant size classes or tissue types have different and distinct nutrient requirements (as seen in responses to growth vs. litterfall responses to N, P and K in the Gigante fertilization experiment; Wright et al. 2011), or because the limited scale and replication of fertilization experiments precludes analyses of the individual species responses that contribute to heterogeneous nutrient limitation (Alvarez-Clare et al. 2013; Dalling et al. 2015).

Mixed Evidence of Phosphorus Limitation at the Local Scale

If P availability alone governs forest growth and productivity then we would predict that, at the community level, N:P would be relatively tightly constrained, with ratios >16 (the Redfield ratio; Koerselman and Mueleman 1996) reflecting selection for efficient P use. However, evidence from community-wide analyses of foliar N:P do not support this prediction, with very wide variation in N:P in lowland forests, but declining N:P with elevation (Fig. 1). In an analysis compiling published datasets of foliar nutrients from neotropical sites, Townsend et al. (2007) found limited evidence for environmental control over N:P. Instead, they found large interspecific variation in N:P among species in lowland forests, suggesting coexisting tree species are likely to differ widely in their nutrient requirements (see also Hättenschwiler et al. 2008; Hedin et al. 2009).

Similarly, analyses of species distributions within large forest dynamics plots show correlations with a multitude of soil nutrients, with no clear preference for P (John et al. 2007). For example, at Barro Colorado Island (BCI), Panama, 40 % of

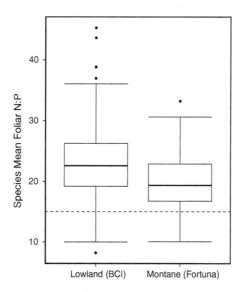

Fig. 1 The distribution of species mean foliar N:P ratios for 281 lowland tree species sampled on Barro Colorado Island (BCI) in the Panama Canal and 91 montane tree species sampled at Fortuna Forest Reserve in western Panama (800–1300 m a.s.l). The hinges of the boxplot represent the interquartile range (IQR), and the whiskers extend to the largest and smallest values within 1.5 * IQR. Community mean N:P ratio was significantly higher on BCI (NP = 22.8) than Fortuna (NP = 19.5; t = 5.65, df = 179.3, P < 0.001). However, variance did not differ significantly between sites (levene test, F = 2.8; df = 1,370; P = 0.093). The line at N:P = 15 represents the Redfield Ratio, which is the empirically derived threshold between N and P limitation in terrestrial ecosystems

the 258 most common tree species in the 50 ha forest dynamics plot show a significant association in their local distribution patterns with at least one principal component axis representing variation in soil chemical variables (John et al. 2007). Analysis of individual soil variables shows a relatively even distribution in the frequency of associations across soil variables (Fig. 2), with as many species showing significant distributional associations with boron (B), calcium (Ca), potassium (K), and zinc (Zn) as with P.

The lack of a clear pattern of association between tree species distributions and soil P could reflect insufficient variability in P availability at the plot level, strong spatial autocorrelation among soil nutrients, or strong similarity among species in P requirements. Despite its location on a relatively flat andesite plateau, nutrient availability within the BCI plot is quite variable. Soil P, extracted using Mehlich III and analyzed using ICP, ranged from 0.45 to 6.8 mg/kg (10th–90th percentiles). Although as a group base cations were strongly positively correlated at BCI (defining the first principal component of soil chemical variation), P concentration was only weakly correlated with base cations and instead loaded most strongly on a second principal component axis. Although it might therefore be tempting to conclude that demographic rates, and therefore distributions of species on BCI are not primarily differentiated by P requirements, species distributions across the Panama Canal area forests suggest otherwise.

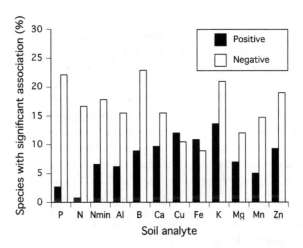

Fig. 2 Frequency of association of 258 tree species with 12 soils variables measured at 300 locations in the 50 ha forest dynamics plot on Barro Colorado Island, Panama. Positive and negative associations indicate that species occurred at sites with higher and lower than expected concentrations ($p < 0.05$) of that variable based on assessment using the Poisson Cluster Method (John et al. 2007)

Evidence for Phosphorus Limitation at Landscape and Regional Scales

Analysis of tree communities in 72 forest plots distributed across the 66 km-wide Isthmus of Panama surrounding BCI has revealed a much stronger imprint of soil P on species distributions. Condit et al. (2013) examined whether the distribution of 550 tree species in this regional species pool showed distributional associations with seven environmental variables measured at each plot, and with dry season moisture availability estimated from a network of rain gauges. This analysis confirmed that dry season rainfall was the strongest environmental predictor of tree distributions, with strong effects on 66 % of species, reflecting a steep gradient in dry season intensity from the Caribbean to Pacific coasts of Panama. More of a surprise, however, was that P availability affected nearly as many species (58 %), revealing a previously unrecognized environmental filter acting on the composition of these forests. Furthermore, species partitioned a gradient of P availability relatively evenly, with 23 % of species showing affinity for high P and 35 % for low P soils.

Phosphorus effects on tree community composition may be particularly strong in central Panama because of the remarkable variation in P availability (resin extractable P varies from <0.1 to 22.8 mg P kg^{-1}; Condit et al. 2013); comparable in magnitude to that of the entire lowland tropics (Condit et al. 2013 and references therein). However, other soil nutrients, notably calcium, were associated with the distribution of 35 % of tree species, raising the possibility that observed associations are at least in part driven by other correlated resources. To test whether species associations reflect differences in growth response to P availability, we have begun measuring seedling growth rates in a pot experiment manipulating P availability. Seedlings of fifteen species with contrasting distributions across the Isthmus were transplanted to a low P soil (0.13 mg kg^{-1} resin extractable P) and fertilized weekly with a full nutrient solution either with or without added P. The magnitude of species growth increase in the P addition treatment was significantly positively correlated with the degree to which species distributions were skewed towards high P soils (determined from the Condit et al. (2013) dataset).

Phosphorus Limitation: Links to Functional and Physiological Traits

Species distributional affinity for P is also likely to be correlated with plant functional traits that potentially impact ecosystem processes. In another pot experiment exploring how Panamanian tree species respond to P availability, Vargas and Lopez (unpublished data), found that *Hura crepitans*, a species with affinity for high P soils, showed significantly higher leaf specific hydraulic conductance (K_L) when grown under high P conditions, suggesting the potential for P availability to influence ecosystem water balance. Higher K_L under high P might be the result of increased xylem vessel diameter or a by-product of shifts in allocation patterns, and reduction in wood density (Goldstein et al. 2013).

More broadly, soil P availability is often correlated with foliar P concentration (Vitousek and Sandford 1983), with effects on decomposition rates and therefore nutrient cycling (reviewed in Cornwell et al. 2008). While foliar P concentration is a plastic trait that responds to P fertilization (e.g., Santiago et al. 2012; Mayor et al. 2014), community wide foliar P values may also reflect the wider distribution patterns of constituent species. We compared foliar N and P concentrations measured in shade leaves of three individuals of each of 137 species collected on BCI (Wright and Turner, unpublished data) with species distributional affinity for P from Condit et al. (2013). Foliar P was significantly positively correlated with distributional affinity for P (Fig. 3; r = 0.45, p < 0.001), and foliar N:P was significantly

Fig. 3 Relationship between foliar N, P, and N:P of shade leaves of 137 tree species collected on Barro Colorado Island, Panama, and the index score of species affinity for P, based on the distribution of the same species across 72 sites across the Isthmus of Panama varying in resin-extractable P. Positive P affinity index scores indicate that species were associated with sites with high resin-extractable P

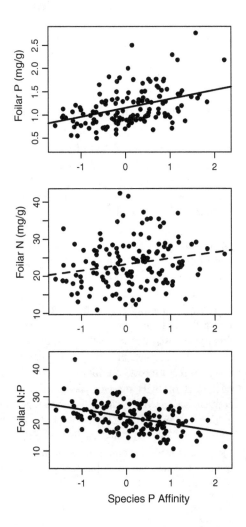

negatively correlated ($r = -0.38$, $p < 0.001$). Foliar N was not significantly corre-lated with P affinity ($r = 0.17$). Interspecific variation in foliar nutrient concentration may, in turn, be expected to correlate with juvenile and adult growth rates, however no significant relationships were found between foliar N, P and N:P with sapling (>1 cm DBH) or juvenile and adult (>10 cm DBH) growth rates of the same 137 species in the BCI 50 ha plot.

Elsewhere, an affinity for low P soils has been found to correlate with a high P use-efficiency (PUE; a measure of biomass increment per unit P). In Queensland, Australia, Gleason et al. (2009) found that tree species that were specialists on P-poor schist-derived soils had twice the PUE of generalist species. High PUE in turn was related to lower wood P concentrations and to higher retention time of P in the canopy, but was unrelated to above-ground net primary productivity (sum of radial growth and litter production). In contrast, along a steeper gradient of soil P availability spanning the eastern slope of the Andes to the highly weathered and P impoverished soils of the Guiana Shield, total soil P correlates positively with foliar P concentration (Fyllas et al. 2009) and wood production, and inversely with wood density (Mercado et al. 2011; Quesada et al. 2012).

Phosphorus Pools in Soil and Biomass

Much of our understanding of the chemistry of P limitation is based on temperate chronosequences (Walker and Syers 1976). These studies generally show that given sufficient time for pedogenesis, a decline in the total P pool size occurs, and is accompanied by a decline in primary mineral P, and a corresponding increase in the proportion of the total P occurring in occluded and organic forms. Similar shifts between inorganic and organic P pools have been observed in the Hawaiian island chronosequence (Crews et al. 1995), and are likely to be more generally reflected in wet tropical forests (although no long-term chronosequence has been identified in species-rich lowland tropical rain forest).

Nonetheless, a traditional view of P cycling in tropical forests holds that the biological uptake of P is tightly coupled with the deposition of labile P via canopy litterfall and relies very little on soil organic and inorganic P pools. According to this hypothesis, nearly all the biologically available P in tropical forests should be sequestered in plant biomass. This view is supported by the observation that while intact tropical forests are highly productive, forested land cleared and converted to pasture is often P-impoverished and frequently abandoned within a few years of conversion (Serrao et al. 1996). Furthermore, Stark and Jordan (1978) provided evidence for "direct-cycling" of P by demonstrating that only 0.1 % of isotopically-labeled P and calcium applied directly soil surface was lost beyond the root mat in a Venezuelan rainforest, indicating that there should be little chance that organic P is leached from the ecosystem, or sequestered to iron or aluminum oxides in the soil.

Recent studies, however, provide increasing evidence that soil organic P pools, which constitute about 25 % of total soil P in tropical forests (Turner and Engelbrecht 2011), are more dynamic and chemically variable than previously thought. For example, total soil organic P concentrations decline markedly (up to 25 %) during the dry season (Turner et al. 2015), while Vincent et al. (2010) showed that organic P contained in DNA and phosphate monoesters declined in response to experimental litter removal and increased in response to litter addition. These studies indicate that there are more persistent pools of accessible organic P in the soil beyond those directly re-assimilated from litter into plant biomass. In broad terms, the pool of organic P available to plants can be considered to constitute three groups: (1) recently senesced plant or microbial material, (2) P sequestered in live microbial biomass and (3) "stable" organic P bound to iron or aluminum oxides. Soil microbial biomass, characterized by low C:P ratios relative to plant biomass, is an important P sink in forested ecosystems, and can represent a large fraction of the total biomass-associated P (Turner et al. 2013). In a lowland tropical forest in Panama, microbial P was estimated to account for approximately two thirds of the total soil organic P (Turner et al. 2015). While microbes may compete with vascular plants for P, microbial sequestration of P prevents the loss of ecosystem P to more permanent geochemical sinks (Oberson et al. 1997), and facilitates the mineralization of organic P to plant available phosphate (Richardson and Simpson 2011). A recent detailed ecosystem P budget for sites representing a five-fold range in total P storage in a Panamanian montane forest reveals that the amount of available P contained in soil microbial and non microbial pools greatly exceeds the amount of P sequestered in foliar and woody plant tissues at all sites (Heineman unpublished data), contrary to the paradigm that labile P is sequestered primarily in plant biomass.

Acquisition of Organic Phosphorus in Tropical Forests

Adaptations of plants to low P environments may reflect alterations both in P use efficiency and in the ability to acquire P from less readily available sources. Turner (2008) suggested that a gradient of biological availability of soil organic P exists, which is determined by the investment cost to plants of producing an array of enzymes required to catalyze the hydrolysis and release of phosphate. At one extreme, plants can take up dissolved phosphate directly from the soil solution or via mycorrhizas. Simple phosphate monoesters, such as glucose 6-phosphate however require hydrolysis using phosphomonoesterase enzymes. Phosphate diesters, including phospholipids and nucleic acids, require both phosphomonoesterase and phosphodiesterase enzymes to release phosphate. Finally, inositol hexakisphosphates require both solubilization from organic matter or clay minerals and hydrolysis by phytase enzymes. Turner (2008) further suggested that differences in the enzymatic capacity of plants confronted with an array of potential P sources could represent an axis of resource partitioning among plant species.

Consistent with the hypothesis that species partition P sources, pools of organic P differ in composition across a gradient in total soil P (Turner and Engelbrecht 2011) with a higher ratio of phosphomonoesters to phosphodiesters in sites with higher total P. Nonetheless, to date, little evidence has emerged to suggest that plants from P impoverished tropical forests differ in their capacity to produce phosphatase enzymes when compared to those from more P-rich sites. In lower montane forest at Fortuna, Panama, Steidinger et al. (2015) measured the production of phosphomonoesterase and phosphodiesterase of excised roots of arbuscular mycorrhizal (AM), ectomycorrhizal (EM) and non-mycorrhizal (NM) plant species. They found only weak evidence for differences in enzymatic activity among these groups, with no difference in activity between AM and EM taxa but significantly higher phosphodiesterase activity in the NM taxon. A second experiment with the same taxa compared growth rates of seedlings infected with mycorrhizas and grown in acid-washed sand fertilized with inorganic phosphate, glucose phosphate, RNA and phytate as the sole P source. Again, AM and EM species did not differ in their response to organic sources of P, but the NM taxon, *Roupala montana* (Proteaceae) was the only species capable of exploiting phytate as a P source. At Fortuna EM and AM taxa, as well as *Roupala* occur in multiple sites that differ in P availability (as determined by resin-extractable P and total P). Although the difference in P source use between *Roupala* and other taxa suggests P source partitioning could occur, its ecological significance in tropical forests requires further exploration, including greater species replication of mycorrhizal groups.

Investment in phosphatase enzymes, however, may explain the abundance of N fixing legumes, which can account for a large fraction of stems and basal area in lowland tropical forests (Losos and Leigh 2004) and on occasions form monodominant stands (Connell and Lowman 1989). The dominance of nitrogen fixers in ecosystems where P is considered to limit forest productivity suggests that increasing N availability may indirectly impact P acquisition. Houlton et al. (2008) pointed out that phosphatase production is nitrogen-intensive (up to 15 % N by mass), and that N_2 fixing plants might therefore be able to increase allocation to phosphatase relative to non-fixers. Evidence generally supports the Houlton hypothesis, with higher phosphatase activity in soil beneath N_2 fixing plants (Houlton et al. 2008), higher root phosphatase activity (and AM infection) in N_2 fixing lowland tropical trees (Nasto et al. 2014), and an increase in the phosphomonoesterase activity of N_2 fixing legumes relative non-fixers with declining soil P along a chronosequence in Western Australia (Png et al. 2014).

Conclusions and Future Directions

The consequences of P limitation from the individual to the whole community-level remains a relatively poorly studied aspect of tropical plant ecophysiology and forest ecology. Just as N availability can limit the productivity of temperate plants under elevated CO_2 (e.g., Long et al. 2004; Norby et al. 2010; but see Drake et al. 2011), chronic P limitation has the potential to constrain the response of tropical forest plants to rising CO_2, particularly if reduced transpiration rates also reduce P uptake (Cernusak et al. 2011). Despite this, experimental data to explore how CO_2 and P interact to influence the growth of tropical plants remains extremely scarce (Cernusak et al. 2013). Likewise, an understanding of how tropical forest plants have adapted to low P environments, either through increases in P-use efficiency or phosphate-acquisition efficiency, can provide insights into how to manage P inputs more efficiently in agricultural settings to maximize crop yields while minimizing P exports from the system (Veneklaas et al. 2012).

References

Alvarez-Clare S, Mack MC, Brooks M (2013) A direct test of nitrogen and phosphorus limitation to net primary productivity in a lowland tropical wet forest. Ecology 94:1540–1551

Baillie IC (1996) Soils of the humid tropics. In: Richards PW (ed) The tropical rain forest. Cambridge University Press, Cambridge, pp 256–285

Benner J, Vitousek PM, Ostertag R (2010) Nutrient cycling and nutrient limitation in tropical montane cloud forest. In: Bruijnzeel LA, Scatena FN, Hamilton LS (eds) tropical montane cloud forest. Cambridge University Press, Cambridge, pp 90–100

Cernusak L, Winter K, Turner BL (2011) Transpiration modulates phosphorus acquisition in tropical tree seedlings. Tree Phys 31:878–885

Cernusak L, Winter K, Dalling JW et al (2013) Tropical forest responses to increasing [CO_2]: current knowledge and opportunities for future research Func. Plant Biol 40:531–551

Condit R, Engelbrecht BMJ, Pino D et al (2013) Species distributions in response to individual soil nutrients and seasonal drought across a community of tropical trees. P Natl Acad Sci 110:5064–5068

Connell JH, Lowman MD (1989) Low-diversity tropical rain forests: some possible mechanisms for their existence. Am Nat 134:19–88

Cornwell WK, Cornelissen JH, Amatangelo K et al (2008) Plant species traits are the predominant control on litter decomposition rates within biomes worldwide. Ecol Lett 11:1065–1071

Crews TE, Kitayama K, Fownes JH et al (1995) Changes in soil-phosphorus fractions and ecosystem dynamics across a long chronosequence in Hawaii. Ecology 76:1407–1424

Dalling JW, Heineman K, González G et al (2015) Geographic, environmental and biotic sources of variation in the nutrient relations of tropical montane forests. J Trop Ecol in press

Drake JE, Gallet-Budynek A, Hofmockel KS et al (2011) Increases in the flux of carbon belowground stimulate nitrogen uptake and sustain the long-term enhancement of forest productivity under elevated CO2. Ecol Lett 14:349–357

Fyllas NM, Patino S, Baker TR et al (2009) Basin-wide variations in foliar properties of Amazonian forest: phylogeny, soils and climate. Biogeosciences 6:2677–2708

Gleason SM, Read J, Ares A et al (2009) Phosphorus economics of tropical rainforest species and stands across soil contrasts in Queensland, Australia: understanding the effects of soil specialization and trait plasticity. Func Ecol 23:1157–1166

Goldstein G, Bucci SJ, Scholz FG (2013) Why do trees adjust water relations and hydraulic architecture in response to nutrient availability? Tree Phys 33:238–240

Grubb PJ (1977) Control of forest growth and distribution on wet tropical mountains: with special reference to mineral nutrition. Ann Rev Ecol Syst 8:83–107

Hättenschwiler S, Aeschlimann B, Coûteaux MM et al (2008) High variation in foliage and leaf litter chemistry among 45 tree species of a neotropical rainforest community. New Phytol 179:165–175

Hedin LO, Brookshire ENJ, Menge DNL et al (2009) The nitrogen paradox in tropical forest ecosystems. Ann Rev Ecol Evol Syst 40:613–635

Houghton RA (2005) Above-ground forest biomass and the global carbon balance. Global Change Bio 11:945–958

Houlton BZ, Wang Y-P, Vitousek PM et al (2008) A unifying framework for dinitrogen fixation in the terrestrial biosphere. Nature 454:327–331

IUSS Working Group WRB (2014) World Reference Base for Soil Resources 2014. International soil classification system for naming soils and creating legends for soil maps. World Soil Resources Reports No. 106. FAO, Rome

John RC, Dalling JW, Harms KE et al (2007) Soil nutrients influence spatial distributions of tropical tree species. P Natl Acad Sci 104:864–869

Koerselman W, Meuleman AFM (1996) The vegetation N:P ratio: a new tool to detect the nature of nutrient limitation. J Appl Ecol 33:1441–1450

Lal R (1986) Conversion of tropical rainforest: agronomic potential and ecological consequences. Adv Agron 39:173–264

Long SP, Ainsworth EA, Rogers A et al (2004) Rising atmospheric carbon dioxide: plants face the future. Ann Rev Plant Biol 55:591–628

Losos EC, Leigh EG (2004) Tropical forest diversity and dynamism: findings from a large-scale plot network. University Chicago Press, Chicago

Mayor JR, Wright SJ, Turner BL (2014) Species-specific responses of foliar nutrients to long-term nitrogen and phosphorus additions in a lowland tropical forest. J Ecol 102:36–44

Mercado LM, Patiño S, Domingues TF et al (2011) Variations in Amazon forest productivity correlated with foliar nutrients and modelled rates of photosynthetic carbon supply. Phil Trans Royal Soc Series B 366:3316–3329

Nasto MK, Alvarez-Clare S, Lekburg Y et al (2014) Interactions among nitrogen fixation and soil P acquisition strategies in lowland tropical forest. Ecol Lett 17:1282–1289

Norby RJ, Warren JM, Iversen CM et al (2010) CO_2 enhancement of forest productivity constrained by limited nitrogen availability. P Natl Acad Sci 107:19368–19373

Nye PH, Greenland DJ (1960) The soil under shifting cultivation. Commonwealth Bureau of Soils, Harpenden UK. Tech Comm 51:73–126

Oberson A, Friesen DK, Morel C et al (1997) Determination of phosphorus released by chloroform fumigation from microbial biomass in high P sorbing tropical soils. Soil Biol Biochem 29:1579–1583

Png GK, Laliberté E, Hayes PE et al (2014) Do N2-fixing plants show higher root phosphatase activity on P-poor soils? In: Mucina L, Price JN, Kalwij JM (eds) Biodiversity and vegetation: patterns, processes, conservation. Kwongan Foundation, Perth, p 255

Quesada CA, Phillips OL, Schwarz M et al (2012) Basin-wide variations in Amazon forest structure and function are mediated by both soils and climate. Biogeosciences 9:2203–2246

Richardson AE, Simpson RJ (2011) Soil microorganisms mediating phosphorus availability. Plant Phys 156:989–996

Sanchez PA, Villachica JH, Bandy DE (1983) Soil fertility dynamics after clearing a tropical rainforest in Peru. Soil Sci Soc Am J 47:1171–1178

Santiago LS, Wright SJ, Harms KE et al (2012) Tropical tree seedling growth responses to nitrogen, phosphorus and potassium addition. J Ecol 100:309–316

Scurlock JMO, Olson RJ (2002) Terrestrial net primary productivity—a brief history and a new worldwide database. Env Rev 10:91–109

Serrao EAS, Nepstad D, Walker R (1996) Upland agricultural and forestry development in the Amazon: sustainability, criticality and resilience. Ecol Econ 18:3–13

Soil Survey Staff (1999) Soil taxonomy: a basic system of soil classification for making and interpreting soil surveys. United States Department of Agriculture-Natural Resources Conservation Service, Lincoln

Stark NM, Jordan CF (1978) Nutrient retention by the root mat of an Amazonian rain forest. Ecology: 434–437

Steidinger BS, Turner BL, Corrales A et al (2015) Variability in potential to exploit different soil organic phosphorus compounds among tropical montane tree species. Func Ecol 29:121–130

Sullivan BW, Alvarez-Clare S, Castle SC et al (2014) Assessing nutrient limitation in complex forested ecosystems: alternatives to large-scale fertilization experiments. Ecology 95:668–681

Tanner EVJ, Vitousek PM, Cuevas E (1998) Experimental investigation of nutrient limitation of forest growth on wet tropical mountains. Ecology 79:10–22

Townsend AR, Cleveland CC, Asner GP et al (2007) Controls over foliar N:P ratios in tropical rain forests. Ecology 88:107–118

Turner BL (2008) Resource partitioning for soil phosphorus: a hypothesis. J Ecol 96:698–702

Turner BL, Lambers H, Condron LM et al (2013) Soil microbial biomass and the fate of phosphorus during long-term ecosystem development. Plant Soil 367:225–234

Turner BL, Engelbrecht BM (2011) Soil organic phosphorus in lowland tropical rain forests. Biogeochemistry 103:297–315

Turner BL, Yavitt JB, Harms KE et al (2015) Seasonal changes in soil organic matter after a decade of nutrient addition in a lowland tropical forest Biogeochemistry (in press)

Veneklaas EJ, Lambers H, Bragg J et al (2012) Opportunities for improving P-use efficiency in crop plants. New Phytol 195:306–320

Vincent AG, Turner BL, Tanner EVJ (2010) Soil organic phosphorus dynamics following perturbation of litter cycling in a tropical moist forest. Eur J Soil Sci 61:48–57

Vitousek PM (1984) Litterfall, nutrient cycling, and nutrient limitation in tropical forests. Ecology 65:285–298

Vitousek PM, Sanford RL (1983) Nutrient cycling in moist tropical forest. Ann Rev Ecol Syst 17:137–167

Vitousek PM, Turner DR, Parton WJ et al (1994) Litter decomposition on the Mauna Loa environmental matrix, Hawaii – patterns, mechanisms, and models. Ecology 75:418–429

Walker TW, Syers JK (1976) The fate of phosphorus during pedogenesis. Geoderm 15:1–19

Wright SJ, Yavitt JB, Wurzburger N et al (2011) Potassium, phosphorus and nitrogen limit forest plants growing on a relatively fertile soil in the lowland tropics. Ecology 92:1616–1625

Tree Nutrient Status and Nutrient Cycling in Tropical Forest—Lessons from Fertilization Experiments

E.J. Sayer and L.F. Banin

Abstract Highly productive tropical forests often occur on nutrient-poor soils. The apparent lack of a relationship between tree growth and site fertility has generated decades of research into which nutrients, if any, limit tropical forest productivity. This chapter looks at the lessons we have learned from several decades of fertilization experiments, which investigate nutrient limitation by measuring changes in growth and productivity in response to the addition of specific nutrients. The enormous diversity of tropical forest ecosystems often confounds attempts to measure a clear ecosystem response to fertilization because tree species' nutrient requirements differ according to life history strategy, adaptation to site fertility, and the life stage of the individuals under study. Importantly, other limiting resources, such as light and water, constrain individual responses to nutrient availability, whereas species interactions such as competition, herbivory, and symbioses can mask growth responses to nutrient amendments. Finally, fertilization changes the timing and balance of nutrient inputs to the forest, whereas litter manipulation studies demonstrate that the combined addition of many different nutrients and organic carbon minimizes nutrient losses. Most fertilization studies have investigated responses to nitrogen and phosphorus additions but there is still no general consensus on nutrient limitation in tropical forests. Future experiments will need to evaluate how the balance of multiple macro- and micronutrients affects tropical forest growth and ecosystem dynamics.

Keywords Belowground biomass · Ecosystem productivity · Life history strategy · Nitrogen fixation · Nutrient limitation · Soil chronosequence · Tree growth

E.J. Sayer (✉)
Lancaster Environment Centre, Lancaster University, Lancaster LA1 4YQ, England
e-mail: e.sayer@lancaster.ac.uk

L.F. Banin
Centre for Ecology and Hydrology, Bush Estate, Midlothian EH26 0QB, Scotland
e-mail: libanin@ceh.ac.uk

© Springer International Publishing Switzerland 2016
G. Goldstein and L.S. Santiago (eds.), *Tropical Tree Physiology*,
Tree Physiology 6, DOI 10.1007/978-3-319-27422-5_13

Nutrient Limitation of Ecosystem Productivity in Tropical Forests

Tropical forests are the most productive of all terrestrial ecosystems and yet large areas of tropical forest occur on nutrient-poor soils (Vitousek and Sanford 1986; Bruijnzeel 1991). Plant communities growing on highly nutrient-limited sites are likely to be well-adapted to nutrient shortage and large proportions of the nutrients available to plants are tied up in the living biomass and recycled with plant litter (e.g. Herrera et al. 1978). The maintenance of such high productivity in tropical forests on infertile soils can therefore be attributed to highly efficient cycling of nutrients in organic matter (e.g. Jordan 1985; Cuevas and Medina 1988) and nutrients from the decomposition of organic matter can make up a large proportion of the nutrients required for plant growth (Bruijnzeel 1991).

The relative infertility of tropical soils and the highly efficient recycling of nutrients in tropical forests have given rise to decades of research into which nutrients, if any, limit productivity in tropical forests (Dalling et al., this volume). The relationship between soil fertility and net primary productivity in tropical forests is uncertain; comparative studies have variously demonstrated that above-ground productivity can be positively (e.g. Quesada et al. 2012), negatively (Proctor 1983) or entirely unrelated to different measures of soil nutrient status (Jordan and Herrera 1981; Proctor 1983). On the one hand, this may be partly due to the challenges in quantifying plant-available nutrients in the soil and the 'efficiency' of nutrient cycling in tropical forests (Vitousek 1984; Vitousek and Sanford 1986), which decouples the simple relationship between soil fertility and plant nutrient acquisition. On the other hand, differences in biomass allocation and species composition can confound results in gradient studies; for example, tree growth was positively related to soil nutrient concentrations in Borneo because of the high density of a particular canopy emergent at the most fertile sites (Paoli et al. 2008).

Fertilization experiments are a useful tool to resolve some of these issues because nutrient limitation can be inferred from a change in the rate of an ecosystem process in response to the addition of a given nutrient (Tanner et al. 1998). Research on the nutrient regulation of plant productivity in the tropics has focused largely on the macronutrients nitrogen (N) and phosphorus (P). An adequate supply of N is essential for plant growth because it is a building block of amino acids, enzymes and nucleic acids (Santiago and Goldstein, this volume, Chap. 14). Phosphorus is also found in nucleic acids and it plays many vital roles in plants, including energy metabolism. Very few tropical studies have investigated the effects of fertilization with other plant macronutrients such as potassium (K), calcium (Ca) or magnesium (Mg), but any nutrient can be said to be 'limiting' when its availability constrains a biological or biochemical process (Tanner et al. 1998).

The concept of limiting nutrients was developed for individual plants or monoculture crops, whereas community- and ecosystem-level responses are likely to vary with species composition (Chapin et al. 1986). The high diversity of plants in tropical forests in particular makes pinpointing nutrient limitation at the ecosystem level especially challenging (Grubb 1989) because not all species are necessarily limited by the same nutrient, two or more nutrients can be co-limiting (Tanner et al. 1998), and apparent limitation by one nutrient may actually be limitation by a different nutrient in disguise, for example when uptake of N is limited by P-availability (Attiwill and Adams 1993). To complicate matters further, simultaneous limitation by different types of resources, such as light, water, or nutrients, is also often the rule (Bloom et al. 1985; Tanner et al. 1998).

During the search for the elusive 'limiting nutrient', fertilization experiments have provided a wealth of valuable information about tropical forest nutrient cycling, plant growth, and species interactions. In this chapter, we draw on the results of more than three decades of fertilization studies conducted in tropical forests to detect patterns in plant responses to altered nutrient supply and identify considerations and constraints for interpreting experimental results. Most of the evidence presented here concerns N- and P-dynamics, for which there is a large body of literature on tropical forests but the same general considerations are likely to apply to other nutrients.

Nutrients, soil development and the extraordinary case of Hawai'i— Theoretically, terrestrial ecosystems will experience a shift from N- to P-limitation over geological time. The primary source of P (and other base cations) in soils is the weathering of bedrock, so their concentrations decline progressively with soil age and development, mainly as a result of erosion and leaching (Walker and Syers 1976). In contrast, N accumulates in soil during the course of soil development, reaching maximum levels in middle-aged soils (Lambers et al. 2008).

Many lowland tropical forest soils are old and highly weathered, particularly those on ancient Precambrian shield geology, and hence they have low availability of P, K and other cations (Grubb 1989; Banin et al. 2015). Conversely, tropical montane forests have low availability and mineralization rates of N (Grubb 1977; Vitousek 1984). Gradient studies also demonstrate that productivity is related to foliar P concentrations in lowland forests (Vitousek 1984) and to foliar N in montane forests (Tanner et al. 1998; Fisher et al. 2013). It is therefore widely accepted that tropical montane forests are more likely to be N-limited whereas lowland forests are more likely to be primarily limited by P (Vitousek and Sanford 1986; Tanner et al. 1998).

The Hawai'ian Long Substrate Age Gradient (LSAG) presents a unique opportunity to test these theories of nutrient availability during soil development in detail. The gradient comprises a chronosequence of primary forest succession at six sites ranging from 300 years to 4.1 million years of age; the relative availability of N is lower at the youngest sites whereas the relative availability of P is lower at the oldest sites (Crews et al. 1995). The sites along the gradient have comparable elevation, rainfall, parent material and species composition (Harrington et al. 2001), and the shift from N- to P-limitation along the gradient is reflected in foliar nutrient concentrations (Crews et al. 1995). Factorial +N and +P fertilization treatments at either end of the gradient have demonstrated N-limitation of forest productivity in the geologically youngest site and P-limitation at the oldest site (Vitousek and Farrington 1997). Unfortunately, another exceptional feature of the Hawaiian LSAG is that the forest is dominated by a single canopy tree species, and is therefore not necessarily representative of the highly diverse forests typical elsewhere in the tropics. Despite this potential limitation for extrapolating results to highly diverse tropical forests, dominance by only one species (in this case *Metrosideros poly-morpha*) makes mechanistic studies possible (Cordell et al. 2001).

The evidence for N-limitation of productivity in tropical montane forests is strong; fertilization with +N increased tree diameter growth and/or litter production in Jamaica, Venezuela and Hawaii (reviewed by Tanner et al. 1998), Ecuador (Homeier et al. 2012), Panama (Adamek et al. 2009) and Peru (Fisher et al. 2013). In contrast, although the concept of P-limitation in lowland tropical forests is widely accepted, the collective results of fertilizer studies are far from conclusive.

Aboveground productivity—Large-scale fertilization experiments in lowland tropical forests have shown no effects of fertilization with +P alone on tree growth in Borneo (Mirmanto et al. 1999) or Cameroon (Newbery et al. 2002) despite very low soil P concentrations. After three years of fertilization in mature forest in Costa Rica, stem growth of small trees doubled with +P additions and a higher percentage of trees increased basal area in +P treatments than in control plots but there were no other community-level responses to fertilization with +N or +P (Alvarez-Clare et al. 2013). Tree growth in young and old Mexican dry forests increased with +P, +N and +NP fertilization; whereas +P and +NP treatments had a greater effect on trunk growth, only fertilization with +NP enhanced litter production (Campo and Vazquez-Yanes 2004). Interestingly, wood and leaf biomass of *Eucalyptus* trees in a Brazilian plantation were greatly enhanced by fertilization with +K and, to a lesser extent, sodium (+Na; Epron et al. 2011). Finally, a factorial fertilization experiment with +N, +P and +K in Panama demonstrated that the addition of each of these three macronutrients enhanced a different component of forest productivity: stem growth of saplings was enhanced by the addition of +NK, whereas seedling height growth increased with +K or +NP fertilization; litterfall increased with +P fertilization, plant investment in fruits and flowers increased with +N, and root biomass of trees and seedlings decreased in response to fertilization with +K or +NK (Kaspari et al. 2008; Yavitt et al. 2011; Wright et al. 2011; Santiago et al. 2012).

Belowground responses—It is possible that the variable responses in aboveground productivity are partly a result of changes in biomass allocation. In theory, plants adapted to nutrient-poor soils should allocate more biomass to roots to improve nutrient acquisition (Chapin 1980), whereas plants on fertile sites should invest a greater proportion of their biomass aboveground. Accordingly, stand-level root biomass should decrease when nutrient limitation is relieved by fertilization. On the other hand, fine roots can proliferate into hotspots of nutrient availability in nutrient-poor soils (St. John 1982), which creates microsites of high root biomass. This provides an alternative approach to assess the effects of fertilization by measuring root growth into microsites spiked with specific nutrients (Cuevas and Medina 1988). Using this method, ingrowth cores containing limiting nutrients represent nutrient hotspots and should therefore have higher root biomass than the surrounding soil (Raich et al. 1994).

A meta-analysis of root biomass responses to fertilization in 45 tropical montane and 52 lowland tropical forests showed that stand-level root production was enhanced by +N fertilization in tropical montane forests and by +P fertilization in lowland tropical forests, and the addition of +NP had a similar effect in both systems (Yuan and Chen 2012). Nonetheless, decreased root biomass has been observed in response to fertilization with +K and +NK in Panama (Wright et al. 2011), +P in Costa Rica (Gower 1987), +N in China (Zhu et al. 2013; Mo et al. 2008) and there were variable responses in root biomass and turnover in response to +N or +P at different sites in Hawai'i (Ostertag 2001).

Similarly, ingrowth core studies demonstrate increased root growth into cores supplemented with +N or +NP in montane forests (Stewart 2000; Graefe et al. 2010), whereas root proliferation into ingrowth cores has been observed in response to +N, +P, +K, +NP, +PK and +Ca fertilization in various lowland tropical forests (Cuevas and Medina 1988; Graefe et al. 2010). These results should be treated with caution because ingrowth cores are often filled with plant growth media made from expanded clays, which represent a source of cations such as K, Ca, or Mg (Raich et al. 1994) and have absorption sites that can interact with added ions (Stewart 2000). In addition, root proliferation into ingrowth cores could represent a response to greater instant availability of inorganic nutrients compared to the surrounding soil, rather than a clear indicator of nutrient limitation.

These contrasting and sometimes contradictory results on the nutrient limitation of above- and belowground productivity at the ecosystem level highlight a number of considerations for interpreting fertilization experiments. Importantly, stand-level responses to a sudden increase in nutrient availability are unlikely to be directly comparable to biomass distribution along natural fertility gradients because adaptation of plants to initial soil fertility affects ecosystem responses to fertilization (Ostertag 2001). Further, the life stage and life history strategies of individual plant species determine biomass allocation and internal demand for nutrients to a large extent. Consequently, several nutrient amendment experiments show highly variable responses of different tree species (Alvarez-Clare et al. 2013; Villagra et al. 2013) and size classes of trees (Wright et al. 2011; Alvarez-Clare et al. 2013). Hence, the interplay of many distinct site- and species characteristics influence the responses of highly diverse tropical forests to nutrient amendments.

Life-History Strategies and Adaptation to Multiple Limiting Resources

A large number of experiments have measured seedling responses to nutrient amendments, which can be conducted on a small scale under controlled conditions. Studies using multiple species have observed highly variable responses to nutrient amendments; although the vast majority of species increased biomass in response to nutrient addition, a much smaller proportion increased relative growth rates (Lawrence 2003). In general, light-demanding species are more likely to increase growth and/or biomass in response to nutrient additions, whereas a larger number of shade-tolerant species increase foliar N and P concentrations (Lawrence 2003; Cai et al. 2008; Tripathi and Raghubanshi 2014). These patterns are fairly clear for pot experiments, whereas field studies with transplanted or naturally occurring seedlings have shown fewer or smaller growth and biomass responses to fertilization (Denslow et al. 1987; Turner et al. 1993). These differences among species and types of experiments highlight the importance of considering other limiting resources and their interactions with nutrient availability.

Light limitation in the understorey—Plants adapted to low-resource environments often have low potential for resource acquisition, invest heavily in defense and storage, and grow slowly even when resources are increased (e.g. Huante et al. 1995). For seedlings growing in the understorey of tropical forests, light availability can be more important than nutrient availability. Adaptation to low light levels often includes greater allocation to leaves, smaller root biomass and low relative growth rates (Cai et al. 2008). Slow-growing, shade-tolerant species have low nutrient requirements when light is limiting (Burslem et al. 1996), so they may not exhibit a strong response to nutrient amendments during the lifetime of an experimental study (Dalling and Tanner 1995). In addition, seedlings grown in low-light conditions invest primarily in increasing leaf area and leaf longevity while reducing biomass allocation to roots; this can preclude a rapid or strong growth response to a sudden increase in nutrient availability (Gunatilleke et al. 1997). Due consideration of species-specific responses to light conditions is also important for designing pot-based studies and interpreting their results: whereas some shade-tolerant understorey species can exhibit a strong positive growth response to nutrient amendments when light limitation is removed (Burslem et al. 1995), others may become photo-inhibited under high light conditions (Fetcher et al. 1996).

Changes in growth or biomass in response to nutrient addition are often minimal or completely absent when seedlings are grown under low-light conditions; instead a strong increase of nutrient concentrations in foliage (Campo and Dirzo 2003; Lawrence 2003) and other plant parts (Raaimakers and Lambers 1996) is observed. Nutrient uptake in excess of requirement is regarded as 'luxury consumption' (e.g. Ostertag 2010; Tripathi and Raghubanshi 2014) but these nutrient stores can enable rapid growth once plants are no longer constrained by light availability, for example when a new gap is created by a treefall (Raaimakers and Lambers 1996). Tropical tree seedlings growing in the understorey are able to respond to very brief increases in light levels such as sunflecks, so higher foliar concentrations of N, P and K in particular could allow understorey plants to maximize photosynthesis even when increased light availability is very sporadic (Pasquini and Santiago 2012). These trade-offs between nutrient-use strategies and light availability can determine competitive outcomes along gap-understorey gradients (Cai et al. 2008), because nutrient storage under low light provides an advantage when the light limitation is removed, whereas fast growth rates are beneficial for competition under high light (Raaimakers and Lambers 1996).

Water stress—Many lowland tropical forests experience periods of low rainfall or drought each year, and there are multiple lines of evidence for important interactions between water and nutrient availability. The relative mobility of different nutrients and the activity of extracellular enzymes vary strongly with soil water content, which for example affects N mineralization rates and the diffusion of nutrients to root surfaces (Cavelier et al. 2000; Cernusak et al. 2010). Adaptation to drought stress can also include greater biomass allocation belowground, which could make plants more responsive to nutrient additions when water availability is low or once drought conditions are alleviated. Hence, total root biomass and root biomass allocation would only decrease in response to increased nutrient supply if the water supply were adequate (Hall et al. 2003). Despite this, few experiments report responses of tropical trees or seedlings to nutrient amendments under drought stress (but see Hall et al. 2003; Burslem et al. 1996).

Nutrient addition can affect the drought resistance of trees by modifying their hydraulic architecture. Rapid growth in response to fertilization was thought to increase the risk of drought-induced embolism as a consequence of lower wood density and larger total leaf surface area, which would result in larger transport vessels and higher rates of transpiration (Goldstein et al. 2013). Surprisingly, an experiment with six species of saplings showed the opposite: growth rates and resistance to drought-induced embolism increased in response to fertilization because changes in wood anatomy and wood density in response to increased leaf surface area and transpiration appear to have mitigated the risk of cavitation (Villagra et al. 2013). Shade-tolerant species in particular were able to increase growth in response to +N and +P fertilization without increasing their vulnerability to drought because they have lower specific leaf conductivity than light-demanding species (Villagra et al. 2013).

Fertilization can reduce transpiration rates in tropical tree seedlings (Winter et al. 2001). An experiment with different species of tree and liana seedlings showed that water use efficiency was positively related to foliar N and negatively related to foliar P concentrations (Cernusak et al. 2010). It is possible that increased water-use efficiency at higher foliar N concentrations decreases transpiration rates, and hence weakens the pressure gradient that transports solutes to root surfaces by mass flow, resulting in lower P uptake (Cernusak et al. 2010). These results are intriguing, because greater water-use efficiency at higher foliar N concentrations may help to explain why many species of tropical trees and seedlings exhibit luxury consumption and storage of N.

Species Interactions

Tropical tree communities are often strongly affiliated with soil type; particular species show strong local associations with certain soils because strong competition excludes species growing in non-optimal habitats (e.g. Russo et al. 2005). Aside from inter- and intraspecific competition for different resources, plant responses to altered nutrient supply can modify or be modified by herbivory and symbioses with mycorrhizal fungi or N-fixing bacteria.

Mycorrhizal associations—The interplay between soil nutrient availability and root symbioses is well established. In tropical forests, plants associated with ecto- and arbuscular mycorrhizal fungi are highly competitive for P (Högberg 1986), so it is particularly striking that mycorrhizal plants rarely respond to experimental additions of P (Denslow et al. 1987; Newbery et al. 2002; Burslem et al. 1995), although the results of some studies suggest that mycorrhizal species may be limited by K, Ca or Mg (Burslem et al. 1995, 1996; Hall et al. 2003).

Although fertilization with inorganic +N and +P can have a negative (Treseder and Vitousek 2001) or no effect on mycorrhizal colonization (Turner et al. 1993; Brearley et al. 2007), mycorrhizas can improve tropical seedling growth by accessing nutrients directly from decomposing litter (Hodge et al. 2001; Brearley et al. 2003). Arbuscular mycorrhizal colonization and growth has been associated with patches of nutrient-rich organic matter in tropical forests (St. John 1982); it is hence conceivable that mycorrhizas are better adapted to access nutrients in organic matter and nutrients added as inorganic fertilizers may not have the same effect (Brearley et al. 2003).

Legumes and N-fixation—Di-nitrogen (N_2) fixation by plants can constitute an important input of N to terrestrial ecosystems when N-availability is low but it is regarded as a 'costly' N-acquisition strategy to plants (Gutschick 1981). The high prevalence of potentially N-fixing leguminous trees in tropical forests, where N is not thought to be limiting, has therefore fueled scientific debate about the ecological and evolutionary advantages of N-fixation (see review by Hedin et al. 2009). N-fixation requires a sufficient supply of P and could therefore be constrained in lowland tropical forests (Vitousek and Howarth 1991; Batterman et al. 2013). Fertilization experiments have contributed evidence to support or refute various hypotheses about how trade-offs between N and P acquisition or investment may explain the abundance of N-fixing plants (henceforth: fixers) in lowland tropical forests. One such theory is that N-fixation occurs in response to low availability of soil N but is down-regulated when N is abundant (e.g. Hedin et al. 2009; Batterman et al. 2013), and experiments show that non-legumes respond more strongly to +N fertilization than legumes when available soil N is low (Tripathi and Raghubanshi 2014). Another theory postulates that N-fixers are able to invest additional N in the production of extracellular phosphatases to acquire P (Houlton et al. 2008; Baribault et al. 2012). However, experimental evidence suggests that plants cannot overcome severe P-limitation by investing N in enzyme production alone (Batterman et al. 2013) and even N-fixing mycorrhizal species can be poor competitors for P (Högberg 1986).

Herbivory—Herbivory can substantially affect the outcome of fertilization experiments. Leaves with higher nutrient concentrations and lower investment in secondary plant compounds or structural carbon are thought to be more susceptible to herbivore attack (Coley and Barone 1996). Consequently, there are trade-offs between increasing growth or photosynthetic capacity in response to fertilization and maintaining defenses against herbivory: firstly, photosynthetic capacity increases with foliar N concentrations (Evans 1989) but higher foliar N also make leaves more palatable to herbivores; secondly, rapid growth in response to fertilization may preclude high investment in chemical or structural defenses (Coley and Barone 1996).

Several experiments have noted that increased herbivory probably masked growth responses to +N fertilization (Andersen et al. 2010), as well as +P and +K (Santiago et al. 2012). Herbivore damage was particularly noticeable in pioneer vegetation (Campo and Dirzo 2003) and fast-growing light-demanding species (Villagra et al. 2013), which invest fewer resources in structural defenses, as well as legumes (Campo and Dirzo 2003) and species associated with more fertile soils (Andersen et al. 2010), which have higher leaf nutrient concentrations. Increased risk of herbivory may also exert selective pressure against luxury consumption of N (Ostertag 2010). This risk could be offset by greater investment in plant chemical defenses but to our knowledge, only a single study has measured increases in phenolic compounds with addition of specific nutrients and found no consistent patterns across light-demanding and shade tolerant species (Denslow et al. 1987).

Life Stages

We would expect large growth responses to fertilization in young forests due to the predominance of fast-growing pioneer vegetation and because tree growth during the years prior to canopy closure (when light availability is high) is very dependent on soil nutrient concentrations (Miller 1981). A study of 10-year and 60-year old stands of secondary tropical dry forest showed that growth rates and litter production were higher in the 10-year old forest compared to the 60-year old stand and greater increases in tree growth and litterfall were also measured in the young forest in response to fertilization (Campo and Vazquez-Yanes 2004).

Strong responses of secondary forest regrowth to nutrient amendments can also be attributed to low soil nutrient availability after land-use for pasture or agricultural crops (Davidson et al. 2004). Accordingly, the productivity of regenerating secondary tropical forests can increase substantially with fertilization: an 85 % increase in annual net primary productivity was observed in plots treated with a complete fertilizer during four years of forest reestablishment (Giardina et al. 2003) and the rates of tree biomass accumulation in a six-year old forest almost doubled after only three years of +N and +NP fertilization, although +P fertilization alone had little effect (Davidson et al. 2004). Similar results were obtained in young dry forest in Mexico, where trunk growth increased substantially after three years of +N or +P fertilization and litterfall increased in response to +NP (Campo and Vazquez-Yanes 2004).

By contrast, the effects of nutrient amendments in mature forests are harder to determine, as they are dominated by slow-growing, shade-tolerant species. Tree size influences individual- and stand-level responses to nutrient amendments because large trees are less likely to be light-limited than smaller trees in the subcanopy (Wright et al. 2011). Species-specific responses of mature tropical trees to nutrient amendments are also expected but this is difficult to test experimentally because of the high diversity of tropical forest trees and the relatively low densities of individuals of the same species. In addition, disturbance (e.g. gap formation) creates areas of forest at different stages of regeneration and as a result, nutrient requirements are likely to be patchy throughout the forest. Seedling experiments are thought to be useful to address this, because the regeneration phase is most likely to influence adult abundance and distribution (Grubb 1977). However, seedlings are subject to very different conditions and constraints than adult trees and hence their responses may not be representative of later life stages.

A direct comparison of tree and seedling responses to fertilization—We are not aware of any direct comparisons of the responses of seedlings and adult trees of the same species to fertilization treatments, but there are legitimate biological reasons to assume that species' responses to nutrient additions will vary according to life stage. We used published data on foliar nutrient concentrations in naturally occurring seedlings and adult trees in a long-term experiment in Panama to examine the responses of three common tropical tree species to fertilization with +N, +P, and +NP (Santiago et al. 2012; Mayor et al. 2013, 2014). The three species included a small subcanopy tree *Heisteria concinna* (Standl.), a medium-sized pioneer tree *Alseis blackiana* (Hemsl.), and a large canopy tree *Tetragastris panamensis* (Engl. Kuntze); all three species have shade-tolerant seedlings that can persist in the understorey (Santiago et al. 2012).

First, we investigated the relationships between seedling and adult foliar nutrients in control plots. We then calculated the proportional response to experimental treatments as log response ratios (Eq. 1) to standardize effect size across species and groups:

$$RR = \ln(Rx/Rc) \tag{1}$$

where Rx is the measured value of the response variable in a given treatment and Rc is the corresponding control value (Santiago and Goldstein, this volume); a response ratio of zero indicates no change in response to a treatment, whereas values greater than or less than zero represent positive and negative responses, respectively (Hedges et al. 1999). The effects of fertilization treatment, species and life stage were determined using linear mixed-effects models in R version 3.1.3. (nlme package; Pinheiro et al. 2015; Development Core Team R 2014) with block as a random effect. Significance of each term was determined by comparing nested models using likelihood ratio tests and AICs to check for model improvement (Pinheiro and Bates 2000).

In unfertilized plots, foliar N and N:P ratios were higher in adults than in seedlings (Fig. 1). Individuals of *Alseis* had the highest foliar concentrations of both nutrients and *Tetragastris* had the lowest. Foliar P concentrations varied little among species but were higher in seedlings than in adults, especially in individuals of *Tetragastris* (Fig. 1).

Although there was no significant overall effect of fertilization with +N, +P or +NP on foliar N concentrations, the response of *Heisteria* seedlings was more positive and more variable than that of adult trees, whereas the other two species showed no notable response. In contrast, foliar P significantly increased in +P and +NP treatments (Fig. 2). Although the interaction terms were not significant in the

Fig. 1 Boxplots of foliar N and P concentrations of adult trees (*dark shading*) and seedlings (*light shading*) of three common tree species in the control plots of a fertilization experiment in lowland tropical forest in Panama, Central America; species are *Alseis blackiana*, *Heisteria concinna* and *Tetragastris Panamensis*

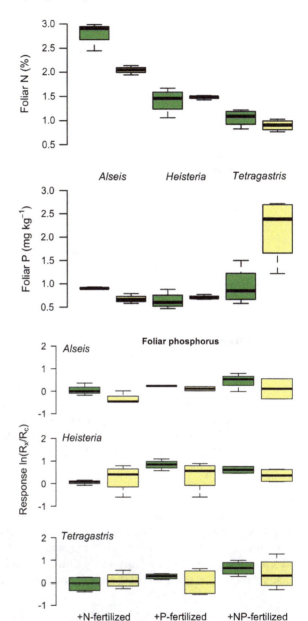

Fig. 2 Boxplots of foliar P concentrations of adult trees (*dark shading*) and seedlings (*light shading*) of three common tree species in a fertilization experiment in lowland tropical forest in Panama, Central America, where +N, +P and +NP are treatments fertilized with nitrogen, phosphorus or both nutrients, respectively; species names are as given in Fig. 1

models, the response to fertilization differed slightly among species and life stages: foliar P varied more amongst seedlings than adult trees and individuals of *Heisteria* displayed the strongest response to fertilization (Fig. 2), whereas foliar N:P ratios tended to decrease in response to +P and +NP fertilization and the fertilization effect was slightly lower in seedlings than in adult trees, especially in individuals of *Heisteria* (Fig. 3).

Fig. 3 Boxplots of foliar NP
concentrations of adult trees
(*dark shading*) and seedlings
(*light shading*) of three
common tree species in a
fertilization experiment in
lowland tropical forest in
Panama, Central America,
where +N, +P and +NP are
treatments fertilized with
nitrogen, phosphorus or both
nutrients, respectively;
species names are as given in
Fig. 1

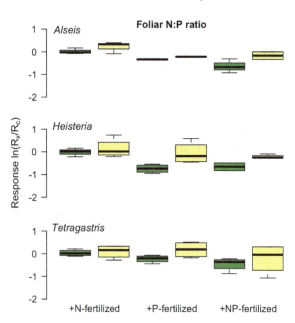

We demonstrate here that the responses of plants to nutrient amendments also
vary with life stage, at least for these three common tropical tree species. Overall,
seedling responses to a given fertilization treatment were more variable, which
suggests stronger constraints on seedling nutrient uptake and storage, e.g. because
of light limitation (see Sect. Light limitation in the understorey), distinct internal
demands for nutrients, less developed root networks, or because seedlings are
outcompeted by adults (Tanner and Barberis 2007). The foliar nutrient concentrations of adult trees used in our analysis were from canopy leaves, which were fully
exposed to sunlight (Mayor et al. 2014), whereas the seedlings experienced light
levels of only *c.* 1 % in the understorey (E.J. Sayer, unpublished data). This alone
could account for the large differences between adult trees and seedlings. However,
it is noteworthy that *Heisteria* displayed the greatest fertilization responses and the
largest differences between seedlings and adult trees, even though subcanopy trees
are often strongly light-limited. This suggests that distinct nutrient demands at
different life stages and by functionally different species may also play a role in
determining foliar nutrient concentrations and the balance of carbon and nutrients.

Litter Manipulation Experiments

Litter manipulation treatments involve the regular removal or addition of the litter
standing crop and fine litterfall (leaves, fruits, flowers and small branches) to disrupt
or enhance the natural forest nutrient cycle (Sayer 2006). Unlike fertilization
experiments, litter manipulation studies do not aim to identify limiting nutrient

elements but instead investigate the importance of nutrient cycling via organic matter (Sayer et al. 2012) and are accordingly discussed separately here. It is important to note that litter manipulation treatments affect several important soil properties, such as soil water content, soil temperature, and habitat space for decomposer organisms (reviewed in Sayer 2006). Furthermore, many of the previously described constraints and interactions affecting plant responses to nutrient amendments will also apply to litter manipulation studies.

Although litter addition can also be regarded as a nutrient addition treatment, two important features distinguish it from fertilization: (1) litter addition treatments supply multiple nutrients in approximate stoichiometric balance, and (2) nutrients are added in combination with organic carbon. This second point is important, because the forest is adapted to cycling nutrients from organic matter and the slow release of many nutrients from decomposing litter minimizes losses from the system (Qualls et al. 1991; Sayer et al. 2012). Litter removal treatments have no parallel in fertilization studies because they effectively disrupt the forests' natural nutrient cycle. This disruption allows us to identify different nutrient cycling strategies. For instance, rapid decreases in the concentrations of N in soil and leaves in response to litter removal could suggest that decomposing organic matter is the principal source of N for plant growth (Sayer and Tanner 2010). On the other hand, it also indicates an inefficient N-cycle, characterized by a lack of mechanisms to mitigate large losses from the system, such as through retranslocation of nutrients before leaf abscission.

Although there are now a number of litter manipulation experiments across the tropics, most have focused on seedling establishment, soil biogeochemistry or decomposition processes. The few experiments investigating tree responses or ecosystem productivity have demonstrated rapid responses to litter addition treatments, including increased stem growth of individual tree species (Villalobos-Vega et al. 2011; see also Sect. The effects of litter manipulation on plant growth.), increased litterfall (Wood et al. 2009; Sayer and Tanner 2010) and changes in foliar nutrient concentrations or nutrient return in litterfall (Tutua et al. 2008; Wood et al. 2009; Sayer and Tanner 2010). Litter removal treatments have had little effect on productivity, possibly because tropical forest adaptation to infertile soils includes efficient nutrient retention mechanisms. Nonetheless, litter removal affected the cycling of N and/or K in most experiments with at least four years of treatments (Tutua et al. 2008; Vasconcelos et al. 2008; Sayer and Tanner 2010), demonstrating the key role of organic matter in minimizing losses of the more mobile nutrient elements.

The effects of litter manipulation on plant growth—Tropical tree seedling responses to litter manipulation have mostly focused on germination, establishment and survival in the field. Litter addition influences seedling establishment by forming a physical barrier and providing a favorable habitat and microclimate for pathogens and herbivores (Sayer 2006), so it is hard to evaluate the effect of litter-derived nutrients on seedling growth in these experiments. Nevertheless, a greenhouse experiment showed that litter addition enhanced seedling growth and biomass via direct uptake of nutrients by mycorrhizal fungi (Brearley et al. 2003).

Few litter manipulation studies to date are sufficiently long-term to assess the effects of treatments on the growth of mature tropical trees (but see Villalobos-Vega et al. 2011). There was no discernable effect of litter manipulation on stand-level tree growth after 4 years of treatments in a lowland tropical forest in Panama (Sayer and Tanner 2010), so we used additional data from the same experiment to explore the individual growth responses of four common tree species to six years of litter removal (L−) and litter addition (L+) treatments. The focal species were *Heisteria concinna* (Standl.) and *Tetragastris panamensis* (Engl. Kuntze), as described in Sect. The effects of litter manipulation on plant growth, and *Simarouba amara* (Aubl.), a fast-growing tree associated with forest gaps, and *Virola sebifera* (Aubl.), a shade-tolerant canopy tree. Experimental treatments began in 2003 and changes in diameter at breast height (dbh) were recorded using dendrometers from 2004 to 2009; all stems had a dbh between 10 and 40 cm at the start of measurements. We produced a linear mixed effects model with relative growth rate as the response variable, treatment and species as fixed effects, and plot and year as random effects. We used the same model simplification approach as described in Sect. 5.1. and the model with the best fit included species, treatment and their interaction term.

Species responded very differently to the treatments. Growth rates of *Simarouba* and *Virola* were highest in the L+ and lowest in the L− plots, and this was most marked in later years (Fig. 4), which suggests that these species have increased growth in response to the nutrients added with the litter. In contrast, the highest growth rates of *Tetragastris* and *Heisteria* were observed in L− treatments and the lowest in control plots (Fig. 4); these counterintuitive results could indicate that these shade-tolerant species are better competitors for fluctuating resources under disturbed conditions.

Our analyses are restricted to these four species because they were sufficiently common in the study forest. Nonetheless, these findings demonstrate non-uniform responses among species to litter and nutrient addition, which perhaps reflect the complexities of life history traits and competitive interactions that occur in diverse, mixed-aged natural forests.

Comparing litter manipulation and inorganic fertilization experiments—The only formal comparison of litter manipulation and fertilization treatments in tropical forest to date suggests that there are substantial differences in the way nutrients are cycled, depending on whether they are added as inorganic fertilizers or in organic material (Sayer et al. 2012). In particular, the dynamics of the most mobile macronutrients N and K differed substantially between experiments at the same study site, even though the fertilization and litter treatments added or removed similar amounts of these nutrients each year. The L+ treatment resulted in much greater availability of inorganic N in the soil and higher N concentrations in litterfall compared to +N fertilization, whereas K concentrations in litterfall decreased more in the L− treatments than they increased with +K fertilization (Sayer et al. 2012). Although the P added with litter in the L+ treatment was only *c.* 12 % of the amount added in the +P-fertilization treatments, most of the P added as fertilizer remained in the soil (*c.* 81 %; Yavitt et al. 2011), whereas increased litterfall in the

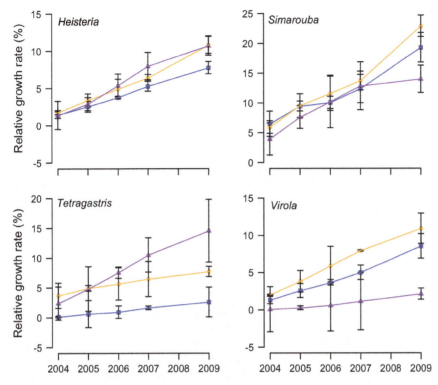

Fig. 4 Relative growth of three common tree species in a litter manipulation experiment in lowland tropical forest in Panama, Central America, showing means and standard errors for $n = 5$ per treatment and species; square symbols denote controls, circles denote litter removal and triangles denote litter addition; species are *Heisteria concinna*, *Simarouba amara*, *Tetragastris panamensis* and *Virola sebifera*

L+ treatment accounted for *c*. 85 % of the P added with litter (Sayer and Tanner 2010; Sayer et al. 2012).

Changes in root biomass distribution were also revealing: whereas root biomass in the uppermost soil horizons decreased in +K and +N fertilized plots (Wright et al. 2011; Yavitt et al. 2011), the distribution of fine roots in L+ plots shifted towards the litter layer and soil surface, probably to take advantage of the greater concentration of nutrients in the forest floor (Sayer 2006). Direct uptake of different nutrients by roots and mycorrhizal hyphae from decomposing litter has been demonstrated clearly by experiments using labeled substrates (Stark and Jordan 1978; Brearley et al. 2003) and a large proportion of the nutrients required for growth may be taken up directly from organic matter on the forest floor (Herrera et al. 1978). Compounds in litter leachate can also increase the availability of nutrients to plants, by limiting sorption of P to clay minerals (Schreeg et al. 2013). Collectively, these results suggest that decomposing litter constitutes an important source of nutrients for plant growth that needs to be considered in studies of nutrient cycling and limitation in tropical forests.

Experimental Limitations

We have demonstrated that differences in life history strategies, species-specific responses to light conditions, water availability, competition, and mutualistic associations need to be taken into account when interpreting the responses of plants to nutrient amendments. The topics we have addressed in this chapter also help us pinpoint the limitations of fertilizer experiments in tropical forests. To explain some discrepancies between pot- and field-based studies, we also need to consider specific 'pot effects', the differences in the mobility of nutrient elements, and the balance and timing of nutrient inputs.

Pot size effects—Although it is certainly possible to use pot experiments to infer plant performance in the field, the influence of pot size is a well-known issue in greenhouse studies. The size of the pot not only determines the amount of water and nutrients available to plants, but the available rooting volume often decides the duration of the study. The 'pot-size effect' is particularly important for fertilization experiments because the different relative mobilities of nitrate and phosphate ions cause a shift from P- to N-limitation during the course of the study, which is determined by the ratio between the soil volume and the length of absorptive roots (Cornforth 1968; Burslem et al. 1995). The highly mobile nitrate ion is delivered to the root surface by diffusion or water flow (Ostertag 2010), whereas the relative immobility of phosphate ions requires roots to forage for P (Yavitt et al. 2011). Consequently, P is more likely to be limiting at the start of a study when root biomass is low, and will become increasingly available as roots grow and access more of the soil in the pot. In contrast, most of the N in the soil is available to plants at the start of the study and it becomes progressively more limiting during the experiment as soil N reserves are depleted. The combination of pot size effects and restricted rooting space could also explain why many fertilization experiments with potted seedlings have not observed the expected changes in plant root:shoot ratios in response to nutrient additions.

Balance and timing of nutrient additions—Differences in the mobility and uptake of nutrient elements also present a problem for field experiments. Substantial losses of mobile elements can occur as soon as supply exceeds demand, even for a short period of time (Vitousek et al. 2010). When added as fertilizer, a large proportion of N in particular can be lost from the system through gas emissions or leaching, rather than taken up by plants (Sayer et al. 2012). Substantial losses of N have been measured in long-term fertilization experiments in montane and lowland tropical forests (Hall and Matson 2003; Koehler et al. 2009; Corre et al. 2010) and as supply continues to exceed demand, the system becomes increasingly 'leaky' (Koehler et al. 2009). In contrast, phosphate is not only immobile in the soil but is also easily sorbed to clay minerals. Hence, many fertilizer experiments often apply large amounts of inorganic P to saturate binding sites and increase the availability of P to plants (Ostertag 2010; Wright et al. 2011). These changes in the ratios of different nutrients can substantially influence the forest nutrient cycle and outcome of an experiment.

Biological stoichiometry is the balance of energy and nutrient elements in living systems. The stoichiometry of C:N:P in particular plays a critical role in a large number of ecosystem properties and processes (Elser et al. 2000 and references therein). Plant growth rates are thought to be related to the balance of N and P, because fast-growing organisms require more P to support protein synthesis relative to N content (termed 'the growth rate hypothesis'; Elser et al. 2000; Matzek and Vitousek 2009). In addition, the decomposition of organic matter is strongly constrained by the stoichiometric requirements of decomposer organisms, which in turn influence the dynamics of carbon and nutrients (Manzoni et al. 2010).

The C:N:P ratios of plant tissues are at least partly determined by physiological constraints and adaptation to nutrient limitation, resulting in a distinct C:N:P ratio within a given forest ecosystem (McGroddy et al. 2004); pulses of fertilizer application or large inputs of specific nutrients not only upset this balance but also alter the timing of nutrient availability, which may be a critical mechanism for maintaining productivity in nutrient-poor tropical forests (Lodge et al. 1994). The potential effects of altering the timing and balance of nutrient inputs is summed up nicely by Newbery et al. (2002): "Forests... are complex, long-lived and highly interconnected systems in which short-term adjustment responses and time lags are to be expected. Adding large quantities of fertilizer to such a presumably near-equilibrium system (in terms of nutrient cycling) is tantamount to a major disturbance."

Conclusions

This chapter contributes to the growing body of evidence for multiple nutrient limitations in tropical forests by demonstrating that plant nutrient demands, and hence response to fertilization, are strongly influenced by life history strategy, life stage, water stress, light availability, mutualistic associations, competition and herbivory, all of which are site- and species-specific.

The interpretation of nutrient amendment experiments also requires due consideration of how different nutrients are transported, stored and cycled. Differential soil availability and internal demand for nutrients by plants can translate into distinct uptake strategies for each nutrient (Ostertag 2001), whereas cost-benefit trade-offs largely determine luxury consumption and storage. The dissimilarities in the biogeochemistry of N and P, combined with plant life-history strategies and adaptation to soil nutrient status could even mean that 'limitation' and 'availability' are qualitatively different for N and P (Harrington et al. 2001; Ostertag 2010).

Plants require at least 17 mineral elements throughout their life cycle (Watanabe et al. 2007) and several fertilizer experiments and physiological studies have demonstrated the importance of other nutrients besides N and P in tropical forest productivity, in particular the macronutrients K and Mg (e.g. Burslem et al. 1995, 1996; Hall et al. 2003; Wright et al. 2011; Santiago et al. 2012). Our understanding of other macro- and micronutrients in tropical ecosystem processes is much less well developed, even though their respective roles in plant physiology and enzyme production are clear. Tropical soils can contain critically low concentrations of many micronutrients (Sobrado 2013) but we lack the experiments to evaluate how this affects tropical forest growth and ecosystem dynamics.

Finally, organic matter plays a crucial role in tropical nutrient cycling, which is underestimated by studies applying inorganic fertilizers and measuring responses in the mineral soil. Aside from being a direct source of nutrients for plant growth, the forest floor helps retain highly mobile elements. Indeed, mineral soil reserves may contribute less than 10 % to a forest's annual N cycle and less than 20 % to annual P cycling (Attiwill and Adams 1993), so nutrients in litter and organic matter may represent a better measure of site fertility than stocks in the mineral soil (Vitousek and Sanford 1986; Tanner et al. 1998).

The accumulation and balance of energy and nutrients in plants underpins the productivity and diversity of ecosystems (Grime 2001). In tropical forests, the high plant diversity and heterogeneity of the ecosystem make it difficult to assess nutrient limitation of tropical forest productivity but fertilizer experiments have nevertheless taught us some valuable lessons about tropical forest nutrient cycling and forest functioning.

References

Adamek M, Corre M, Hölscher D (2009) Early effect of elevated nitrogen input on aboveground net primary production of a tropical lower montane rain forest, Panama. J Trop Ecol 25:637–647

Alvarez-Clare S, Mack MC, Brooks M (2013) A direct test of nitrogen and phosphorus limitation to net primary productivity in a lowland tropical wet forest. Ecology 94:1540–1551

Andersen K, Corre M, Turner BL, Dalling JW (2010) Plant-soil associations in lower montane tropical forest: physiological acclimation and herbivore-mediated responses to nitrogen addition. Funct Ecol 24:1171–1180

Attiwill PM, Adams MA (1993) Nutrient cycling in forests. New Phytol 124:561–582

Banin LF, Phillips OL, Lewis SL (2015) Tropical Forests. In: Peh KS-H, Corlett RT, Bergeron Y (eds) Routledge handbook of forest ecology. Routledge

Baribault TW, Kobe RK, Finley AO (2012) Tropical tree growth is correlated with soil phosphorus, potassium, and calcium, though not for legumes. Ecol Monogr 82:189–203

Batterman SA, Wurzburger N, Hedin LO (2013) Nitrogen and phosphorus interact to control tropical symbiotic N_2 fixation: a test in *Inga punctata*. J Ecol 101:1400–1408

Bloom AJ, Chapin FS, Mooney HA (1985) Resource limitation in plants—an economic analogy. Ann Rev Ecol Syst 16:363–392

Brearley FQ, Press MC, Scholes JD (2003) Nutrients obtained from leaf litter can improve the growth of dipterocarp seedlings. New Phytol 160:101–110

Brearley FQ, Scholes JD, Press MC, Palfner G (2007) How does light and phosphorus fertilisation affect the growth and ectomycorrhizal community of two contrasting dipterocarp species? Plant Ecol 192:237–249

Bruijnzeel LA (1991) Nutrient input-output budgets of tropical forest ecosystems: a review. J Trop Ecol 7:1–24

Burslem DFRP, Grubb PJ, Turner IM (1996) Responses to simulated drought and elevated nutrient supply among shade-tolerant tree seedlings of lowland tropical forest in Singapore. Biotropica 28:636–648

Burslem DFRP, Grubb PJ, Turner IM (1995) Responses to nutrient addition among shade-tolerant tree seedlings of lowland tropical rain-forest in Singapore. J Ecol 83:113–122

Cai ZQ, Poorter L, Han Q, Bongers F (2008) Effects of light and nutrients on seedlings of tropical *Bauhinia* lianas and trees. Tree Physiol 28:1277–1285

Campo J, Dirzo R (2003) Leaf quality and herbivory responses to soil nutrient addition in secondary tropical dry forests of Yucatan, Mexico. J Trop Ecol 19:525–530

Campo J, Vazquez-Yanes C (2004) Effects of nutrient limitation on aboveground carbon dynamics during tropical dry forest regeneration in Yucatán, Mexico. Ecosystems 7:311–319

Cavelier J, Tanner EVJ, Santamaría J (2000) Effect of water, temperature and fertilizers on soil nitrogen net transformations and tree growth in an elfin cloud forest of Colombia. J Trop Ecol 16:83–99

Cernusak LA, Winter K, Turner BL (2010) Leaf nitrogen to phosphorus ratios of tropical trees: experimental assessment of physiological and environmental controls. New Phytol 185:770–779

Chapin FS (1980) The mineral nutrition of wild plants. Ann Rev Ecol Syst 11:233–260

Chapin FS, Vitousek PM, van Cleve K (1986) The nature of nutrient limitation in plant communities. Am Nat 127:48–58

Coley PD, Barone JA (1996) Herbivory and plant defences in tropical forests. Ann Rev Ecol Syst 27:305–335

Cordell S, Goldstein G, Meinzer FC, Vitousek PM (2001) Morphological and physiological adjustment to N and P fertilization in nutrient-limited *Metrosideros polymorpha* canopy trees in Hawaii. Tree Physiol 21:43–50

Cornforth IS (1968) Relationship between soil volume used by roots and nutrient accessibility. J Soil Sci 19:291–301

Corre MD, Veldkamp E, Arnold J, Wright SJ (2010) Impact of elevated N input on soil N cycling and losses in old-growth lowland and montane forests in Panama. Ecology 91:1715–1729

Crews TE, Kitayama K, Fownes JH, Riley RH, Herbert DA, Mueller-Dombois D, Vitousek PM (1995) Changes in soil phosphorus fractions and ecosystem dynamics across a long chronosequence in Hawaii. Ecology 76:1407–1424

Cuevas E, Medina E (1988) Nutrient dynamics within Amazonian forests. II fine root growth, nutrient availability and leaf litter decomposition. Oecologia 76:222–235

Dalling JW, Tanner EVJ (1995) An experimental study of regeneration on landslides in montane rain forest in Jamaica. J Ecol 83:55–64

Davidson EA, Carvalho CJR, Vieira ICG, Figueiredo RD, Moutinho P, Ishida FY, dos Santos MTP, Guerrero JB, Kalif K, Saba RT (2004) Nitrogen and phosphorus limitation of biomass growth in a tropical secondary forest. Ecol Appl 14:150–163

Denslow JS, Vitousek PM, Schultz JC (1987) Bioassays of nutrient limitation in a tropical rain forest soil. Oecologia 74:370–376

Elser JJ, Sterner RW, Gorokhova E, Fagan WF, Markow TA, Cotner JB, Harrison JF, Hobbie SE, Odell GM, Weider LW (2000) Biological stoichiometry from genes to ecosystems. Ecol Lett 3:540–550

Epron D, Laclau J-P, Almeida JCR, Gonçalves JLM, Ponton S, Sette CR Jr, Delgado-Rojas JS, Bouillet J-P, Nouvellon Y (2011) Do changes in carbon allocation account for the growth response to potassium and sodium applications in tropical *Eucalyptus* plantations? Tree Physiol 32:667–679

Evans JR (1989) Photosynthesis and nitrogen relationships in leaves of C_3 plants. Oecologia 78:9–19

Fetcher N, Haines BL, Cordero RA, Lodge DJ, Walker LR, Fernandez DS, Lawrence WT (1996) Responses of tropical plants to nutrients and light on a landslide in Puerto Rico. J Ecol 84:331–341

Fisher JB, Malhi Y, Torres IC, Metcalfe DB, van de Weg MJ, Meir P, Silva-Espejo JE, Huaranca Hasce W (2013) Nutrient limitation in rainforests and cloud forests along a 3,000-m elevation gradient in the Peruvian Andes. Oecologia 172:889–902

Goldstein G, Bucci, SJ, Scholz FG (2013) Why do trees adjust water relations and hydraulic architecture in response to nutrient availability? Tree Physiol 33:238–240

Giardina CP, Ryan MG, Binkley D, Fownes JH (2003) Primary production and carbon allocation in relation to nutrient supply in a tropical experimental forest. Glob Change Biol 9:1438–1450

Gower ST (1987) Relations between mineral nutrient availability and fine root biomass in two Costa Rican tropical wet forests: a hypothesis. Biotropica 19:171–175

Graefe S, Hertel D, Leuschner C (2010) N, P and K limitation of fine root growth along an elevation transect in tropical mountain forests. Acta Oecol 36:537–542

Grime JP (2001) Plant strategies, vegetation processes and ecosystem properties, 2nd edn. Wiley, Chichester

Grubb PJ (1977) Control of forest growth and distribution on wet tropical mountains—with special reference to mineral nutrition. Ann Rev Ecol Syst 8:83–107

Grubb PJ (1989) The role of mineral nutrients in the tropics: a plant ecologist's view. In: Proctor J (ed) Mineral nutrients in tropical forest and savanna ecosystems. Blackwell, Oxford

Gunatilleke CVS, Gunatilleke IAUN, Perera GAD, Burslem DFRP, Ashton PMS, Ashton PS (1997) Responses to nutrient addition among seedlings of eight closely related species of *Shorea* in Sri Lanka'. J Ecol 85:301–311

Gutschick VP (1981) Evolved strategies in nitrogen acquisition by plants. Am Nat 118:607–637

Hall SJ, Matson PA (2003) Nutrient status of tropical rain forests influences soil N dynamics after N additions. Ecol Monogr 73:107–129

Hall JS, Ashton MS, Berlyn GP (2003) Seedling performance of four sympatric *Entandrophragma* species (Meliaceae) under simulated fertility and moisture regimes of central African rain forest. J Trop Ecol 19:55–66

Harrington RA, Fownes JH, Vitousek PM (2001) Production and resource-use efficiencies in N- and P- limited tropical forests: a comparison of responses to long-term fertilization. Ecosystems 4:646–657

Hedges LV, Gurevitch J, Curtis PS (1999) The meta-analysis of response ratios in experimental ecology. Ecology 80:1150–1156

Hedin LO, Brookshire ENJ, Menge D, Barron AR (2009) The nitrogen paradox in tropical forest ecosystems. Ann Rev Ecol Syst 40:613–635

Herrera R, Merida T, Stark NM, Jordan CF (1978) Direct phosphorus transfer from leaf litter to roots. Naturwissenschaften 65:208–209

Hodge A, Campbell CD, Fitter AH (2001) An arbuscular mycorrhizal fungus accelerates decomposition and acquires nitrogen directly from organic material. Nature 413:297–299

Högberg P (1986) Soil nutrient availability, root symbioses and tree species composition in tropical Africa: a review. J Trop Ecol 2:359–372

Homeier J, Hertel D, Camenzind T, Cumbicus NL, Maraun M, Martinson GO, Nohemy Poma L, Rillig MC, Sandmann D, Scheu S, Veldkamp E, Wilcke W, Wullaert H, Leuschner C (2012) Tropical Andean forests are highly susceptible to nutrient inputs—rapid effects of experimental N and P addition to an Ecuadorian montane forest. PLoS ONE 7:e47128

Houlton BZ, Wang Y-P, Vitousek PM, Field CB (2008) A unifying framework for dinitrogen fixation in the terrestrial biosphere. Nature 454:327–330

Huante P, Rincon E, Chapin FS (1995) Responses to phosphorus of contrasting successional tree-seedling species from the tropical deciduous forest of Mexico. Funct Ecol 9:760–766

Jordan CF (1985) Nutrient cycling in tropical forest ecosystems. Wiley, Chichester

Jordan CF, Herrera R (1981) Tropical rain forests—are nutrients really critical? Am Nat 117: 167–180

Kaspari M, Garcia MN, Harms KE, Santana M, Wright SJ, Yavitt JB (2008) Multiple nutrients limit litterfall and decomposition in a tropical forest. Ecol Lett 11:35–43

Koehler B, Corre MD, Veldkamp E, Wullaert H, Wright SJ (2009) Immediate and long-term nitrogen oxide emissions from tropical forest soils exposed to elevated nitrogen input. Glob Change Biol 15:2049–2066

Lambers H, Raven JA, Shaver GR, Smith SE (2008) Plant nutrient-acquisition strategies change with soil age. Trends Ecol Evol 23:95–103

Lawrence D (2003) The response of tropical tree seedlings to nutrient supply: meta-analysis for understanding a changing tropical landscape. J Trop Ecol 19:239–250

Lodge DJ, McDowell WH, McSwiney CP (1994) The importance of nutrient pulses in tropical forests. Trends Ecol Evol 9:384–387

Manzoni S, Trofymow JA, Jackson RB, Porporato A (2010) Stoichiometric controls dynamics of carbon, nitrogen, and phosphorus in decomposing litter. Ecol Monogr 80:89–106

Matzek V, Vitousek PM (2009) N: P stoichiometry and protein: RNA ratios in vascular plants: an evaluation of the growth-rate hypothesis. Ecol Lett 12:765–771

Mayor JR, Wright SJ, Turner BL (2013) Data from: Species-specific responses of foliar nutrients to long-term nitrogen and phosphorus additions in a lowland tropical forest. Dryad Digital Repository. doi:10.5061/dryad.257b9

Mayor JR, Wright SJ, Turner BL (2014) Species-specific responses of foliar nutrients to long-term nitrogen and phosphorus additions in a lowland tropical forest. J Ecol 103:36–44

McGroddy ME, Daufresne T, Hedin LO (2004) Scaling of C:N:P: stoichiometry in forest ecosystems worldwide: implications of terrestrial Redfield-type ratios. Ecology 85:2390–2401

Miller HG (1981) Forest fertilization: some guiding concepts. Forestry 54:157–167

Mirmanto E, Proctor J, Green J, Nagy L, Suriantata (1999) Effects of nitrogen and phosphorus fertilization in lowland evergreen rainforest. Phil Trans Roy Soc B 354:1825–1829

Mo J, Zhang W, Zhu W, Gundersen P, Fang Y, Li D, Wang H (2008) Nitrogen addition reduces soil respiration in a mature tropical forest in southern China. Glob Change Biol 14:403–412

Newbery DM, Chuyong GB, Green JJ, Songwe NC, Tchuenteu F, Zimmermann L (2002) Does low phosphorus supply limit seedling establishment and tree growth in groves of ectomyc-orrhizal trees in a central African rainforest? New Phytol 156:297–311

Ostertag R (2001) Effects of nitrogen and phosphorus availability on fine-root dynamics in Hawaiian montane forests. Ecology 82:485–499

Ostertag R (2010) Foliar nitrogen and phosphorus accumulation responses after fertilization: an example from nutrient-limited Hawaiian forests. Plant Soil 334:85–98

Pasquini SC, Santiago LS (2012) Nutrients limit photosynthesis in seedlings of a lowland tropical forest tree species. Oecologia 168:311–319

Paoli GD, Curran LM, Slik JWF (2008) Soil nutrients affect spatial patterns of aboveground biomass and emergent tree density in southwestern Borneo. Oecologia 155:287–299

Pinheiro JC, Bates DM (2000) Mixed-effects models in S and S-plus. Springer, New York

Pinheiro J, Bates D, DebRoy S, Sarkar D, Core Team R (2015) nlme: linear and nonlinear mixed effects models. R package version 3:1–122

Proctor J (1983) Mineral nutrients in tropical forests. Prog Phys Geog 7:422–431

Qualls RG, Haines BL, Swank WT (1991) Fluxes of dissolved organic nutrients and humic substances in a deciduous forest. Ecology 72:254–266

Quesada CA, Phillips OL, Schwarz M, Czimczik CI, Baker TR, Patino S, Fyllas NM, Hodnett MG, Herrera R, Almeida S, Davila EA, Arneth A, Arroyo L, Chao KJ, Dezzeo N, Erwin T, Fiore A, Higuchi N, Coronado EH, Jimenez EM, Killeen T, Lezama AT, Lloyd G, Lopez-Gonzalez G, Luizao FJ, Malhi Y, Monteagudo A, Neill DA, Vargas PN, Paiva R, Peacock J, Penuela MC, Cruz AP, Pitman N, Priante N, Prieto A, Ramirez H, Rudas A, Salomao R, Santos AJB, Schmerler J, Silva N, Silveira M, Vasquez R, Vieira I, Terborgh J, Lloyd J (2012) Basin-wide variations in Amazon forest structure and function are mediated by both soils and climate. Biogeosciences 9:2203–2246

Development Core Team R (2014) R: a language and environment for statistical computing. R Foundation for Statistical Computing, Vienna

Raaimakers D, Lambers H (1996) Response to phosphorus supply of tropical tree seedlings: a comparison between a pioneer species *Tabirira obtusa* and a climax species *Lecythis corrugata*. New Phytol 132:97–102

Raich JW, Riley RH, Vitousek PM (1994) Use of root-ingrowth cores to assess nutrient limitations in forest ecosystems. Can J For Res 24:2135–2138

Russo SE, Davies SJ, King DA, Tan S (2005) Soil-related performance variation and distributions of tree species in a Bornean rain forest. J Ecol 93:879–889

Santiago LS, Wright SJ, Harms KE, Yavitt JB, Korine C, Garcia MN, Turner BL (2012) Tropical tree seedling growth responses to nitrogen phosphorus and potassium addition. J Ecol 100:309–316

Sayer EJ, Tanner EVJ (2010) Experimental investigation of the importance of litterfall in lowland semi-evergreen tropical forest nutrient cycling. J Ecol 98:1052–1062

Sayer EJ (2006) Using experimental litter manipulation to assess the roles of leaf litter in the functioning of forest ecosystems. Biol Rev 81:1–31

Sayer EJ, Tanner EVT, Wright SJ, Yavitt JB, Harms KE, Powers JS, Kaspari M, Garcia MN, Turner BL (2012) Comparative assessment of lowland tropical forest nutrient status in response to fertilization and litter manipulation. Ecosystems 15:387–400

Schreeg LA, Mach MC, Turner BL (2013) Leaf litter inputs decrease phosphate sorption in a strongly weathered tropical soil over two time scales. Biogeochem 113:507–524

Sobrado MA (2013) Soil and leaf micronutrient composition in contrasting habitats in podzolized sands of the Amazon region. Am J Plant Sci 4:1918–1923

St. John TV (1982) Response of tree roots to decomposing organic matter in two lowland Amazonian rain forests. Can J For Res 13:346–349

Stark NM, Jordan CF (1978) Nutrient retention by the root mat of an Amazonian rain forest. Ecology 59:434–437

Stewart CG (2000) A test of nutrient limitation in two tropical montane forests using ingrowth cores. Biotropica 32:369–373

Tanner EVJ, Barberis IM (2007) Trenching increased growth, and irrigation increased survival of tree seedlings in the understorey of a semi-evergreen rain forest in Panama. J Trop Ecol 23:257–268

Tanner EVJ, Vitousek PM, Cuevas E (1998) Experimental investigation of nutrient limitation of forest growth on wet tropical mountains. Ecology 79:10–22

Treseder KK, Vitousek PM (2001) Effects of soil nutrient availability on investment in acquisition of N and P in Hawaiian rain forests. Ecology 82:946–954

Tripathi SN, Raghubanshi AS (2014) Seedling growth of five tropical dry forest species in relation to light and nitrogen gradients. J Plant Ecol 7:250–263

Turner IM, Brown ND, Newton AC (1993) The effect of fertilizer application on dipterocarp seedling growth and mycorrhizal infection. For Ecol Manage 57:329–337

Tutua SS, Xu ZH, Blumfield TJ, Bubb KA (2008) Long-term impacts of harvest residue management on nutrition, growth and productivity of an exotic pine plantation of sub-tropical Australia. For Ecol Manage 256:741–748

Vasconcelos SS, Zarin DJ, Machado Araújo M, Rangel-Vasconcelos LGT, Reis de Carvalho CJ, Staudhammer CL, Oliveira FA (2008) Effects of seasonality, litter removal and dry-season irrigation on litterfall quantity and quality in eastern Amazonian forest regrowth, Brazil. J Trop Ecol 24:27–38

Villagra M, Campanello PI, Montti Goldstein G (2013) Removal of nutrient limitations in forest gaps enhances growth rate and resistance to cavitation in subtropical canopy tree species differing in shade tolerance. Tree Physiol 33:285–296

Villalobos-Vega R, Goldstein G, Haridasan M, Franco AC, Miralles-Wilhelm F, Scholz FG, Bucci SJ (2011) Leaf litter manipulations alter soil physicochemical properties and tree growth in a Neotropical savanna. Plant Soil 346:385–397

Vitousek PM, Farrington H (1997) Nutrient limitation and soil development: experimental test of a biogeochemical theory. Biogeochem 37:63–75

Vitousek PM, Howarth RW (1991) Nitrogen limitation on land and in the sea: how can it occur? Biogeochem 13:87–115

Vitousek PM (1984) Litterfall, nutrient cyclling and nutrient limitation in tropical forests. Ecology 65:285–298

Vitousek PM, Sanford RL (1986) Nutrient cycling in moist tropical forest. Ann Rev Ecol Syst 17:137–167

Vitousek PM, Porder S, Houlton BZ, Chadwick OA (2010) Terrestrial phosphorus limitation: mechanisms, implications, and nitrogen-phosphorus interactions. Ecol Appl 20:5–15

Walker TW, Syers JK (1976) The fate of phosphorus during pedogenesis. Geoderma 15:1–19

Watanabe T, Broadley MR, Jansen S, White PJ, Takada J, Satake K, Takamatsu T, Tuah SJ, Osaki M (2007) Evolutionary control of leaf element composition in plants. New Phytol 174:516–523

Winter K, Aranda J, Garcia M, Virgo A, Paton SR (2001) Effect of elevated CO_2 and soil fertilization on whole-plant growth and water use in seedlings of a tropical pioneer tree *Ficus insipida* Willd. Flora 196:458–464

Wood TE, Lawrence D, Clark DA, Chazdon RL (2009) Rain forest nutrient cycling and productivity in response to large-scale litter manipulation. Ecology 90:109–121

Wright SJ, Yavitt JB, Wurzburger N, Turner BL, Tanner EVJ, Sayer EJ, Santiago LS, Kaspari M, Hedin LO, Harms KE, Garcia MN, Corre MD (2011) Potassium, phosphorus or nitrogen limit root allocation, tree growth and litter production in a lowland tropical forest. Ecology 92: 1616–1625

Yavitt JB, Harms KE, Garcia MN, Mirabello MJ, Wright SJ (2011) Soil fertility and fine root dynamics in response to 4 years of nutrient (N, P, K) fertilization in a lowland tropical moist forest, Panama. Austral Ecol 36:433–445

Yuan ZY, Chen HYH (2012) A global analysis of fine root production as affected by soil nitrogen and phosphorus. Proc Roy Soc B 279:3796–3802

Zhu F, Yoh M, Gilliam FS, Lu X, Mo J (2013) Nutrient limitation in three lowland tropical forests in southern China receiving high nitrogen deposition: insights from fine root responses to nutrient additions. PLoS ONE 8:e82661

Is Photosynthesis Nutrient Limited in Tropical Trees?

Louis S. Santiago and Guillermo Goldstein

Abstract Tropical forests play an enormously important role in the global cycling of carbon. However, the extent to which nutrients limit the potential for tropical trees to increase carbon gain as atmospheric carbon dioxide rises, a phenomenon known as the carbon-concentration feedback, is uncertain. This chapter addresses our current state of knowledge on nutrient limitation of photosynthesis in tropical trees, summarizing and synthesizing the results of over 20 studies on photosynthetic responses to nutrient manipulation experiments. Our results indicate that nutrient limitation of photosynthesis is widespread, but that contrasting species and ecosystems vary in their responses, with savannah trees showing the least response at the leaf scale. Second, although photosynthesis is strongly limited by N in particular species, N limitation of photosynthesis is modest compared to P limitation of photosynthesis when considering all of the available literature. Finally, alleviation of nutrient limitation through addition of combined nutrient treatments produces the strongest photosynthetic responses, highlighting the potentially complex stroichiometric interactions among elements. The authors discuss several ways forward for resolving questions regarding the potential and limits of tropical trees to influence key carbon cycling processes and thus improve our ability to forecast global responses and feedbacks to climate change.

L.S. Santiago (✉)
Department of Botany & Plant Sciences, University of California,
2150 Batchelor Hall, Riverside, CA 92521, USA
e-mail: santiago@ucr.edu

L.S. Santiago
Smithsonian Tropical Research Institute, Ancon, Panama, Republic of Panama

G. Goldstein
Laboratorio de Ecología Funcional, Departamento de Ecología Genética y Evolución,
Instituto IEGEBA (CONICET-UBA), Facultad de Ciencias Exactas y naturales,
Universidad de Buenos Aires, Buenos Aires, Argentina
e-mail: gold@bio.miami.edu

G. Goldstein
Department of Biology, University of Miami, Coral Gables, FL 33146, USA

© Springer International Publishing Switzerland 2016
G. Goldstein and L.S. Santiago (eds.), *Tropical Tree Physiology*,
Tree Physiology 6, DOI 10.1007/978-3-319-27422-5_14

Keywords Nitrogen · Phosphorus · Potassium Meta-analysis · Fertilization · Atmospheric carbon-dioxide

Introduction

Tropical forests are responsible for approximately one-third of terrestrial gross primary production (Beer et al. 2010), and are thought to be especially sensitive to climate change (Phillips et al. 2010), suggesting that their responses to climate change and feedbacks to the global climate system are likely to be substantial. But one outstanding question is whether tropical trees can exploit the increasing atmospheric supply of CO_2 through enhanced photosynthetic carbon income (the concentration-carbon feedback) (Bonan and Levis 2010). Understanding the trajectory of plant photosynthetic responses to increasing CO_2 is not straightforward because nutrient availability has the potential to limit the concentration-carbon feedback in tropical vegetation such that photosynthesis may not continue to increase with atmospheric CO_2 concentration (Santiago 2015). Predictions are further complicated by the fact that soil substrates vary greatly in the tropics. For example, the Amazon lies along a dramatic east–west nutrient availability gradient (Quesada et al. 2010), and there are often strong changes in nutrient availability with elevation (Dalling et al., this volume). So how does the formidable diversity of trees combine with heterogeneity of soil types and elemental availability in the tropics to limit photosynthetic exploitation of increasing atmospheric CO_2? To address this question, this synthesis focuses on photosynthetic responses of tropical trees to nutrients and how such information is critical to inform Earth system models to understand potential climate change feedbacks.

There are multiple ways in which nutrients could mediate the concentration-carbon feedback. The first is simply that plants increase leaf-level photosynthetic rates per area with increasing atmospheric CO_2 to the degree allowed by nutrient availability (Bloom et al. 1985; Field and Mooney 1986) (Fig. 1a). The second is that as atmospheric CO_2 increases, plants allocate increased carbon income to growth such that leaf-level photosynthesis per area remains the same, but the greater leaf surface area associated with enhanced growth leads to increased photosynthesis at the whole-plant scale (Poorter and Remkes 1990) (Fig. 1b). A third scenario is one in which increasing CO_2 causes a decrease in stomatal conductance such that internal CO_2 concentration and photosynthetic rate remain the same, but water use efficiency, defined as photosynthetic carbon gain per unit water loss, increases (Fig. 1c) (Farquhar and Richards 1984; Cernusak et al. 2007a, b). In all of these scenarios we expect nutrients to mediate physiological responses to changing atmospheric CO_2 (Raven et al. 2004; Cernusak et al. 2007a, b), such that tropical trees growing on the most fertile soils would have the greatest increase, or least reduction in photosynthetic carbon assimilation. All of these potential responses to atmospheric CO_2 could also have larger scale impacts because of the strong bearing

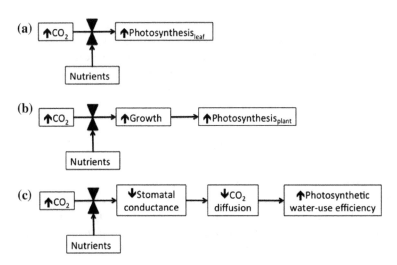

Fig. 1 Possible ways for nutrients to mediate responses of tropical plants to increasing CO_2

of tropical forests on the global carbon cycle and climate system (Friedlingstein et al. 2006; Bonan and Levis 2010). Yet the only way to determine the extent to which nutrients limit the carbon-concentration feedback in tropical trees is to determine the identity of elements that limit photosynthesis and the degree of that limitation.

There are also several valid eco-physiological reasons why tropical woody species may not show greater photosynthesis on more fertile soils. First, several studies have now reported increased herbivory following nutrient addition and an increase in foliar nutrient concentration that appears to make leaves more palatable to herbivores (Campo and Dirzo 2003; Santiago et al. 2012). In some cases, the increase in herbivory can be strong enough to mask plant growth responses to nutrient addition (Andersen et al. 2010). Second, there is evidence that as atmospheric CO_2 increases, some plants reduce stomatal conductance such that photosynthetic rates remain the same, but water loss goes down and water use efficiency goes up (Cernusak et al. 2013). Because nutrient availability can interact with plant regulation of water loss, especially when transpiration rates are linked to uptake of key elements through mass flow in the transpiration stream (Cramer et al. 2009), reductions in stomatal conductance with increasing CO_2 could reduce nutrient uptake through mass flow and worsen nutrient limitation of photosynthesis. In other cases, plants in poor soils have been shown to exhibit greater transpiration rates than plants in fertile soils and it is thought that the benefit is increased rates of absorption of nutrients through mass flow (Cramer et al. 2008). In such cases, how plants in poor versus fertile soils might alter their photosynthetic rates with changing atmospheric CO_2 is unknown. Finally, if increasing atmospheric CO_2 concentration indeed increases photosynthetic carbon uptake, at some point a buildup of carbohydrate products of photosynthesis could lead to down regulation of photosynthetic rate in a process known as feedback inhibition (Delucia et al. 1985). Thus careful evaluation of the study design, including plant-animal interactions and other physiological processes such as nutrient allocation to leaves and interactions with transpiration will increase the depth of interpretation and predictive power of experimental studies.

How Is Nutrient Limitation Studied?

Nutrient limitation of plant productivity and photosynthesis has been studied through a variety of approaches. Perhaps the most effective has been through nutrient addition experiments in which responses to fertilization relative to control treatments are interpreted as limitation by that nutrient, of if there is no response, the conclusion is that the added nutrient was not limiting (Vitousek 2004). Many of these are stand-level fertilization experiments consisting of large (40 × 40 m) plots receiving individual or combined addition of nitrogen (N), phosphorus (P), potassium (K) or micronutrients (Tanner et al. 1992; Vitousek et al. 1993; Mirmanto et al. 1999; Newbery et al. 2002; Campo and Vázquez-Yanes 2004; Davidson et al. 2004; Bucci et al. 2006; Lu et al. 2010; Wright et al. 2011; Álvarez-Clare et al. 2013; Fisher et al. 2013). Such large experiments are able to evaluate nutrient limitation at multiple scales including leaf physiological processes, growth and ecosystem-level processes such as nutrient cycling and litterfall. Yet, because canopy access poses substantial challenges and because ecosystem science is usually the focus of these studies, measurements of leaf-scale physiological measurements are only available from five of the known large-scale tropical fertilization experiments (Cordell et al. 2001; Saraceno 2006; Lu et al. 2007; Pasquini and Santiago 2012).

Other approaches to manipulating nutrient availability in tropical trees include litter removal and addition experiments (Sayer and Tanner 2010; Sayer, this volume) or removal of vegetation to increase nutrient availability of target individuals. Such process-based nutrient manipulations do not directly reveal the identities of limiting elements. However, when foliar nutrient responses are included, information on the relative importance of key elements can sometimes be determined (Lewis and Tanner 2000; Barberis and Tanner 2005). At smaller scales, fertilizer has also been applied directly to individual plants in the field (e.g. Yavitt and Wright 2008), or in pots (Burslem 1996; Gunatilleke et al. 1997; Lawrence 2003; Cai et al. 2008). In many cases, such experiments have been successful in determining which elements limit growth and physiological processes in tropical seedlings or saplings.

When stand-scale experimental manipulation of nutrient availability is impractical, basic questions about plant nutritional effects on physiology have been approached along environmental gradients that vary in soil nutrient availability (Vitousek et al. 1992; Santiago et al. 2004; Santiago and Mulkey 2005; Baltzer et al. 2008; Fyllas et al. 2009). In these in situ studies, by evaluating how foliar nutrient concentration or physiological characteristics vary with soil type it is possible to generate hypotheses regarding nutrient limitation that can be further tested experimentally. For example, in Hawaii, the gradual shift from N-limitation of ecosystem productivity in very young soils to P-limitation of ecosystem productivity in very old soils creates opportunities to evaluate the relative importance of these elements for photosynthesis and growth (Kitayama et al. 1997; Cordell et al. 2001). In lowland tropical forest in Panama, increasing precipitation leads to species replacement of dry-season deciduous tree species with evergreen species, consistent with the idea that evergreeness conserves nutrients by minimizing the cost of leaf replacement and providing an advantage for evergreen species in wetter sites with leached soils (Santiago and Mulkey 2005). Even broad ecological patterns, such as in Brazilian savannahs, where patches of closed canopy dry season deciduous forest appear on calcium-rich outcrops (Haridasan 1992), provide fodder for new hypotheses and experiments. Thus the interplay between observation and experiment has continued to be an important process in understanding nutrient limitation in the tropics.

Direct analysis of foliar ratios of nitrogen-to-phosphorus (N:P) is also considered a method for inferring relative nutrient limitation of N versus P using the guideline that N:P ratios below 12 indicate N-limitation, ratios above 16 indicate P-limitation and values between 12 and 16 indicate co-limitation by N and P (Aerts and Chapin 2000). It is difficult for foliar N:P data alone to establish limitation, yet it seems to be an excellent measure of relative limitation. Previously, most researchers used the eco-physiological standard of the youngest mature leaves to study N:P, but recently, stem, root, and older leaf N:P have been shown to be more responsive indicators of soil nutrient availability than new leaf N:P, because new leaves appear to act as nutrient reservoirs during active growth, allowing maintenance of optimal N:P ratios in recently produced, physiologically active leaves (Schreeg et al. 2014). Overall, combining soil and plant observational approaches with experiments offer our greatest potential for unraveling complex interactions between nutrient availability, increasing atmospheric CO_2 and photosynthetic performance in tropical forests.

The Physiological Basis for Elemental Limitation of Photosynthesis

The proteins of the Calvin cycle and thylakoids make up the majority of N in leaves, thus N is known to be a strong determinant of photosynthesis (Field and Mooney 1986; Wright et al. 2004). Therefore, as leaf N concentration increases with experimental fertilization (Ostertag 2009; Wright et al. 2011; Pasquini and Santiago 2012; Santiago et al. 2012), maximum photosynthetic rate is expected to increase as well. N addition in shade leaves is expected to increase N allocation toward light harvesting proteins of the thylakoid membrane of the chloroplast, including the pigment-protein complexes and components of the electron transport chain (Evans 1989). In sun leaves, the maximum rate of CO_2 assimilation (A_{max}) is expected to increase with N addition, but in any case, there should be correspondence between light harvesting and carboxylation because carboxylation capacity and electron transport are tightly coupled (Wullschleger 1993; Pasquini and Santiago 2012). The increase in foliar N with fertilization is usually modest, owing to the metabolic costs of storing N in plant tissue (Ostertag 2009). Thus the expected enhancement of photosynthesis with N addition should occur over a relatively small range of leaf N values.

Leaf P is a component of polyphosphates and phospholipids and is necessary for activity of the Calvin cycle and rubisco regeneration (Marschner 1995). Within the biochemical reactions of carbon assimilation, activation through phosphorylation is ubiquitous, highlighting the potential for P to limit or promote rapid photosynthetic reactions. P addition has been shown to stimulate A_{max}, but reduce A_{max} per unit leaf P (Cordell et al. 2001), because of the ability of leaves to sequester large amounts of P in vacuoles (Sinclair and Vadez 2002; Ostertag 2009). In most soils of lowland tropical forests, P is thought to be the most limiting element to productivity (Dalling, this volume), so understanding P limitation of photosynthesis is fundamental for predicting plant responses to climate change.

K^+ is the most abundant ion in plant cells and is critical for numerous biochemical functions including osmoregulation, photosynthesis, cell extension, oxidative phosphorylation and protein activation (Evans and Sorger 1966; Morgan 1984; Marschner 1995; Santiago and Wright 2007). Therefore, K addition has the potential to increase photosynthesis through efficient biochemistry and stomatal control, but of the macronutrients, it is the least studied and most rare to find in fertilization experiments on tropical trees. Currently, there is only one stand-scale natural forest fertilization study that has incorporated K, and growth of seedlings, saplings and poles, but not emergent trees, as well as aspects of photosynthesis have been shown to be limited by K (Wright et al. 2011; Pasquini and Santiago 2012). K is also unique in that it has no organic phase, so unlike N and P which can be bound to organic complexes within leaf litter, K leaches quickly (Schreeg et al. 2013), potentially promoting shortages in availability.

Nutrient Addition Studies with Tropical Trees

Of the nutrient addition studies on tropical plants, we have identified 19 studies on 46 species that measured A_{max} on plants fertilized with N, P, K or a combination of these and micronutrient elements (Table 1). These include studies on plants in pots in controlled greenhouse settings, as well as plants in the field, growing in large stand-level fertilized or control plots. We only included studies that report A_{max} for individual species at multiple nutrient levels, but did not consider proxy variables

Table 1 Summary of species, locations, experimental details and references for studies involving responses of maximum photosynthetic rate to nutrient addition in tropical tree species

Location	Species	Treatment	Study type	References
Brasília, Brazil	*Blepharocalyx salicifolius*	N, P, Combined	field	Saraceno (2006)
Brasília, Brazil	*Caryocar brasiliense*	N, P, Combined	field	Saraceno (2006)
Brasília, Brazil	*Ouratea hexasperma*	N, P, Combined	field	Saraceno (2006)
Brasília, Brazil	*Qualea parviflora*	N, P, Combined	field	Saraceno (2006)
Dinghushan, China	*Cryptocarya chinensis*	N	field	Lu et al. (2007)
Dinghushan, China	*Cryptocarya concinna*	N	field	Lu et al. (2007)
Dinghushan, China	*Randia canthioides*	N	field	Lu et al. (2007)
Dinghushan, China	*Cryptocarya concinna*	N, P, Combined	field	Zhu et al. (2014)
Dinghushan, China	*Randia canthioides*	N, P, Combined	field	Zhu et al. (2014)
Freetown, Sierra Leone	*Entandrophragma angolense*	Combined	pot	Riddoch et al. (1991)
Freetown, Sierra Leone	*Nauclea diderrichii*	Combined	pot	Riddoch et al. (1991)
Fujian, China	*Eucalyptus dunnii*	P	pot	Wu et al. (2014)
Fujian, China	*Eucalyptus grandis*	P	pot	Wu et al. (2014)
Gamboa, Panama	*Ficus insipida*	Combined	pot	Cernusak et al. (2007a, b)
Gamboa, Panama	*Platymiscium pinnatum*	Combined	pot	Cernusak et al. (2009)
Gamboa, Panama	*Swietenia macrophylla*	Combined	pot	Cernusak et al. (2009)
Gamboa, Panama	*Tectona grandis*	Combined	pot	Cernusak et al. (2009)
Gigante, Panama	*Alesis blackiana*	N, P, K	field	Pasquini and Santiago (2012)
Guangzhou, China	*Acmena acuminatissima*	N	chamber	Liu et al. (2012)
Guangzhou, China	*Castanopsis hystrix*	N	chamber	Liu et al. (2012)
Guangzhou, China	*Ormosia pinnata*	N	chamber	Liu et al. (2012)
Guangzhou, China	*Schima superba*	N	chamber	Liu et al. (2012)
Hawaii, USA	*Metrosideros polymorpha*	N, P, Combined	field	Cordell et al. (2001)
Queensland, Australia	*Auranticarpa ilicifolia*	P	pot	Bloomfield et al. (2014)
Queensland, Australia	*Cryptocarya triplinervis*	P	pot	Bloomfield et al. (2014)
Queensland, Australia	*Flindersia bourjotiana*	P	pot	Bloomfield et al. (2014)
Queensland, Australia	*Guioa lasioneura*	P	pot	Bloomfield et al. (2014)
Queensland, Australia	*Hymenosporum flavum*	P	pot	Bloomfield et al. (2014)

(continued)

Table 1 (continued)

Location	Species	Treatment	Study type	References
Queensland, Australia	*Neolitsea dealbata*	P	pot	Bloomfield et al. (2014)
Queensland, Australia	*Syzygium wilsonii*	P	pot	Bloomfield et al. (2014)
Queensland, Australia	*Flindersia brayleyana*	Combined	pot	Thompson et al. (1988)
Queensland, Australia	*Argyrodendron trifoliolatum*	Combined	pot	Thompson et al. (1992)
Queensland, Australia	*Argyrodendron sp*	Combined	pot	Thompson et al. (1992)
Queensland, Australia	*Flindersia brayleyana*	Combined	pot	Thompson et al. (1992)
Queensland, Australia	*Toona australis*	Combined	pot	Thompson et al. (1992)
Recife, Brazil	*Anadenanthera colubrina*	P	pot	Oliveira et al. (2014)
Recife, Brazil	*Prosopis juliflora*	P	pot	Oliveira et al. (2014)
Sabah, Malaysia	*Dryobalanops lanceolata*	Combined	field	Bungard et al. (2002)
Sabah, Malaysia	*Shorea johorensis*	Combined	field	Bungard et al. (2002)
San Antonio, Costa Rica	*Cedrela odorata*	Combined	pot	Carswell et al. (2000)
Sao Paulo, Brazil	*Cariniana legalis*	Combined	pot	de Oliveira et al. (2012)
Sao Paulo, Brazil	*Croton urucurana*	Combined	pot	de Oliveira et al. (2012)
Twin Cays, Belize	*Rhizophora mangle*	N, P	field	Lovelock et al. (2006)
Xishuangbanna, China	*Bauhinia aurea*	N	pot	Cai et al. (2008)
Xishuangbanna, China	*Bauhinia claviflora*	N	pot	Cai et al. (2008)
Xishuangbanna, China	*Bauhinia monandra*	N	pot	Cai et al. (2008)
Xishuangbanna, China	*Bauhinia purpurea*	N	pot	Cai et al. (2008)
Xishuangbanna, China	*Bauhinia tenuiflora*	N	pot	Cai et al. (2008)

Reference (Oliveira et al. 2014)

related to photosynthetic rate, such as foliar nutrient concentration or specific leaf area. Whereas most of these studies used simple fertilizer additions, four created Hoagland or Ingestad solutions to match desired levels of nutrient availability (Thompson et al. 1988, 1992; Riddoch et al. 1991; Carswell et al. 2000), but in an attempt to be inclusive we considered all of these studies due to the limited availability of such information in the literature. Within Table 1, there is also variation in light availability among studies so we used the highest light availability if there were multiple levels, yet some studies (Lu et al. 2007; Pasquini and Santiago 2012), were done on understory plants in the field, so very low light conditions prevailed. Similarly, in studies where multiple levels of nutrient addition were performed, we utilized treatments with the highest level of nutrient addition except for the study of Lu et al. (2007), in which the goal of the study was to simulate N deposition to the point of toxicity. Here, we used the lowest level of fertilization, which was similar to the high nutrient treatments in other studies. If multiple time points were given in the paper, we used the latest time point to represent the values produced under a longer period of experimental treatment. Finally, we restricted our analysis to woody species and included lianas in addition to trees, thus three of the *Bauhinia* species in Table 1 are lianas (*B. claviflora, B. tenuiflora, B. aurea*).

Our data show that overall there have been enhancements of A_{max} with nutrient addition in tropical trees (Fig. 2). Of the studies on nutrient limitation of A_{max} in this synthesis, 22 included control and high N treatments, 21 included control and high P treatments, and 24 included control and combined elemental treatments of various kinds, for a total of 67 species × site combinations (Table 1). The raw data shows that mean A_{max} trended higher with N, P or combined nutrient addition (Fig. 2). However, to evaluate the overall effect of nutrient addition on photosynthesis, we used the *ln*-transformed response ratio as our main effect size metric: $RR_x = \ln (E/C)$, where E is A_{max} in the enriched or high nutrient treatment, C is A_{max} in the un-enriched control or low nutrient treatment (Elser et al. 2007). We then tested statistically whether RR_x for each treatment (RR_N, RR_P, $RR_{Combined}$), was significantly different from zero. In comparisons across multiple experimental systems where species, soils and environmental conditions vary considerably, the analysis of change relative to control is generally considered to be more meaningful than absolute differences between means (Elser et al. 2007).

Fig. 2 Mean values of maximum photosynthetic rate under control versus nutrient addition treatments for tropical tree species fertilized with N, P, or combined nutrient treatments

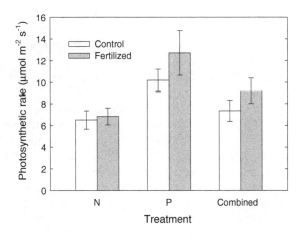

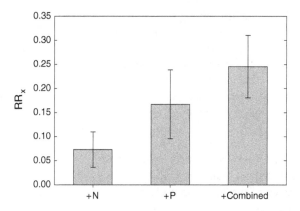

Fig. 3 Responses of area-based maximum photosynthetic rate to single enrichment of N, P or to combined elemental enrichment in tropical woody species. Data are natural-log transformed response ratios (RR_x) in which photosynthesis in the enriched treatment is divided by its value in the control treatment and then *ln*-transformed (Elser et al. 2007). For example, a value of 0.5 indicates a value in the manipulated treatment that is approximately 1.6 times its value in the control. *Sample sizes* +N (22), +P (21) and +Combined (24). *Error bars* indicate plus or minus one standard error

Our synthesis indicates that nutrient limitation of photosynthesis exists across the woody tropical species studied thus far (Fig. 3). However, the magnitude of the response varied strongly with the element in question. Across the 22 species × site combinations for N, photosynthesis trended higher with N addition, but mean RR_N was not significantly different from zero at the 0.05 significance level ($p = 0.061$ for t-test of $RR_N = 0$). Mean RR_P was significantly different from zero, indicating that photosynthesis significantly increased with P across the 21 species × site combinations ($p = 0.03$ for t-test of $RR_P = 0$). Thus our results across studies are consistent with the idea that P is the most widespread limiting nutrient in tropical forests. Whereas N produced substantial species-specific responses at some sites, with RR_N values up to 0.50 indicating an increase of approximately 1.6 times its value in the control, the magnitude of photosynthetic response to N was muted compared to P, which showed RR_P values as high as 0.90 indicating an increase of 2.45 times its value in the control treatment.

Our data also indicate that the 24 species × site combinations with combined manipulation of N, P, K or micronutrients produce the greatest response across all studies ($p = 0.0009$ for t-test of $RR_{Combined} = 0$; Fig. 3). $RR_{Combined}$ ranged as high as 0.92 indicating an increase of 2.51 times its value in the control. The highest significance in the case of combined nutrient addition leads to one of two possible conclusions. The first is that synergistic effects of multiple elemental manipulation has a stronger impact on alleviating nutrient limitation of photosynthesis than single element manipulation, consistent with the theory of multiple limiting nutrients (Kaspari et al. 2008). This theory suggests that plants allocate to uptake of limiting resources until all resources simultaneously limit a physiological process such that addition of any resource increases the rate of that process to the same degree (Bloom et al. 1985). Alternatively, the greatest response to combined nutrient addition could result from the fact that manipulating multiple elements leads to a greater chance that the single most limiting element is included in the study. Overall, the high values for $RR_{Combined}$ support the idea that the stoichiometric demands of fundamental biochemical processes enable co-limitation by multiple elements (Elser et al. 2007; Kaspari et al. 2008; Wright et al. 2011; Santiago et al. 2012).

Although our mean RR_x values were significantly different from zero or nearly so in all cases, a closer look at individual species × site combinations reveals that RR_x values were negative in 21 out of 67 cases. Negative RR_x values were fairly distributed across nutrient treatments, with eight negative RR_N values, six negative RR_P values, and seven negative $RR_{Combined}$ values. Furthermore, 11 out of the 21 negative RR_x values were from a single study, in which 11 out of the 12 RR_x values in the study were negative (Saraceno 2006). The study by Saraceno (2006) is in many ways unique because it was conducted in Brazilian cerrado sensu stricto vegetation, an open canopy savannah ecosystem, whereas most of the other studies include tree species of closed canopy rainforest ecosystems. Cerrado regions are known for nutrient poor soils that are acidic with a low cation exchange capacity and high concentrations of aluminum (Furley and Ratter 1988). Additionally, light availability is likely to be higher in cerrado because of the open canopy and rainfall is more seasonal than in many closed canopy rainforest ecosystems. Whereas most rainforest species increase photosynthetic rate with the addition of nutrients, cerrado trees nearly uniformly decrease photosynthetic rate (Saraceno 2006), and reduce both nighttime and daytime transpiration with the addition of nutrients (Bucci et al. 2006; Scholz et al. 2007).

Several studies have directly addressed the responses of tropical trees to elevated atmospheric CO_2 by experimentally combining elevated CO_2 treatments with nutrient addition. Application of soil fertilizer substantially accelerated biomass accumulation of two tropical tree species at elevated CO_2, but elevated CO_2 had no significant effect on biomass accumulation in unfertilized communities (Winter et al. 2001), providing support for nutrient limitation of the concentration-carbon feedback. However, relatively little data is available on photosynthetic responses to nutrients at high CO_2 (Fig. 4). In one study from southeast China, N fertilization did not modify the effects of CO_2 on photosynthesis (Liu et al. 2012). Yet, in two other studies from the neotropics, greater combined nutrient availability increased the magnitude of the response of photosynthesis to elevated CO_2 (Carswell et al. 2000; de Oliveira et al. 2012), providing support for nutrient limitation of the concentration-carbon feedback. The results of these studies are consistent with the conclusion that N limitation of photosynthesis in tropical trees is modest compared to P, and combined nutrient treatments produce the greatest responses (Fig. 4).

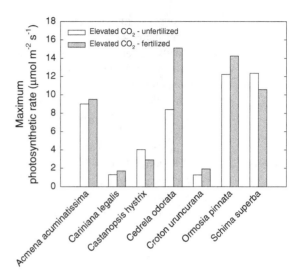

Fig. 4 Maximum photosynthetic rate at elevated CO_2 for tropical trees in control and fertilized treatments

Conclusions and Future Directions

We sought to determine whether nutrients limit the concentration-carbon feedback, the ability of tropical trees to increase rates of photosynthetic carbon gain as atmospheric CO_2 increases. Our evaluation of the available literature indicates several key patterns. First, nutrient limitation of photosynthesis is widespread in rainforest trees and appears to manifest itself differently, if at all, in savannah trees. Second, although photosynthesis is strongly limited by N in particular species, N limitation of photosynthesis is modest compared to P limitation of photosynthesis when considering all of the available literature. Finally, alleviation of nutrient limitation through addition of combined nutrient treatments produces the strongest photosynthetic responses, highlighting the potentially complex stroichiometric interactions among elements.

Currently, our predictions for how tropical forest ecosystems will respond to climate change and how their responses will feedback to the global climate system are limited by model realism. Several models still represent photosynthesis as continuously increasing in proportion to atmospheric CO_2 (Bonan and Levis 2010), illustrating the need for improved representation of photosynthesis in Earth system models (Ali et al. 2015). To build on the current state of knowledge regarding nutrient limitation of photosynthesis in tropical trees, an expansion of approaches will be necessary. First, the focus on N and P needs to be broadened to include other potentially limiting elements. Only one long-term tropical fertilization study has added K in a factorial design, but it reported substantial K limitation of key ecosystem processes (Wright et al. 2011). Also, given the diversity of tropical soils, one wonders if there are other elements beyond N, P, and K that deserve focused study. Second, better correspondence between large-scale field experiments and key greenhouse experiments designed to test specific mechanisms are needed to pinpoint physiological responses and refine their role in processes at larger scales. Finally, a greater distinction between nutrient limitation of photosynthetic productivity between early and late successional forests is needed to elucidate the potential carbon sequestration of secondary forest during regeneration. Secondary forests are becoming a larger proportion of the total tropical land area, and their rapid growth during early succession has been implicated as a potential major route for carbon absorption by terrestrial vegetation (Cernusak et al. 2013). This potential must be determined to improve our predictive power. Overall, understanding the potential and limits to the key carbon cycling processes of tropical trees will improve our ability to forecast global responses and feedbacks to climate change.

Acknowledgments We gratefully acknowledge Alex Barron, Sandra Bucci, Walt Carson, Sarah Pasquini, Jennifer Powers, Fabian Scholz, Jim Sickman, Ben Turner, Peter Vitousek, Klaus Winter, Joe Wright, and Nina Würzburger for insightful discussions; Xiankai Lu and Keith Bloomfield for contributing raw data; the Smithsonian Tropical Research Institute and the University of California, Riverside, for logistical support; and a University of California Faculty Fellowship to Santiago.

References

Aerts R, Chapin FS (2000) The mineral nutrition of wild plants revisited: a re-evaluation of processes and patterns. Adv Ecol Res 30:1–66

Ali AA, Xu C, Rogers A, McDowell NG, E. MB, Fisher R, Wullschleger SD, Reich PB, Vrugt JA, Baurle WL, Santiago LS, Wilson CJ (2015) Global scale environmental control of plant photosynthetic capacity. Ecol Appl 25:2349–2365

Álvarez-Clare S, Mack MC, Brooks M (2013) A direct test of nitrogen and phosphorus limitation to net primary productivity in a lowland tropical wet forest. Ecology 94:1540–1551

Andersen KM, Corre MD, Turner BL, Dalling JW (2010) Plant-soil associations in a lower montane tropical forest: physiological acclimation and herbivore-mediated responses to nitrogen addition. Funct Ecol 24:1171–1180

Baltzer JL, Davies SJ, Bunyavejchewin S, Noor NSM (2008) The role of desiccation tolerance in determining tree species distributions along the Malay-Thai Peninsula. Funct Ecol 22:221–231

Barberis IM, Tanner EVJ (2005) Gaps and root trenching increase tree seedling growth in Panamanian semi-evergreen forest. Ecology 86:667–674

Beer C, Reichstein M, Tomelleri E, Ciais P, Jung M, Carvalhais N, Rodenbeck C, Arain MA, Baldocchi D, Bonan GB, Bondeau A, Cescatti A, Lasslop G, Lindroth A, Lomas M, Luyssaert S, Margolis H, Oleson KW, Roupsard O, Veenendaal E, Viovy N, Williams C, Woodward FI, Papale D (2010) Terrestrial gross carbon dioxide uptake: global distribution and covariation with climate. Science 329:834–838

Bloom AJ, Chapin FS III, Mooney HA (1985) Resource limitation in plants-an economic analogy. Annu Rev Ecol Syst 16:363–392

Bloomfield KJ, Farquhar GD, Lloyd J (2014) Photosynthesis-nitrogen relationships in tropical forest tree species as affected by soil phosphorus availability: a controlled environment study. Funct Plant Biol 41:820–832

Bonan GB, Levis S (2010) Quantifying carbon-nitrogen feedbacks in the Community Land Model (CLM4). Geophys Res Lett 37(7):L07401

Bucci SJ, Scholz FG, Goldstein G, Meinzer FC, Franco AC, Campanello PI, Villalobos-Vega R, Bustamante M, Miralles-Wilhelm F (2006) Nutrient availability constrains the hydraulic architecture and water relations of savannah trees. Plant Cell Environ 29:2153–2167

Bungard RA, Zipperlen SA, Press MC, Scholes JD (2002) The influence of nutrients on growth and photosynthesis of seedlings of two rainforest dipterocarp species. Funct Plant Biol 29:505–515

Burslem DFRP (1996) Differential responses to nutrients, shade and drought among tree seedlings of lowland tropical forest in Singapore. In: Swaine MD (ed) The ecology of tropical forest tree seedlings. UNESCO, Paris, pp 211–244

Cai ZQ, Poorter L, Han Q, Bongers F (2008) Effects of light and nutrients on seedlings of tropical *Bauhinia* lianas and trees. Tree Physiol 28:1277–1285

Campo J, Dirzo R (2003) Leaf quality and herbivory responses to soil nutrient addition in secondary tropical dry forests of Yucatan, Mexico. J Trop Ecol 19:525–530

Campo J, Vázquez-Yanes C (2004) Effects of nutrient limitation on aboveground carbon dynamics during tropical dry forest regeneration in Yucatán, Mexico. Ecosystems 7:311–319

Carswell FE, Grace J, Lucas ME, Jarvis PG (2000) Interaction of nutrient limitation and elevated CO_2 concentration on carbon assimilation of a tropical tree seedling (*Cedrela odorata*). Tree Physiol 20:977–986

Cernusak LA, Aranda J, Marshall JD, Winter K (2007a) Large variation in whole-plant water-use efficiency among tropical tree species. New Phytol 173:294–305

Cernusak LA, Winter K, Aranda J, Turner BL, Marshall JD (2007b) Transpiration efficiency of a tropical pioneer tree (*Ficus insipida*) in relation to soil fertility. J Exp Bot 58:3549–3566

Cernusak LA, Winter K, Turner BL (2009) Physiological and isotopic ($\delta^{13}C$ and $\delta^{18}O$) responses of three tropical tree species to water and nutrient availability. Plant Cell Environ 32:1441–1455

Cernusak LA, Winter K, Dalling JW, Holtum JAM, Jaramillo C, Korner C, Leakey ADB, Norby RJ, Poulter B, Turner BL, Wright SJ (2013) Tropical forest responses to increasing atmospheric CO_2: current knowledge and opportunities for future research. Funct Plant Biol 40:531–551

Cordell S, Goldstein G, Meinzer FC, Vitousek PM (2001) Regulation of leaf life-span and nutrient-use efficiency of *Metrosideros polymorpha* trees at two extremes of a long chronosequence in Hawaii. Oecologia 127:198–206

Cramer MD, Hoffmann V, Verboom GA (2008) Nutrient availability moderates transpiration in *Ehrharta calycina*. New Phytol 179:1048–1057

Cramer MD, Hawkins HJ, Verboom GA (2009) The importance of nutritional regulation of plant water flux. Oecologia 161:15–24

Davidson EA, de Carvalho CJR, Vieira ICG, Figueiredo RD, Moutinho P, Ishida FY, dos Santos MTP, Guerrero JB, Kalif K, Saba RT (2004) Nitrogen and phosphorus limitation of biomass growth in a tropical secondary forest. Ecol Appl 14:S150–S163

de Oliveira EAD, Approbato AU, Legracie JR, Martinez CA (2012) Soil-nutrient availability modifies the response of young pioneer and late successional trees to elevated carbon dioxide in a Brazilian tropical environment. Environ Exp Bot 77:53–62

Delucia EH, Sasek TW, Strain BR (1985) Photosynthetic inhibition after long-term exposure to elevated levels of atmospheric carbon dioxide. Photosynth Res 7:175–184

Elser JJ, Bracken MES, Cleland EE, Gruner DS, Harpole WS, Hillebrand H, Ngai JT, Seabloom EW, Shurin JB, Smith JE (2007) Global analysis of nitrogen and phosphorus limitation of primary producers in freshwater, marine and terrestrial ecosystems. Ecol Lett 10:1135–1142

Evans JR (1989) Photosynthesis and nitrogen relationships in leaves of C_3 plants. Oecologia 78:9–19

Evans HJ, Sorger GJ (1966) Role of mineral elements with emphasis on the univalent cations. Ann Rev Plant Physiol 17:47–76

Farquhar GD, Richards RA (1984) Isotopic composition of plant carbon correlates with water-use efficiency of wheat genotypes. Aust J Plant Physiol 11:539–552

Field C, Mooney HA (1986) The photosynthesis-nitrogen relationship in wild plants. In: Givnish TJ (ed) On the economy of plant form and function. University Press, Cambridge, pp 25–55

Fisher JB, Malhi Y, Torres IC, Metcalfe DB, van de Weg MJ, Meir P, Silva-Espejo JE, Huasco WH (2013) Nutrient limitation in rainforests and cloud forests along a 3,000-m elevation gradient in the Peruvian Andes. Oecologia 172:889–902

Friedlingstein P, Cox P, Betts R, Bopp L, Von Bloh W, Brovkin V, Cadule P, Doney S, Eby M, Fung I, Bala G, John J, Jones C, Joos F, Kato T, Kawamiya M, Knorr W, Lindsay K, Matthews HD, Raddatz T, Rayner P, Reick C, Roeckner E, Schnitzler KG, Schnur R, Strassmann K, Weaver AJ, Yoshikawa C, Zeng N (2006) Climate-carbon cycle feedback analysis: results from the C^4MIP model intercomparison. J Clim 19:3337–3353

Furley PA, Ratter JA (1988) Soil resources and plant communities of the Central Brazilian cerrado and their development. J Biogeogr 15:97–108

Fyllas NM, Patiño S, Baker TR, Nardoto GB, Martinelli LA, Quesada CA, Paiva R, Schwarz M, Horna V, Mercado LM, Santos A, Arroyo L, Jiménez EM, Luizão FJ, Neill DA, Silva N, Prieto A, Rudas A, Silviera M, Vieira ICG, Lopez-Gonzalez G, Malhi Y, Phillips OL, Lloyd J (2009) Basin-wide variations in foliar properties of Amazonian forest: phylogeny, soils and climate. Biogeosciences 6:2677–2708

Gunatilleke CVS, Gunatilleke IAUN, Perera GAD, Burslem DFRP, Ashton PMS, Ashton PS (1997) Responses to nutrient addition among seedlings of eight closely related species of *Shorea* in Sri Lanka. J Ecol 85:301–311

Haridasan M (1992) Observations on soils, foliar nutrient concentrations and floristic composition of cerrado sensu stricto and cerradão communities in central Brazil. In: Pa F, Ratter J (eds) Nature and dynamics of forest-savanna boundaries. Chapman & Hall, London, pp 171–184

Kaspari M, Garcia MN, Harms KE, Santana M, Wright SJ, Yavitt JB (2008) Multiple nutrients limit litterfall and decomposition in a tropical forest. Ecol Lett 11:35–43

Kitayama K, Pattison R, Cordell S, Webb D, Mueller-Dombois D (1997) Ecological and genetic implications of foliar polymorphism in *Metrosideros polymorpha* Gaud. (Myrtaceae) in a habitat matrix on Mauna Loa. Hawaii. Annals Botany 80:491–497

Lawrence D (2003) The response of tropical tree seedlings to nutrient supply: meta-analysis for understanding a changing tropical landscape. J Trop Ecol 19:239–250

Lewis SL, Tanner EVJ (2000) Effects of above- and belowground competition on growth and survival of rain forest tree seedlings. Ecology 81:2525–2538

Liu JX, Zhang DQ, Zhou GY, Duan HL (2012) Changes in leaf nutrient traits and photosynthesis of four tree species: effects of elevated CO2, N fertilization and canopy positions. J Plant Ecol 5:376–390

Lovelock CE, Ball MC, Choat B, Engelbrecht BMJ, Holbrook NM, Feller IC (2006) Linking physiological processes with mangrove forest structure: phosphorus deficiency limits canopy development, hydraulic conductivity and photosynthetic carbon gain in dwarf *Rhizophora mangle*. Plant Cell Environ 29:793–802

Lu X, Mo J, Li D, Zhang W, Fang Y (2007) Effects of simulated N deposition on the photosynthetic and physiologic characteristics of dominant understorey plants in Dinghushan Mountain of subtropical China. J Beijing For Univ 29:1–9

Lu X, Mo J, Gilliam FS, Zhou G, Fang Y (2010) Effects of experimental nitrogen additions on plant diversity in an old-growth tropical forest. Glob Change Biol 16:2688–2700

Marschner H (1995) Mineral nutrition in higher plants. Academic Press, London

Mirmanto E, Proctor J, Green J, Nagy L, Suriantata (1999) Effects of nitrogen and phosphorus fertilization in a lowland evergreen rainforest. Philos Trans R Soc Lond Ser B-Biol Sci 354:1825–1829

Morgan JM (1984) Osmoregulation and water-stress in higher-plants. Ann Rev Plant Physiol Plant Mol Biol 35:299–319

Newbery DM, Chuyong GB, Green JJ, Songwe NC, Tchuenteu F, Zimmermann L (2002) Does low phosphorus supply limit seedling establishment and tree growth in groves of ectomyc-orrhizal trees in a central African rainforest? New Phytol 156:297–311

Oliveira MT, Medeiros CD, Frosi G, Santos MG (2014) Different mechanisms drive the performance of native and invasive woody species in response to leaf phosphorus. supply during periods of drought stress and recovery. Plant Physiol Biochem 82:66–75

Ostertag R (2009) Foliar phosphorus accumulation in relation to leaf traits: an example in a tropical wet forest in Hawaii. S Afr J Bot 75:415–415

Pasquini SC, Santiago LS (2012) Nutrients limit photosynthesis in seedlings of a lowland tropical forest tree species. Oecologia 168:311–319

Phillips OL, van der Heijden G, Lewis SL, Lopez-Gonzalez G, Aragao L, Lloyd J, Malhi Y, Monteagudo A, Almeida S, Davila EA, Amaral I, Andelman S, Andrade A, Arroyo L, Aymard G, Baker TR, Blanc L, Bonal D, de Oliveira ACA, Chao KJ, Cardozo ND, da Costa L, Feldpausch TR, Fisher JB, Fyllas NM, Freitas MA, Galbraith D, Gloor E, Higuchi N, Honorio E, Jimenez E, Keeling H, Killeen TJ, Lovett JC, Meir P, Mendoza C, Morel A, Vargas PN, Patino S, Peh KSH, Cruz AP, Prieto A, Quesada CA, Ramirez F, Ramirez H, Rudas A, Salamao R, Schwarz M, Silva J, Silveira M, Slik JWF, Sonke B, Thomas AS, Stropp J, Taplin JRD, Vasquez R, Vilanova E (2010) Drought-mortality relationships for tropical forests. New Phytol 187:631–646

Poorter H, Remkes C (1990) Leaf area ratio and net assimilation rate of 24 wild species differing in relative growth rate. Oecologia 83:553–559

Quesada CA, Lloyd J, Schwarz M, Patino S, Baker TR, Czimczik C, Fyllas NM, Martinelli L, Nardoto GB, Schmerler J, Santos AJB, Hodnett MG, Herrera R, Luizao FJ, Arneth A, Lloyd G, Dezzeo N, Hilke I, Kuhlmann I, Raessler M, Brand WA, Geilmann H, Moraes JO, Carvalho FP, Araujo RN, Chaves JE, Cruz OF, Pimentel TP, Paiva R (2010) Variations in chemical and physical properties of Amazon forest soils in relation to their genesis. Biogeosciences 7:1515–1541

Raven JA, Handley LL, Wollenweber B (2004) Plant nutrition and water use efficiency. In: Bacon MA (ed) Water use efficiency in plant biology. CRC Press, Boca Raton, pp 171–197

Riddoch I, Lehto T, Grace J (1991) Photosynthesis of tropical tree seedlings in relation to light and nutrient supply. New Phytol 119:137–147

Santiago LS (2015) Nutrient limitation of eco-physiological processes in tropical trees. Trees 29:1291–1300

Santiago LS, Mulkey SS (2005) Leaf productivity along a precipitation gradient in lowland Panama: patterns from leaf to ecosystem. Trees 19:349–356

Santiago LS, Wright SJ (2007) Leaf functional traits of tropical forest plants in relation to growth form. Funct Ecol 21:19–27

Santiago LS, Kitajima K, Wright SJ, Mulkey SS (2004) Coordinated changes in photosynthesis, water relations and leaf nutritional traits of canopy trees along a precipitation gradient in lowland tropical forest. Oecologia 139:495–502

Santiago LS, Wright SJ, Harms KE, Yavitt JB, Korine C, Garcia MN, Turner BL (2012) Tropical tree seedling growth responses to nitrogen, phosphorus and potassium addition. J Ecol 100:309–316

Saraceno MIS (2006) Efeitos da fertilização a longo prazo no metabolism fotossintético, nas características foliares e no crescimento em árvores do cerrado. In: Departamento de Ecologia, vol. Mestre em Ecologia. Universidade de Brasília, Instituto de Ciências Biológicas, Brasília, Brazil, p 65

Sayer EJ, Tanner EVJ (2010) Experimental investigation of the importance of litterfall in lowland semi-evergreen tropical forest nutrient cycling. J Ecol 98:1052–1062

Scholz FG, Bucci SJ, Goldstein G, Meinzer FC, Franco AC, Miralles-Wilhelm F (2007) Removal of nutrient limitations by long-term fertilization decreases nocturnal water loss in savanna trees. Tree Physiol 27:551–559

Schreeg LA, Mack MC, Turner BL (2013) Nutrient-specific solubility patterns of leaf litter across 41 lowland tropical woody species. Ecology 94:94–105

Schreeg LA, Santiago LS, Wright SJ, Turner BL (2014) Stem, root, and older leaf N: P ratios are more responsive indicators of soil nutrient availability than new foliage. Ecology 95:2062–2068

Sinclair TR, Vadez V (2002) Physiological traits for crop yield improvement in low N and P environments. Plant Soil 245:1–15

Tanner EVJ, Kapos V, Franco W (1992) Nitrogen and phosphorus fertilization effects on Venezuelan montane forest trunk growth and litterfall. Ecology 73:78–86

Thompson WA, Stocker GC, Kriedemann PE (1988) Growth and photosynthetic response to light and nutrients of *Flindersia brayleyana* F Muell a rainforest tree with broad tolerance to sun and shade. Aust J Plant Physiol 15:299–315

Thompson WA, Huang LK, Kriedemann PE (1992) Photosynthetic response to light and nutrients in sun-tolerant and shade-tolerant rainforest trees. II. Leaf gas exchange and component processes of photosynthesis. Aust J Plant Physiol 19:19–42

Vitousek PM (2004) Nutrient cycling and limitation. Princeton University Press, Princeton

Vitousek PM, Aplet GH, Turner DR, Lockwood JJ (1992) The Mauna Loa environmental matrix: foliar and soil nutrients. Oecologia 89:372–382

Vitousek PM, Walker LR, Whiteaker LD, Matson PA (1993) Nutrient limitations to plant growth during primary succession in Hawaii Volcanoes National Park. Biogeochemistry 23:197–215

Winter K, Garcia M, Gottsberger R, Popp M (2001) Marked growth response of communities of two tropical tree species to elevated CO2 when soil nutrient limitation is removed. Flora 196:47–58

Wright IJ, Reich PB, Westoby M, Ackerly DD, Baruch Z, Bongers F, Cavender-Bares J, Chapin T, Cornelissen JHC, Diemer M, Flexas J, Garnier E, Groom PK, Gulias J, Hikosaka K, Lamont BB, Lee T, Lee W, Lusk C, Midgley JJ, Navas M-L, Niinemets Ü, Oleksyn J, Osada N, Poorter H, Poot P, Prior L, Pyankov V, Roumet C, Thomas SC, Tjoelker MG, Veneklaas EJ, Villar R (2004) The worldwide leaf economics spectrum. Nature 428:821–827

Wright SJ, Yavitt JB, Wurzburger N, Turner BL, Tanner EVJ, Sayer EJ, Santiago LS, Kaspari M, Hedin LO, Harms KE, Garcia MN, Corre MD (2011) Potassium, phosphorus, or nitrogen limit root allocation, tree growth, or litter production in a lowland tropical forest. Ecology 92:1616–1625

Wu P-F, Ma X-Q, Tigabu M, Huang Y, Zhou L-L, Cai L, Hou X-L, Oden PC (2014) Comparative growth, dry matter accumulation and photosynthetic rate of seven species of Eucalypt in response to phosphorus supply. J For Res 25:377–383

Wullschleger SD (1993) Biochemical limitations to carbon assimilation in C$_3$ plants—a retrospective analysis of the A/C$_i$ curves from 109 species. J Exp Bot 44:907–920

Yavitt JB, Wright SJ (2008) Seedling growth responses to water and nutrient augmentation in the understorey of a lowland moist forest, Panama. J Trop Ecol 24:19–26

Zhu F, Lu X, Mo J (2014) Phosphorus limitation on photosynthesis of two dominant understory species in a lowland tropical forest. J Plant Ecol 7:526–534

Part V
Carbon Economy and Allocation of Tropical Trees and Forests

Facing Shortage or Excessive Light: How Tropical and Subtropical Trees Adjust Their Photosynthetic Behavior and Life History Traits to a Dynamic Forest Environment

Guillermo Goldstein, Louis S. Santiago, Paula I. Campanello,
Gerardo Avalos, Yong-Jiang Zhang and Mariana Villagra

Abstract Light is critical for plant establishment, growth, and survival in wet tropical forests. The objective of this chapter is to analyze paradigms of photosynthetic performance and life history traits of tropical forest trees to contrasting light environments across the forest floor, gaps and upper canopy. Physiological and morphological plasticity as well as genetically fixed adaptive traits are analyzed, including leaf optical properties and photoprotection from high irradiance. Photosynthetic adaptations to contrasting light environments of closely related species are discussed. This approach has the advantage among comparative studies

G. Goldstein (✉)
Laboratorio de Ecología Funcional, Departamento de Ecología Genética y Evolución,
Instituto IEGEBA (CONICET-UBA), Facultad de Ciencias Exactas y naturales,
Universidad de Buenos Aires, Buenos Aires, Argentina
e-mail: goldstein@ege.fcen.uba.ar

G. Goldstein
Department of Biology, University of Miami, Coral Gables, FL 33146, USA

L.S. Santiago
Department of Botany & Plant Sciences, University of California,
2150 Batchelor Hall, Riverside, CA 92521, USA
e-mail: santiago@ucr.edu

L.S. Santiago
Smithsonian Tropical Research Institute, Apartado Postal, 0843-03092 Balboa,
Republic of Panama

P.I. Campanello · M. Villagra
Laboratorio de Ecología Forestal y Ecofisiología, Instituto de Biología Subtropical,
CONICET, FCF, Universidad Nacional de Misiones, Posadas, Argentina
e-mail: pcampanello@yahoo.com

M. Villagra
e-mail: marian.villagra@gmail.com

© Springer International Publishing Switzerland 2016
G. Goldstein and L.S. Santiago (eds.), *Tropical Tree Physiology*,
Tree Physiology 6, DOI 10.1007/978-3-319-27422-5_15

of adaptations across species in that genetic relationships among species are known. Species-specific variations in maximum photosynthetic rates, which reflect the degree of adaptation to growth irradiance, are shown to be gradual, suggesting that classification into two distinct functional groups in terms of light requirements is somewhat arbitrary. Trees growing in gaps or in the upper canopy rely strongly on biochemical mechanisms to dissipate excess energy and to avoid damage to the light reaction centers and photosystems. Consistent with their high photosynthetic capacity, light demanding species are capable of plastic changes in hydraulic architecture, such as increases in hydraulic conductivity under high irradiance, which makes them more competitive in open habitats.

Keywords Electron transport rate · Hydraulic architecture · Leaf pigments · Photoinhibition · Photosynthesis · Plasticity · Shade tolerant trees

Introduction

Light is one of the most important environmental determinants of plant establishment, growth, and survival in wet tropical and subtropical forests (Osunkoya and Ash 1991; Nicotra et al. 1999). Light varies spatially and temporally in these forests and consequently trees have developed various adaptations to cope with dynamic light regimes. Canopy trees are exposed to high irradiance, evaporative demand and temperature, which can induce nonstomatal limitations to photosynthesis through the effects of deactivation of photosynthetic enzymes and the reduction of photochemistry (Krause et al. 1995; Ishida et al. 1999; Demmig-Adams and Adams 2006). In the understory of tropical rainforests, light is considered the most limiting resource, with understory plants often receiving less than 1 % of the ambient light that reaches the canopy (Björkman and Ludlow 1972; Chazdon and Fetcher 1984). Forest gaps have a different light regime compared to the upper canopy and the forest understory (Canham et al. 1990). At a regional scale, a lack of solar radiation in tropical and subtropical forests during the rainy season appears to limit plant

G. Avalos
Escuela de Biología, Universidad de Costa Rica, San Pedro, San José 11501-2060,
Costa Rica
e-mail: gerardo.avalos@ucr.ac.cr

G. Avalos
The School for Field Studies, Center for Sustainable Development Studies,
100 Cummings Center, Suite 534-G, Beverly, MA 01915-6239, USA

Y.-J. Zhang
Department of Organismic and Evolutionary Biology, 16 Divinity Avenue,
Cambridge, MA 02138, USA
e-mail: yongjiangzhang@oeb.harvard.edu

carbon assimilation and growth due to heavy cloud cover associated with the Intertropical Convergence Zone (Wright and vanSchaik 1994; Graham et al. 2003).

Light in the forest understory consists of diffuse radiation that is randomly interspersed by short duration light flecks (Chazdon and Fetcher 1984; Canham et al. 1990), and depends on canopy cover at any point in the forest (Fig. 1). Light flecks occur when movement of leaves in the canopy or changes in the angle of the sun allow direct light penetration for intermittent periods of time (Lüttge 2008). The diffuse light in the understory has different spectral characteristics compared to upper canopy light because certain wavebands are differentially absorbed as light passes through leaves within the canopy. The red-to-far red ratio (R:FR) ranges from 1.10 in full sun to as low as 0.10 under an intact forest canopy (Lee 1987). Such changes in R:FR have profound effects on plant development and growth (e.g. Ballaré et al. 1997). Under strongly light-limiting conditions of the forest understory, plants are expected to maximize light capture in the most efficient way by obtaining light resources at the lowest costs in terms of construction and maintenance (Givnish 1988).

Tropical trees are often classified in terms of succession and light requirements as early-successional, light demanding, fast growing, pioneer species at one side of the spectrum and late successional, shade tolerant, slow growing species at the other. During the establishment stage, light requiring species tend to grow in gaps under relatively high irradiance levels, whereas shade tolerant species can become established in the shaded forest understory. When trees reach the canopy, they are exposed to environmental conditions very different from where they spend the first part of their life cycles. Light requiring and shade tolerant species, however, may not be two completely distinct functional groups (Ellis et al. 2000; Gilbert et al. 2006). The objective of this chapter is to analyze paradigms regarding physiological plasticity and adaptation of tropical forest trees to the contrasting light environments

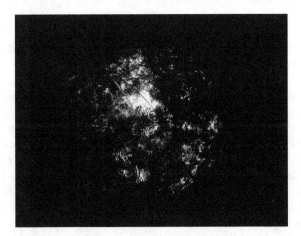

Fig. 1 Vertical fish-eye camera view from the forest floor to the canopy of a lowland tropical forest in Gigante, Panama (*Photograph* Sarah Pasquini). The picture shows the low light levels that plants experience on the forest floor and shows the small openings of the canopy that allow direct solar radiation to reach the understory of the forest for short periods of time (*light flecks*)

provided by the forest understory, treefall gaps, and the upper canopy. A substantial improvement in the understanding of light responses from tropical trees during the last two decades, particularly in high light environments, is in part associated with accessing the upper canopy, often 30–60 m above the forest floor. The first construction tower crane to gain access to the canopy was established in a seasonal dry tropical forest in Panama in 1990. Since then, other cranes have been established in a variety of tropical forests around the world.

The Paradigm of Forest Gap Dynamics

Regeneration of trees in rainforests tends to occur when treefall gaps initiate a canopy opening. These gaps, where most of the incoming light is direct solar radiation, are filled with previously suppressed or newly arriving seedlings and saplings. This process may have important effects on forest structure, composition, and dynamics (Whitmore 1975; Hartshorn 1978; Runkle 1981; Brokaw 1985). In fact, mature forests have been described as shifting mosaics of different-sized patches at contrasting stages of gap closure (Watt 1947; Whitmore 1975; Oldeman 1978; Bormann and Likens 1979). In addition to higher light availability, tropical forest treefall gaps are also less humid, hotter, and usually have greater soil nutrient and water availability than surrounding forest understory (Vitousek and Denslow 1986; Pearcy 1987; Meinzer et al. 1995). Gap formation has been suggested to play a major role in maintaining tree diversity in species-rich tropical and subtropical forests because it provides habitat heterogeneity that can be partitioned by species with contrasting light demands (Denslow 1980). However, Hubbell et al. (1999) found that species richness of seedlings was greater in gaps than in forest understory, but species richness per stem was similar in gaps and in non-gap sites suggesting that gap phase regeneration does not fully explain species richness of the forest. Species richness is divided by the number of stems because stem density in gaps is substantially higher than stem density in the forest understory.

Some plant species can be used as "biological clocks" to estimate the temporal dynamics of gap openings (Zalamea et al. 2012). This is important because treefall gaps tend to close fast and a few years after gap formation it can be difficult to recognize where and when the gap opening occurred. *Astrocaryum mexicanun* is an understory palm species that endures tree falls when gaps are formed. The palm bends under falling trees, but rapidly recovers vertical growth. Based on the age-constant rate of stem elongation, the vertical stems of the palm were used to determine the age of the gap opening (Martinez-Ramos et al. 1988). Using this morphological feature of the palm, it was estimated that the turnover rate of gap openings in a Mexican wet tropical forest is about 47 years, providing evidence that gap openings and thus the release of suppressed seedling and saplings of forest trees occur frequently in time, and spatially at short distances within the forest. Species of the genus *Cecropia* can also be used to date previous disturbances or gap formation in tropical forest because the scars of fallen leaves, inflorescences and branches remain visible along the trunk. The high annual periodicity in branching and alternation of long and short nodes make it possible to distinguish fast and slow growth periods, allowing the gap history of particular sites to be quantified (Zalamea et al. 2008).

In tropical and subtropical forests, growth of particular plant groups can alter the gap phase regeneration process. For example, in some disturbed forest, native bamboos that are present in the understory can colonize treefall gaps and inhibit tree regeneration, preventing recovery of disturbed forests (Campanello et al. 2007). There is also evidence that lianas can send treefall gaps on an alternate successional pathway, whereby gap-phase regeneration is dominated by lianas and stalled in a low leaf area index canopy state for many years (Schnitzer et al. 2000).

Developmental Plasticity of Physiological and Morphological Traits Under Contrasting Light Regimes

Tropical and subtropical plants or plant parts, such as leaves, have the ability to adjust their morphology and physiology to a particular light environment. For example, there are substantial differences between sun exposed and shade exposed leaves, and it is well documented that leaves developed in the sun are smaller, thicker, have more layers of palisade cells, and have higher photosynthetic and respiration rates than shade leaves (Pearcy 1987; Vogelmann et al. 1996). Optical properties such as reflectance, absorptance, transmittance, and pigment composition of leaves also change depending on the growing light environment (Poorter et al. 2000). At the crown level, crown architecture, including branching patterns and degree of self-shading, as well as allometry, including key ratios such as height-to-diameter, and supporting-to-non-supporting tissue mass, can also change depending on light regime (Hallé et al. 1978; Kohyama and Hotta 1990; King 1991; Kitajima 1996; Valladares 1999; Valladares et al. 2000).

Tropical and subtropical tree species have developed species-specific morphological and physiological characteristics enabling them to optimize the capture of scarce solar radiation in the dynamic light environment of wet forests. Plants growing in the forest understory assimilate up to 80 % of their carbon uptake during short duration high intensity light flecks (Chazdon 1988). Responses to low light therefore involve rapid photosynthetic induction in response to sudden increases in light (Chazdon and Pearcy 1986; Montgomery and Givnish 2008). Maximizing carbon gain during light flecks depends on the photosynthetic induction response once a leaf is illuminated and on the ability to continue photosynthetic activity immediately after a light fleck strikes a leaf (post-illumination CO_2 assimilation) (Sharkey et al. 1986; Valladares et al. 1997; Montgomery and Givnish 2008; Santiago and Dawson 2014). Maximizing carbon gain also depends on the ability to maintain induction during diffuse-light periods, which is affected by light regime and overhead canopy structure. Frequent light flecks will enhance carbon gain because photosynthetic induction is a function of Calvin cycle enzymes and build-up of metabolite pools (Sassenrath-Cole et al. 1994).

Drastic changes in solar radiation produced by a gap opening in a forest can subject understory plants to substantial changes in physiological performance. The success of the trees under the new light conditions will depend on their phenotypic plasticity. Light requiring species grow fast and use substantial amounts of light resources before the gap is closed. Therefore these tree species depend on physiological and morphological attributes enhancing light capture and carbon assimilation. Phenotypic plasticity varies greatly among different species. Generally, high light requiring trees exhibit larger differences in photosynthetic rates and specific leaf area between high irradiance and low irradiance growing conditions compared to shade tolerant species (Fig. 2). In this figure *B. densiflora* and *C. fissillis* are light demanding while *L. leucanthus* and *B. ridelianum* are more shade tolerant species.

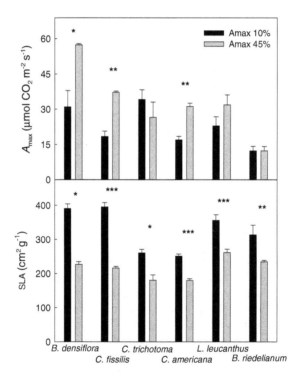

Fig. 2 Maximum assimilation rate (A_{max}, n = 3) in the *upper panel* and specific leaf area (SLA, n = 6–10 depending on species) in the *bottom panel* for saplings of six tree species, differing in their tolerance to the shade, growing under 10 and 45 % growth irradiances (*black* and *gray bars*, respectively). *Bars* indicate the mean + SE. Values of A_{max} were obtained from light response curves with an infrared gas analyzer (CIRAS I, PPSystems). T-tests were performed to assess differences between growth irradiances within each species (in the case of A_{max} values obtained from an asymptotic exponential regression of light response curves were compared). Significant differences at *$P < 0.05$ %, ** $P < 0.01$ *, and *** $P < 0.005$ are indicated. From Campanello et al. (2011)

Are Shade Tolerant and Light Demanding Tree Species Two Distinct Functional Groups?

In contrast to acclimation, which occurs within the lifetime of one individual, adaptation represents gradual genetic change over several generations that result in new fixed characteristics (Lambers et al. 1998). With regard to photosynthetic adaptations to light environments, the traits that characterize adaptation to high or low light tend to be similar to the traits that characterize acclimation to high or low light described above. Thus tropical and subtropical tree species adapted to high light tend to exhibit relatively small and thick leaves with high rates of photosynthesis and respiration, whereas trees adapted to low light tend to exhibit opposite traits. The capacity to survive long periods under deep shade involves trade-offs in which increased physiological performance under shade results in a pattern of resource use that favors growth and survival under limited light, but at the cost of partially restricting capacity to respond to sudden light increases, and specifically, to grow and survive under high light conditions (Kitajima 1994; Wright et al. 2010). Many types of tropical trees, including light demanding species, can spend relatively long periods of time under shade, but to complete their regeneration trajectory they must survive the lack of light on the forest floor, at least temporarily. Most pioneer tree species can only regenerate in open conditions, such as large gaps and forest edges, but many can survive variable periods under deep shade (Popma and Bongers 1988; Clark and Clark 1992). Long-lived pioneers are finally excluded from the gap phase regeneration process due to shade conditions prevalent in late successional stages, and are replaced by long-lived tree species that started regeneration as shade-tolerants (Gómez-Pompa and Vázquez-Yanes 1981).

Regardless of the physiological and morphological adjustments that the leaves of each species experience under different light environments, species-specific differences can be observed even when they are growing under similar environmental conditions. Maximum photosynthetic rates of leaves exposed to full sun in the canopy can reflect their light requirements because species-specific differences in photosynthetic performance are maintained even if measurements are done under the same environmental conditions. Figure 3 depicts photosynthetic rates of twenty one species of tropical trees accessible from the canopy crane in San Lorenzo, Panama (Santiago and Wright 2007), ranked in order of maximum mass-based photosynthetic rate and classified into shade tolerant or light demanding species according to Santiago (2010). The data show that classification into shade tolerant and light demanding species approximates the order of rates, with only one shade tolerant species having a rate greater than any light demanding species (Fig. 3). Additionally, across all species, light demanding species as a group had significantly greater maximum rates of photosynthesis than shade tolerant species ($t = 4.90$; $P < 0.0001$). Yet, a close view of the differences among species also demonstrates that species fall along a continuum, and except for the much higher rates in *Ochroma* and *Cecropia*, there are no major jumps in the rates across species. Similarly, among trees in a subtropical forest in Argentina, species with higher light requirements also tend to have elevated maximum electron transport

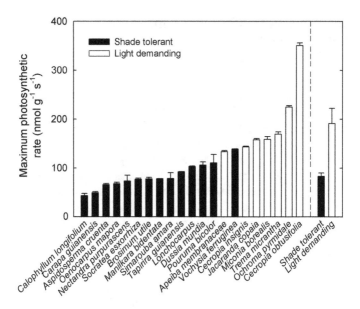

Fig. 3 Mean maximum photosynthetic rate (A_{max}) per unit mass (±1 SE) of tree species measured at the Fort Sherman Canopy Crane, San Lorenzo, Panama, and categorized into *shade tolerant* (canopy) or *light demanding* (pioneer) species according to Santiago (2010) and Santiago and Wright (2007). Species are ranked in order of increasing values from *left* to *right*. The two far *right bars* represent the means (±1 SE) of all *shade tolerant* and *light demanding* species; means are significantly different ($t = 4.90$; $P < 0.0001$)

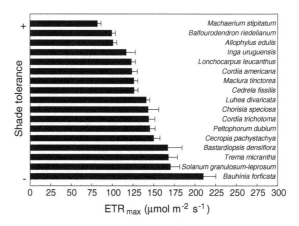

Fig. 4 Maximum electron transport rate (ETR_{max}) estimated from light response curves obtained in tree saplings of 17 canopy species growing in the field at 100 % irradiance. Species are ranked according to their shade tolerance. *Bars* are means + 1SE, n = 4 per species from Campanello et al. (2011)

rates of photosystem II (ETR_{max}) compared to shade tolerant tree species (Fig. 4). Again, it is notable that species-specific variation in ETR_{max} is gradual, ranging from 75 to 225 μmol m^{-2} s^{-1}, suggesting that any classification of these species into two distinct functional groups regarding light adaptations would be arbitrary.

Photosynthetic Adaptations to Contrasting Light Environments of Closely Related Species

Photosynthetic adaptations to light have rarely been studied in large numbers of closely related taxa (Robichaux and Pearcy 1980, 1984; Lei and Lechowicz 1997), and even less so for plants whose distributions along light gradients have been quantified and whose phylogenetic relationships to each other are known (Givnish et al. 2004; Santiago and Kim 2009). A study on static photosynthetic light responses of 11 *Lobelia* species provides support for predictions of physiological performance based on economic theory under different growth irradiance (Givnish et al. 2004). Among Hawaiian lobelias, maximum photosynthetic rate per unit area increases with average photosynthetic photon flux density (PPFD), as do dark respiration rates. Species growing under high light environments exhibit high respiration rates and therefore require more light to achieve a zero net rate of carbon uptake and reach maximum photosynthetic rates compared to species growing in darker habitats (Givnish et al. 2004). Yet, species growing under low light environments also achieve higher photosynthetic rates at high PPFD, however the maximum rates are not as high as species from open habitats. These results suggest an "adaptive cross-over" in photosynthetic light response, in which shade tolerant species outperform light requiring species at low light levels with low respiration, whereas light requiring species have higher photosynthetic performance at high light levels.

It is interesting that the instantaneous light compensation points (the PPFD at which net CO_2 assimilation is zero) in the Givnish et al. (2004) study range, depending on the species, from 0.1 to 1.2 % of full sunlight, which is much lower than the ecological or whole plant compensation points of 1.1–29 % of maximum incoming solar radiation. Ecological compensation points take into account nighttime leaf respiration, leaf construction costs amortized over leaf lifetimes, and allocation to non-leaf tissue, assuming identical rates of turnover for leaf and non leaf tissue (Givnish 1988; Givnish et al. 2004). The ecological compensation points are substantially closer to the lower limits of PPFD actually experienced by the lobelia species growing along a gradient of different light conditions, suggesting that the ecological compensation point can better explain the distribution of plants under different light conditions than the more traditional instantaneous leaf compensation points obtained from light response curves.

Adjustment in Leaf Optical Properties in Response to Changes in Light Conditions

Leaves of shade adapted tropical tree species have higher specific leaf area, synthesize less chlorophyll per unit area and use less chlorophyll for capturing the same amount of PPFD compared with leaves of high light requiring tree species (Lee

1986). The leaf anatomical properties correlated with increased photosynthetic efficiency are palisade cells with equal dimension having more chloroplasts on their abaxial surfaces. This dense layer of chloroplasts maximizes light capture efficiency. The leaves of high light requiring trees have columnar palisade cells making the chloroplast less efficient in PPFD absorption, but allowing light to reach chloroplasts in the spongy mesophlyll (Lee 1986).

Leaf accumulation of anthocyanins is also commonly regarded as a photoprotection mechanism under high irradiance. Anthocyanins can strongly absorb blue-green light (Harborne 1988; Barnes et al. 2000). The presence of anthocyanins in leaves can function as a 'sunscreen' to reduce the photons captured by chlorophylls (Smillie and Hetherington 1999; Feild et al. 2001; Gould et al. 2002; Oberbauer and Starr 2002; Gould 2004; Hughes et al. 2005; Hughes and Smith 2007). Anthocyanins are commonly produced during leaf expansion, especially in tropical plants, and appear to protect expanding leaves from photo-damage when the photosynthetic apparatus is not fully developed (Zhang et al. 2013b). In addition to photoprotection, anthocyanins may also function to backscatter light for understory plants under light-limiting conditions. Anthocyanins are prevalent on abaxial leaf surfaces of mature leaves in some understory plants of light requiring tropical tree species. Lee (1986) hypothesized that the anthocyanin rich cell layers located in the abaxial leaf surfaces can backscatter transmitted light back to the chloroplasts of shade adapted tree species growing in the forest understory.

Photoinhibition and Photoprotection of Tropical Canopy Trees Exposed to High Irradiances

Excessive light energy can cause transient reduction in the photosystem II (PSII) efficiency and may consequently limit CO_2 assimilation in sun-exposed leaves of the upper canopy, which may have an impact on plant growth. At midday, the reduction in photosynthetic rate due to midday decline in stomatal conductance that most species experience (Tenhunen et al. 1987; Zhang et al. 2013a), and high temperatures can induce a decline in enzyme activity, subjecting canopy leaves to excessive light energy. Photoinhibition may result in a decline in photosynthetic rates when leaves are exposed to high light levels that exceed their photon requirements for photosynthesis. Excessive light can place considerable stress on the leaf and result in damage to the photosynthetic apparatus. In plants that have been exposed to high light, the efficiency of photosynthesis decreases because of an increase in the number of photons absorbed per molecule of CO_2 assimilated. This decline in the quantum yield of PSII can cause both short- and long-term reductions in CO_2 assimilation depending on the extent of the photoprotective recovery and repair processes (Krause 1988). Among species growing under different light conditions, species restricted to high light habitats exhibit a higher capacity to tolerate and recover from light stress than shade tolerant species (e.g. Kamaluddin and Grace 1992; Johnson et al. 1993).

Shade tolerant trees, however, growing in small size gaps where light levels are intermediate between full sun and full shade, can avoid reductions in photosynthetic rates when exposed to high light levels. Acclimation capacity to respond to high light may be a requirement of shade tolerant plants that need to efficiently utilize short periods of high light levels in the forest understory or in small gap openings. Shade tolerant species growing in small gaps of subtropical wet forest in NE Argentina are less photoinhibited and recover faster than higher light requiring species (Villagra 2012). For example, shade tolerant *Balfourodendron riedelianum* and *Euterpe edulis* growing in small gaps recovered 90 % of their maximum PS II efficiency, measured as Fv/Fm (the ratio of fluorescence under the measuring light and under a light saturation pulse) (Fig. 5). The light requiring tree species *Cedrela fissilis* and *Maclura tinctoria* growing under similar light regimes, on the other hand only recovered 80 % of their maximum Fv/Fm. It does appear that shade tolerant tree species growing at intermediate light levels inside the forest have the capacity to dissipate excess light and to utilize incoming radiation for photosynthesis instead of undergoing strong photoinactivation and/or downregulation of the PSII, which might reduce carbon gain.

Some plants avoid absorbing excess radiation through leaf movements, leaf folding, chloroplast re-orientation and increased reflectance. In particular, light requiring canopy tree species in subtropical forests can fold their leaves when exposed to high light (Huang et al. 2012). Leaf folding reduces light absorptance and consequently helps to reduce the potential damage of PSII. Most plants, however, rely strongly on biochemical mechanisms to dissipate excess energy and to avoid damage to photosystems (Bilger and Björkman 1990; Johnson et al. 1993; Fetene et al. 1997; Demmig-Adams 1998). One of the principal means of photo-protection occurs through xanthophyll-cycle-dependent energy dissipation under high light. Photoprotective mechanisms are critical for individuals growing in sun-exposed sites and for shaded individuals during intense light flecks. Depending on the species and the habitat of origin, such plants use only 10–50 % of peak irradiance for photosynthesis (Demmig-Adams and Adams 1996).

Despite the potential negative effects of excess light, considerable inactivation of PSII in some species can occur under high light conditions without a significant effect on carbon assimilation. Under high light conditions, the energy absorbed exceeds the downstream capacity to use that energy (Lee et al. 1999; Kornyeyev et al. 2006, Adams et al. 2007). *Dipterocarpus retusus* is one of the tallest tropical tree species in the world reaching up to 72 m in height in SW China. The leaves of this species exhibit photosynthetic depression at midmorning in the upper canopy, and are able to reduce photochemistry and increase heat dissipation during the dry season (Zhang et al. 2009). Photorespiration in this species plays an important role

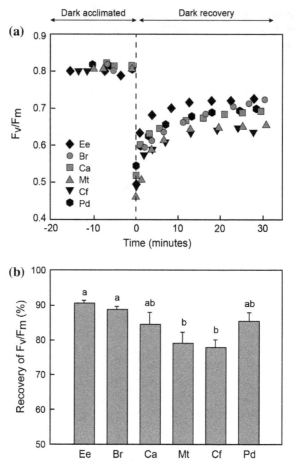

Fig. 5 The *Fv/Fm* ratios of five subtropical saplings during dark-acclimation (Maximum efficiency of PSII) and after exposure to saturating light levels (**a**) and the percentage Fv/Fm dark recovery (**b**) after 30 min (*bars* are means (n = 3)). Fv/Fm of dark-adapted leaves was measured three times at 5 min intervals after which a saturating beam of actinic light (2,000 μmol m^{-2} s^{-1}) was applied for 20 min (*dashed line*). Recovery of Fv/Fm was measured at a photon flux density of 20 μmol m^{-2} s^{-1} over a 30 min period. Different letters indicate significant differences among species (LSD-Fisher, $P < 0.05$). Abbreviations are: Ee, *Euterpe edulis*; Br, *Balfourodendron riedelianum*; Ca, *Cordia americana*; Mt, *Maclura tinctoria*; Cf, *Cedrela fissilis*; and Pd, *Peltophorum dubium*. Adapted from Villagra (2012)

in photoprotection. The sustained photosynthetic depression in the uppermost-canopy leaves of these giant trees could be a protective response to prevent excessive water loss and to avoid catastrophic leaf hydraulic dysfunction without a large negative effect on photosynthetic gas exchange (Zhang et al. 2009).

Coordination Between Photosynthetic Capacity and Water-Transport Efficiency in Tree Species Growing at Different Irradiances

The ability of plants to allocate resources among functions in a coordinated manner can determine competitive success in a tropical forest where substantial and unpredictable changes in light environments occur as a result of the opening and closure of canopy gaps. High water-transport capacity should allow maintenance of high leaf water potentials, stomatal conductance and consequently support high photosynthetic carbon assimilation rates (Santiago et al. 2004), especially in environments with high evaporative demand and solar radiation as commonly found in canopy openings. A large investment in an efficient water transport system in the understory of the forest would be non-adaptive as the light levels and evaporative demand conditions are relatively low and the size of the trees is small.

Dominant subtropical tree species in northern Argentina exhibit different degrees of coordination between photosynthetic capacity and water transport efficiency depending on their light requirements. Slow growing shade tolerant species exhibit lower maximum electron transport rate (*ETR*) than fast growing species. A positive correlation between ETR_{max} and maximum leaf hydraulic conductivity has been observed in high light requiring species growing at different light levels such as *Cedrela fissilis*, *Patagonula Americana* and *Cordia trichotoma*: they increase both ETR_{max} and hydraulic conductivity with increased growth irradiance (Fig. 6). This coordinated change in photosynthesis and hydraulic conductivity suggests that the

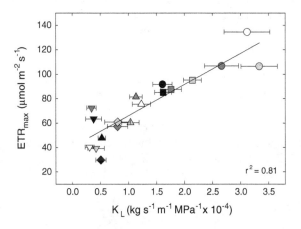

Fig. 6 Maximum electron transport rate of PSII (ETR_{max}) versus leaf specific hydraulic conductivity (K_L). A linear regression was fitted to the data. *Symbols* represent mean values ± SE for K_L and maximum electron transport rates obtained from light response curves for each species and light treatment. Irradiances of 10, 30, 40 and 65 % of full sun are indicated by *black*, *dark gray*, *light gray* and *open symbols* respecievely. Species are *Balfourodendron riedelianum* (*Triangle down*), *Cedrela fissilis* (*circle*), *Cordia trichotoma* (*square*), *Lonchocarpus leucanthus* (*diamond*) and *Patagonula Americana* (*triangle up*). Adapted from Campanello et al. (2008)

high light requiring species are capable of plastic changes in hydraulic architecture and water transport efficiency in response to increases in light availability resulting from the formation of canopy gaps which makes them more competitive in open habitats. In contrast, the hydraulic conductivity of slower-growing and more shade tolerant species such as *Balfourodendron riedelianum* and *Lonchocarpus leucanthus* does not increase with increasing growth irradiance (Fig. 6).

Summary

Irradiance is extremely variable in forest ecosystems with large spatial and temporal changes, particularly in tropical and subtropical forests. A very low percentage of the incoming irradiance reaches the forest floor, consisting of diffuse radiation that is randomly punctuated by short duration light flecks. Consequently understory plants face a shortage of light energy. Understory plants maximize light capture in the most efficient way, i.e., at the lowest costs in terms of construction and maintenance resulting in relatively low photosynthetic capacity. High light demanding species, on the other hand, exhibit high photosynthetic capacity. Even though carbon assimilation increases asymptotically with increasing PPFD, excessive light can result in photoinhibition of the photosynthetic process. Canopy leaves may avoid absorbing excess radiation through leaf movements, leaf folding, chloroplast re-orientation and increased reflectance. Most tree species, however, rely strongly on biochemical mechanisms to dissipate excess energy and to avoid damage to the photosystems. Maximum photosynthetic rate per unit area increases with changes in light growing conditions, from shaded understory to gaps and upper canopy trees, consistent with increases in dark respiration, leaf mass and photosynthetic capacity per unit area. High water-transport capacity allows maintenance of high carbon assimilation in environments with high evaporative demand and solar radiation, commonly found in canopy openings and in the upper canopy. The ecological compensation point better predicts the distribution of species along gradients in light growing condition than the instantaneous photosynthetic points obtained from traditional light response curves with leaves. Shade adapted and light requiring tree species from tropical and subtropical rainforests tend to have contrasting life history traits and leaf photosynthetic characteristics. Most of the species, however fall on intermediate places along a continuum of light adaptations.

References

Adams WW III, Watson AM, Mueh KE, Amiard V, Turgeon R, Ebbert V, Logan BA, Combs AF, Demmig-Adams B (2007) Photosynthetic acclimation in the context of structural constraints to carbon export from leaves. Photosynth Res 94:455–466

Ballaré CL, Scopel AL, Sanchez RA (1997) Foraging for light: photosensory ecology and agricultural implications. Plant Cell Environ 20:820–825

Barnes PW, Searles PS, Ballare CL, Ryel RJ, Caldwell MM (2000) Non-invasive measurements of leaf epidermal transmittance of UV radiation using chlorophyll fluorescence: field and laboratory studies. Physiol Plant 109:274–283

Bilger W, Björkman O (1990) Role of xanthophyll cycle in photoprotection elucidated by measurements of light-induced absorbency changes, fluorescence and photosynthesis in leaves of *Hedera canariensis*. Photosynth Res 25:173–185

Björkman O, Ludlow M (1972) Characterization of the light climate on the floor of a Queensland rainforest. Carnegie Inst Wash Yearb 71:85–91

Bormann FH, Likens GE (1979) Pattern and process in a forested ecosystem. Springer-Verlag, New York, USA

Brokaw NVL (1985) Gap-phase regeneration in a tropical forest. Ecology 66:682–687

Campanello PI, Gatti MG, Ares A, Montti L, Goldstein G (2007) Tree regeneration and microclimate in a liana and bamboo-dominated semideciduous Atlantic Forest. For Ecol Manage 252:108–117

Campanello PI, Gatti MG, Goldstein G (2008) Coordination between water-transport efficiency and photosynthetic capacity in canopy tree species at different growth irradiances. Tree Physiol 28:85–94

Campanello PI, Gatti MG, Ares A, Montti L, Villara M, Goldstein G (2011) Ser o no ser tolerante a la sombra: economía de agua y carbono en especies arbóreas del Bosque Atlántico (Misiones, Argentina). Ecología Austral 21:285–300

Canham CD, Denslow JS, Platt WJ, Runkle JR, Spies TA, White PS (1990) Light regimes beneath closed canopies and tree fall gaps in temperate and tropical forests. Can J Forest Res-Rev Can Rech Forestiere 20:620–631

Chazdon RL (1988) Sunflecks and their importance to forest understory plants. Adv Ecol Res 18:1–63

Chazdon RL, Fetcher N (1984) Light environments of tropical forests. In: Medina E, Mooney HA, Vázquez-Yanes C (eds) Physiological Ecology of Plants of the Wet Tropics. Dr. W Junk Publishers, The Hague, pp 553–564

Chazdon RL, Pearcy RW (1986) Photosynthetic responses to light variation in rainforest species. II. Carbon gain and photosynthetic efficiency during lightflecks. Oecologia 69:524–531

Clark DA, Clark DB (1992) Life history diversity of canopy and emergent trees in a neotropical rain forest. Ecol Monogr 62:315–344

Demmig-Adams B (1998) Survey of thermal energy dissipation and pigment composition in sun and shade leaves. Plant Cell Physiol 39:474–482

Demmig-Adams B, Adams WW (1996) The role of xanthophyll cycle carotenoids in the protection of photosynthesis. Trends Plant Sci 1:21–26

Demmig-Adams B, Adams WW (2006) Photoprotection in an ecological context: the remarkable complexity of thermal energy dissipation. New Phytol 172:11–21

Denslow JS (1980) Gap partitioning among tropical rainforest trees. Biotropica 12:47–55

Ellis AR, Hubbell SP, Potvin C (2000) In situ field measurements of photosynthetic rates of tropical tree species: a test of the functional group hypothesis. Can J Bot-Rev Can Botanique 78:1336–1347

Feild TS, Lee DW, Holbrook NM (2001) Why leaves turn red in autumn. The role of anthocyanins in senescing leaves of red-osier dogwood. Plant Physiol 127:566–574

Fetene M, Nauke P, Lüttge U, Beck E (1997) Photosynthesis and photoinhibition in a tropical alpine giant rosette plant, *Lobelia rhynchopetalum*. New Phytol 137:453–461

Gilbert B, Wright SJ, Muller-Landau HC, Kitajima K, Hernández A (2006) Life history trade-offs in tropical trees and lianas. Ecology 87:1281–1288

Givnish TJ (1988) Adaptation to sun and shade: A whole-plant perspective. Aust J Plant Physiol 15:63–92

Givnish TJ, Montgomery RA, Goldstein G (2004) Adaptive radiation of photosynthetic physiology in the Hawaiian lobeliads: light regimes, static light responses, and whole-plant compensation points. Am J Bot 91:228–246

Gómez-Pompa A, Vázquez-Yanes C (1981) Successional studies of a rain forest in Mexico. In: Forest Succession. Springer, pp 246–266

Gould KS (2004) Nature's Swiss army knife: the diverse protective roles of anthocyanins in leaves. J Biomed Biotechnol:314–320

Gould KS, Vogelmann TC, Han T, Clearwater MJ (2002) Profiles of photosynthesis within red and green leaves of *Quintinia serrata*. Physiol Plant 116:127–133

Graham EA, Mulkey SS, Kitajima K, Phillips NG, Wright SJ (2003) Cloud cover limits net CO_2 uptake and growth of a rainforest tree during tropical rainy seasons. Proc Natl Acad Sci USA 100:572–576

Hallé F, Oldeman RAA, Tomlinson PB (1978) Tropical trees and forests: an architectural analysis. Springer Science and Business Media, 2012. Springer-Verlag, Berlin

Harborne JB (1988) The flavonoids: recent advances

Hartshorn GS (1978) Treefalls and tropical forest dynamics. In: Tomlinson PB, Zimmerman MH (eds) Tropical Trees as Living Systems. Cambridge University Press, Cambridge, England, pp 617–683

Huang W, Zhang S-B, Cao K-F (2012) Evidence for leaf fold to remedy the deficiency of physiological photoprotection for photosystem II. Photosynth Res 110:185–191

Hubbell SP, Foster RB, O'Brien ST, Harms KE, Condit R, Wechsler B, Wright SJ, de Lao SL (1999) Light-gap disturbances, recruitment limitation, and tree diversity in a neotropical forest. Science 283:554–557

Hughes NM, Smith WK (2007) Attenuation of incident light in *Galax urceolata* (Diapensiaceae): Concerted influence of adaxial and abaxial anthocyanic layers on photoprotection. Am J Bot 94:784–790

Hughes NM, Neufeld HS, Burkey KO (2005) Functional role of anthocyanins in high-light winter leaves of the evergreen herb *Galax urceolata*. New Phytol 168:575–587

Ishida A, Toma T, Marjenah (1999) Limitation of leaf carbon gain by stomatal and photochemical processes in the top canopy of Macaranga conifera, a tropical pioneer tree. Tree Physiol 19:467–473

Johnson GN, Young AJ, Scholes JD, Horton P (1993) The dissipation of excess excitation energy in Brittish plant species. Plant Cell Environ 16:673–679

Kamaluddin M, Grace J (1992) Photoinhibition and light acclimation in seedlings of *Bischofia javanica*, a tropical forest tree from Asia. Ann Bot 69:47–52

King DA (1991) Correlations between biomass allocation, relative growth rate and light environment in tropical forest saplings. Funct Ecol 5:485–492

Kitajima K (1994) Relative importance of photosynthetic traits and allocation patterns as correlates of seedling shade tolerance of 13 tropical trees. Oecologia 98:419–428

Kitajima K (1996) Ecophysiology of tropical tree seedlings. In: Mulkey SS, Chazdon RL, Smith AP (eds) Tropical forest plant ecophysiology. Chapman and Hall, New York, pp 559–596

Kohyama T, Hotta M (1990) Significance of allometry in tropical saplings. Funct Ecol 4:515–521

Kornyeyev D, Logan BA, Tissue DT, Allen RD, Holaday AS (2006) Compensation for PSII photoinactivation by regulated non-photochemical dissipation influences the impact of photoinactivation on electron transport and CO_2 assimilation. Plant Cell Physiol 47:437–446

Krause GH (1988) Photoinhibition of photosynthesis: an evaluation of damaging and protective mechanisms. Physiol Plant 74:566–574

Krause GH, Virgo A, Winter K (1995) High susceptibility to photoinhibition of young leaves of tropical forest trees. Planta 197:583–591

Lambers H, Chapin FS III, Pons TL (1998) Plant Physiological Ecology. Springer-Verlag, New York

Lee DW (1986) Unusual strategies of light absorption in rainforest herbs. In: Givnish TJ (ed) The Economics of Plant Form and Function. Cambridge University Press, Cambridge, UK, pp 105–131

Lee DW (1987) The spectral distribution of radiation in 2 neotropical rainforests. Biotropica 19:161–166

Lee HY, Chow WS, Hong YN (1999) Photoinactivation of photosystem II in leaves of *Capsicum annuum*. Physiol Plant 105:377–384

Lei TT, Lechowicz MJ (1997) The photosynthetic response of eight species of Acer to simulated light regimes from the centre and edges of gaps. Funct Ecol 11:16–23

Lüttge U (2008) Physiological Ecology of Tropical Plants, 2nd edn. Springer-Verlag, Berlin

Martinez-Ramos M, Alvarez-Buylla E, Sarukhan J, Piñero D (1988) Treefall age determination and gap dynamics in a tropical forest. J Ecol 76:700–716

Meinzer FC, Goldstein G, Jackson P, Holbrook NM, Gutiérrez MV, Cavelier J (1995) Environmental and physiological regulation of transpiration in tropical forest gap species: the influence of boundary layer and hydraulic conductance properties. Oecologia 101:514–522

Montgomery RA, Givnish TJ (2008) Adaptive radiation of photosynthetic physiology in the Hawaiian lobeliads: dynamic photosynthetic responses. Oecologia 155:455–467

Nicotra AB, Chazdon RL, Iriarte SVB (1999) Spatial heterogeneity of light and woody seedling regeneration in tropical wet forests. Ecology 80:1908–1926

Oberbauer SF, Starr G (2002) The role of anthocyanins for photosynthesis of Alaskan arctic evergreens during snowmelt. In: Gould KS, Lee DW (eds) Anthocyanins in leaves. advances in botanical research, vol 37. Academic Press, New York, pp 129–145

Oldeman RAA (1978) Architecture and energy exchange of dicotyledonous trees in the forest. In: Tomlinson PB, Zimmerman MH (eds) Tropical Trees as Living Systems. Cambridge University Press, Cambridge, England, pp 535–560

Osunkoya OO, Ash JE (1991) Acclimation to a change in light regime in seedlings of 6 Australian rainforest tree species. Aust J Bot 39:591–605

Pearcy RW (1987) Photosynthetic gas exchange responses of Australian tropical forest trees in canopy, gap and understory micro-environments. Funct Ecol 1:169–178

Poorter L, Kwant R, Hernandez R, Medina E, Werger MJA (2000) Leaf optical properties in Venezuelan cloud forest trees. Tree Physiol 20:519–526

Popma J, Bongers F (1988) The Effect of canopy gaps on growth and morphology of seedlings of rain forest species. Oecologia 75:625–632

Robichaux RH, Pearcy RW (1980) Environmental characteristics, field water relations, and photosynthetic responses of C_4 Hawaiian *Euphorbia* species from contrasting habitats. Oecologia 47:99–105

Robichaux RH, Pearcy RW (1984) Evolution of C_3 and C_4 plants along an environmental moisture gradient: Patterns of photosynthetic differentiation in Hawaiian *Scaevola* and *Euphorbia* species. Am J Bot 71:121–129

Runkle JR (1981) Gap regeneration in some old growth forests of the eastern United States. Ecology 62:1041–1051

Santiago LS (2010) Can growth form classification predict litter nutrient dynamics and decomposition rates in lowland wet forest? Biotropica 42:72–79

Santiago LS, Dawson TE (2014) Light-use efficiency of California redwood forest understory plants along a moisture gradient. Oecologia 174:351–363

Santiago LS, Kim S-C (2009) Correlated evolution of leaf shape and physiology in the woody *Sonchus* alliance (Asteraceae: Sonchinae) in Macaronesia. Int J Plant Sci 170:83–92

Santiago LS, Wright SJ (2007) Leaf functional traits of tropical forest plants in relation to growth form. Funct Ecol 21:19–27

Santiago LS, Goldstein G, Meinzer FC, Fisher JB, Machado K, Woodruff D, Jones T (2004) Leaf photosynthetic traits scale with hydraulic conductivity and wood density in Panamanian forest canopy trees. Oecologia 140:543–550

Sassenrath-Cole GF, Pearcy RW, Steinmaus S (1994) The role of enzyme activation state in limiting carbon assimilation under variable light conditions. Photosynth Res 41:295–302

Schnitzer SA, Dalling JW, Carson WP (2000) The impact of lianas on tree regeneration in tropical forest canopy gaps: evidence for an alternative pathway of gap-phase regeneration. J Ecol 88:655–666

Sharkey TD, Seemann JR, Pearcy RW (1986) Contribution of metabolites of photosynthesis to postillumination CO_2 assimilation in response to lightflecks. Plant Physiol 82:1063–1068

Smillie RM, Hetherington SE (1999) Photoabatement by anthocyanin shields photosynthetic systems from light stress. Photosynthetica 36:451–463

Tenhunen JD, Pearcy RW, Lange OL (1987) Diurnal variations in leaf conductance and gas exchange in natural environments. In: Zeiger E, Farquhar GD, Cowan IR (eds) Stomatal Function. University Press, Stanford, USA, Stanford, pp 323–351

Valladares F (1999) Architecture, ecology, and evolution of plant crowns. Handbook of Functional Plant Ecology. Marcel Dekker, New York, pp 121–194

Valladares F, Allen MT, Pearcy RW (1997) Photosynthetic responses to dynamic light under field conditions in six tropical rainforest shrubs occurring along a light gradient. Oecologia 111:505–514

Valladares F, Wright SJ, Lasso E, Kitajima K, Pearcy RW (2000) Plastic phenotypic response to light of 16 congeneric shrubs from a Panamanian rainforest. Ecology 81:1925–1936

Villagra M (2012) Plasticidad morfológica y fisiológica de especies arbóreas del Bosque Atlántico en respuesta a la disponibilidad de luz y nutrientes. In: Facultad de Ciencias Exactas y Naturales, vol. PhD. Universidad de Buenos Aires, Argentina

Vitousek PM, Denslow JS (1986) Nitrogen and Phosphorus Availability in Treefall Gaps of a Lowland Tropical Rain-Forest. J Ecol 74:1167–1178

Vogelmann TC, Nishio JN, Smith WK (1996) Leaves and light capture: Light propagation and gradients of carbon fixation within leaves. Trends Plant Sci 1:65–70

Watt AS (1947) Pattern and process in the plant community. J Ecol 35:1–22

Whitmore TC (1975) Tropical rainforests of the Far East. Clarendon, Oxford, England

Wright SJ, vanSchaik CP (1994) Light and the phenology of tropical trees. Am Nat 143:192–199

Wright SJ, Kitajima K, Kraft NJB, Reich PB, Wright IJ, Bunker DE, Condit R, Dalling JW, Davies SJ, Diaz S, Engelbrecht BMJ, Harms KE, Hubbell SP, Marks CO, Ruiz-Jaen MC, Salvador CM, Zanne AE (2010) Functional traits and the growth-mortality trade-off in tropical trees. Ecology 91:3664–3674

Zalamea P-C, Stevenson PR, Madrinan S, Aubert P-M, Heuret P (2008) Growth pattern and age determination for *Cecropia sciadophylla* (Urticaceae). Am J Bot 95:263–271

Zalamea P-C, Heuret P, Sarmiento C, Rodriguez M, Berthouly A, Guitet S, Nicolini E, Delnatte C, Barthelemy D, Stevenson PR (2012) The Genus *Cecropia*: A Biological Clock to Estimate the Age of Recently Disturbed Areas in the Neotropics. Plos One 7

Zhang JL, Meng LZ, Cao KF (2009) Sustained diurnal photosynthetic depression in uppermost-canopy leaves of four dipterocarp species in the rainy and dry seasons: does photorespiration play a role in photoprotection? Tree Physiol 29:217–228

Zhang Y-J, Meinzer FC, Qi J-H, Goldstein G, Cao K-F (2013a) Midday stomatal conductance is more related to stem rather than leaf water status in subtropical deciduous and evergreen broadleaf trees. Plant Cell Environ 36:149–158

Zhang Y-J, Yang Q-Y, Lee DW, Goldstein G, Cao K-F (2013b) Extended leaf senescence promotes carbon gain and nutrient resorption: importance of maintaining winter photosynthesis in subtropical forests. Oecologia 173:721–730

Carbon Economy of Subtropical Forests

Yong-Jiang Zhang, Piedad M. Cristiano, Yong-Fei Zhang,
Paula I. Campanello, Zheng-Hong Tan, Yi-Ping Zhang,
Kun-Fang Cao and Guillermo Goldstein

Abstract Compared to tropical and temperate forests, subtropical forests have received little attention in physiological and ecological studies until now, and the contribution of this ecosystem type to the global carbon cycle has not been fully assessed. In this chapter we discuss results on the carbon balance of subtropical forests at different spatial and temporal scales, analyze the potential limitation of seasonal low temperatures and water deficits on physiological processes of subtropical trees, and characterize the uniqueness of subtropical forest ecosystems in terms of carbon economy. Results from multiple techniques and scales were included in the carbon balance assessment. The largest two regions with subtropical forests are located in Asia and South America. The net ecosystem carbon gain of

Y.-J. Zhang (✉)
Department of Organismic and Evolutionary Biology, Harvard University,
Cambridge, MA 02138, USA
e-mail: yongjiangzhang@oeb.harvard.edu

P.M. Cristiano
Laboratorio de Ecología Funcional, Departamento de Ecología Genética y Evolución,
Instituto IEGEBA (CONICET-UBA), Facultad de Ciencias Exactas y Naturales, Universidad
de Buenos Aires, Buenos Aires, Argentina
e-mail: piedad@ege.fcen.uba.ar

Y.-F. Zhang
Department of Geological Sciences, John A. and Katherine G. Jackson School
of Geosciences, University of Texas at Austin, Austin, Texas, USA
e-mail: yongfei@utexas.edu

P.I. Campanello
Laboratorio de Ecología Forestal y Ecofisiología, Instituto de Biología Subtropical,
CONICET, FCF, Universidad Nacional de Misiones, Posadas, Argentina
e-mail: pcampanello@yahoo.com

Z.-H. Tan · Y.-P. Zhang
Key Laboratory of Tropical Forest Ecology, Xishuangbanna Tropical Botanical Garden,
Chinese Academy of Sciences, Mengla, Yunnan 666303, China
e-mail: yipingzh@xtbg.ac.cn

© Springer International Publishing Switzerland 2016
G. Goldstein and L.S. Santiago (eds.), *Tropical Tree Physiology*,
Tree Physiology 6, DOI 10.1007/978-3-319-27422-5_16

subtropical forests in these two regions, which have annual precipitations larger than 800 mm, is probably neither strongly limited by soil water availability nor by seasonal low temperatures. Relatively low evapotranspiration in the winter/dry season and high soil water-holding capacity help maintain good water availability for trees in most subtropical forests. High solar radiation, light penetration and low ecosystem respiration in winter may compensate for the negative effects of low temperatures on gross photosynthesis. Therefore, subtropical forests in many areas can assimilate carbon in excess of respiration throughout the year and they are, probably, among the largest terrestrial carbon sinks across terrestrial ecosystems worldwide. In addition, because leaf and ecosystem respiration respond to temperature changes to a larger extent compared to ecosystem carbon assimilation, a negative relationship between net ecosystem carbon gain and mean annual temperature was found in Asian subtropical and tropical forests. This relationship suggests that global warming may weaken the carbon sink strength of these forest ecosystems. These results indicate the important contribution of subtropical forests to the global carbon cycle and the potentially negative response of these forests to global warming. We hope this information will promote additional physiological and ecological research and conservation in subtropical forests.

Keywords Carbon sink · Climate change · Evapotranspiration · Respiration · Water deficit

Introduction

Compared to tropical and temperate forests, the ecological processes of subtropical forests and the physiology of subtropical trees are less-studied. The largest two regions with subtropical forests are located in China and Argentina, with some forests also occurring in North America, Africa and Australia (Fig. 1). In general, subtropical forests have received relatively little attention in terms of physiological

K.-F. Cao
State Key Laboratory for Conservation and Utilization of Subtropical Agro-Bioresources, and College of Forestry, Guangxi University, Nanning 530004, China
e-mail: caokf@xtbg.ac.cn

G. Goldstein
Laboratorio de Ecología Funcional, Departamento de Ecología Genética y Evolución, Instituto IEGEBA (CONICET-UBA), Facultad de Ciencias Exactas y naturales, Universidad de Buenos Aires, Buenos Aires, Argentina
e-mail: gold@bio.miami.edu; goldstein@ege.fcen.uba.ar

G. Goldstein
Department of Biology, University of Miami, Coral Gables, FL 33146, USA

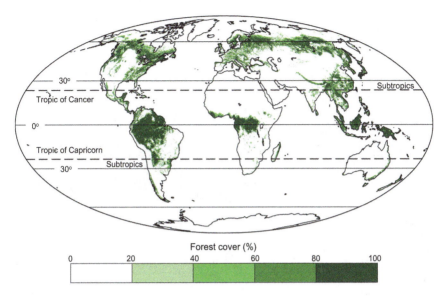

Fig. 1 A world forest cover map. Global forest cover derived from MODIS (Moderate Resolution Imaging Spectroradiometer) with a yearly temporal resolution and a spatial resolution of 0.05°. *Product type* vegetation type data (MOD12C1) for 2010. The subtropical zones are indicated

and ecological research until recently (Tan et al. 2012; Zhang et al. 2013b; Cristiano et al. 2014; Yu et al. 2014). Different from the 'tropics' that are clearly defined by diurnal air temperature variations being substantially larger than seasonal temperature changes, the Subtropical region is not a well-defined area. Subtropical climatic regimes also occur at high elevations within the tropics. A large portion of the world's deserts are also located within the subtropics due to the development of the subtropical ridge, which is a significant belt of high pressure situated around 30° latitude in the Northern or Southern Hemisphere. While the limits of subtropical zones toward the equator (23.4° latitude) are widely accepted, the poleward limits of the subtropics are controversial and lack a universal agreement; latitudes around 30°–35°, and even 40° have been used to define the northern and southern limits of the subtropics (Corlett 2013). Unfortunately, the subtropics have been considered by some authors as a transition zone between the tropical and temperate zones and not as a distinct climatic zone. In previous studies related to carbon stocks and net primary productivity (NPP) of worldwide forests, the subtropical forests were often included together with temperate forests (Pan et al. 2011, 2013).

Subtropical forests have unique ecological characteristics compared to tropical and temperate forests. Regardless of the latitudinal limits, subtropical climates are those characterized by seasonal air temperature ranges being relatively similar to diurnal temperature changes. Compared to typical tropical rainforests with little climatic seasonality and year-round high temperatures, subtropical forests have mild winters (Tan et al. 2012; Zhang et al. 2013a; Cristiano et al. 2014) (Fig. 2), which potentially negatively affect leaf photosynthetic performance (Zhang et al. 2013a)

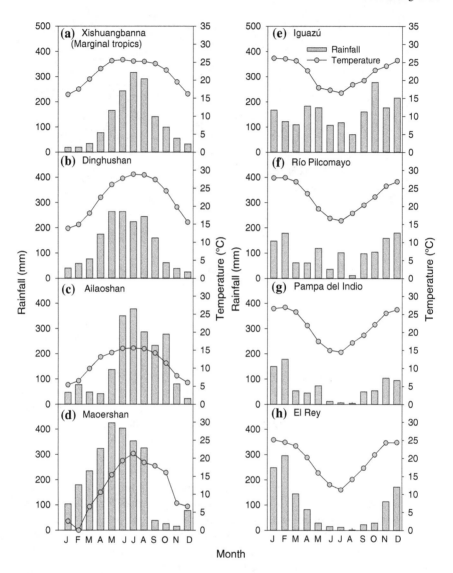

Fig. 2 Seasonal dynamics in average air temperature and precipitation of subtropical forests and a forest in the marginal Asian tropics (Xishuangbanna). *Circles* are monthly average air temperatures; *bars* are average monthly precipitation. Xishuangbanna data from Xishuangbanna Tropical Rainforest Ecosystem Station, China. Ailaoshan data from Ailaoshan Station for Subtropical Forest Ecosystem Studies, China and (Qiu and Xie 1998). Dinghushan data from Huang et al. (2011), and Maoershan data from Bai et al. (2010). Iguazú, Río Pilcomayo, Pampa del Indio and El Rey data from Meteorological Stations in Argentina (Cristiano et al. 2014; Cristiano unpublished data)

and ecosystem carbon gain. However, there is some evidence that mild winters have little impact on tree growth, as depicted by high Normalized Difference Vegetation Index (NDVI) values observed throughout the year, and by high ecosystem net carbon assimilations measured during winters on subtropical forests (Tan et al. 2012; Cristiano et al. 2014). Further, plant species composition and leaf and flower phenology of subtropical forests tend to be more similar to tropical forests compared to temperate forests (Qiu and Xie 1998).

Subtropical forests are a strong carbon sink potentially contributing greatly to global forest carbon sequestration. A few recent studies have suggested that these forests assimilate carbon all-year-round with high canopy photosynthetic rates and high ecosystem carbon sequestration (Tan et al. 2011, 2012; Cristiano et al. 2014; Yu et al. 2014). For instance, an old-growth subtropical forest in Ailaoshan, Southwest (SW) China is the largest carbon sink (~ 9 t ha^{-1} year^{-1}; estimated by eddy covariance method) among all Asian ecosystems studied so far, including tropical, subtropical, and temperate forests, croplands, wetlands, and grasslands (Chen et al. 2013). Further, because the subtropics are geographically in the transition zone between tropical and temperate zones, understanding the ecological and physiological mechanisms controlling the carbon sink strength of subtropical forests could provide some hints for predicting the future of temperate forests during global warming.

Despite the potentially important role of subtropical forests in the global carbon cycle, their carbon economy and the controlling environmental, ecological, and physiological factors are poorly-understood, and have attracted little attention until now. In this chapter, we summarize published results on leaf and ecosystem level carbon exchange of subtropical forests (Table 1 and those in Chen et al. 2013), analyze the limiting effects of water and temperature on carbon economy of subtropical forests, and assess the potential for subtropical forests to be among the largest terrestrial carbon sinks across all terrestrial ecosystems. Here we focus on

Table 1 Major subtropical forest ecosystems in Southern China and Northern Argentina, including a tropical forest at the northern limit of the Asian tropics, used in this chapter, and their location, elevation, mean annual temperatures (MAT), and mean annual precipitation (MAP)

Forest name	Latitude	Longitude	Elevation (m)	MAT (°C)	MAP (mm)
China					
Xishuangbanna (Marginal tropics)	21° 41′ N	101° 25′ E	570	21.7	1560
Dinghushan	23° 10′ N	112° 32′ E	240	21.0	1956
Ailaoshan	24° 32′ N	101° 01′ E	2460	11.3	1931
Maoershan	25° 50′ N	101° 49′ E	1500	10.8	2510
Argentina					
Iguazú	26° 25′ S	54° 37′ W	193	21.0	2000
Río Pilcomayo	25° 10′ S	58° 90′ W	57	22.6	1260
Pampa del Indio	26° 15′ S	60° 00′ W	80	21.5	827
El Rey	24° 40′ S	64° 40′ W	1500	19.5	1251

Asian and South American subtropical forests because they account for the largest proportion of worldwide subtropical forests (Fig. 1). Although the poleward limits for subtropical forests could vary depending on regional geography (e.g. elevation) and climate (Zhang 1989; Ni and Song 1997), here we adopted the poleward limits of 30.0° as this is the most conventional and convenient definition (Corlett 2013).

Water Limitation of Carbon Economy

Does water availability limit carbon gain of subtropical forests? Since many subtropical forests are characterized by a seasonal decrease in monthly precipitation (Fig. 2), seasonal water deficits could be a limiting factor for ecosystem carbon gain. Water deficits negatively affect leaf photosynthesis through lowering stomatal conductance and by limiting CO_2 supply to the Calvin Cycle (Flexas et al. 2001). At a global scale, ecosystem net primary productivity (NPP) is positively correlated with annual precipitation across different terrestrial ecosystems (Lieth 1975; Michaletz et al. 2014). The most humid south American subtropical forests showed year-round high precipitation with no apparent dry season, thus canopy photosynthesis and ecosystem carbon assimilation of these subtropical forests in South America is probably not substantially influenced by seasonal water deficits (Cristiano et al. 2014). These humid subtropical forests are located in northeastern Argentina (Misiones Province) and also along the eastern slopes of the Andes Mountains in northwestern Argentina, a region locally known as Las Yungas. Both areas are rainy, humid, and have mild winters, with some eastern slope Andean subtropical forests having a dry season with substantial fog (El Rey). Between these two humid forests, there are large regions with seasonally humid subtropical forests (also known as humid Chaco) and dry subtropical forests (also known as arid Chaco). We have included in this chapter information on the humid subtropical forests and the seasonally dry subtropical forests (Fig. 2 right panels). Arid Chaco is not included in the current analysis because it has regular frosts during the winter and its annual precipitation is lower than 800 mm. Arid Chaco also has strong floristic affinities with temperate ecosystems, while seasonally humid Chaco has strong floristic affinities with neotropical forests (Pennington et al. 2000).

Influenced by the Asian summer monsoon (the East Asian Monsoon and Indian Monsoon), the Asian subtropics are characterized by a warm and wet summer and a cold and dry winter. During the dry season, precipitation of the Asian subtropical forests can be low (Dinghushan, Ailaoshan, and Moershan, Fig. 2). For example, Ailaoshan has 6 months with rainfall <100 mm (Fig. 2c). No evidence of water deficits, however, has been found in the dry season in high elevation subtropical forests in Southwest China (Ailaoshan, see Table 1). The dry season predawn leaf water potentials of all 10 tree species studied are close to zero, ranging from −0.07 to −0.16 MPa (Zhang et al. 2013a). Further, even during a once-in-a century regional drought event (∼half year without rains; 2009–2010), trees in the

Ailaoshan subtropical forest showed no symptoms of drought stress and still maintained predawn leaf water potential very close to zero (Qi et al. 2012, 2013). Indeed, the soil layers where the roots are mainly distributed were able to maintain high water availability (soil water potential > -0.5 MPa) during the driest period of the 2009–2010 drought event. High capacity of the Ailsoshan subtropical forest in buffering rainfall shortages has been attributed to its high soil water-holding capacity, which is distinctly higher than that of some plantations, shrub lands, and secondary forests in this region (Qi et al. 2012). Although it is still unknown whether other subtropical forests have high soil water-holding capacity and can effectively buffer the effects of seasonal rainfall shortages, it is very likely that water availability is not a strong limiting factor on net ecosystem carbon exchange (NEE) of most Asian and Argentinean subtropical forests.

Limited impact of seasonal rainfall shortage on NEE of subtropical forests is supported by the seasonal dynamics in NDVI of these forests. The NDVI is a spectral index that estimates the amount of green biomass and thus is related to the rate of ecosystem net CO_2 assimilation level and net productivity. Satellite-based NDVI (or the enhanced vegetation index, EVI) has been used to assess canopy carbon exchange (Cristiano et al. 2014) and predict forest NPP or NEE (Xiao et al. 2004; Campos et al. 2013). The NDVI of some Asian subtropical forests in the driest months (e.g. December to April for Ailaoshan, September to November for Maoershan, Fig. 2) is higher than that of the rainy season (Fig. 3a). The most humid Argentinean subtropical forest (Iguazú, 2000 mm annual precipitation) exhibits a constantly high NDVI pattern while the other three subtropical forests of Northern Argentina with seasonal declines in precipitation exhibit only small declines in NDVI during the winter/dry season (Fig. 3b). The decrease could be the result of lower temperature effects on photosynthesis, decrease in leaf area index during this period, and/or drought effects on plant metabolism. Notably, the dry season coincides with the winter in both the Asian and Argentinean subtropics. Since evapotranspiration is positively correlated with air temperature (Kosa 2009, and Fig. 4), winter evapotranspiration is much lower than that of the summer for subtropical forests (Fig. 4a). The winter/dry season evapotranspiration of four Argentinian subtropical forests is only about 23–53 % of that of the summer/wet season (Fig. 4a). Lower winter ecosystem evapotranspiration could help in water conservation during the winter/dry season, and thus partially compensate for the negative effects of rainfall shortages on tree carbon assimilation.

Is Seasonal Low Temperature a Limiting Factor?

Is the carbon gain of subtropical forests limited by seasonal low temperatures? The answer is yes and no. Compared to typical temperate forests with strong seasonal changes in air temperatures, winter is generally mild and the seasonality of temperature or NDVI is weaker (Fig. 5). Trees in subtropical forests are exposed to

Fig. 3 MODIS NDVI seasonal dynamics (Normalized Difference Vegetation Index) of subtropical forests and a tropical forest at the marginal tropics (Xishuangbanna). *Error bars* represent standard errors of the values. Xishuangbanna, Ailaoshan, Maoershan data from Zhang and Zhang (unpublished data). Iguazú data from Cristiano et al. (2014). Río Pilcomayo, Pampa del Indio and El Rey data from Cristiano (unpublished data)

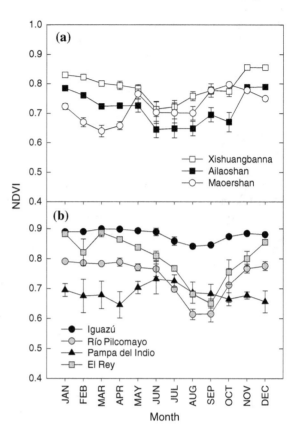

mild seasonal low temperatures, which can negatively affect leaf photosynthesis in several ways. Low temperatures can directly inhibit photosynthetic enzyme activities (Holaday et al. 1992; Kingston Smith et al. 1997), increase the probability of photoinhibition (Levitt 1980; Allen and Ort 2001; Huang et al. 2010), reduce stomatal conductance (Allen and Ort 2001; Zhang et al. 2014), and increase carbohydrate concentration (high sink effect) in leaves consequently inhibiting photosynthesis (Azcon-Bieto 1983). In forests with subzero temperatures, freeze-thaw cycles could cause xylem embolism (blocking of xylem conduits by air bubbles) (Davis et al. 1999) slowing down water transport and consequently limiting photosynthesis (Taneda and Tateno 2005). Based on field measurements, leaf net photosynthetic rates of well-irrigated trees and shrubs at the northern limit of Asian tropics declined 28–44 % in the cool season, and that of trees from three warm subtropical forests declined by 20–30 % (Table 2). The winter decline in leaf net photosynthetic rate could be as high as 81 % in cold subtropical forests such as that in Maoershan, China, which has an average air temperature of 0 °C in the coldest month (elevation 1500 m; Table 2).

Fig. 4 **a** Monthly average evapotranspiration estimated using the MODIS product MOD16A2, for four subtropical forests in Argentina, and **b** monthly average evapotranspiration in relation to mean monthly air temperature. The relationship between evapotranspiration and air temperature was fitted by a linear regression. Data from Cristiano et al. (unpublished data)

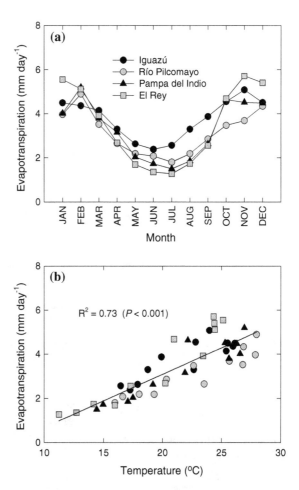

The negative effects of seasonal low temperatures on ecosystem carbon gain coincide with a decrease in total leaf surface area during the winter. The leaf area index (LAI) of an Asian and a South American subtropical forest declined significantly in winter (Fig. 6). The winter decline in LAI is mainly the result of leaf drop of winter-deciduous trees (Cristiano et al. 2014). The subtropical forests in South America and Asia are evergreen dominated forests mixed with some deciduous tree species (occasionally dominated by deciduous species) (Bai et al. 2010; Chen et al. 2013; Zhang et al. 2013b; Cristiano et al. 2014).

Subtropical forests have ecosystem and leaf level mechanisms to compensate for the negative effects of seasonal low temperatures on leaf photosynthesis. First of all, although leaf drop of deciduous trees decreases total leaf surface area, it can increase the light penetration through the canopy and increase the light interception of lower canopy leaves, sub-canopy trees, as well as understory saplings and shrubs (Miyazawa and Kikuzawa 2005; Cristiano et al. 2014). Second, increased leaf

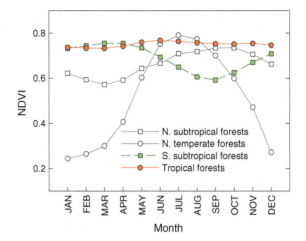

Fig. 5 Seasonal dynamics in MODIS NDVI (Normalized Difference Vegetation Index) of forests of different climatic zones. Forests were identified according to MODIS vegetation type data (Moderate Resolution Imaging Spectroradiometer; *product type* vegetation type data (MOD12C1) for 2010). Data from Zhang and Zhang (unpublished data)

Table 2 Summer and winter leaf net photosynthetic rates of trees from a marginal tropical forest (Xishuangbanna) and three subtropical forests (Dinghushan, Ailaoshan, Maoershan) in Southern China. Values are means ±SEs

	Leaf net photosynthetic rate (μmol m^{-2} s^{-1})		Percent decline (%)
	Summer	Winter	
Xishuangbanna (native species, n = 3)	12.6 ± 1.4	7.3 ± 2.1	44
Xishuangbanna (introduced species, n = 9)	9.7 ± 1.0	7.2 ± 1.6	28
Dinghushan (native species, n = 4)	8.0 ± 0.1	6.4 ± 0.4	20
Ailaoshan (native species, n = 10)	11.4 ± 0.4	7.9 ± 0.3	30
Maoershan (native species, n = 4)	10.0 ± 0.5	1.9 ± 0.3	81

Data source Xishuangbanna (Zhang et al. 2014), Dinghushan (Zhang and Ding 1996), Ailaoshan (Zhang 2012; Zhang et al. 2013a), Maoershan (Bai et al. 2010)

chlorophyll concentrations of trees could partially compensate for the negative effects of low temperatures on LAI and leaf photosynthetic capacity as observed in a South American subtropical forest (Cristiano et al. 2014). Third, in the case of Asian subtropical forests, solar radiation is higher and sunshine duration is longer in autumn and winter seasons compared to summer owing to less frequent rain and fog events, which can also compensate for the negative effects of seasonal low temperatures on photosynthesis (Zhang et al. 2013b). Lastly, in humid subtropical forests in Argentina, winter daytime temperatures are within the range of optimum temperature for CO_2 assimilation of trees (Goldstein, unpublished information).

Fig. 6 Seasonal dynamics in leaf area index (LAI) of two subtropical forests. **a** Field measured LAI in summer and winter of Ailaoshan and Iguazú subtropical forests. **b** Average monthly MODIS LAI of Iguazú. Data source; Ailaoshan data from Qi et al. (2013), Iguazú data from Cristiano et al. (2014)

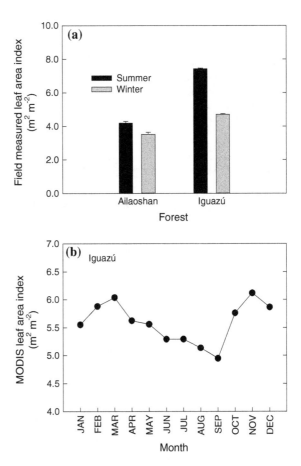

To summarize, the leaf photosynthesis and gross ecosystem carbon exchange (GEE) of subtropical forests are negatively affected by seasonal low temperatures, but subtropical forests have many tree and ecosystem level adaptations to compensate for low temperature effects. Then the question is to what extent trees and ecosystems can maintain net carbon gain during winters. The trees in the high elevation Ailaoshan subtropical forest (SW China), where average monthly winter temperature drops to 5 °C, maintained considerably high winter carbon assimilation rates (5.4–8.8 mol m^{-2} s^{-1}) and the estimated winter carbon gain (by photosynthetic light response curves and PPFD dynamics) accounts for ∼30 % of the yearly net carbon gain of the evergreen broadleaf tree species (Zhang 2012; Zhang et al. 2013b). Indeed, substantial winter carbon gain by the evergreen broadleaf trees, which is the dominant growth form in these forests, helps them to maintain high net yearly carbon gain compared to deciduous species. Interestingly, a deciduous species extends the leaf senescence process to facilitate photosynthetic carbon assimilation during part of the winter period (Zhang et al. 2013b), possibly as a

Fig. 7 Seasonal dynamics in net ecosystem CO_2 assimilation of two Asian subtropical forests and one forest located at the northern limit of the Asian tropics. Data for Ailaoshan and Dinghushan subtropical forests are from Tan et al. (2012), and data for Xishuangbanna tropical-subtropical forest are from Zhang et al. (2010). *Error bars* are standard errors of the values in each month

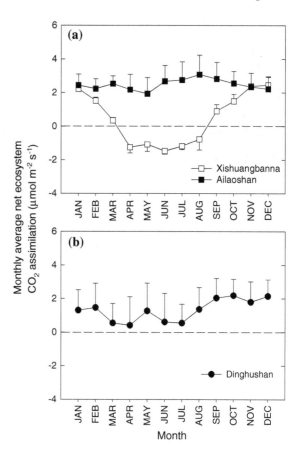

result of the selective pressures favoring year round carbon assimilation. At the ecosystem level, some subtropical forests can maintain high winter NEE comparable to that of the summer season (Tan et al. 2011, 2012) (Fig. 7). The NEE of the Ailaoshan subtropical forest is constantly high throughout the year (Fig. 7a), while the NEE of the Dinghushan subtropical forest is even higher in winter than in summer (Fig. 7b).

Why Are Subtropical Forests Among the Largest Carbon Sinks?

High year-round leaf photosynthesis and net ecosystem carbon exchange or carbon gains (NEE) (Fig. 7) suggest that subtropical forests are large carbon sinks. This prediction has been supported by eddy covariance data. At least in Asia, NEE of subtropical forests are among the highest of Asian ecosystems and are generally higher than that for tropical and temperate forests (Fig. 8) (Chen et al. 2013; Yu et al. 2014). In general, NEE of Asian forests increases with latitude from 0 to 28° N and declines as latitude further increases from 28° N (Fig. 8a). Therefore the maximum NEE of Asian forests is achieved in the subtropical zone (indicated as the shaded area of Fig. 8a).

Different responses of respiration and photosynthesis to temperature may explain the high carbon sink strength of subtropical forests. Plant tissue and soil respiration are very sensitive to changes in temperature (Atkin and Tjoelker 2003; Cavaleri et al. 2008; Tan et al. 2012; Slot et al. 2013). Across Asian tropical and subtropical forests, both gross ecosystem carbon exchange (GEE) and net ecosystem respiration increase exponentially with increasing mean annual temperature (MAT; Fig. 9). However, ecosystem respiration increases at a faster rate with increasing MAT than GEE, resulting in a negative relationship between NEE as a function of increasing MAT (Fig. 9). The relationship between NEE and MAT is well-described by a linear regression; NEE decreases as MAT increases and approaches zero at a MAT of 25.8 °C (Fig. 9). Above this temperature threshold the forests are sources (rather than sinks) of the main greenhouse gas (CO_2). This pattern is also supported by

Fig. 8 Net annual ecosystem carbon gains of Asian forests in relation to latitude (**a**) and mean annual temperature (**b**). Data from Tan et al. (2012) and Chen et al. (2013). All forest types were included. The *solid black lines* indicate Gaussian equations fitted to the data; $y = a + b \times e^{-0.5\,[(x-c)/d]^2}$. The *blue* and *red break lines* are 95 % confidence bands and 95 % prediction bands, respectively. The *shaded area* in (**a**) indicates the subtropical zone

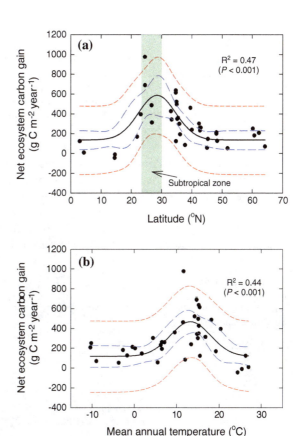

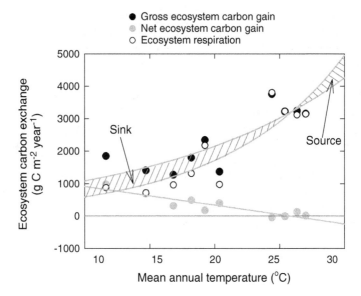

Fig. 9 Gross ecosystem carbon gain, net ecosystem carbon gain, and ecosystem respiration of Asian subtropical and tropical forests in relation to mean annual temperature (MAT). Data from Tan et al. (2012) and Chen et al. (2013). Subtropical forests include Ailaoshan, Dinghushan, Chilanshan, Qianyanzhou, and Huitong. Tropical forests include Pasoh, Lambir, Sakaerat, Mae Klong, and Xishuangbanna. *Solid symbols* are gross ecosystem carbon gain; *grey circles* are net ecosystem carbon gain; *open circles* are ecosystem respiration. The relationships between gross ecosystem carbon gain and MAT, and between ecosystem respiration and MAT were fitted with exponential regressions. The relationship between net ecosystem carbon gain and MAT was fitted with a linear regression

field measurements in Ailaoshan. Low temperatures in winter caused a 30 % decline in leaf photosynthetic rate of trees in Ailaoshan (Table 2), but at the same time resulted in a 62.9, 38.1 and 46.6 % decline in leaf, stem and soil respiration, respectively (Table 3). Higher declines in tree and soil respiration than photosynthesis contribute to the high winter carbon sequestration of the Ailaoshan subtropical forest. In addition, the young forest ages and high nitrogen deposition are

Table 3 Summer and winter respiration of a subtropical forest on Mount. Ailaoshan, Southwest China

	Respiration rate (μmol m^{-2} s^{-1})		Percent decline (%)
	Summer	Winter	
Leaf dark respiration	1.03	0.38	62.9
Tree trunk respiration	0.85	0.53	38.1
Soil respiration	4.37	2.33	46.6

Data source leaf respiration data from Zhang (2012), tree trunk respiration data from Hu (2010), and soil respiration data from Tan et al. (2012)

also factors contributing to the high NEE of the Asian subtropical forests (Yu et al. 2014). Compared to Asian tropical and temperate forests, Asian subtropical forests are generally younger and show significantly higher nitrogen deposition, both of which are positively related with forest net productivity (Yu et al. 2014).

In contrast to the Asian subtropical forests that have constantly high NEE throughout the year (Tan et al. 2012), a forest at the northern limit of the Asian tropics (Xishugangbanna) is a carbon sink in the cool season, but a carbon source during the warm season (Zhang et al. 2010) (Fig. 7), resulting in its lower annual net carbon gain (168.83 g C m^{-2} year^{-1}) compared to the subtropical forest in the same region (Ailaoshan; 976.15 g C m^{-2} year^{-1}). Indeed, Xishugangbanna has higher NDVI than Ailaoshan does in all seasons (Fig. 3a). It also has higher annual GEE (2342.67 g C m^{-2} year^{-1}) than Ailaoshan has (1848.33 g C m^{-2} year^{-1}). However, high warm season temperatures result in a much higher annual ecosystem respiration in Xishuangbanna (2173.83 g C m^{-2} year^{-1}) than in Ailaoshan (871.68 g C m^{-2} year^{-1}) (Zhang et al. 2010), which weakens its carbon sink strength. Therefore, respiration is the main factor determining the strength of carbon sinks of the Asian subtropical and tropical forests.

A global data synthesis has shown that the net ecosystem primary productivity (NPP) estimated by forest inventory (including stem, root and foliage growth) is positively correlated with MAT (Michaletz et al. 2014). It is possible that the forest inventory method has not taken the soil carbon stock into account, whose respiration accounts for a large proportion of the forest respiration and it is a very sensitive process which increases substantially as temperature increases (Tan et al. 2012). The forest inventory method may underestimate the carbon sink strength of the subtropical forests, e.g., the NPP of Ailaoshan subtropical forest estimated by forest inventory is 3–4 t ha^{-1} year^{-1} less than the NEE estimated by eddy covariance (Tan et al. 2011). More mechanistic studies are needed to address these differences and to better understand processes that are responsible for the differences.

Potential Responses of Subtropical Forests to Warming

Warming may weaken the carbon sink strength of subtropical forests. In the Asian tropics and subtropics, forest net carbon exchange or carbon gain (NEE) is negatively correlated with mean annual temperature (MAT; Fig. 9). Therefore, predicted global warming in the future (IPCC 2007) probably will reduce the net carbon sequestration strength of the subtropical forests. Since the ecosystem respiration dominates the forest NEE responses to MAT, higher increases in ecosystem respiration than in carbon assimilation will weaken the carbon sink strength and even turn the carbon sinks into sources (Fig. 9). According to the temperature response of respiration and carbon assimilation or GEE, the optimum MATs for NEE are between 11 and 15 °C, and further increase in MAT above 25.8 °C will turn the forests from carbon sinks to sources (Fig. 9). Although NEE of a forest can acclimate to changes in temperature (Yuan et al. 2011), the NEE of the forests used for NEE versus MAT analysis in this chapter are values of trees acclimated to the temperature regimes at each specific location. Thus the negative relationship between NEE and MAT found in Asian subtropical and tropical forests (Fig. 9) suggests that the temperature related decline has a strong potential to affect the carbon sink strength of these forests.

These predictions are consistent with tree growth studies in Australian subtropical and temperate forests. The optimum MAT for growth of Australian subtropical and temperate eucalyptus forests is around 11 °C. Because the current MAT of most Australian eucalyptus forests is above 11 °C, it is predicted that warming will decrease their growth and carbon sink strength (Bowman et al. 2014). It is possible that global warming may shift the temperate forests bordering the subtropics into subtropical forests and turn them into substantial carbon sinks. In addition, it is still unknown whether potentially positive effects of increased atmospheric CO_2 concentration and nitrogen deposition related to anthropogenic climate change would be able to compensate for the negative effects of warming on ecosystem net carbon gain.

Final Remarks

By analyzing the current results on carbon balance of subtropical forests, we hope to provide a better assessment on the potentially large contribution of subtropical forests to the global carbon cycle. We acknowledge that leaf, tree and ecosystem level mechanistic studies have only been conducted in a few forests of the subtropical region, preventing us from getting a comprehensive carbon economy map for all subtropical forest ecosystems. With the information obtained so far, however, we can conclude that high photosynthesis and carbon sequestration and relatively low respiration explain why subtropical forests are among the largest terrestrial carbon sinks of the world. A higher sensitivity of respiration than photosynthesis to low temperatures contributes to their high carbon sink strength throughout the year. Therefore, the subtropical forests may highly impact the global carbon cycle, calling for global attention on subtropical forests and indicating the importance of preserving these ecosystems for their buffering power of industrial carbon emissions.

References

Allen DJ, Ort DR (2001) Impacts of chilling temperatures on photosynthesis in warm-climate plants. Trends Plant Sci 6:36–42

Atkin OK, Tjoelker MG (2003) Thermal acclimation and the dynamic response of plant respiration to temperature. Trends Plant Sci 8:343–351

Azcon-Bieto J (1983) Inhibition of photosynthesis by carbohydrates in wheat leaves. Plant Physiol 73:681–686

Bai K, Jiang D, Cao K, Wan X, Liao D (2010) Photosynthetic response to seasonal temperature changes in evergreen and deciduous broad-leaved trees in montane forests of Ailao mountain and Maoer mountain. Acta Ecologica Sinica 30:0905–0913

Bowman DMJS, Williamson GJ, Keenan RJ, Prior LD (2014) A warmer world will reduce tree growth in evergreen broadleaf forests: evidence from Australian temperate and subtropical eucalypt forests. Glob Ecol Biogeogr 23:925–934

Campos GEP, Moran MS, Huete A, Zhang YG, Breslöff C, Huxman TE, Eamus D, Bosch DD, Buda AR, Gunter SA, Scalley TH, Kitchen SG, McClaran MP, McNab WH, Montoya DS, Morgan JA, Peters DPC, Sadler EJ, Seyfried MS, Starks PJ (2013) Ecosystem resilience despite large-scale altered hydroclimatic conditions. Nature 494:349–352

Cavaleri MA, Oberbauer SF, Ryan MG (2008) Foliar and ecosystem respiration in an old-growth tropical rain forest. Plant Cell Environ 31:473–483

Chen Z, Yu GR, Ge JP, Sun XM, Hirano T, Saigusa N, Wang QF, Zhu XJ, Zhang YP, Zhang JH, Yan JH, Wang HM, Zhao L, Wang YF, Shi PL, Zhao FH (2013) Temperature and precipitation control of the spatial variation of terrestrial ecosystem carbon exchange in the Asian region. Agric For Meteorol 182:266–276

Corlett RT (2013) Where are the Subtropics? Biotropica 45:273–275

Cristiano P, Madanes N, Campanello P, di Francescantonio D, Rodríguez S, Zhang Y-J, Carrasco L, Goldstein G (2014) High NDVI and potential canopy photosynthesis of south american subtropical forests despite seasonal changes in leaf area index and air temperature. Forests 5:287–308

Davis SD, Sperry JS, Hacke UG (1999) The relationship between xylem conduit diameter and cavitation caused by freezing. Am J Bot 86:1367–1372

Flexas J, Gulias J, Jonasson S, Medrano H, Mus M (2001) Seasonal patterns and control of gas exchange in local populations of the Mediterranean evergreen shrub *Pistacia lentiscus* L. Acta Oecol 22:33–43

Holaday AS, Martindale W, Alred R, Brooks AL, Leegood RC (1992) Changes in activities of enzymes of carbon metabolism in leaves during exposure of plants to low temperature. Plant Physiol 98:1105–1114

Hu W-Y (2010) Stem respiration of dominant tree species and epiphytic bryophyte respiration in a montane evergreen broad-leaved forest in Ailao mountains, China. Chinese Academy of Sciences, Beijing

Huang W, Zhang SB, Cao KF (2010) The different effects of chilling stress under moderate light intensity on photosystem II compared with photosystem I and subsequent recovery in tropical tree species. Photosynth Res 103:175–182

Huang YH, Zhou GY, Tang XL, Jiang H, Zhang DQ, Zhang QM (2011) Estimated soil respiration rates decreased with long-term soil microclimate changes in successional forests in Southern China. Environ Manag 48:1189–1197

IPCC (2007) Climate change 2007: the physical science basis. In: Solomon S, Qin D, Manning M, Chen Z, Marquis M, Averyt K, Tignor M, Miller H (eds) Contribution of working group I to the fourth assessment report of the intergovernmental panel on climate change. Cambridge University Press, Cambridge, p 996

Kingston Smith AH, Harbinson J, Williams J, Foyer CH (1997) Effect of chilling on carbon assimilation, enzyme activation, and photosynthetic electron transport in the absence of photoinhibition in maize leaves. Plant Physiol 114:1039–1046

Kosa P (2009) Air temperature and actual evapotranspiration correlation using Landsat 5 TM satellite imagery. Kasetsart J Nat Sci 43:605–611

Levitt J (1980) Responses of plants to environmental stresses. Physiological ecology, 2d edn. Academic Press, New York

Lieth H (1975) Modeling the primary productivity of the world. Primary productivity of the biosphere 14:237–263

Michaletz ST, Cheng DL, Kerkhoff AJ, Enquist BJ (2014) Convergence of terrestrial plant production across global climate gradients. Nature 512:3

Miyazawa Y, Kikuzawa K (2005) Winter photosynthesis by saplings of evergreen broad-leaved trees in a deciduous temperate forest. New Phytol 165:857–866

Ni J, Song Y (1997) Relationship between climate and distribution of main species of subtropical evergreen broad-leaved forest in China. Acta Phytoeclogica Sinica 21:115–129

Pan YD, Birdsey RA, Fang JY, Houghton R, Kauppi PE, Kurz WA, Phillips OL, Shvidenko A, Lewis SL, Canadell JG, Ciais P, Jackson RB, Pacala SW, McGuire AD, Piao SL, Rautiainen A, Sitch S, Hayes D (2011) A large and persistent carbon sink in the world's forests. Science 333:988–993

Pan YD, Birdsey RA, Phillips OL, Jackson RB (2013) The structure, distribution, and biomass of the world's forests. Annu Rev Ecol Evol Syst 44(44):593–622

Pennington RT, Prado DE, Pendry CA (2000) Neotropical seasonally dry forests and quaternary vegetation changes. J Biogeogr 27:261–273

Qi J-H, Zhang Y-J, Zhang Y-P, Liu Y-H, Yang Q-Y, Song L, Gong H-D, Lu Z-Y (2012) Water-holding capacity of an evergreen broadleaf forest in Ailao mountain and its functions in mitigating the effects of Southwest China drought. Acta Ecologica Sinica 32:1692–1702

Qi JH, Zhang YJ, Zhang YP, Liu YH, Lu ZY, Wu CS, Wen HD (2013) The impacts of the Southwest China drought on the litterfall and leaf area index of an evergreen broadleaf forest on Ailao mountain. Acta Ecologica Sinica 32:1692–1702

Qiu X-Z, Xie S-C (1998) Studies on the forest ecosystem in Ailao mountains, Yunnan, China. Yunnan Sciences and Technology Press, Kunming

Slot M, Wright SJ, Kitajima K (2013) Foliar respiration and its temperature sensitivity in trees and lianas: in situ measurements in the upper canopy of a tropical forest. Tree Physiol 33:505–515

Tan ZH, Zhang YP, Schaefer D, Yu GR, Liang NS, Song QH (2011) An old-growth subtropical Asian evergreen forest as a large carbon sink. Atmos Environ 45:1548–1554

Tan Z-H, Zhang Y-P, Liang N, Hsia Y-J, Zhang Y-J, Zhou G-Y, Li Y-L, Juang J-Y, Chu H-S, Yan J-H, Yu G-R, Sun X-M, Song Q-H, Cao K-F, Schaefer DA, Liu Y-H (2012) An observational study of the carbon-sink strength of East Asian subtropical evergreen forests. Environ Res Lett 7:044017

Taneda H, Tateno M (2005) Hydraulic conductivity, photosynthesis and leaf water balance in six evergreen woody species from fall to winter. Tree Physiol 25:299–306

Xiao XM, Zhang QY, Braswell B, Urbanski S, Boles S, Wofsy S, Berrien M, Ojima D (2004) Modeling gross primary production of temperate deciduous broadleaf forest using satellite images and climate data. Remote Sens Environ 91:256–270

Yu GR, Chen Z, Piao SL, Peng CH, Ciais P, Wang QF, Li XR, Zhu XJ (2014) High carbon dioxide uptake by subtropical forest ecosystems in the East Asian monsoon region. Proc Natl Acad Sci U S A 111:4910–4915

Yuan W, Luo Y, Liang S, Yu G, Niu S, Stoy P, Chen J, Desai AR, Lindroth A, Gough CM, Ceulemans R, Arain A, Bernhofer C, Cook B, Cook DR, Dragoni D, Gielen B, Janssens IA, Longdoz B, Liu H, Lund M, Matteucci G, Moors E, Scott RL, Seufert G, Varner R (2011) Thermal adaptation of net ecosystem exchange. Biogeosciences 8:1453–1463

Zhang S (1989) Comment on the northern boundary of the central subtropical evergreen broad-leaved forest zone in Shanghai area. Acta Phytoecologica et Geobotanica Sinica 13:93–95

Zhang YJ (2012) Water and carbon balances of evergreen and deciduous broadleaf trees from a subtropical cloud forest in Southwest China. University of Miami, Miami

Zhang Z, Ding M (1996) Biomass and efficiency of radiation utilization in monsoon evergreen broadleaved forest in Dinghushan biosphere reserve. Acta Ecologica Sinica 16:525–534

Zhang YP, Tan ZH, Song QH, Yu GR, Sun XM (2010) Respiration controls the unexpected seasonal pattern of carbon flux in an Asian tropical rain forest. Atmos Environ 44:3886–3893

Zhang YJ, Cao KF, Goldstein G (2013a) Winter photosynthesis of evergreen broadleaf trees from a montane cloud forest in subtropical China photosynthesis research for food, fuel and the future. Springer, pp 812–817

Zhang YJ, Yang QY, Lee DW, Goldstein G, Cao KF (2013b) Extended leaf senescence promotes carbon gain and nutrient resorption: importance of maintaining winter photosynthesis in subtropical forests. Oecologia 173:721–730

Zhang YJ, Holbrook NM, Cao KF (2014) Seasonal dynamics in photosynthesis of woody plants at the northern limit of the Asian tropics: potential role of fog in maintaining tropical rainforests and agriculture in Southwest China. Tree Physiol 34:1069–1078

The Ecophysiology of Leaf Lifespan in Tropical Forests: Adaptive and Plastic Responses to Environmental Heterogeneity

Sabrina E. Russo and Kaoru Kitajima

Abstract Leaf lifespan, the time from leaf expansion to shedding, exhibits wide variation and is a key integrator of relationships with photosynthetic rate, leaf mass per area (LMA), and leaf nitrogen among coexisting tropical tree species. We present a hierarchical view of sources of variation in leaf lifespan in tropical forests, emphasizing the importance of substantial within-species variation, which has rarely been addressed. Interspecific variation in leaf lifespan is positively correlated with LMA, varying from short-lived, low-LMA leaves to long-lived, high-LMA leaves of species associated with resource-rich versus resource-depleted habitats, respectively. Phenotypic responses of leaf lifespan and LMA to light show counter-gradient variation: with acclimation to shade, leaf lifespan increases, and LMA decreases, but both increase with adaptation to shade. In contrast, phenotypic responses to soil fertility are predicted to show co-gradient variation: both leaf lifespan and LMA increase with declining fertility both inter- and intraspecifically. We present new data analyses supporting these predictions, but the interactive effects of light and soil resources can produce complex phenotypic responses. Future studies of leaf lifespan should devote more attention to within-species variation to better quantify and explain how leaf lifespan is central to trade-offs generating the contrasting ecological strategies of tropical tree species.

S.E. Russo (✉)
School of Biological Sciences, University of Nebraska, Lincoln, USA
e-mail: srusso2@unl.edu

K. Kitajima
Graduate School of Agriculture, Kyoto University, Kyoto, Japan
e-mail: kaoruk@kais.kyoto-u.ac.jp

K. Kitajima
Smithsonian Tropical Research Institute, Balboa, Panama

© Springer International Publishing Switzerland 2016
G. Goldstein and L.S. Santiago (eds.), *Tropical Tree Physiology*,
Tree Physiology 6, DOI 10.1007/978-3-319-27422-5_17

357

Introduction

Leaf lifespan, the duration of time between when a leaf is first expanded and when it is senesced from the plant, differs greatly among species, among individual plants, and also among leaves on a plant (Chabot and Hicks 1982). Since the leaf is the principal photosynthetic organ of higher plants, its lifespan determines how long it will return photosynthetically fixed carbon to the plant (Kikuzawa and Lechowicz 2011). Plant growth and survival are critically dependent upon cumulative net photosynthetic carbon gain, which in turn depends strongly not only on insolation, but also on the availability of nutrients, especially nitrogen, and water in soil (Field 1983). Thus, leaf lifespan and nutrient allocation patterns are functionally linked (Ackerly and Bazzaz 1995; Hikosaka 2005), making leaf lifespan a critically important trait mediating the carbon and nutrient economies of plants that ultimately translate into fitness variation in relation to environmental heterogeneity.

Leaf lifespan is one axis of variation in the worldwide leaf economic spectrum (WLES), which describes a spectrum of coordinated leaf functional trait variation ranging from fast-growing species that produce short-lived, structurally inexpensive leaves with high nutrient concentrations and high photosynthetic productivity to slow-growing species that produce longer-lived, structurally expensive leaves that have lower nutrient concentrations and photosynthetic rates (Reich et al. 1991, 1992, 1997; Wright et al. 2004). Recent analyses of leaf lifespan using increasingly larger databases tend to focus on site and species-level means, neglecting large within-species variation, even though it can be substantial (Westoby et al. 2002). As a result, our knowledge of within-species variation in leaf lifespan in relation to differences in resource availability and other plant functional traits, at both the individual and leaf-levels, is comparatively rudimentary. The ability of a plant to respond to environmental shifts through acclimation will in part dictate responses to climate change, as well as determine patterns of species distribution along environmental gradients (e.g., Vanderwel et al. 2015). Furthermore, an integrated understanding of how multiple sources of natural selection operate on leaf lifespan in relation to the evolution of diverse plant ecological strategies (Donovan et al. 2011) requires quantitative estimates of how leaf lifespans change with environmental variation.

In this review, we seek to call attention to within-species variation in leaf lifespan, some of which can be understood as optimal plastic responses. We focus on tropical forests, where tree species display a wide range of leaf lifespans, including very long-lived leaves. First, we present a hierarchical view of the sources of variation in leaf lifespan and the dynamic underlying physiological mechanisms that influence how lifespan affects a plant's carbon and nutrient economies. Then, we discuss within-versus among-species variation in leaf lifespan and leaf mass per area (LMA) that can be related to variation in light availability and soil fertility from the perspective of theories on optimal leaf lifespan. We show that the direction and strength of the relationship of leaf lifespan with LMA differ among versus within-species, depending on the type of environmental factors considered. In the last section, we discuss knowledge gaps and research questions that are worth pursuing toward a more mechanistic understanding of leaf lifespan in the carbon and nutrient economies of the whole-tree.

Theories of Optimal Leaf Lifespan

Cost-benefit theories of leaf lifespan have a long history (e.g., Chabot and Hicks 1982). Here, we limit our review to the essential ideas needed for understanding the key elements of optimization of leaf function that have resulted in the global diversity and distribution of leaf lifespans and leaf habits. The fundamental question addressed by optimal leaf lifespan models is, for how long should a tree retain its leaves in a given environment? The optimal answer depends on the costs of leaf construction and maintenance, as well as costs associated with leaf turnover, versus the benefits that the leaf provides, namely photosynthetically fixed carbon and nutrient storage. Table 1 summarizes potential key costs and benefits, which will be described throughout this chapter. The cost of leaf construction is the total cost of acquiring all energy and materials required to build a leaf and its supporting organs (e.g., stem), as well as the cost of the maintaining molecules that make up the leaf (e.g., respiration). However, given the difficulty of assessing some of these costs, what is generally quantified is a minimum leaf construction cost, estimated as the total chemical bonding energy in organic molecules multiplied with a factor for biosynthetic pathway costs (Williams et al. 1987; Poorter et al. 2006). Per unit dry mass, this minimum biosynthetic cost of leaf construction may not differ much among species (Griffin 1994). Hence, leaf construction cost per unit area is approximated by LMA, and leaves with higher LMA require either fast photosynthetic rates or long lifespan to pay back construction costs and generate a net carbon gain.

Since the target of natural selection is individuals, optimization of carbon gain relative to carbon and nutrient costs must be considered at the whole-plant level, even though ecophysiological analyses, including many that we review here, often treat leaves as the unit of study. Leaves are expensive to manufacture: large amounts of limited resources such as nitrogen and phosphorus, as well as carbon to construct systems for structural support, vascular transport, and belowground resource uptake, must be allocated for their construction (Givnish 1988; Williams et al. 1989; Kikuzawa and Ackerly 1999; Reich et al. 2009). Thus, how many leaves a plant should maintain at a given time reflects a dynamic optimization of maximizing benefit, i.e., photosynthetic income, relative to costs of carbon and nutrient allocation for construction and maintenance. It is dynamic, because as a plant produces new leaves or is overtopped by neighbors, old leaves become shaded and less productive (e.g., Mooney et al. 1981). Moreover, aging results in the decrease of net photosynthesis per unit area and photosynthetic nitrogen use efficiency (PNUE) (Field and Mooney 1983; Sobrado 1994; Kitajima et al. 1997b, 2002). In addition, the cost of making and keeping a leaf is not fixed, as it is influenced by variation in structural and chemical defense (McKey 1974;

Table 1 Potential trade-offs that are relevant for cost-benefit models of leaf lifespan

Property	Benefit	Cost
Slower leaf turnover	• Leaf construction costs are infrequently incurred • Nutrient resorption and translocation costs are infrequently incurred • Nutrient and carbon losses are minimized, and leaves may store nutrients and carbohydrates, contributing to better nutrient retention and nutrient use efficiency • Cost of allocation to roots for uptake of belowground resources is reduced • Slower development of self-shading enables leaf to remain near its maximal productivity for longer time period	• Leaf area of tree does not always maximize light interception • Slower height and crown growth rate limit competitive ability • Slower leaf turnover delays the response to spatio-temporal fluctuations of light • Lost opportunity cost due to less optimal allocation to maximize the compounding interest of photosynthetic production
Slower A_{max}	• Lower nutrient demands alleviate need for extensive belowground investment • Slower transpiration rates reduce need for extensive water uptake • Photosynthetic machinery requires less maintenance respiration	• Slower rate of return of photosynthetic carbon requires longer lifespan to repay initial carbon construction cost • Plants fail to benefit from compounded interest associated with high A_{max}
Greater investment in defense	• Durability is enhanced through greater resistance to damage agents, such as physical forces and natural enemies • Functional deterioration with leaf age may be slower *via* slower accumulation of damage to the leaf	• Larger construction cost per unit leaf area means longer payback time

The benefits and costs at the plant level of three key properties associated with extending leaf lifespan for evergreen tropical species are summarized. Abbreviations: A_{max}, maximum rate of photosynthesis at the leaf level

Coley 1983), lost-opportunity costs of not allocating nutrients to newer leaves or non-leaf tissues (Harper 1989; Westoby et al. 2000), and costs of stem and root tissues to support leaves (Kikuzawa and Ackerly 1999; Givnish 2002). Thus, dynamic optimization of leaf lifespan involves processes operating at both the leaf and whole-plant levels, and it is adaptive for plants to adjust leaf phenotype and lifespan to acclimate to spatio-temporal variation in the environment and resource availability (Ackerly and Bazzaz 1995).

Sources of Variation in Leaf Lifespan

Here, we present a hierarchical view of the total phenotypic variation in leaf lifespan across all individuals, which can be partitioned into mechanisms operating at different levels of organization (Fig. 1).

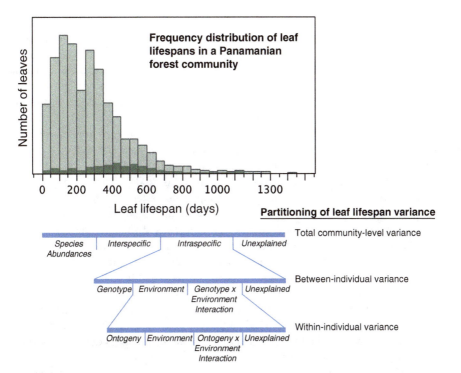

Fig. 1 Conceptual diagram of how the total variation in leaf lifespan within a forest community is hierarchically partitioned among different sources of variation at different levels of biological organization (*lower portion*), along with an example of leaf-level variation in estimates of leaf lifespan of seedlings across 58 species in a Panamanian forest community (*upper portion*). The part of the histogram in *darker green* indicates data for one species, *Virola surinamensis* (Myristicaceae), demonstrating substantial within-species variation in leaf lifespan, some of which reflects plasticity in response to light. Different species in a community contribute disparate amounts to the total variance of the community, depending on their intraspecific variation

Interspecific Variation

The covariation of leaf lifespan with other leaf functional traits defines key axes in the WLES describing interspecific variation in species ecological strategies (Reich et al. 1991, 1992, 1997; Wright et al. 2004). Tropical tree species vary enormously in their leaf lifespans, from pioneer species that exchange leaves within several weeks (e.g., *Heliocarpus appendiculatus*, Ackerly and Bazzaz 1995) to shade tolerant species that retain leaves over multiple years (e.g., 25 years reported for *Araucaria* by Molisch (1928), cited by Chabot and Hicks (1982); 12 years reported for a dicot tree sapling in Panama, pers. comm., P.D. Coley). Although phylogenetic history can constrain evolutionary changes, it is widely accepted that interspecific variation in leaf lifespan, and the shape of the WLES axes, arise through natural selection (Donovan et al. 2011), which has produced a diversity of convergent solutions, even within a given tropical forest community (e.g., Reich et al. 1991). Overall, we can interpret the trait syndromes associated with leaf lifespan as being evolutionary answers to optimization problems posed by complex, interacting trade-offs related to the carbon and nutrient economies of plants and their consequences for fitness in heterogeneous environments.

One of the widely-reported patterns from tropical forest tree communities is that species with shorter leaf lifespans tend to have faster maximum rates of photosynthesis and higher leaf nitrogen concentrations per unit leaf mass (Williams et al. 1989; Reich et al. 1991, 1992, 1997; Wright et al. 2004). They also generally occupy more productive habitats that can support the faster growth rates that make such a strategy advantageous, such as moist soils with high nutrient availability and irradiance. Since rapid shoot growth also causes self-shading, decreasing the insolation and productivity of older leaves (Hikosaka 1996; Ackerly 1999; Yamada et al. 2000), the rate of decline in photosynthetic rate with leaf age tends to be faster for species with shorter-lived leaves (Kitajima et al. 1997a, b, 2002). In such productive environments, greater whole-plant photosynthetic income may be gained by reallocating nutrients in aging leaves to support production of new leaves and rapid height growth (Field 1983; deJong 1995; Hikosaka 2005; Marty et al. 2010), and there would be little to gain by investing in structural durability beyond the minimal need to achieve the short optimal leaf lifespan.

Conversely, long-lived leaves with high LMA are generally found on slower-growing tree species that persist in less productive habitats, such as nutrient-depleted, well-drained soils or the shaded understory, where diurnal photosynthetic carbon gain is constrained by light, nutrient, or water availability. Longer leaf lifespan is advantageous because it prolongs the time over which such high carbon construction costs can be recouped (Chabot and Hicks 1982; Poorter et al. 2006). Moreover, in these habitats, allocation of limited resources to roots may constrain allocation to leaf construction (Bryant et al. 1983; Poorter et al. 2012), and leaves may have lower nutrient concentrations per unit mass (Reich et al. 1991, 1992, 1997; Wright et al. 2004). Longer lifespans are also selected because they reduce the nutrient loss associated with leaf turnover, and thus increase the whole-plant retention time of expensive-to-acquire nutrients (Monk 1966; Small 1972; Chapin 1980; Aerts and de Caluwe 1994). Likewise, defense (often carbon-based structural and chemical defenses, rather than nitrogen-based chemical defense, on infertile soils; Bryant et al. 1983) to avoid damage and premature leaf loss from herbivory or other hazards should also be favored (Janzen 1974; McKey 1979; Coley and Barone 1996).

In summary, variation among species in leaf lifespan should be viewed as an important part of the functional variation underlying the interspecific trade-off between growth and survival rates, which represents plant species' ecological strategies spanning fast growth and low survival to slow growth and high survival

(Kitajima 1994; Kobe 1999; Hubbell 2001; Kitajima and Myers 2008; Russo et al. 2008). Indeed, interspecific variation in leaf lifespan is positively correlated with survival rate in shade for seedlings and saplings in neotropical forests (Poorter and Bongers 2006; Kitajima and Poorter 2010; Kitajima et al. 2013) and saplings in Bornean rain forests (Russo, unpub. data).

Intraspecific Variation

Many studies, including those cited above, focus on interspecific variation, comparing mean or median lifespan of species, ignoring large variation within species (Fig. 1). Tree species with evergreen leaf habits should have evolved the capacity to produce leaves with varying lifespans, given that leaf structural and biochemical traits show such ecological plasticity (Valladares et al. 2007) and that plasticity in leaf lifespan enables trees to respond to environmental changes to maintain positive net carbon gain. The total variance within a population of a phenotypic trait such as leaf lifespan can be partitioned into four sources, plus unexplained variance (Fig. 1): (1) variation attributable to genes, (2) variation attributable to the environment, (3) variation attributable to ontogeny, (4) variation attributable to genotype-by-environment interaction (genetic variation for phenotypic plasticity).

Within-species variation in leaf lifespan may arise because the genotypes in the population differ in their leaf lifespan, and the relative proportion of these genotypes may differ among populations. Environmental heterogeneity can also be a significant source of variation in leaf lifespan both between habitat types and between microenvironments within a habitat due to differences in forest canopy structure, microtopography, or tree-size, which all influence access to above- and belowground resources (Weiner 1990). Even individual leaves and branches on a tree experience contrasting environments. In seasonally dry tropical forests, leaves produced in the early wet season function for a longer time under lower light availability of the cloudy rainy season, whereas those produced prior to the dry season can achieve higher productivity under a brighter sky, but are limited in maximum leaf lifespan due to dry-season deciduousness (Kitajima et al. 1997a). The shapes of leaf survival curves, which show the proportion of leaves remaining versus leaf age, demonstrate that leaf-level variation in lifespan can be substantial, even within a species (Fig. 2). When most leaves have similar lifespans, the survival curve shows low mortality before declining dramatically, which coincides with a short phase of synchronous senescence (Fig. 2a). Alternatively, the survival curve can decline more gradually, indicating steady mortality from early to late leaf ages, which reflects greater variation leaf lifespan among individual leaves (Fig. 2b).

Phenotypic plasticity is the capacity of a genotype to produce different phenotypes in different environments (Sultan 1995). It is considered by some to be a trait under selection that is favored when the environment is highly and unpredictably variable (Via and Lande 1985) or when it enables a plant to take maximal

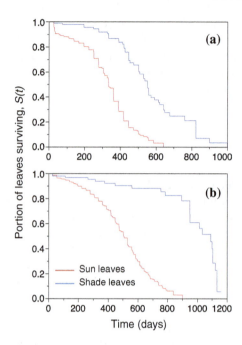

Fig. 2 Examples from two Panamanian tree species to demonstrate how the shapes of leaf survivorship curves differ between environments and species. Panels show Kaplan-Meier estimates of the survival function for sun leaves in the canopy (*red*) and shade leaves in the understory (*blue*). In the case of *Bombacopsis sessilis* (Bombacaceae) (**a**), sun and shade leaves show similar magnitudes of leaf-level variation in leaf lifespan across contrasting light environments, but for *Virola surinamensis* (Myristicaceae) (**b**), the survival function is steeper for shade than for sun leaves, indicating less variance in observed leaf lifespans among individual leaves in the former

advantage of ephemeral pulses of resources, and so faster-growing species are thought to have greater plasticity (Bazzaz 1979; Lambers and Poorter 1992; Valladares et al. 2000). Quantifying phenotypic plasticity at the individual level is challenging because genetic clones are rarely available for wild tree species, hence, environmentally induced phenotypic variation in leaf lifespan is measured at the population level using maternal siblings grown in contrasting environments. Even so, studies of plasticity in leaf lifespan of tropical species are rare (Ackerly and Bazzaz 1995; Kitajima et al. 2013). However, they still yield insights because even though the variation due to genotype and genetic variance for phenotypic plasticity are unknown, total phenotypic variation can be partitioned into what is explained by the environment versus all other sources (Whitman and Agrawal 2009).

Ontogenetic plasticity, in which the phenotype depends on an individual's developmental stage, also contributes to within-species variation. The tissue density and toughness of the leaf lamina increase from saplings to adults in tropical trees (Kitajima and Poorter 2010), and in a Malaysian forest, Osada et al. (2001) found that more sunlit leaves at taller heights within tree crowns had shorter lifespans relative to more shaded leaves at shorter heights. Because the environment changes dramatically with tree size in closed canopy tropical forests, the proportion of this variation due to purely environmental versus developmental influences is unknown.

Leaf Structure and Nitrogen: Key Aspects of Leaf Construction Cost and Lifespan

LMA and nitrogen concentrations vary with leaf lifespan among species, as described in the WLES, and also plastically within species (Ackerly and Bazzaz 1995). It is overly simplistic to consider LMA only as a measure of structural defense or to assume that a fixed proportion of nitrogen resides only in the carboxylation enzymes. It is therefore important to consider a leaf as a heterogeneous structure (Terashima et al. 2011). In mesic tropical forests, trees with needle-shaped leaves are rare. A common design that simultaneously allows for efficient harvest of light energy under a shaded canopy while enabling CO_2 uptake is for a plant to have multiple layers of thin and flat leaf blades arranged in its crown (Horn 1971; Halle et al. 1978; Hikosaka 2005). The leaf blade is a complex structure consisting of photosynthetic and vascular cells sandwiched between cuticles that provide protection from desiccation and physical damage. Within the leaf blade, metabolically active molecules reside in cells surrounded by cell walls of different thickness and mechanical properties (Onoda et al. 2015). There are many sources of structural and mechanical variation in the leaf blade that contribute to its toughness and density, two correlates of leaf lifespan (Coley 1983; Kitajima and Poorter 2010; Westbrook et al. 2011; Kitajima et al. 2012).

When a plant should shed a leaf is a function of nutrient, as well as carbon balance (Field 1983; Aerts and Chapin 2000). Shorter leaf lifespan should be favored when the cost of acquiring nutrients is low, whereas the reverse is true in nutrient-depleted soils, and only by accounting for the cost of nutrient acquisition can models accurately predict an evergreen leaf habit in highly seasonal, nutrient-limited, boreal environments (Givnish 2002). Consistent with these predictions, in an Argentine subtropical forest, leaf lifespans of seedlings of five tree species grown in high-light gaps declined with N and P fertilization, although some only marginally so (Villagra et al. 2013). Similar results were seen by Cordell et al. (2001) for *Metrosideros polymorpha* (Myrtaceae), but only on N-limited substrates. On P-limited substrates, fertilization with N and P had no effect on leaf lifespan. Thus, the effects of nutrient limitation on leaf lifespan are likely to depend upon the nutrient in question.

In some leaf lifespan models that consider costs and benefits only in terms of carbon, the predicted optimal time of leaf shedding is the time at which the leaf can no longer off-set its own carbon costs of maintenance respiration (Monsi and Saeki 2005; translated from Monsi and Saeki 1953). Consideration of optimum nitrogen allocation strategy in the plant canopy, however, is critical (Hirose 2005). For example, Oikawa et al. (2006) showed that this prediction is met for an herb when it was grown under high nitrogen availability, but that under low nitrogen availability, leaves were shed despite still having positive net carbon gain. When nitrogen availability limits the maximum leaf area within the plant canopy, the effects of leaf area index and self-shading are less pronounced, such that the oldest leaves at lower positions may be receiving sufficient light to allow C-gain to remain positive up until the time of leaf shedding (Ackerly and Bazzaz 1995). To maximize net carbon gain over all leaves in an entire, heterogeneously illuminated plant canopy, nitrogen should be allocated so that the most sun-lit leaves contain the greatest photosynthetic nitrogen concentrations, and such differential nitrogen allocation is expected to be more pronounced in species with a steep self-shading gradient (Hikosaka 2005; Hirose 2005). In a tropical forest canopy, *Cecropia* species with a low leaf area index (<1) had a shallower nitrogen gradient compared to species with greater LAI and self-shading (Kitajima et al. 2002, 2005). Thus, there are complicated interactions between insolation at the top of the crown and soil nutrient availability that influence LAI and the steepness of the self-shading gradient, which together affect differential nitrogen allocation among leaves and leaf lifespan. Moreover, how these processes interact to affect whole-plant C-gain is likely to vary among species with different ecological strategies.

One of the most commonly used, but deceptively simple, ecological measures of leaf structure is LMA (Osnas et al. 2013). However, LMA is a complex trait that can be decomposed into lamina thickness and density, which play different functional roles (Witkowski and Lamont 1991; Poorter et al. 2009). Sun-exposed leaves have multiple layers of elongated palisade mesophyll cells for more thorough absorption of high irradiance, resulting in greater lamina thickness and high LMA (Goldstein et al. this volume; Givnish 1988; Terashima et al. 2011). Greater LMA of sun leaves compared to shade leaves also involves change in the abundance of thick-walled vascular cells to meet high transpirational water demand and consequent increases in tissue density (Poorter et al. 2009). In contrast to these sun-shade acclimation responses, adaptation to high versus low-light habitats exhibit different directional responses: the leaves of light-demanding tree species have lower LMA than those of shade-tolerant tree species (Kitajima 1994; Walters and Reich 1999; Rozendaal et al. 2006; Markesteijn et al. 2007). This interspecific trend in LMA can be understood as adaptation to resource-limited environments, such as the shaded forest understory or infertile soils, in which longer time is required for paying back leaf construction cost (Mooney and Gulmon 1982; Williams et al. 1989).

Selection for Leaf Lifespan—What Is the Role of Defense?

If leaf lifespan is merely a function of how quickly a leaf wears out under the bombardment of attacks from herbivores and physical stresses, it may be reasonable to hypothesize a positive association between defense and leaf lifespan. But, a casual walk in a tropical forest reveals many "holey" leaves that exhibit extensive damage (e.g., 11-year old leaves in a Bornean rain forest Fig. 3). Coley (1983) evaluated saplings of 46 tropical tree species in Panama under a standardized

Fig. 3 Portraits of leaves that are greater than 11 years old on saplings of five tree species in the mixed dipterocarp forest of Lambir Hills National Park, Borneo, **a** *Polyalthia sarawakensis* (Annonaceae), **b** *Syzygium cf. grande* (Myrtaceae), **c** *Shorea laxa* (Dipterocarpaceae), **d** *Knema galleata* (Myristicaceae), **e** *Dipterocarpus globosus* (Dipterocarpaceae)

environment of treefall gaps to examine which leaf traits best explain differences in herbivory of young and mature leaves, leaf lifespan, and inherent differences in growth rates. The results show that interspecific variation in herbivory rates of young leaves is not explainable by most putative defense traits except for toughness. In contrast, differences between species in herbivory rates of mature leaves, although they are overall much lower than in young leaves, can be explained by toughness, cellulose contents, and pubescence, but not by carbon-based chemical defense of tannins and phenols. Similar results are reported for 24 species from the same Panamanian forest in Kitajima et al. (2012), in which leaf toughness is measured as fracture toughness and work-to-shear, along with their material bases.

Interestingly, these putative structural and mechanical defenses explain species differences in leaf lifespan and growth rates better than herbivory rate does. Thus, there is a paradox. Leaf lifespan, rather than herbivory, is associated with leaf functional traits associated with physical defense. But, a cafeteria experiment with a generalist herbivore has shown a clear negative correlation between toughness and herbivory rate ($r = -0.78$), which were as strong as positive correlations of toughness with leaf lifespan ($r = 0.88$) and sapling survival rate ($r = 0.78$) (Kitajima and Poorter 2010). Perhaps, specialist herbivores are responsible for herbivory in the field, but they may be influenced differently by different chemical defenses, such that comparison of a broad range of species does not reveal significant association between herbivory and putative leaf defense traits.

Recent comparative studies also shed light on how LMA is linked to leaf structural properties through evolution and acclimation. In comparative analysis of subcanopy leaves of 197 species, Westbrook et al. (2011) asked, "what makes leaves tough?" from an evolutionary perspective, using structural equation models with phylogenetic independent contrasts. The results showed lamina density and cellulose per unit dry mass as two alternative paths to evolutionarily increase leaf fracture toughness, which was greater for species with high survival in shade. Their results also suggested that lamina thickness evolved independently of density and cellulose per unit mass. Perhaps, fracture toughness and its material bases should be viewed as the physical robustness necessary to set the upper limit for maximum lifespan. For a given leaf, realized leaf lifespan, however, may be shorter than this potential maximum. Even if a leaf is built to last for five years, it is perhaps not adaptive for a sapling to keep it that long. Particularly when a canopy opening forms, it may be more beneficial to senesce old leaves, translocate nitrogen and phosphorus to new leaves at higher and sunnier positions, and thereby extend the main stem.

For a given leaf, LMA is a product of lamina thickness and tissue density (dry mass per volume). Both lamina thickness and density may have plastic responses to light and nutrient availability as described in later sections with respect to LMA. Higher light availability in treefall gaps is associated with thicker, denser, and tougher leaves, but leaf lifespan is shorter, and herbivory is more common in gap seedlings than in understory seedlings (Kitajima et al. 2012). Hence, it is overly simplistic to interpret LMA as a putative structural defense trait. For interspecific variation, the tissue-density aspect of LMA enhances leaf lifespan, but the lamina thickness appears to be of lesser importance. For intraspecific variation associated with sun-shade gradients, leaf lifespan may decrease while LMA increases from shade to sun. Such counter-gradient variation (Lusk et al. 2008) is also found for putative chemical defense as well (Coley 1993); for temperate tree saplings, Shure and Wilson (1993) found that within species, acclimation to larger gap size (i.e., more light) resulted in higher tannin concentrations, but among species, higher tannin was associated with adaptation to shade.

Empirical Tests of Conceptual Models for Acclimation and Adaptation to Varying Resources

As discussed above, acclimation to variable light availability versus adaptive specialization to contrasting irradiance habitats may exhibit different relationships of leaf lifespan with LMA and other leaf functional traits. The interaction of nitrogen and light availability is also key to understanding the optimization of leaf lifespan. Unfortunately, much less is known about how acclimation and adaptation exhibit co-variation of leaf lifespan and LMA within and among species in relation to soil fertility variation in tropical forests. In this section, we first present conceptual predictions of covariation of leaf lifespan and LMA (Fig. 4). We then test them with empirical data on plasticity of species specialized to habitats with contrasting resource availabilities, using data on leaves from Panama (seedlings grown under contrasting light environments) and Borneo (seedlings grown under contrasting light and soil environments). Our goal is to interpret these patterns in adaptive ecological plasticity with respect to the resource economic strategies of tree species.

Acclimation Versus Adaptation to Light Availability: Counter-Gradient Variation

As discussed in the previous section, the among-species correlation in leaf lifespan and LMA is positive (WLES), whereas plastic covariation within species is negative. Such counter-gradient variation is summarized graphically for light (Fig. 4a): acclimation to shade causes a decrease in LMA (Arrows 1 and 3 in Fig. 4a) and an increase in lifespan (Arrows 2 and 4 in Fig. 4a) within species, whereas between species, longer leaf lifespan is associated with higher LMA (positive slope of across species in the same light environment). This conceptual model is similar to the one by Lusk et al. (2008), but our model is more explicit about variation in the acclimation response of each species, depending on its resource economic strategy. More specifically, we predict that the degree of plasticity in leaf lifespan is greater, but that of LMA is smaller, for species with longer maximum leaf lifespan (shallower slopes for the gray arrows representing plastic covariance toward the right-hand side of Fig. 4a). For species under selection to have rapid leaf turnover in a resource-rich environment, it may be infeasible to reduce leaf lifespan beyond a

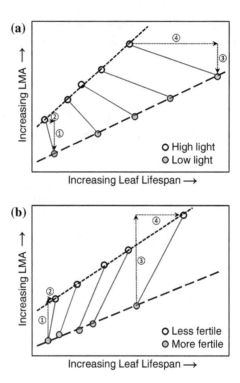

Fig. 4 Conceptual models of how leaf mass per area (LMA) covaries with leaf lifespan across and within species in terms of phenotypic responses to **a** contrasting light regimes (*open symbols* high light, *closed symbols* low light) and **b** contrasting soil types (*open symbols* fertile, *closed symbols* infertile). Each pair of *circles* connected by a *solid gray line* corresponds to an individual species and its phenotypic reaction norm. The interspecific relationship of LMA versus lifespan in a given environment is shown as *dotted* or *broken lines*: in high or low light (*short* and *long-dash*, respectively) and in more or less fertile soil types (*long* and *short-dash*, respectively). *Small circled numbers* indicate directional changes in LMA and lifespan, and the mechanisms involved for each are described in the main text

certain minimum, due to ever-increasing costs associated with leaf-turnover, even though they may exhibit a high degree of plasticity in LMA under higher light availability. At the other end of the LMA-lifespan spectrum, species selected to have structural defense to enable longer leaf lifespans have more latitude for plastic lifespan responses, which may result in greater shifts in response to light environment. There is also likely to be an upper limit on LMA, since light attenuates within thick mesophyll layers, and severely decreased lamina area may reduce photosynthetic surface area. Thus, the degrees of plasticity in LMA and leaf lifespan may differ across species with different ecological strategies, influencing the slope of the interspecific LMA-lifespan relationship (Fig. 4a).

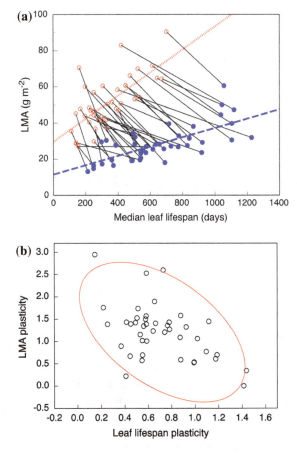

Fig. 5 **a** Interspecific variation and plasticity of leaf lifespan and LMA across seedlings of 41 Panamanian tropical tree species. Seedlings were grown from seeds in common gardens in treefall gaps (*red, open circles*) or understory common gardens or shade house (0.8 % of total daily PPFD in both; *blue, closed circles*). Each *solid line* connecting two data points from a given species shows the direction of phenotypic co-variation in LMA and leaf lifespan. The *dotted* and *broken lines* indicate the interspecific relationship. Species with estimates of median leaf lifespan exceeding 1300 days were not included, nor were species that were raised or could survive only in one of the environments. See Kitajima et al. (2013) for further details. **b** Relationship between plasticity in LMA and leaf lifespan derived from the data shown in (**a**), with the 95 % confidence *ellipse* in *red*. The plasticity index was calculated as the difference in trait values between contrasting light environments divided by the geometric mean of those values

We tested these predictions using data on the LMA and leaf lifespans of seedlings from 41 Panamanian tree species that were experimentally grown in replicated common gardens in gaps and shaded understory. These species represent a range of shade-tolerance strategies, from light-demanding pioneers to shade tolerant late-successional species (Kitajima et al. 2012, 2013). Overall, the results shown in

Fig. 5a support the conceptual model predictions (Fig. 4a). Within-species, acclimation to shade involved an increase of leaf lifespan and decrease of LMA (negative correlation). Among species, longer leaf lifespans were associated with
increased LMA under each standardized light environment (positive correlation).
Also, light-demanding species with short leaf lifespan exhibited large degrees of
plasticity in LMA, but relatively smaller changes in leaf lifespan between the two
light environments. Consistent with our conceptual model, plasticity in LMA
declined with increasing plasticity in leaf lifespan ($r = -0.491$, $P = 0.001$; Fig. 5b).

Acclimation Versus Adaptation to Soil Resource Availability: Co-gradient Variation

Among and within tropical forests, soil properties vary greatly in terms of availability of mineral nutrients, such as nitrogen and phosphorous, and water, even
under the same climate (Ashton 2015). Co-variation of leaf lifespan and LMA in
relation to natural soil gradients is widely demonstrated both in temperate and
tropical ecosystems (e.g., Cordell et al. 2001; Wright et al. 2002), although in many
such studies, variation due to acclimation versus adaptation are confounded. While
we recognize that different types of belowground resources may produce varying
responses (e.g., low-rainfall vs. low-nutrient sites, Wright et al. 2002; nitrogen vs.
phosphorus, Cordell et al. 2001), here we develop a general conceptual model for
inter- and intraspecific variation in leaf lifespan and LMA due to soil fertility.

Variation in soil fertility is expected to produce co-gradient variation (Fig. 4b),
unlike the case for insolation (Fig. 4a), since lower soil fertility should be associated
with increases in both LMA (Arrows 1 and 3 in Fig. 4b) and leaf lifespan (Arrows 2
and 4 in Fig. 4b) both within and between species. Species with contrasting soil
associations across fertility gradients experience and are presumably adapted to
different soil nutrient and moisture regimes, and so we expect them to differ in
LMA and leaf lifespan plasticity, causing the slope of the interspecific
LMA-lifespan relationship to vary across soil habitat types, analogous to
light-related patterns.

Patterns of within-species variation in leaf lifespan with soil fertility, however,
can be quite inconsistent across studies, often with apparently non-adaptive phenotypic responses, such as reduced leaf lifespan in less fertile soils (Aerts and
Caluwe 1995; Richardson et al. 2010; Pornon et al. 2011). Pornon et al. (2011)
proposed a conceptual model to explain these counter-intuitive soil-related
responses, in which they consider that sink activity due to growth accesses
endogenous, more than exogenous, nitrogen when soil nitrogen is very low,
accelerating leaf senescence (Marty et al. 2009; Pornon et al. 2011). However, this
logic only makes sense if a plant species does not sufficiently down-regulate sink
activity in resource-depleted environments via regulation of leaf production and
growth rates, which would reduce demand for nitrogen in the first place. Although

height competition is less intense in nutrient-depleted soils, down regulation of height growth may not be an evolutionarily stable strategy in some circumstances (Anten 2005). There is ample evidence that growth rates vary depending on resource availability, but there are likely to be limits to the amount by which a species can adjust inherent variation in growth rate (sensu Lambers and Poorter 1992).

Interactions of Light and Soil Resource Availability on Leaf Lifespan Variation

Both light and soil resources vary in time and space, and in the forest understory, they are often interdependent (Coomes and Grubb 2000; Russo et al. 2012), and so it makes sense to examine their interactive effects on acclimation of LMA and leaf lifespan. We examined these patterns using data from seedlings of 13 Bornean tree species that were reciprocally transplanted into contrasting light and soil environments in forest experimental plots. Leaf lifespan was estimated on seedlings of each species sown on clay and sandy loam soils in gaps and understory. All study species are shade tolerant, but differ in soil specialization, ranging from species associated with nutrient-depleted, well-drained sandy loam soil, to clay soil with greater nutrient concentrations and water-holding capacity, and generalist species associated with both soil types (Davies et al. 2005). Moreover, the sandy loam specialists have slower diameter growth and higher survival rates compared to clay specialists (Russo et al. 2005).

The patterns among the Bornean species across light and soil treatments (Fig. 6) were considerably more complex than the pattern due to insolation alone (Fig. 5). As predicted by our conceptual models (Fig. 4), the results show counter-gradient variation for acclimation versus adaptation of LMA and leaf lifespan with light environment, and co-gradient variation with contrasting soil environments. However, the direction and magnitude of plasticity of both traits depended on the soil and light environment, as well as the species' soil specialization. Among species, increases in leaf lifespan were associated with increases in LMA, and, consistent with our conceptual model, the slopes of these relationships depended upon the environment. Within-species, acclimation to shade produced an increase in leaf lifespan and a decrease in LMA (negative slopes of the black lines in Fig. 6a, b). These relationships were most consistently seen on sandy loam (Fig. 6b), which exhibited counter-gradient selection and resembled patterns for the Panamanian species (Fig. 5a). On the more fertile clay (Fig. 6a), LMA decreased for all species with acclimation to shade, as expected. However, leaf lifespan on clay showed variable patterns with acclimation to shade, increasing or showing little change, as expected for clay specialists, but counterintuitively decreasing for sandy loam

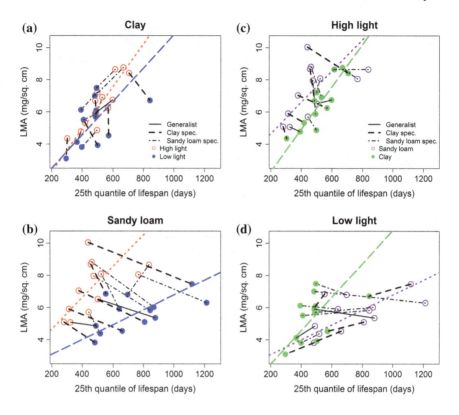

specialists and generalists. One explanation is that the sandy loam specialist seedlings may have accelerated leaf senescence in order to reduce carbohydrate consumption (Sevanto et al. 2014), since the understory on clay is shadier than the understory on their home sandy loam soil to which they are presumably adapted (Russo et al. 2012).

Within-species, acclimation from sandy loam to clay produced a decrease in leaf lifespan and either little change or a decrease in LMA (Fig. 6c, d), co-gradient variation that is most consistently seen in the low light understory (Fig. 6d). In gaps, however, patterns were complex, with some species exhibiting counter-gradient selection or little change in leaf lifespan across contrasting soils (Fig. 6c). These disparate responses could be related to whether species are N or P-limited. On a P-limited substrate, Cordell et al. (2001) found no change in the leaf lifespan of *M. polymorpha* in response to fertilization with N, P, or their combination, in contrast to an N-limited substrate, on which leaf lifespan declined with fertilization. Although responses of individual species varied, at a given LMA, leaf lifespan was always longer for seedlings in low light especially on sandy loam (Fig. 6a, b). However, at a given LMA, leaf lifespan was on average longer for seedlings on sandy loam than on clay soil only in low light (Fig. 6d); in high light, leaf lifespan

◀ **Fig. 6** Interspecific variation and plasticity of leaf lifespan and LMA of seedlings of 13 tropical, shade-tolerant tree species in the Dipterocarpaceae specializing on sandy loam soil, clay soil, or neither (generalists). Seedlings were grown from newly germinated seeds in a field reciprocal transplant experiment in plots in Bornean forest (Lambir Hills National Park, Malaysia) located on sandy loam (*purple, open circles*) or clay (*green, closed circles*) and in treefall gaps (*red, open circles*) or shaded understory (*blue, closed circles*). Responses in the four environmental regimes are in each panel, comparing plots in high versus low light (**a** and **b**) or sandy loam versus clay (**c** and **d**), holding either soil or light environment constant, respectively. The two data points for each species are connected by a *black line* showing the phenotypic reaction norm, with the dashing pattern indicating soil specialization (*solid*, generalist; *long dash*, clay specialist; *dot-dash*, sandy loam specialist). *Colored dotted* and *broken lines* (colors correspond to the symbol colors for the different environments) indicate the interspecific relationship of LMA with the 25th quantile of leaf lifespan within each soil or light environment. LMA was determined for all true leaves at approximately 1.5 years after seeds were sown. Species' 25th quantile leaf lifespan was determined for marked true leaves for up to 1200 days. The 25th quantile of lifespan was used because the median lifespan could not be estimated for all species based on the study duration (3.25 years) and long leaf lifespans. Study species are shade-tolerant and either canopy or emergent tree species. Congeneric species of contrasting soil associations include the following: sandy loam specialists, *Dryobalanops aromatica, Hopea beccariana, Dipterocarpus globosus, Shorea beccariana, Shorea laxa, Vatica nitens*; clay specialists, *Dryobalanops lanceolata, Hopea dryobalanoides, Dipterocarpus palembanicus, Shorea macrophylla, Shorea xanthophylla*; and generalists, *Anisoptera grossivenia* and *Dipterocarpus acutangulus*. Soil associations are based on Poisson cluster model analyses in Davies et al. (2005; see also Russo et al. 2005), with stems of specialists being significantly aggregated on that soil type and generalists showing no significant aggregation due to soil type

at a given LMA was shorter for seedlings on sandy loam than on clay (Fig. 6c). This unexpected response resembles findings from Reich et al. (1999) and Wright et al. (2002) in relation to water availability: at a given LMA, species associated with lower rainfall sites had shorter leaf lifespan than those at wetter sites. For these Bornean species, the LMA required to achieve a given lifespan is higher on sandy loam, potentially owing to lower soil moisture (Russo et al. 2010). Higher LMA may confer better tolerance to low soil water potentials experienced at low-rainfall sites or well drained soils (Niinemets 2001; Wright et al. 2002). Thus, soil-related plasticity in leaf lifespan when light is not limiting may depend upon the particular belowground resource in question, as well as variation among species in which resource is most limiting to growth.

As with the results from Panama, plasticity in LMA and leaf lifespan were negatively correlated ($r = -0.56$, $p < 0.001$; Fig. 7). However, on the resource-depleted habitats, the mean plasticity index for leaf lifespan (sandy loam: 0.50 and low light 0.45) was on overage two to four-fold higher than that for LMA (sandy loam: 0.26 and low light 0.11), whereas in the more productive environments, the mean plasticity indices for leaf lifespan and LMA were more comparable (lifespan, 0.15 and 0.20, and LMA, 0.19 and 0.14, for clay and high light, respectively), suggesting that in resource-depleted habitats leaf lifespan may be extended by means other than increasing LMA.

Fig. 7 Relationship between plasticity in leaf mass per area (LMA) and leaf lifespan for the data shown in Fig. 6. The plasticity index was calculated as the difference in trait values between contrasting soil or light environments (with the other factor held constant) divided by the geometric mean of those values

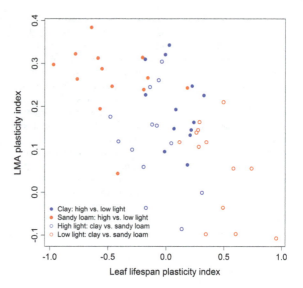

Towards a More Mechanistic Understanding of Leaf Lifespan—What Remains Unknown?

Leaf lifespan is central to the carbon and nutrient economies of trees, with direct and indirect links to individual growth and survival. Yet, even very basic information about leaf lifespan in tropical trees is still lacking. Here, we highlight some important knowledge gaps impeding a more mechanistic understanding of the role of leaf lifespan in whole-tree carbon and nutrient dynamics of tropical species.

How Long Is the Longest Leaf Lifespan among Tropical Broad-leaved Species?

Although there are several methods for estimating leaf lifespan (see Kikuzawa and Lechowicz 2011, Chap. 1 for a review) accurate estimates of the distribution of leaf lifespans for evergreen tropical tree species are difficult to obtain. Estimates of leaf lifespan are improved when the data are not censored, i.e., when lifespan of all leaves are followed from their birth to death (Dungan et al. 2003). But data collection campaigns over long time periods for a sufficient number of leaves are logistically challenging, especially for slow-growing species in aseasonal or resource-poor tropical environments. Leaf lifespans of shade-tolerant juveniles in tropical moist forests can be typically five or more years (e.g., Kursar and Coley 1993; King 1994). In saplings of shade-tolerant tree species in an ever-wet Bornean rain forest growing on nutrient-depleted soils, for 14 out of 31 species, even the

25th percentile of the leaf lifespan distribution was not well estimated after 6.5 years of censuses, suggesting that leaf lifespans of these species may routinely exceed 10 years (Russo, unpub. data; Fig. 3). Thus, many published leaf lifespans for evergreen tropical species may not only fail to encompass the full range of within- and between-tree variation, but they may also be underestimates.

Why Are Seemingly Unproductive Leaves Retained?

Juvenile trees of shade tolerant species in closed-canopy tropical forest often survive for long periods of time under deep shade. Given the aged appearance of their often self-shaded leaves, which themselves are frequently covered with lichens and epiphylls (Fig. 3), it is hard to imagine that such leaves are contributing significantly to the tree's carbon economy. So, why are they retained? These leaves may have such low respiratory maintenance costs that they can still achieve positive net carbon balance, in which case, there may be little advantage to senescing the leaf, particularly if the new leaf would not intercept appreciably more insolation as is the case for shaded juveniles. New and young leaves are more likely to be attacked by herbivores (Coley 1983; Coley et al. 2005), and thus replacing old leaves could present a significant risk of resource drain. It is also possible that these leaves are retained because they function to store nutrients, for example, nitrogen in disused proteins, that will be used to produce a flush of new leaves when the growing environment improves. Carbon and nitrogen isotope labelling studies (e.g., Pornon and Lamaze 2007) hold promise for identifying the sources of nitrogen used to form new leaves and the leaf-level photosynthesis-respiration balance of older leaves.

Are Leaves Senesced When Leaf Photosynthetic Function Declines to Zero?

How rapidly photosynthetic rates and PNUE decline with leaf age is a key parameter in theoretical models of optimal leaf lifespan because it strongly influences photosynthetic productivity through the lifetime of the leaf (Kikuzawa 1991, 1995; Ackerly 1999; Escudero and Mediavilla 2003). At least three classes of mechanism likely cause the decline in function with leaf age: self-shading, age-related deterioration per se, and withdrawal of nitrogen from the lamina. With the exception of a few studies (Kitajima et al. 1997b, 2002), how much variation in the age-related decline in leaf function exists among and within tropical tree species, and the environmental controls over this process, remain largely unmeasured.

Estimation of decline-rate functions can be accomplished with either a chronosequence approach, in which leaf position substitutes for exact leaf age, or with a repeated-measurements approach, in which photosynthetic parameters are repeatedly estimated on the same leaf over its lifetime (Kitajima et al. 2002; Osada et al. 2015). The chronosequence approach is less labor intensive, but it ignores within-plant and within-species variation the leaf lifespan or decline-rate function, which can be substantial, as discussed earlier. This results in substantial bias in estimating functional decline rate with leaf age, and mismatch of empirical data to theoretical predictions of leaf lifespan as to whether leaves are senesced when photosynthetic function declines to zero (Osada et al. 2015). Importantly, the bias worsens as the leaf-level variance in lifespan increases. Given the substantial variation in survival time among individual leaves within-species and even within plants (Figs. 1 and 2), we call attention to the importance of considering variance in age-related changes in leaf function and demography in future tests of optimal leaf lifespan models with empirical data.

Models of Leaf Lifespan—What Are the Key Trade-Offs?

The internal dynamics of carbon and nutrients in trees are complex, are determined by multiple functional traits, and often involve latent physiological and biochemical processes. As a result, the mechanisms by which leaf lifespan affects these dynamics are difficult to quantify with empirical studies. Models of leaf-lifespan that only consider leaf-level processes cannot account for key processes affecting leaf lifespan that are operating at the whole-tree level. Many optimal leaf lifespan models differ in fundamental ways in terms of which mechanisms and physiological processes they incorporate. As a result, each is more appropriate in some ecological contexts than in others. For example, the whole-tree model of Givnish (2002) includes a parameter controlling the fractional carbon allocation to leaves versus roots, and not accounting for this yields predictions of the distribution of deciduous versus evergreen leaf habits in seasonal environments that do not match those found in nature. Somewhat paradoxically, simpler models that do not explicitly consider whole-plant nutrient allocation strategies, in contrast, can also give rough approximation of the prevalence of deciduous versus evergreen habits (Kikuzawa et al. 2013). Although there is no complete consensus as to what are the critical parameters that explain leaf lifespan variation from an ecophysiological perspective, there is a general recognition that trade-offs in resource allocation strategies are key to mechanistic understanding. In Table 1, we summarize some of the critical leaf-level and whole-tree trade-offs in function that mechanistic models of leaf lifespan should capture. Whether we want to explain global variation in leaf habit or the biological mechanisms governing leaf lifespan variation in a forest community in relation to resource availability, dynamic resource allocation models that incorporate these major physiological processes and that accurately capture these critical trade-offs are needed. Development of such models is particularly exciting in species-rich and functionally diverse tropical tree communities.

Acknowledgements Bornean and Panamanian leaf trait data were collected under the US National Science Foundation (NSF) award DEB-0919136 to SER and IBN-0093033 to KK, respectively. The manuscript preparation was initiated while SER was supported by a Short-term Fellowship (S-14181) from the Japanese Society for Promotion of Science.

Literature Cited

Ackerly D (1999) Self-shading, carbon gain and leaf dynamics: a test of alternative optimality models. Oecologia 119:300–310

Ackerly DD, Bazzaz FA (1995) Leaf dynamics, self-shading and carbon gain in seedlings of a tropical pioneer tree. Oecologia 101:289–298

Aerts R, Caluwe Hd (1995) Interspecific and intraspecific differences in shoot and leaf lifespan of four *carex* species which differ in maximum dry matter production. Oecologia 102:467–477

Aerts R, Chapin FS (2000) The mineral nutrition of wild plants revisited: a re-evaluation of processes and patterns. Adv Ecol Res 30:1–67

Aerts R, de Caluwe H (1994) Nitrogen use efficiency of *carex* species in relation to nitrogen supply. Ecology 75:2362–2372

Anten NPR (2005) Optimal photosynthetic characteristics of individual plants in vegetation stands and implications for species coexistence. Ann Bot 95:495–506

Ashton P (2015) On the forests of tropical asia: lest the memory fade. The University of Chicago Press, Chicago

Bazzaz FA (1979) The physiological ecology of plant succession. Annu Rev Ecol Syst 10:351–371

Bryant JP, Chapin FS III, Klein DR (1983) Carbon/nutrient balance of boreal plants in relation to vertebrate herbivory. Oikos 40:357–368

Chabot BF, Hicks DJ (1982) The ecology of leaf life spans. Annu Rev Ecol Syst 13:229–259

Chapin FSI (1980) Mineral nutrition of wild plants. Annu Rev Ecol Evol 11:233–260

Coley PD (1983) Herbivory and defensive characteristics of tree species in a lowland tropical forest. Ecol Monogr 53:209–234

Coley PD (1993) Gap size and plant defenses. Trends Ecol Evol 8:1–2

Coley PD, Barone JA (1996) Herbivory and plant defences in tropical forests. Annu Rev Ecol Syst 27:305–335

Coley PD, Lokvam J, Rudolph K, Bromberg K, Sackett TE, Wright L, Brenes-Arguedas T, Dvorett D, Ring S, Clark A, Baptiste C, Pennington RT, Kursar TA (2005) Divergent defensive strategies of young leaves in two species of *inga*. Ecology 86:2633–2643

Coomes DA, Grubb PJ (2000) Impacts of root competition in forests and woodlands: a theoretical framework and review of experiments. Ecol Monogr 70:171–207

Cordell S, Goldstein G, Meinzer FC, Vitousek PM (2001) Regulation of leaf life-span and nutrient-use efficiency of metrosideros polymorpha trees at two extremes of a long chronosequence in hawaii. Oecologia 127:198–206

Davies SJ, Tan S, LaFrankie JV, Potts MD (2005) Soil-related floristic variation in the hyperdiverse dipterocarp forest in Lambir hills, Sarawak. In: Roubik DW, Sakai S, Hamid A (eds) Pollination ecology and rain forest diversity, Sarawak studies. Springer, New York, pp 22–34

deJong TJ (1995) Why fast-growing plants do not bother about defence. Oikos 74:545–548

Donovan LA, Maherali H, Caruso CM, Huber H, de Kroon H (2011) The evolution of the worldwide leaf economics spectrum. Trends Ecol Evol 26:88–95

Dungan RJ, Duncan RP, Whitehead D (2003) Investigating leaf lifespans with interval-censored failure time analysis. New Phytol 158:593–600

Escudero A, Mediavilla S (2003) Decline in photosynthetic nitrogen use efficiency with leaf age and nitrogen resorption as determinants of leaf life span. J Ecol 91:880–889

Field C (1983) Allocating leaf nitrogen for the maximization of carbon gain: leaf age as a control on the allocation program. Oecologia 56:341–347

Field C, Mooney HA (1983) Leaf age and seasonal effects on light, water, and nitrogen use efficiency in a california shrub. Oecologia 56:348–355

Givnish T (1988) Adaptation to sun and shade: a whole-plant perspective. Aust J Plant Physiol 15:63–92

Givnish T (2002) Adaptive significance of evergreen vs. deciduous leaves: solving the triple paradox. Silva Fenn 36

Griffin KL (1994) Calorimetric estimates of construction cost and their use in ecological studies. Funct Ecol 8:551–562

Halle F, Oldeman RAA, Tomlinson PB (1978) Tropical trees and forests: an architectural analysis. Springer-Verlag, New York

Harper JL (1989) The value of a leaf. Oecologia 80:53–58

Hikosaka K (1996) Effects of leaf age, nitrogen nutrition and photon flux density on the organization of the photosynthetic apparatus in leaves of a vine (ipomoea tricolor cav.) grown horizontally to avoid mutual shading of leaves. Planta 198:144–150

Hikosaka K (2005) Leaf canopy as a dynamic system: ecophysiology and optimality in leaf turnover. Ann Bot 95:521–533

Hirose T (2005) Development of the monsi–saeki theory on canopy structure and function. Ann Bot 95:483–494

Horn HS (1971) The adaptive geometry of trees. Princeton University Press, Princeton, NJ

Hubbell SP (2001) The unified neutral theory of biodiversity and biogeography. Princeton University Press, Princeton

Janzen DH (1974) Tropical blackwater rivers, animals, and mast fruiting by the dipterocarpaceae. Biotropica 6:69–103

Kikuzawa K (1991) A cost-benefit analysis of leaf habit and leaf longevity of trees and their geographical pattern. Am Nat 138:1250–1263

Kikuzawa K (1995) The basis for variation in leaf longevity of plants. Plant Ecol 121:89–100

Kikuzawa K, Ackerly D (1999) Significance of leaf longevity in plants. Plant Species Biol 14:39–45

Kikuzawa K, Lechowicz MJ (2011) Ecology of leaf longevity. Springer

Kikuzawa K, Onoda Y, Wright IJ, Reich PB (2013) Mechanisms underlying global temperature-related patterns in leaf longevity. Global Ecol Biogeogr 22:982–993

King DA (1994) Influence of light level on the growth and morphology of saplings in a panamanian forest. Am J Bot 81:948–957

Kitajima K (1994) Relative importance of photosynthetic traits and allocation patterns as correlates of seedling shade tolerance of 13 tropical trees. Oecologia 98:419–428

Kitajima K, Myers JA (2008) Seedling ecophysiology: strategies towards achievement of positive carbon balance. In: Leck MA, Parker VT, Simpson RL (eds) Seedling ecology and evolution. Cambridge University Press, Cambridge, pp 172–188

Kitajima K, Poorter L (2010) Tissue-level leaf toughness, but not lamina thickness, predicts sapling leaf lifespan and shade tolerance of tropical tree species. New Phytol 186:708–721

Kitajima K, Mulkey SS, Wright SJ (1997a) Seasonal leaf phenotypes in the canopy of a tropical dry forest: photosynthetic characteristics and associated traits. Oecologia 109:490–498

Kitajima K, Mulkey SS, Wright SJ (1997b) Decline of photosynthetic capacity with leaf age in relation to leaf longevities for five tropical canopy tree species. Am J Bot 84:702–708

Kitajima K, Mulkey SS, Samaniego M, Wright SJ (2002) Decline of photosynthetic capacity with leaf age and position in two tropical pioneer tree species. Am J Bot 89:1925–1932

Kitajima K, Mulkey SS, Wright SJ (2005) Variation in crown light utilization characteristics among tropical canopy trees. Ann Bot 95:535–547

Kitajima K, Llorens A-M, Stefanescu C, Timchenko MV, Lucas PW, Wright SJ (2012) How cellulose-based leaf toughness and lamina density contribute to long leaf lifespans of shade-tolerant species. New Phytol 195:640–652

Kitajima K, Cordero RA, Wright SJ (2013) Leaf life span spectrum of tropical woody seedlings: effects of light and ontogeny and consequences for survival. Ann Bot 112:685–699

Kobe RK (1999) Light gradient partitioning among tropical tree species through differential seedling mortality and growth. Ecology 80:187–207

Kursar TA, Coley PD (1993) Photosynthetic induction times in shade-tolerant species with long and short-lived leaves. Oecologia 93:165–170

Lambers H, Poorter H (1992) Inherent variation in growth rate between higher plants: a search for physiological causes and ecological consequences. Adv Ecol Res 34:187–261

Lusk CH, Reich PB, Montgomery RA, Ackerly DD, Cavender-Bares J (2008) Why are evergreen leaves so contrary about shade? Trends Ecol Evol 23:299–303

Markesteijn L, Poorter L, Bongers F (2007) Light-dependent leaf trait variation in 43 tropical dry forest tree species. Am J Bot 94:515–525

Marty C, Lamaze T, Pornon A (2009) Endogenous sink–source interactions and soil nitrogen regulate leaf life-span in an evergreen shrub. New Phytol 183:1114–1123

Marty C, Lamaze T, Pornon A (2010) Leaf life span optimizes annual biomass production rather than plant photosynthetic capacity in an evergreen shrub. New Phytol 187:407–416

McKey D (1974) Adaptive patterns in alkaloid physiology. Am Nat 108:305–320

McKey D (1979) The distribution of secondary compounds within plants. In: Rosenthal GA, Janzen DH (eds) Herbivores, their interaction with secondary plant metabolites. Academic Press, Boston, pp 55–133

Molisch H (1928) The Longevity of Plants. NY: E H Fulling p 226

Monk CD (1966) An ecological significance of evergreenness. Ecology 47:504–505

Monsi M Saeki T (1953) Uber den lichtfackor in den pflanzengesellschaften und seine bedeutung fur die stoffproduktion. Japanese Journal of Botany 14:22–52

Monsi M, Saeki T (2005) On the factor light in plant communities and its importance for matter production. Ann Bot 95:549–567 (Translated from Monsi and Saeki 1953)

Mooney HA, Gulmon SL (1982) Constraints on leaf structure and function in reference to herbivory. Bioscience 32:198–206

Mooney HA, Field C, Gulmon SL, Bazzaz FA (1981) Photosynthetic capacity in relation to leaf position in desert versus old-field annuals. Oecologia 50:109–112

Niinemets Ü (2001) Global-scale climatic controls of leaf dry mass per area, density, and thickness in trees and shrubs. Ecology 82:453–469

Oikawa S, Hikosaka K, Hirose T (2006) Leaf lifespan and lifetime carbon balance of individual leaves in a stand of an annual herb, *xanthium canadense*. New Phytol 172:104–116

Onoda Y, Schieving F, Anten NPR (2015) A novel method of measuring leaf epidermis and mesophyll stiffness shows the ubiquitous nature of the sandwich structure of leaf laminas in broad-leaved angiosperm species. J Exp Bot

Osada N, Takeda H, Furukawa A, Awang M (2001) Leaf dynamics and maintenance of tree crowns in a malaysian rain forest stand. J Ecol 89:774–782

Osada N, Oikawa S, Kitajima K (2015) Implications of life span variation within a leaf cohort for evaluation of the optimal timing of leaf shedding. Funct Ecol 29:308–314

Osnas JLD, Lichstein JW, Reich PB, Pacala SW (2013) Global leaf trait relationships: mass, area, and the leaf economics spectrum. Science 340:741–744

Poorter L, Bongers F (2006) Leaf traits are good predictors of plant performance across 53 rain forest species. Ecology 87:1733–1743

Poorter H, Pepin S, Rijkers T, de Jong Y, Evans JR, Korner C (2006) Construction costs, chemical composition and payback time of high- and low-irradiance leaves. J Exp Bot 57:355–371

Poorter H, Niinemets Ü, Poorter L, Wright IJ, Villar R (2009) Causes and consequences of variation in leaf mass per area (lma): a meta-analysis. New Phytol 182:565–588

Poorter H, Niklas KJ, Reich PB, Oleksyn J, Poot P, Mommer L (2012) Biomass allocation to leaves, stems and roots: meta-analyses of interspecific variation and environmental control. New Phytol 193:30–50

Pornon A, Lamaze T (2007) Nitrogen resorption and photosynthetic activity over leaf life span in an evergreen shrub, rhododendron ferrugineum, in a subalpine environment. New Phytol 175:301–310

Pornon A, Marty C, Winterton P, Lamaze T (2011) The intriguing paradox of leaf lifespan responses to nitrogen availability. Funct Ecol 25:796–801

Reich PB, Uhl C, Walters MB, Ellsworth DS (1991) Leaf life-span as a determinant of leaf structure and function among 23 amazonian tree species. Oecologia 86:16–24

Reich PB, Walters MB, Ellsworth DS (1992) Leaf life-span in relation to leaf, plant, and stand characteristics among diverse ecosystems. Ecol Monogr 62:365–392

Reich PB, Walters MB, Ellsworth DS (1997) From tropics to tundra: global convergence in plant functioning. P Natl Acad Sci USA 94:13730–13734

Reich PB, Ellsworth DS, Walters MB, Vose JM, Gresham C, Volin JC, Bowman WD (1999) Generality of leaf trait relationships: A test across six biomes. Ecology 80:1955–1969

Reich PB, Falster DS, Ellsworth DS, Wright IJ, Westoby M, Oleksyn J, Lee TD (2009) Controls on declining carbon balance with leaf age among 10 woody species in australian woodland: do leaves have zero daily net carbon balances when they die? New Phytol 183:153–166

Richardson SJ, Peltzer DA, Allen RB, McGlone MS (2010) Declining soil fertility does not increase leaf lifespan within species: evidence from the franz josef chronosequence, new zealand. N Z J Ecol 34:306–310

Rozendaal DMA, Hurtado VH, Poorter L (2006) Plasticity in leaf traits of 38 tropical tree species in response to light; relationships with light demand and adult stature. Funct Ecol 20:207–216

Russo SE, Davies SJ, King DA, Tan S (2005) Soil-related performance variation and distributions of tree species in a bornean rain forest. J Ecol 93:879–889

Russo SE, Brown P, Tan S, Davies SJ (2008) Interspecific demographic trade-offs and soil-related habitat associations of tree species along resource gradients. J Ecol 96:192–203

Russo SE, Cannon WL, Elowsky C, Tan S, Davies SJ (2010) Variation in leaf stomatal traits of 28 tree species in relation to gas exchange along an edaphic gradient in a bornean rain forest. Am J Bot 97:1109–1120

Russo SE, Zhang L, Tan S (2012) Covariation between understorey light environments and soil resources in bornean mixed dipterocarp rain forest. J Trop Ecol 28:33–44

Sevanto S, McDowell NG, Dickman LT, Pangle R, Pockman WT (2014) How do trees die? A test of the hydraulic failure and carbon starvation hypotheses. Plant, Cell Environ 37:153–161

Shure DJ, Wilson LA (1993) Patch-size effects on plant phenolics in successional openings of the southern appalachians. Ecology 74:55–67

Small E (1972) Photosynthetic rates in relation to nitrogen recycling as an adaptation to nutrient deficiency in peat bog plants. Can J Botany 50:2227–2233

Sobrado MA (1994) Leaf age effects on photosynthetic rate, transpiration rate and nitrogen content in a tropical dry forest. Physiol Plant 90:210–215

Sultan SE (1995) Phenotypic plasticity and plant adaptation. Acta Bot Neerl 44:363–383

Terashima I, Hanba YT, Tholen D, Niinemets Ü (2011) Leaf functional anatomy in relation to photosynthesis. Plant Physiol 155:108–116

Valladares F, Wright SJ, Lasso E, Kitajima K, Pearcy RW (2000) Plastic phenotypic response to light of 16 congeneric shrubs from a Panamanian rainforest. Ecology 81:1925–1936

Valladares F, Gianoli E, Gómez JM (2007) Ecological limits to plant phenotypic plasticity. New Phytol 176:749–763

Vanderwel MC, Slot M, Lichstein JW, Reich PB, Kattge J, Atkin OK, Bloomfield KJ, Tjoelker MG, Kitajima K (2015) Global convergence in leaf respiration from estimates of thermal acclimation across time and space. New Phytol

Via S, Lande R (1985) Genotype-environment interaction and the evolution of phenotypic plasticity. Evolution 39:505–522

Villagra M, Campanello PI, Bucci SJ, Goldstein G (2013) Functional relationships between leaf hydraulics and leaf economic traits in response to nutrient addition in subtropical tree species. Tree Physiol 33:1308–1318

Walters MB, Reich PB (1999) Low-light carbon balance and shade tolerance in the seedlings of woody plants: do winter deciduous and broad-leaved evergreen species differ? New Phytol 143:143–154

Weiner J (1990) Asymmetric competition in plant populations. Trends Ecol Evol 5:360–364

Westbrook JW, Kitajima K, Burleigh JG, Kress WJ, Erickson DL, Wright SJ (2011) What makes a leaf tough? Patterns of correlated evolution between leaf toughness traits and demographic rates among 197 shade-tolerant woody species in a neotropical forest. Am Nat 177:800–811

Westoby M, Warton D, Reich PB (2000) The time value of leaf area. Am Nat 155:649–656

Westoby M, Falster DS, Moles AT, Vesk PA, Wright IJ (2002) Plant ecological strategies: some leading dimensions of variation between species. Annu Rev Ecol Syst 33:125–159

Whitman DW, Agrawal AA (2009) What is phenotypic plasticity and why is it important? In: Whitman DW, Ananthakrishnan TN (eds) Phenotypic plasticity of insects. Science Publishers

Williams K, Percival F, Merino J, Mooney HA (1987) Estimation of tissue construction cost from heat of combustion and organic nitrogen content. Plant, Cell Environ 10:725–734

Williams K, Field CB, Mooney HA (1989) Relationships among leaf construction cost, leaf longevity, and light environment in rain-forest plants of the genus *piper*. Am Nat 133:198–211

Witkowski ETF, Lamont BB (1991) Leaf specific mass confounds leaf density and thickness. Oecologia 88:486–493

Wright IJ, Westoby M, Reich PB (2002) Convergence towards higher leaf mass per area in dry and nutrient-poor habitats has different consequences for leaf life span. J Ecol 90:534–543

Wright IJ, Reich PB, Westoby M, Ackerly DD, Baruch Z, Bongers F, Cavender-Bares J, Chapin T, Cornelissen JHC, Diemer M, Flexas J, Garnier E, Groom PK, Gulias J, Hikosaka K, Lamont BB, Lee T, Lee W, Lusk C, Midgley JJ, Navas ML, Niinemets U, Oleksyn J, Osada N, Poorter H, Poot P, Prior L, Pyankov VI, Roumet C, Thomas SC, Tjoelker MG, Veneklaas EJ, Villar R (2004) The worldwide leaf economics spectrum. Nature 428:821–827

Yamada T, Okuda T, Abdullah M, Awang M, Furukawa A (2000) The leaf development process and its significance for reducing self-shading of a tropical pioneer tree species. Oecologia 125:476–482

The Effects of Rising Temperature on the Ecophysiology of Tropical Forest Trees

Martijn Slot and Klaus Winter

Abstract The response of tropical trees to rising temperatures represents a key uncertainty that limits our ability to predict biosphere-atmosphere feedbacks in a warming world. We review the current understanding of temperature effects on the ecophysiology of tropical trees from organelle to biome level, where we distinguish between short-term responses, acclimation, and adaptation. We present new data on short-term temperature responses of photosynthesis and dark respiration, and temperature acclimation of photosynthesis. We also compare new field and laboratory-obtained photosynthesis-temperature response data. We identify several priority study areas. (1) Acclimation: We need to better understand photosynthetic acclimation, for example to determine whether the adjustment of the thermal optimum of photosynthesis (T_{Opt}) is consistently negated by a decrease in photosynthesis at T_{Opt}, as we observed. (2) Growth: Whereas tropical seedlings may grow better with warming, canopy trees reportedly grow worse; we do not currently know what explains these contrasting temperature effects. (3) Reproduction: Tropical trees may be close to reproductive temperature thresholds, as heat sterility in crops occurs in the upper 30 °C range. Nonetheless, the temperature sensitivity of tropical tree reproduction is virtually unstudied. (4) Mortality: How does heat-induced atmospheric drought (high leaf-to-air vapor pressure deficit) affect tropical tree mortality? (5) Stomatal behavior: What is the specific role of temperature in the induction of midday-stomtal closure on sunny days? Better knowledge in these areas will improve our ability to predict carbon fluxes in tropical forests experiencing ongoing warming.

M. Slot (✉) · K. Winter
Smithsonian Tropical Research Institute, Apartado 0843-03092,
Balboa, Ancón, Republic of Panama
e-mail: SlotM@si.edu; martijnslot78@gmail.com

© Springer International Publishing Switzerland 2016
G. Goldstein and L.S. Santiago (eds.), *Tropical Tree Physiology*,
Tree Physiology 6, DOI 10.1007/978-3-319-27422-5_18

385

Introduction

Tropical forests cover only 15 % of the planet's terrestrial surface (Pan et al. 2013), yet they account for more than one-third of its net primary productivity (NPP) (Saugier et al. 2001) and two-thirds of its plant biomass (Pan et al. 2013). Given this disproportionally large contribution to the global carbon cycle, it is important that we increase our understanding of the effects of global warming on tropical forest trees. Recent analyses have suggested that global variation in temperature and precipitation have no direct effect on global patterns of NPP after accounting for stand age and biomass (Michaletz et al. 2014), but this does not mean that changes in climatic variables will not affect NPP within a given biome. The coming decades will see ongoing warming in tropical regions, leading to unprecedented temperature regimes (Diffenbaugh and Scherer 2011) that currently do not support closed-canopy forests (Wright et al. 2009). Tropical regions have previously experienced warming—most notably leading up to the Paleocene-Eocene Thermal Maximum ($\sim$ 56 million years ago) when temperatures rose by 3–5 °C—but even the most rapid of such historical warming events took place over thousands to tens of thousands of years, timescales during which gradual changes in species composition, adaptive responses, and speciation are possible (Jaramillo et al. 2010). Current warming, on the other hand, occurs over the lifetime of individual trees, necessitating a high degree of thermal plasticity to maintain long-term growth and survival.

Warming effects on the physiology of woody plants have been studied primarily in mid- and high-latitiude ecosystems. Many of these effects are expected to be broadly generalizable, given the universal effects of temperature on enzyme kinetics, and the common principles of the biochemical pathways of both photosynthesis and mitochondrial respiration, the key drivers of terrestrial biosphere-atmosphere carbon exchange. However, there are several important differences between tropical and higher latitude ecosystems that might differentially affect plant responses to climate warming.

First, seasonal temperature fluctuations are minimal in the tropics, and such thermally stable conditions may not favor evolution of the capacity to acclimate to temperature changes (Janzen 1967; Cunningham and Read 2003). If tropical species have limited thermal plasticity, current warming may be highly detrimental to their performance. Second, for most of the past 2.6 million years tropical regions have experienced lower temperatures than today. Natural selection under such conditions would not have favored heat-protective traits (Corlett 2011). Temperate and boreal climates have also been cooler over this period, but current warming is more likely to expose tropical vegetation to temperatures near the thermal limit of photosynthesis than higher latitude vegetation. Indeed, tropical forests might be close to their thermal optimum already (Doughty and Goulden 2008; but see Lloyd and Farquhar 2008). Third, warming does not lengthen the growing season in the tropics as it does at higher latitutes (Menzel and Fabian 1999; Menzel et al. 2006).

This means that a longer growing period will not compensate for any negative effect of temperature on carbon uptake in the tropics. Given the above three points, temperature relationships of tropical trees deserve special attention.

Here we review the current understanding of the effects of elevated temperature on the ecophysiology of tropical trees and forests. We present new data on instantaneous effects and growth-temperature effects on seedling gas exchange; we compare laboratory-based measurements with in situ measurements of temperature effects on foliar physiology; we review experimental, observational, and modeling studies; and we identify areas of study that should be prioritized to improve our ability to predict carbon fluxes in gradually warming tropical forests. We will focus on temperature responses at the ecophysiological level, emphasizing the effects of warming in the non-damaging temperature range. For information on molecular responses to heat stress we refer readers to several reviews on this subject that have been published in the past decade (e.g., Wahid et al. 2007; Allakhverdiev et al. 2008; Ashraf and Harris 2013). Our primary concern here is the effect that warming has on tropical lowlands, which are currently the warmest ecosystems that support closed-canopy forests (Wright et al. 2009). The ~ 15 % of tropical forests that are montane are beyond the scope of this chapter.

We distinguish between different temporal scales at which temperature may affect tropical trees, and between the organizational scales at which these temperature response processes operate (Table 1). We will start our review at the organelle and leaf level—with a strong focus on photosynthesis and respiration—then work our way up through the whole-tree level out to stand-, ecosystem-, and biome-level temperature effects. Along the way we go from a highly mechanistic understanding of organelle- and leaf-level processes—based on foundational work on temperate species and backed up by a large body of experimental work—to more speculation as we attempt to predict temperature effects at higher organizational scales and over longer time periods for which no experimental work exists to date.

Leaf-Level Temperature Effects

Photosynthesis

Net photosynthesis (A_{Net}) increases with short-term warming before reaching a maximum CO_2 assimilation rate (A_{Opt}) at optimum temperature (T_{Opt}), beyond which net CO_2 uptake rates decline and eventually drop to zero at the upper CO_2 compensation point (Box 1). We will first discuss the short-term temperature response over non-harmful, reversible temperature ranges. After briefly addressing heat damage, we discuss temperature responses at timescales over which acclimation may occur.

Table 1 Temperature responses (+ and − indicate the direction of change) across organizational and temporal scales

Organizational scale		Biophysical response	Acclimation	Adaptation	Adaptation/Species turnover	Species turnover/speciation
	Biome					Possible temperature/drought-induced collapse of closed-canopy forest. If closed canopy forest is lost: big changes in rainfall patterns, climate feedbacks; risk of runaway warming
	Ecosystem		+ Mineralization in soils could benefit tree growth in nitrogen limited ecosystems; − NPP/GPP ratio[a]	+ Isoprene emission[h]; − NPP/GPP ratio[a]	Savanna or desert species may take over. If closed canopy forest is lost: big changes in hydraulics, rainfall patterns, climate feedbacks; risk of runaway warming	
	Tree	+ Respiration of all organs (however: root and bulk stem temperatures are well buffered and do not track instantaneous air temperature changes); − Pollen viability	− Respiratory capacity of all organs; species-specific interactions with mycorrhizae may either strengthen or weaken the acclimation response[b]; − Reproductive efficiency; − Stem diameter growth of canopy trees; Seedling/sapling growth unaffected	Possible change in biomass allocation and leaf display; + Leaf cooling traits for adapted species; + Isoprene emission[h]	If closed canopy forest is lost: increased abundance of savanna or desert species that are better adapted to drought	
	Leaf	Photosynthesis peaks near local mean temperature and then declines; + Respiration; + Isoprene emission; + Monoterpene emission	+ T_{Opt} of photosynthesis; Constructive adjustment: + A_{Opt}; Detractive adjustment: − A_{Opt}; − Respiratory capacity; + Isoprene emission; Change in morphology, e.g. − leaf mass per area[c]; + Mesophyll conductance[d]	+ Heat tolerance; − Low-temperature tolerance; + Isoprene emission[h]; Respiration homeostatic with pre-warming		

(continued)

Table 1 (continued)

	Biophysical response	Acclimation	Adaptation	Adaptation/Species turnover	Species turnover/speciation
Sub-leaf/Organelle	+ Monoterpene synthesis + Protein turnover + Membrane fluidity + Photorespiration − Electron transport rate − CO_2 solubility	+ Heat tolerance of photosynthetic machinery + Thermostability of Rubsico activase or different isoforms being expressed[e,f] + Electron transport capacity + Membrane rigidity Change in membrane composition (e.g., increased ratio of saturated to unsaturated fatty acids)[g]			If closed canopy forests are converted into savanna/desert ecosystems: increased abundance of C_4 and CAM species
	Instantaneous	*Days–months*	*Years–decades*	*Decades–decennia*	*Decennia–Millenia*
		Temporal scale			

Ecosystem and biome-level predictions for long timescales are speculative

[a]Zhang et al. (2014), [b]Fahey et al. (2015), [c]Cheesman and Winter (2013b), [d]Makino et al. (1994), [e]Salvucci and Crafts-Brander (2004), [f]Wang et al. (2010), [g]Murakami et al. (2000), [h]Sharkey and Monson (2014)

Direct Effects: CO$_2$ Fixation

At a given light level, photosynthesis is limited by one of three factors: (1) the capacity of ribulose 1,5-bisphosphate carboxylase/oxygenase (Rubisco) to carboxylate ribulose bisphosphate (RuBP) (Rubisco-limited, or RuBP-carboxylation-limited photosynthesis); (2) the capacity to regenerate RuBP through the Calvin cycle and the thylakoid reactions (RuBP-regeneration-limited photosynthesis); or (3) the capacity for triose phosphate use (TPU) through starch and sucrose synthesis, where low TPU limits the regeneration of inorganic phosphate necessary for photophosphorylation (TPU-limited photosynthesis) (Harley and Sharkey 1991; von Caemmerer 2000). At light levels that saturate photosynthesis, RuBP-regeneration capacity generally reflects the maximum electron transport capacity (J_{max}), whereas the rate of RuBP-carboxylation-limited photosynthesis represents the maxium capacity for Rubisco carboxylase activity (V_{Cmax}). TPU limitation at current atmospheric CO$_2$ concentrations only occurs at low temperatures (Sage and Kubien 2007), and is expected to have minimal impact on photosynthesis in lowland tropical forest trees (but see Ellsworth et al. 2015 for potential TPU limitation driven by foliar phosphorus limitation). At current ambient CO$_2$ concentrations, RuBP-caboxylation limitation is common at intermediate temperatures, whereas RuPB-regeneration can become limiting at high temperatures (Sage and Kubien 2007). High temperature can also impede Rubisco functioning through its effect on Rubsico activase, the enzyme that promotes the dissociation of inhibitory sugar phosphates from the active site of Rubisco, thereby activating it (Portis 2003). Rubisco activase has lower heat tolerance than Rubisco itself, so high-temperature impairment of Rubisco activity may represent reduced activation of Rubsico, rather than reduced functioning of activated Rubisco (Salvucci and Crafts-Brandner 2004).

Rubsico can not only carboxylate RuBP, it can also oxygenate it, leading to photorespiration, a process that results in loss rather than gain of CO$_2$. Photorespiration, although associated with reduced carbon gain, appears to play a beneficial role under high temperatures by maintaining electron flow and preventing photo-oxidation (Osmond and Björkman 1972; Kozaki and Takeba 1996), and by providing a substrate for the synthesis of isoprene (Jardine et al. 2014), a volatile organic compound believed to be associated with thermoprotection of photosynthesis (see "Volatile organic compounds and thermoprotection"). Increasing temperature promotes photorespiration in two ways. First, the solubility of CO$_2$ decreases more strongly with temperature than that of O$_2$, so proportionally more O$_2$ can reach the active sites of Rubisco and stimulate RuBP oxygenation. Second, the relative specificity of Rubisco for CO$_2$ compared to O$_2$ decreases with increasing temperature, so at higher temperatures, higher CO$_2$ concentrations are needed to achieve a given RuBP-carboxylation rate (von Caemmerer and Quick 2000).

Under current ambient CO$_2$ concentrations, photosynthesis is controlled by TPU capacity at low temperature ($< \sim 20$ °C for the tropical species *Ipomoea batatas* (L.) Lam. (sweet potato; Sage and Kubien 2007). At intermediate temperatures RuBP carboxylation becomes the rate-limiting step. With further warming ($> \sim 32$ °C for *I. batatas*) increased photorespiration reduces photosynthetic efficiency, and the electron transport processes of RuBP regeneration become limiting. Ultimately, at temperatures exceeding the thermal tolerance of Rubisco activase, Rubisco activation may become limiting. Increasing CO_2 concentrations will reduce photorespiration, thus increasing the rates of both RuBP-regeneration-limited photosynthesis and RuBP-caboxylation-limited photosynthesis, without affecting TPU-limited photosynthesis. RuBP-caboxylation-limited photosynthesis is more strongly stimulated by CO_2 than RuBP-regeneration limited photosynthesis. Consequently, at higher CO_2, there is a shift in what controls net photosynthesis at a given temperature, with control exerted by TPU extending to higher temperatures than at low CO_2, and RuBP regeneration starting to limit photosynthesis at lower temperatures, resulting in a reduced or disappearing role of RuBP-caboxylation-limitation in constraining net photosynthesis (Sage and Kubien 2007). In other words, at elevated CO_2, J_{max} exerts stronger control over the temperature response of photosynthesis than V_{Cmax}. The remainder of our discussion will focus primarily on temperature effects at current ambient CO_2 concentration (~ 400 ppm).

Figure 1a illustrates how temperature affects net CO_2 assimilation at 21 % versus 2 % O_2—i.e., with and without photorespiration—in two early-successional tropical tree species, *Ochroma pyramidale* (Cav. ex Lam.) Urb. and *Ficus insipida* Willd., and in the late-successional species *Calophyllum longifolium* Willd. Because photorespiration is proportionally higher at higher temperatures (Slot et al. 2016), T_{Opt} of net photosynthesis is higher at 2 % O_2, similar to what has been observed under elevated CO_2 (e.g., Berry and Björkman 1980). *C. longifolium* has the lowest rates of photosynthesis regardless of temperature, consistent with its conservative growth strategy. *O. pyramidale* and *F. insipida*, in contrast, are fast growing species with higher T_{Opt} and A_{Opt} values. They germinate in forest gaps where irradiance levels as well as temperatures are higher than in the understory where *C. longifolium* typically regenerates.

Direct Effects: Thermodamage

One of the most temperature-sensitive components of plants is photosystem II (PSII), a protein complex located in the thylakoid membrane of the chloroplast. PSII efficiency can be readily determined using chlorophyll *a*-fluorescence techniques (e.g., Krause and Weis 1991). Thermodamage assessment from chlorophyll fluorescence measurements correlates well with classical analysis of post-heating necrosis (Krause et al. 2010, 2013). The critical temperature for irreparable leaf damage is about 52 or 53 °C, depending on whether thermal damage is determined in the dark or in the light (Krause et al. 2015). Such temperatures are only a few degrees above leaf temperatures that already occur occasionally in situ in sun-exposed leaves of tropical forest trees (Krause et al. 2010; Fig. 2). There is some variation across species in reported thermotolerance, but this may in part

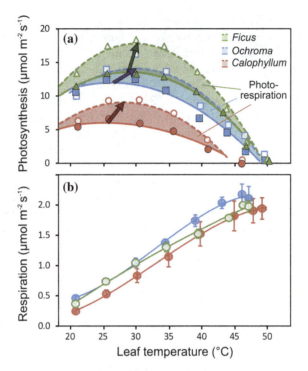

Fig. 1 Temperature response *curves* of net photosynthesis under photorespiratory (21 % O_2, *dashed lines, closed symbols*), and non-photorespiratory conditions (2 % O_2, *solid lines, open symbols*) **a**, and dark respiration **b** of recently fully expanded leaves of *Calophyllum longifolium* (*circles*), *Ficus insipida* (*triangles*), and *Ochroma pyramidale* (*squares*) seedlings grown outdoors under full natural radiation in Panama (mean annual temperature 27 °C). *Arrows* indicate increases in T_{Opt} and A_{Opt} with removal of photorespiration. Non-photorespiratory conditions were created by mixing air entering the cuvette with nitrogen gas at a 1:9.5 ratio. After passing the mixture through soda lime, CO_2 was added to generate a CO_2 concentration of 400 ppm. Measurements were made on attached leaves (n = 3–6) in a temperature-controlled Walz cuvette (Walz GmbH, Eiffeltrich, Germany) attached to an LI-6252 infrared gas analyzer (Licor). Photosynthesis *curves* were fit as: net photosynthesis $= b \times (T_{Leaf} - T_{Min}) \times \left(1 - e^{c \times (T_{Leaf} - T_{Max})}\right)$. T_{Min} and T_{Max} are the hypothetical low- and high-temperature CO_2 compensation points, and b and c are constants. All four variables were estimated using a non-linear solver function. Respiration fits are 3rd order polynomials. For clarity only means are shown in **a**; in **b** means ± 1 SEM are shown. Modified after Slot et al. (2016)

reflect differences in methodology used to assess thermotolerance. In a study comparing tropical and temperate rainforest tree species in Australia, the leaf temperature at which irreversible damage occurred was independent of growth temperature and biome of origin (Cunningham and Read 2006), suggesting minimal acclimation and phylogeographic predisposition to thermal damage.

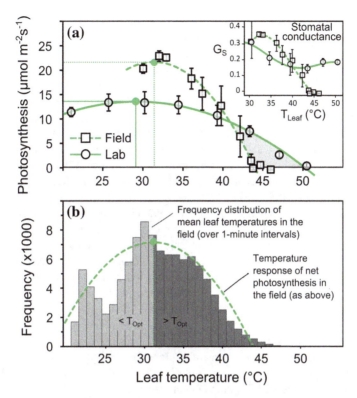

Fig. 2 Comparison of field and laboratory based measurements of net photosynthesis versus leaf temperature (T_{Leaf}) of *Ficus insipida* **a**, and the frequency distribution of daytime T_{Leaf} for this species in the field **b** based on 1 month (Jan–Feb, dry season) of continuous monitoring of 6 leaves with thermocouple wires touching the abaxial surfaces. The *inset* in **a** shows the temperature response of stomatal conductance (G_S). Lab measurements were made as described in Fig. 1. Field measurements were made with an LI-6400 (Licor) between 8 a.m. and noon on a typical dry season day. The leaves of the ~30 m tall study tree were accessed using a canopy crane. The response *curve* of field-based measurements is overlaid on the frequency distribution of T_{Leaf} in **b** to illustrate the fact that T_{Opt} occurs at the most frequently occurring T_{Leaf}. *Error bars* represent 1 SEM (n = 4–6). Modified after Slot et al. (2016)

Indirect Effects: CO$_2$ Fixation

Photosynthetic carbon uptake may also be affected by temperature via leaf-to-air vapor pressure deficit (VPD). Warm air can accomodate more water vapor than cool air, so at a given absolute moisture content of the air, VPD increases with leaf temperature, potentially causing stomatal closure. It is very difficult with standard gas-exchange equipment to keep VPD constant over a large range of leaf temperatures. One study that appears to have successfully controlled for VPD reported T_{Opt} values of 28–29 °C for two rainforest tree species in Costa Rica measured in situ (Vargas and Cordero 2013), values that are comparable to those reported in Fig. 1, in which VPD was not controlled. These observations also roughly correspond to decreasing A_{Net} values at temperatures $>30\ °C$ that Doughty and Goulden (2008) observed on three species measured in situ in the Brazilian Amazon, and with observations of Cheesman and Winter (2013a) of decreasing A_{Net} when T_{Leaf} was $>27–30\ °C$ in three tropical tree species grown in controlled-environment chambers in Panama. Whereas Vargas and Cordero (2013) maintained VPD < 2.0 kPa at leaf temperatures above T_{Opt}, VPD in the experiment depicted in Fig. 1 increased from about 2.0 kPa at T_{Opt} to 8.0 kPa at the upper CO_2 compensation point. Nonetheless, these two studies show similar decreases in net photosynthesis at temperatures above $\sim 30\ °C$. Insufficient data on VPD-controlled temperature responses of A_{Net} of tropical trees complicate the interpretation of the role of VPD-induced stomatal closure in determining T_{Opt} and photosynthetic decline above T_{Opt}.

In upper-canopy tree leaves of tropical forest canopies, the combination of high temperature, high irradiance and high VPD can lead to pronounced midday depressions of stomatal conductance and CO_2 uptake (e.g., Zotz et al. 1995; Cernusak et al. 2013; Goldstein et al. this volume, Santiago et al. this volume). Figure 2 shows changes in net CO_2 uptake of *F. insipida* leaves measured in situ in the upper-canopy of a tropical forest in Panama as photosynthetic photon flux density (PPFD), temperature and VPD all increase from morning to noon on a sunny day. Compared to laboratory-measured seedling leaves (see Fig. 1 for methods), A_{Opt} of canopy leaves was much higher because these leaves were acclimated to higher PPFD and had higher leaf mass per area. T_{Opt} was also higher in the field, probably owing to generally higher leaf temperatures in situ. Whereas stomatal conducantance and photosynthesis in the field decreased to zero when leaf temperatures reached $45\ °C$, in the laboratory net photosynthesis remained positive up to $50\ °C$ and conductance—after a marked decrease between 30 and $45\ °C$— started to increase above $45\ °C$ (Fig. 2). This re-opening of stomata at high temperatures in the laboratory was also observed for *O. pyramidale* and *C. longifolium* (Slot et al. 2016). Nonetheless, in exposed canopy leaves, high heat load caused by high solar radiation can become so stressful that photosynthetic carbon uptake may take place mainly during morning hours. During the course of the day depicted in Fig. 2a, 50 % of the time leaf temperatures exceeded T_{Opt} of *F. insipida* (Fig. 2b), stomata were partially closed, and A_{Net} was reduced. Depending on leaf orientation, the extent to which midday depression occurs may vary considerably, even between neighboring leaves, or within leaves if they are undulated (e.g., *Cecropia*; K. Winter, unpublished data). The variability in leaf-level irradiance and leaf temperature makes stomatal conductance particularly challenging to model and predict with high spatial and temporal resolution. Given the frequent occurrence of midday stomatal reductions in tropical trees on sunny days, further understanding of this phenomenon is warranted to better predict global carbon fluxes in a warming world.

Acclimation

Thermal acclimation is a biochemical, physiological, or morphological adjustment by individual plants in response to a change in the temperature regime, which results in an alteration in the short-term response to temperature (Smith and Dukes 2013). Thermal acclimation of photosynthesis commonly leads to a shift of the thermal optimum of photosynthesis towards the new growth temperature (Box 1). This shift results from a change in relative contribution of RuBP carboxylation and RuBP regeneration in controlling net photosynthesis. The optimum temperature of RuBP regeneration is higher than that of RuBP carboxylation (Kirschbaum and Farquhar 1984; Hikosaka et al. 2006). Warm-acclimated plants tend to have a lower J_{max}/V_{Cmax} ratio than cool-grown control plants (e.g., Bernacchi et al. 2003). When J_{max}/V_{Cmax} is low, RuBP regeneration—with its relatively high T_{Opt}— exerts greater control over net photosynthesis than when J_{max}/V_{Cmax} is high in cool-grown plants, and as a result T_{Opt} of net photosynthesis is higher in warm-acclimated plants (Hikosaka et al. 2006). If the J_{max}/V_{Cmax} ratio is not reduced by high temperature, T_{Opt} may still increase if the activation energy of V_{Cmax} increases more with growth temperature than that of J_{max} (Hikosaka et al. 2006).

Consistent with thermal acclimation, T_{Opt} of *C. longifolium* seedlings grown in controlled-environment chambers increased with growth temperature (Fig. 3), but

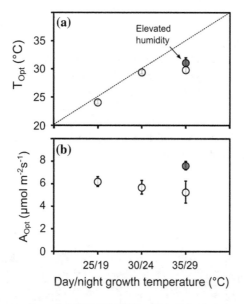

Fig. 3 The optimal temperature of photosynthesis (T_{Opt}) (**a**), and net photosynthesis at T_{Opt} (A_{Opt}) (**b**) plotted against growth temperature of *Calopyllum longifolium* seedlings (mean ± 1 SEM). Seedlings were grown in individual pots (2.8 l) in commercial potting soil in growth cabinets (Environmental Growth Chambers, Chagrin Falls, OH, USA) at day/night temperatures of 25/19, 30/24, 35/29, 35/29 + elevated RH, and 39/33 °C. Plants from the 39/33 °C did not have enough live leaves for photosynthesis measurements (see Fig. 4). Net photosynthesis was measured with an LI-6400 (Licor) at 20, 25, 30, 35, and 40 °C on one leaf per seedling (n = 6/treatment) after equilibration for at least 45 min in growth cabinets set to the target temperature. T_{Opt} was calculated as described in the caption of Fig. 1. In one 35/29 °C cabinet relative humidity was increased to ~90 % (as opposed to ~45 %) by maintaining a pot of water near boiling point in the cabinet during the light period

there is a limit to this adjustment. Plants grown at 25/19 °C (day/night) and 30/24 °C had T_{Opt} values close to their daytime growth temperature, whereas plants grown at 35/29 °C had a T_{Opt} of 30 °C, suggesting that, at least for this late-successional species, 30 °C represents the maximum T_{Opt} at current CO_2 concentrations. While T_{Opt} shifted towards the growth temperature, A_{Opt} decreased with increasing growth temperature. Way and Yamori (2014) call this *detractive* adjustment to warming, as opposed to *constructive* adjustment (see Box 1). The exception to this pattern was the 35/29 °C treatment associated with elevated relative humidity (~ 90 % as opposed to ~ 45 %), for which both T_{Opt} and A_{Opt} were highest among all treatments. This may indicate that part of the decrease in A_{Opt} relative to growth temperature resulted from temperature effects on VPD, with higher stomatal conductance being maintained when plants were grown at elevated relative humidity.

Kositsup et al. (2009) grew *Hevea brasiliensis* Müll. Arg. seedlings at 18 and 28 °C and also found that T_{Opt} adjusted to growth temperature. They further observed that A_{Net}, V_{Cmax} and J_{max} all increased with increased growth temperature, suggesting constructive acclimation. However, 18 °C is probably sub-optimal for lowland tropical species such as *H. brasiliensis*, and the observed adjustments may indicate that 28 °C is closer to the optimum temperature for this species, rather than indicate strong constructive thermal acclimation. Doughty (2011) warmed leaves of trees and lianas (woody vines) in the Brazilian Amazon and found no acclimation of A_{Net}. The decrease in photosynthesis with warming was assigned to leaf damage caused by occasionally very high leaf temperatures of experimentally warmed leaves during sunny spells. Similarly, *C. longifolium* plants grown in controlled-environment chambers at 39/33 °C (day/night) showed severe leaf damage and photosynthesis could not be measured (Fig. 4).

Dusenge et al. (2015) reported that V_{Cmax} was similar between leaves of trees from two tropical rainforest sites in Rwanda differing in temperature regime,

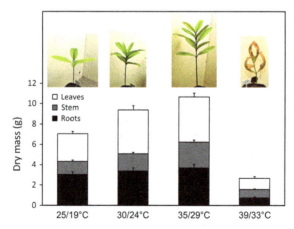

Fig. 4 The effect of growth temperature on total biomass (mean ± 1 SEM of 6 seedlings per treatment) and biomass allocation in *Calophyllum longifolium* seedlings, with photographs of representative plants of each treatment at final harvest

whereas J_{max} was significantly lower at the warmer site, resulting in a reduced J_{max}/V_{Cmax}, in line with observations on temperate species. The physiological plasticity notwithstanding, A_{Net} was reduced at the warmer site compared to the cool site, suggesting detractive adjustment, similar to observations on *C. longifolium* (Fig. 3). We still know little about the thermal acclimation potential of V_{Cmax} and J_{max} of lowland tropical species, and to date no study has compared the acclimation potential of V_{Cmax} and J_{max} of tropical and temperate tree species. Such experimental studies are needed to better predict the effect of long-term warming on the photosynthetic properties of tropical forest vegetation.

Adaptation

There is at least one comparative study that points to fundamental differences in the physiological properties of species adapted to temperate versus tropical conditions. Cunningham and Read (2002) reported higher T_{Opt} values for tropical than temperate species at a given growth temperature. Furthermore, the temperature response curve for temperate species was wider, i.e., temperate species had photosynthesis rates of >80 % of A_{Opt} over a larger temperature range. These results are consistent with the absence of strong selection for a broad temperature range of photosynthesis for tropical species. We found that T_{Opt} of tropical species is lower than growth temperature when growth temperatures exceed ~ 30 °C (Fig. 3). The narrower curves of tropical species and the apparent limit to increasing A_{Opt}, despite a small increase in T_{Opt}, suggest that significant warming in the tropics may cause a decline in photosynthetic carbon uptake.

Dark Respiration

Temperature Response

About 30 % of the carbon fixed by tropical forests through photosynthesis is released back into the atmosphere by foliar respiration (Chambers et al. 2004; Malhi 2012). Respiration is vital for plant growth and survival, supporting biosynthesis, cellular maintenance and repair, but from a carbon-balance perspective respiration represents a loss.

In the short term, dark respiration increases with temperature, peaks with a $T_{Opt} > 50$ °C, and then steeply declines. The initial increase is often assumed to be exponential, and can be expressed as a Q_{10} value—the proportional increase in respiration with 10 °C warming (Box 1). The general assumption is that respiration rates double for every 10 °C warming, resulting in a Q_{10} of 2.0, although in reality a wide range of Q_{10} values—from ~ 1.4 to 4.2— has been reported (Atkin et al. 2005, and references therein). Leaf respiration of tropical forest species is at least as sensitive to short-term temperature increase as that of temperate species, with Q_{10}

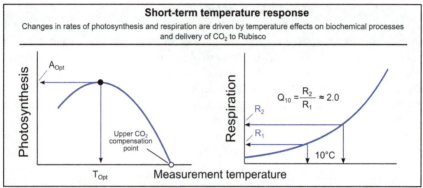

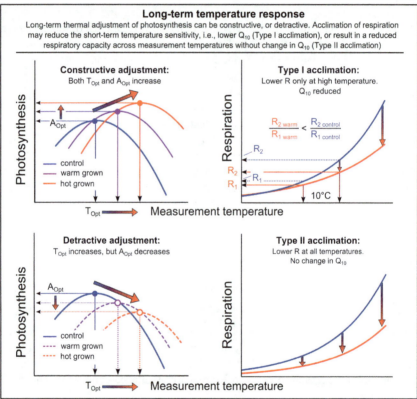

Box 1. Photosynthesis and respiration are highly sensitive to temperature, but the effects depend on the timescale of exposure. Long term warming (e.g., weeks to months) may modify the short-term response through acclimation, i.e., the biochemical, or structural adjustment by individual plants in response to a new temperature regime

values ranging from 1.5 to 4.1 (Meir et al. 2001; Cavaleri et al. 2008; Slot et al. 2013, 2014b), and averaging above 2.0 (Slot et al. 2013, 2014b). Figure 1b shows respiration-temperature response curves for leaves of *C. longifolium*, *F. insipida*, and *O. pyramidale*. Notably, the response-curves (n = 3–6) are not exponential. Over a relatively narrow temperature range it is difficult to distinguish different shapes of temperature response curves, but the examples in Fig. 1b cover a range of $\sim 25\ ^\circ$C, and an exponential curve is clearly not an appropriate approximation. We do not know whether the data shown in Fig. 1b reflect a uniquely tropical phe-nomenon, but a recent study with *Oryza sativa* L. (rice) reported similarly linear temperature responses of respiration over a $\sim 14\ ^\circ$C range (Peraudeau et al. 2015). In contrast, measurements over a broad temperature range on the temperate *Eucalyptus pauciflora* suggest a good exponential fit up to critical temperatures around 50 °C (O'Sullivan et al. 2013), whereas Hüve et al. (2012) showed con-siderable variation in the shape of the response curve over a $\sim 25\ ^\circ$C range across three temperate herbs. Modeling temperature responses using an exponential Q_{10} or Arrhenius-type fit when a linear fit better describes the actual response could lead to overestimation of respiration rates at high temperatures. Future research will need to assess how widely distributed the non-exponential temperature responses of dark respiration are among tropical species.

Acclimation

Nighttime temperatures increase more rapidly than daytime temperatures (e.g., Easterling et al. 1997), potentially leading to increased respiration costs in tropical forest systems. However, acclimation of respiration might mitigate this. Respiration at a given temperature is generally decreased in warm-acclimated plants compared to control plants, either through down-regulation of respiratory capacity at all tempera-tures, or because the short-term temperature sensitivity in warm-acclimated plants is reduced, leading to lower rates at intermediate to high temperatures (Atkin and Tjoelker 2003; see Box 1). Recently, we showed that upper-canopy leaves of tropical canopy trees and lianas acclimate to nighttime warming (Slot et al. 2014a). Acclimation has also been shown in seedlings of tropical tree species grown under different temperature regimes in controlled-environment chambers and open-top chambers (Cheesman and Winter 2013b). In fact, when accounting for the degree and duration of warming there is no indication that tropical species differ from species from other biomes in their capacity for thermal acclimation of respiration (Slot and Kitajima 2015).

At the whole-plant level, respiratory acclimation can be affected by interactions with mycorrhizal symbionts (Atkin et al. 2009). The only study to date that has addressed these interactions for tropical tree species indicated a strong species-specific effect of mycorrhizal colonization of roots on acclimation of whole-plant respiration to nighttime warming (Fahey et al. 2016). The host species-specific nature of the effects of mycorrhizae may complicate generalizations of acclimation trends in vegetation models, and calls for more research into the mechanisms behind these interactions.

Adaptation

Respiration rates at ambient temperature vary considerably across biomes, with rates increasing twofold from the Arctic to the tropics (Atkin et al. 2015). This doubling over a ~ 20 °C range in growing season temperature represents a considerable deviation from the much steeper instantaneous temperature response. Respiration standardized to 25 °C is threefold lower in the tropics than in the Arctic (Atkin et al. 2015), suggesting significant downregulation of respiratory capacity in warm-adapted species. These global patterns may reflect adaptation to different climates, but model simulations by Vanderwel et al. (2015) show that accounting for acclimation to local temperature regimes produces very similar patterns in simulated respiration across global temperature gradients. Given that the capacity for acclimation is common across the globe (Slot and Kitajima 2015) and sufficient to explain global patterns, it appears that there may not have been further selective pressure leading to adaptive temperature responses of respiration.

Volatile Organic Compounds and Thermoprotection

Volatile organic compounds (VOC) can contribute significantly to the carbon emissions from forests. The most common VOCs are isoprenes and monoterpenes, and the emission of both increases with short-term warming in the tropics (Keller and Lerdau 1999; Bracho-Nuñez et al. 2013). Isoprene emission is hypothesized to protect photosynthesis from short, high-temperature episodes (Sharkey and Yeh 2001; Sasaki et al. 2007) by helping stabilize thylakoid membranes (Velikova et al. 2011), quenching reactive oxygen species (Velikova et al. 2012; Jardine et al. 2013, 2014), and/or increasing heat dissipation (Sasaki et al. 2007; Pollastri et al. 2014). In the short term (minutes to hours), isoprene emission increases exponentially with temperature, both in temperate (Sharkey and Monson 2014) and tropical trees (Keller and Lerdau 1999), but above 38–40 °C emissions may decrease (Tingey et al. 1979; Alves et al. 2014; Jardine et al. 2013, 2014). T_{Opt} of isoprene emission is ~ 10 °C higher than T_{Opt} of net photosynthesis. Consequently, at high temperatures isoprene synthesis requires carbon sources other than recent photosynthates. Photorespiration may provide such an alternative source at temperatures exceeding T_{Opt} of photosynthesis (Jardine et al. 2014). Controlled-environment studies with temperate species have shown that plants acclimated to higher growth temperature have a higher capacity for isoprene release than plants grown at moderate temperatures (Tingey et al. 1979; Monson et al. 1992), but there is currently no evidence to suggest that warm-adapted tropical species have higher emission rates than temperate species.

Despite its potential importance in thermoprotection, isoprene emission does not occur in all plants. It only occurs in perennial plants with C_3 photosynthesis (Loreto and Fineschi 2015), but without a clear phylogenetic distribution, due in part to frequent loss and secondary acquisition of the trait (Monson et al. 2013). In the tropics isoprene emission is equally common in deciduous and evergreen plants, whereas emission is comparatively rare in temperate evergreens (Loreto and Fineschi 2015). Tambunan et al. (2006) studied isoprene emission in 42 tropical tree species and reported high rates in 4 species, very low rates (<10 µg g^{-1} h^{-1}) in 28 species, and non-detectable rates in 10 species. Of the 51 tropical tree and lianas studied by Keller and Lerdau (1999), only 15 emitted isoprene at detectable rates. Even for the latter species, the total contribution of emission of isoprene to the carbon balance was relatively small, with an overall mean rate of 24 nmol m^{-2} s^{-1} (~ 65 µg g^{-1} h^{-1} at mean leaf mass per area of 90 g m^{-2}) at 30 °C and PPFD of 1000 µmol m^{-2} s^{-1}, which is ~ 500 times lower than the mean photosynthesis rate. It is likely that long-term warming will increase rates of isoprene emission from leaves of isoprene-emitting species (Sharkey and Monson 2014), but given the thermoprotective role these emissions appear to play, the loss of isoprene carbon is probably more than compensated for by the maintenance or recovery of photo-synthetic functioning during episodic heat.

Whole-Plant Temperature Effects

Biomass Accumulation

Two meta-analyses suggest that plant growth decreases with warming when starting temperatures are already high (Way and Oren 2010; Lin et al. 2010); in both studies tropical species were underrepresented. Figure 4 shows the effect of combined day and nighttime warming on growth of *C. longifolium*. Growth was optimal at 35/29 ° C (day/night), i.e., at temperatures well above those in its natural habitat in the forest understory. Several studies show that nighttime warming stimulates growth in seedlings of tropical tree species (Esmail and Oelbermann 2011; Cheesman and Winter 2013a, b). These seedling responses contrast with results obtained for tropical canopy trees, which suggest a negative correlation between nighttime temperature and tree growth (Clark et al. 2003, 2010, 2013; Vlam et al. 2014; Anderegg et al. 2015a). A recent study on *Eucalyptus* species across a temperature gradient in Australia reported that, compared to small trees, big trees show a dis-proportionally strong decline in growth with increasing mean annual temperature (Prior and Bowman 2014). Perhaps ontogenetic differences in temperature sensi-tivity govern the contrasting temperature effects on tropical canopy trees and seedlings. Leaf area ratio (leaf area divided by total plant mass) decreases with ontogeny, so canopy trees have proportionally less photosynthetic material with which to support the maintenance of non-photosynthetic tissue than seedlings. If the capacity for thermal acclimation of root and stem respiration is smaller than that of leaf respiration, then warming would increase the respiration load in canopy trees more strongly than in seedlings. There is no indication that acclimation of root respiration is lower than that of leaf respiration in temperate species (Loveys et al. 2003), but little is known about thermal acclimation of root and stem respiration in the tropics beyond the observation by Fahey et al (2016) that acclimation of whole-plant- and root respiration varied greatly among species.

In in situ studies of growth-temperature relationships—using latitudinal or elevational temperature gradients, or inter-annual trends in temperature—temperature is one of multiple factors that change, and patterns of biological response cannot be assigned to temperature differences alone. Controlled-environment chambers, on the other hand, enable the isolated study of temperature effects. However, in controlled-environment studies with potted seedlings, soil temperature may track changes in air temperature more closely than in the field, and soil warming itself may increase growth (Königer and Winter 1993; Holtum and Winter 2014). Forest soils are thermally very stable, so soil processes may explain at least part of the observed contrasting responses of seedlings and canopy trees to nighttime warming. Soil warming may increase nutrient mineralization rates (Jarvis and Linder 2000), thereby stimulating growth. However, we found that nighttime warming stimulated growth of *O. pyramidale* seedlings equally in fertilized and non-fertilized soils (Slot and Winter unpublished data). Altered source-sink relationships (Pilkington et al. 2015) may provide an alternative, though not yet tested explanation for the observed responses.

Mortality

A quantitative trait locus (QTL) for thermotolerance has been identified for the comparatively heat-tolerant *Oryza glaberrima* Steud. (African rice), the overexpression of which in *O. sativa* L., *Arabidopsis thaliana* L., and *Festuca arundinacea* Schreb., significantly reduced their mortality following a 12 day 38/35 °C treatment (Li et al. 2015). The QTL is associated with degradation of cytotoxic denatured proteins that accumulate during heat exposure, suggesting an important role for the proteins encoded by the QTL in mitigating warming-induced cell death and plant mortality. Several other factors may play a role in warming-induced mortality. For example, increased rates of respiration and decreased rates of photosynthesis could lead to carbon starvation. Atmospheric drought caused by temperature-induced increase in VPD may also kill plants. It is hard to disentangle drought and temperature effects, as soil moisture availability may prevent warming-induced carbon starvation by enabling transpirational leaf cooling and maintenance of a positive carbon balance during heat events (Bauweraerts et al. 2014), and carbon reserves in the form of soluble carbohydrates may prevent drought-induced hydraulic failure (O'Brien et al. 2014). Drought is a common agent of mortality in the tropics (e.g., Condit et al. 1995; Allen et al. 2010), subject of observational (e.g., Condit et al. 1995; Phillips et al. 2010) and experimental studies (e.g., Slot and Poorter 2007; Nepstad et al. 2007), but atmospheric drought, such as occurs during warming, is much less studied (Breshears et al. 2013).

In the *C. longifolium* study depicted in Fig. 4, daytime temperature of 39 °C and relative humidity of ~29 % was lethal for most seedlings. We do not know whether the well-watered seedlings died from cytotoxicity, carbon starvation, or hydraulic failure, but the fact that T_{Opt} and A_{Opt} at 35/29 °C were higher under elevated relative humidity (Fig. 3), shows that atmospheric drought can negatively impact the maintenance of a positive carbon balance, and potentially survival. Tropical sapling mesocosms grown outdoors in Panama, with daytime warming of ~ 5 °C did not negatively impact biomass accumulation and photosynthetic traits as long as plants had access to soil water (Slot and Winter unpublished data), further highlighting the interconnected nature of heat and drought as agents of plant mortality.

Elevated CO_2 improves both the carbon balance and water use efficiency of plants, but a recent greenhouse study on *Eucalyptus radiata* A.Cunn. ex DC suggests that negative effects of elevated temperature on drought-induced mortality are not alleviated by elevated CO_2, despite higher leaf-level water use efficiency in the high CO_2 treatment (Duan et al. 2014). Thus, drought-induced mortality may occur in parts of the tropics where warming is accompanied by decreased precipitation (Olivares et al. 2015). Furthermore, abiotic climatic changes are likely to interact with biotic sources of mortality such as insect outbreaks (Anderegg et al. 2015b). For example, warming can affect phytochemistry and the feeding efficiency of invertebrate herbivores (Jamieson et al. 2015), potentially leading to increased leaf loss and plant mortality.

Reproduction

There have not been any detailed experimental studies on how rising temperatures affect reproduction of tropical vegetation, although the reproductive phase is one of the most temperature-sensitive parts of the lifecycle of plants. Pollen production, pollen viability and pollen tube growth all decrease with warming in a range of crop species (Sage et al. 2015, and references therein). There is considerable variation in temperature sensitivity across genotypes and species, but no clear relationship between ambient growth environment and thermal sensitivity (Prasad et al. 2006). Pollen viability of the tropical tree *Mangifera indica* L. (mango) decreases above 33 °C (Issarakraisila and Considine 1994), which is not an unusually high temperature in the tropics. Many tropical species flower for only short periods and risk reproductive failure if flowering coincides with anomalously high temperatures. Surprisingly, flower production correlated positively with temperature both seasonally and interannually in tropical forests in Panama and Puerto Rico (Pau et al. 2013). Increased flower production could compensate for decreased pollen viability but detailed study is required to better understand temperature effects on tropical tree reproduction and to disentangle direct warming effects on plant fertility, and warming effects on reproductive allocation.

Temperature may also affect reproduction through its effect on pollinators. For example, fig-pollinating wasps—which develop inside developing figs—die at temperatures only a few degrees above current ambient temperatures (Patiño et al. 1994). Transpirational cooling helps maintain fruit temperatures in non-lethal ranges, but further warming could have major negative consequences for fig reproduction.

Exposure to high temperature is necessary to break physical dormancy of seeds of many species. This is especially common for species that form seed banks, as the high temperature is associated with high light conditions favorable for seedling establishment of these early-successional species. More intense heatwaves (Meehl and Tebaldi 2004) could potentially trigger germination under conditions not favorable for seedling establishment, leading to mortality. The potential effects of climate change on the seed ecology of tropical forest trees is not well understood (Walck et al. 2011).

Stand-, Ecosystem-, and Biome-Level Temperature Effects

An important aspect of tropical forests is their very high species diversity. All trees ultimately use—and often directly compete for—the same resources, and species coexistence is at least in part maintained by niche specialization along resource gradients (e.g., Kitajima and Poorter 2008; Condit et al. 2013). This means that climate warming may shift the competitive balance within a forest community if rising temperature differentially affects mortality and reproduction across species. For example, warming may increase soil mineralization rates in the tropics (Salinas et al. 2011). Tropical tree species vary widely in their phosphorus affinity (Condit et al. 2013), and temperature-induced changes in nutrient availability potentially contribute to shifts in species composition. This may result in reduced diversity, or a change in the dominant species and functional types associated with a particular nutrient regime. Climate-induced changes in species composition will affect forest growth (Coomes et al. 2014). Systematic changes in growth rates of trees may also cause turnover rates in tropical forests to increase (Phillips and Gentry 1994). Nevertheless, changes in tree growth rates and turnover do not necessarily change ecosystem carbon storage if the size structure of trees within the ecosystem is maintained (Körner 2009).

Tropical ecosystems have lower ratios of NPP over gross primary productivity (GPP) than most higher latitude ecosystems (Zhang et al. 2009, 2014). Furthermore, rising temperatures decrease the NPP/GPP ecosystem carbon use efficiency (Zhang et al. 2014) suggesting that, despite observed thermal acclimation capacity at the leaf level, warming causes ecosystem-level respiration cost to increase. Soil respiration is a major component of ecosystem respiration, and soil respiration rates in tropical forests are higher than in any other ecosystem in the world (Raich and Schlesinger 1992). The capacity of tropical soil respiration to acclimate to higher temperatures is unknown.

Concluding Remarks

Because most research on temperature effects on the ecophysiology of plants has been done on non-tropical species, much is still to be discovered about how tropical trees will respond to climate change. We have a fairly sound understanding of general principles of temperature responses at the organelle and leaf level (Table 1), but we do not know whether tropical trees differ systematically from non-tropical species in how warming affects leaf and whole-plant performance. For example, more research is needed to establish the generality of detractive acclimation of photosynthesis in tropical species, as observed in *C. longifolium* (Fig. 3). We have shown that the commonly assumed exponential increase of respiration with temperature may not apply to tropical leaves over wide temperature ranges, a finding that requires further confirmation. At this point, we can only speculate on the whole tree-level effects of warming, apart from noting that warming is associated with atmospheric drought and may affect reproduction. At the stand level, warming may exacerbate midday reductions in stomatal conductance. A better mechanistic understanding of the causes and consequences of midday stomatal closure will improve our ability to translate understanding based on individual leaf responses to stand-level photosynthesis simulations.

In the decades to come the tropics will experience further increases in atmospheric CO_2 concentrations and warming, and significant alterations in the timing and amount of rainfall (Diffenbaugh and Scherer 2011; Mora et al. 2013). The vulnerability of tropical forests to climate change is under much debate (e.g., Clark 2004; Lloyd and Farquhar 2008, Lewis et al. 2009; Booth et al. 2012; Randerson 2013). Because of the large contribution of tropical forests to the global carbon cycle, a better understanding of how these forests respond to climate change drivers will improve our ability to predict future climate and biogeochemical cycling at the global scale (Booth et al. 2012; Huntingford et al. 2013; Piao et al. 2013). Information on thermal acclimation of photosynthetic parameters V_{Cmax} and J_{Max} will be crucial to improve predictive models De Kauwe et al. (2016). We can learn a lot about plant-temperature interactions from elevation gradients in the tropics (e.g., Malhi et al. 2010), but these gradients do not reveal how plants at the high-temperature end of the gradient will respond to further warming, and that is precisely the key in determining the fate of tropical lowland forests in a warming world. As global warming generates novel climate regimes in tropical latitudes, maintenance of current forest structure requires processes of thermal acclimation to occur in conjuction with adjustments to changing precipitation regimes (Mora et al. 2013)—and possibly changes in cloud cover—under gradually rising ambient CO_2 concentrations and increasing atmospheric deposition of nutrients (Hietz et al. 2011). Clearly, in situ warming experiments of entire forest segments are needed to better assess the effects of rising temperatures on tropical vegetation (Cavaleri et al. 2015).

Acknowledgements This work was supported by the Smithsonian Tropical Research Institute. M.S. was recipient of a CTFS-Forest-GEO postdoctoral fellowship. Milton Garcia assisted with in situ canopy measurements.

References

Allakhverdiev SI, Kreslavski VD, Klimov VV, Los DA, Carpentier R, Mohanty P (2008) Heat stress: an overview of molecular responses in photosynthesis. Photosynth Res 98:541–550

Allen CD, Macalady AK, Chenchouni H, Bachelet D, McDowell N, Vennetier M, Kitzberger T, Rigling A et al (2010) A global overview of drought and heat-induced tree mortality reveals emerging climate change risks for forests. For Ecol Man 259:660–684

Alves EG, Harley P, Gonçalves JFC, da Silva CE, Moura KJ (2014) Effects of light and temperature on isoprene emission at different leaf developmental stages of *Eschweilera coriacea* in central Amazon. Acta Amazonica 44:9–18

Anderegg WR, Ballantyne AP, Smith WK, Majkut J, Rabin S, Beaulieu C, Birdsey R, Dunne JP et al (2015a) Tropical nighttime warming as a dominant driver of variability in the terrestrial carbon sink. Proc Nal Acad Sci USA 112:15591–15596

Anderegg WRL, Hicke JA, Fisher RA, Allen CD, Aukema J, Bentz B, Hood S, Lichstein JW et al (2015b) Tree mortality from drought, insects, and their interactions in a changing climate. New Phytol 208:674–683

Ashraf M, Harris PJC (2013) Photosynthesis under stressful environments: an overview. Photosynthetica 51:163–190

Atkin OK, Tjoelker MG (2003) Thermal acclimation and the dynamic response of plant respiration to temperature. Trends Plant Sci 8:343–351

Atkin OK, Bruhn D, Tjoelker MG (2005) Response of plant respiration to changes in temperature: mechanisms and consequences of variations in Q_{10} values and acclimation. In: Lambers H, Ribas-Carbo M (eds) Plant respiration: from cell to ecosystem. Springer, Dordrecht, pp 95–135

Atkin OK, Sherlock D, Fitter A, Jarvis S, Hughes J, Campbell C, Hurry V, Hodge A (2009) Temperature dependence of respiration in roots colonized by arbuscular mycorrhizal fungi. New Phytol 182:188–199

Atkin OK, Bloomfield KJ, Reich PB, Tjoelker MG, Asner GP, Bonal D, Bönisch G, Bradford M et al (2015) Global variability in leaf respiration among plant functional types in relation to climate and leaf traits. New Phytol 206:614–636

Bauweraerts I, Ameye M, Wertin TM, McGuire MA, Teskey RO, Steppe K (2014) Water availability is the decisive factor for the growth of two tree species in the occurrence of consecutive heat waves. Agr Forest Meteorol 189:19–29

Bernacchi CJ, Pimentel C, Long SP (2003) In vivo temperature response functions of parameters required to model RuBP-limited photosynthesis. Plant Cell Environ 26:1419–1430

Berry J, Björkman O (1980) Photosynthetic response and adaptation to temperature in higher plants. Annu Rev Plant Physiol 31:491–543

Booth BBB, Jones CD, Collins M, Totterdell IJ, Cox PM, Sitch S, Huntingford C, Betts RA, Harris GR, Lloyd J (2012) High sensitivity of future global warming to land carbon cycle processes. Environ Res Lett 7:024002

Bracho-Nuñez A, Knothe NM, Welter S, Staudt M, Costa WR, Liberato MAR, Piedade MTF, Kesselmeier J (2013) Leaf level emissions of volatile organic compounds (VOC) from some Amazonian and mediterranean plants. Biogeosci 10:5855–5873

Breshears DD, Adams HD, Eamus D, McDowell NG, Law DJ, Will RE, Williams AP, Zou CB (2013) The critical amplifying role of increasing atmospheric moisture demand on tree mortality and associated regional die-off. Front Plant Sci 4:266

von Caemmerer S (2000) Biochemical models of leaf photosynthesis (2). CSIRO Publishing, Collingwood, Australia

von Caemmerer S, Quick WP (2000) Rubisco: physiology in vivo. In: Leegood RC, Sharkey TD, von Caemmerer S (eds) Photosynthesis. Springer, Dordrecht, pp 85–113

Cavaleri MA, Oberbauer SF, Ryan MG (2008) Foliar and ecosystem respiration in an old-growth tropical rain forest. Plant Cell Environ 31:473–483

Cavaleri MA, Reed SC, Smith WK, Wood TE (2015) Urgent need for warming experiments in tropical forests. Glob Change Biol 21:2111–2121

Cernusak LA, Winter K, Dalling JW, Holtum JA, Jaramillo C, Körner C, Leakey ADB, Norby RJ et al (2013) Tropical forest responses to increasing atmospheric CO_2: current knowledge and opportunities for future research. Funct Plant Biol 40:531–551

Chambers JQ, Tribuzy ES, Toledo LC, Crispim BF, Higuchi N, Dos Santos J, Araújo AC, Kruijt B et al (2004) Respiration from a tropical forest ecosystem: partitioning of sources and low carbon use efficiency. Ecol Appl 14:S72–S88

Cheesman AW, Winter K (2013a) Growth response and acclimation of CO_2 exchange characteristics to elevated temperatures in tropical tree seedlings. J Exp Bot 64:3817–3828

Cheesman AW, Winter K (2013b) Elevated night-time temperatures increase growth in seedlings of two tropical pioneer tree species. New Phytol 197:1185–1192

Clark DA (2004) Sources or sinks? The responses of tropical forests to current and future climate and atmospheric composition. Philos Trans Roy Soc B 359:477–491

Clark DA, Piper SC, Keeling CD, Clark DB (2003) Tropical rain forest tree growth and atmospheric carbon dynamics linked to interannual temperature variation during 1984–2000. Proc Natl Acad Sci USA 100:5852–5857

Clark DA, Clark DB, Oberbauer SF (2013) Field-quantified responses of tropical rainforest aboveground productivity to increasing CO_2 and climatic stress, 1997–2009. J Geophys Res-Biogeo 118:783–794

Clark DB, Clark DA, Oberbauer SF (2010) Annual wood production in a tropical rain forest in NE Costa Rica linked to climatic variation but not to increasing CO_2. Glob Change Biol 16:747–759

Condit R, Hubbell SP, Foster RB (1995) Mortality rates of 205 neotropical tree and shrub species and the impact of a severe drought. Ecol Monogr 65:419–439

Condit R, Engelbrecht BM, Pino D, Pérez R, Turner BL (2013) Species distributions in response to individual soil nutrients and seasonal drought across a community of tropical trees. Proc Natl Acad Sci USA 110:5064–5068

Coomes DA, Flores O, Holdaway R, Jucker T, Lines ER, Vanderwel MC (2014) Wood production response to climate change will depend critically on forest composition and structure. Glob Change Biol 20:3632–3645

Corlett RT (2011) Impacts of warming on tropical lowland rainforests. Trends Ecol Evol 27:145–150

Cunningham SC, Read J (2002) Comparison of temperate and tropical rainforest tree species: photosynthetic responses to growth temperature. Oecologia 133:112–119

Cunningham SC, Read J (2003) Do temperate rainforest trees have a greater ability to acclimate to changing temperatures than tropical rainforest trees? New Phytol 157:55–64

Cunningham SC, Read J (2006) Foliar temperature tolerance of temperate and tropical evergreen rain forest trees of Australia. Tree Physiol 26:1435–1443

De Kauwe MG, Lin YS, Wright IJ, Medlyn BE, Crous KY, Ellsworth DS, Maire V, Prentice IC et al (2016) A test of the 'one-point method' for estimating maximum carboxylation capacity from field-measured, light-saturated photosynthesis. New Phytol In Press. doi:10.1111/nph.13815

Diffenbaugh NS, Scherer M (2011) Observational and model evidence of global emergence of permanent, unprecedented heat in the 20th and 21st centuries. Clim Change 107:615–624

Doughty CE (2011) An in situ leaf and branch warming experiment in the Amazon. Biotropica 43:658–665

Doughty CE, Goulden ML (2008) Are tropical forests near a high temperature threshold? J Geophys Res Biogeosci 113:G00B07

Duan H, Duursma RA, Huang G, Smith RA, Choat B, O'Grady AP, Tissue DT (2014) Elevated [CO_2] does not ameliorate the negative effects of elevated temperature on drought-induced mortality in Eucalyptus radiata seedlings. Plant Cell Environ 37:1598–1613

Dusenge ME, Wallin G, Gårdesten J, Niyonzima F, Adolfsson L, Nsabimana D, Uddling J (2015) Photosynthetic capacity of tropical tree species in relation to leaf nutrients, successional group identity and growth temperature. Oecologia 177:1183–1194

Easterling DR, Horton B, Jones PD, Peterson TC, Karl TR, Parker DE, Salinger MJ, Razuvayev V et al (1997) Maximum and minimum temperature trends for the globe. Science 277:346–366

Ellsworth DS, Crous KY, Lambers H, Cooke J (2015) Phosphorus recycling in photorespiration maintains high photosynthetic capacity in woody species. Plant Cell Environ 38:1142–1156

Esmail S, Oelbermann M (2011) The impact of climate change on the growth of tropical agroforestry tree seedlings. Agrofor Syst 83:235–244

Fahey C, Winter K, Slot M, Kitajima K (2016) Influence of arbuscular mycorrhizal fungi on whole-plant respiration and thermal acclimation of tropical tree seedlings. Ecol & Evol. In Press. doi:10.1002/ece3.1952

Harley PC, Sharkey TD (1991) An improved model of C_3 photosynthesis at high CO_2: reversed O_2 sensitivity explained by lack of glycerate reentry into the chloroplast. Photosyn Res 27:169–178

Hietz P, Turner BL, Wanek W, Richter A, Nock CA, Wright SJ (2011) Long-term change in the nitrogen cycle of tropical forests. Science 334:664–666

Hikosaka K, Ishikawa K, Borjigidai A, Muller O, Onoda Y (2006) Temperature acclimation of photosynthesis: mechanisms involved in the changes in temperature dependence of photosynthetic rate. J Exp Bot 57:291–302

Holtum JA, Winter K (2014) Limited photosynthetic plasticity in the leaf-succulent CAM plant *Agave angustifolia* grown at different temperatures. Funct Plant Biol 41:843–849

Huntingford C, Zelazowski P, Galbraith D, Mercado LM, Sitch S, Fisher R, Lomas M, Walker AP et al (2013) Simulated resilience of tropical rainforests to CO_2-induced climate change. Nat Geosci 6:268–273

Hüve K, Bichele I, Ivanova H, Keerberg O, Pärnik T, Rasulov B, Tobias M, Niinemets Ü (2012) Temperature responses of dark respiration in relation to leaf sugar concentration. Physiol Plant 144:320–334

Issarakraisila M, Considine JA (1994) Effects of temperature on pollen viability in mango cv. 'Kensington'. Ann Bot-London 73:231–240

Jamieson MA, Schwartzberg EG, Raffa KF, Reich PB, Lindroth RL (2015) Experimental climate warming alters aspen and birch phytochemistry and performance traits for an outbreak insect herbivore. Glob Change Biol 21:268–2710

Janzen DH (1967) Why mountain passes are higher in the tropics. Am Nat 101:233–249

Jaramillo C, Ochoa D, Contreras L, Pagani M, Carvajal-Ortiz H, Pratt LM, Krishnan S, Cardona A et al (2010) Effects of rapid global warming at the Paleocene-Eocene boundary on neotropical vegetation. Science 330:957–961

Jardine K, Meyers K, Abrell L, Alves EG, Serrano AMY, Kesselmeier J, Karl T, Guenther A et al (2013) Emissions of putative isoprene oxidation products from mango branches under abiotic stress. J Exp Bot 64:3669–3679

Jardine K, Chambers JQ, Alves EG, Teixeira A, Garcia S, Holm J, Niguchi N, Abrell L et al (2014) Dynamic balancing of isoprene carbon sources reflects photosynthetic and photorespiratory responses to temperature stress. Plant Physiol 166:2051–2064

Jarvis P, Linder S (2000) Constraints to growth of boreal forests. Nature 405:904–905

Keller M, Lerdau M (1999) Isoprene emission from tropical forest canopy leaves. Glob Biogeochem Cycles 13:19–29

Kirschbaum MUF, Farquhar GD (1984) Temperature dependence of whole-leaf photosynthesis in *Eucalyptus pauciflora* Sieb. ex Spreng. Funct Plant Biol 11:519–538

Kitajima K, Poorter L (2008) Functional basis for resource niche partitioning by tropical trees. In: Carson WP, Schnitzer SA (eds) Tropical forest community ecology. Blackwell, Oxford, pp 172–188

Königer M, Winter K (1993) Growth and photosynthesis of *Gossypium hirsutum* L. at high photon flux densities: effects of soil temperatures and nocturnal air temperatures. Agronomie 13:423–431

Körner C (2009) Responses of humid tropical trees to rising CO_2. Annu Rev Ecol Evol Syst 40:61–79

Kositsup B, Montpied P, Kasemsap P, Thaler P, Améglio T, Dreyer E (2009) Photosynthetic capacity and temperature responses of photosynthesis of rubber trees (*Hevea brasiliensis* Müll. Arg.) acclimate to changes in ambient temperatures. Trees 23:357–365

Kozaki A, Takeba G (1996) Photorespiration protects C_3 plants from photooxidation. Nature 384:557–560

Krause GH, Weis E (1991) Chlorophyll fluorescence and photosynthesis: the basics. Annu Rev Plant Biol 42:313–349

Krause GH, Winter K, Krause B, Jahns P, García M, Aranda J, Virgo A (2010) High-temperature tolerance of a tropical tree, *Ficus insipida*: methodological reassessment and climate change considerations. Funct Plant Biol 37:890–900

Krause GH, Cheesman AW, Winter K, Krause B, Virgo A (2013) Thermal tolerance, net CO_2 exchange and growth of a tropical tree species, *Ficus insipida*, cultivated at elevated daytime and nighttime temperatures. J Plant Physiol 170:822–827

Krause GH, Winter K, Krause B, Virgo A (2015) Light-stimulated heat tolerance in leaves of two neotropical tree species, *Ficus insipida* and *Calophyllum longifolium*. Funct Plant Biol 42:42–51

Lewis SL, Lloyd J, Sitch S, Mitchard ET, Laurance WF (2009) Changing ecology of tropical forests: evidence and drivers. Annu Rev Ecol Evol Syst 40:529–549

Lin D, Xia J, Wan S (2010) Climate warming and biomass accumulation of terrestrial plants: a meta-analysis. New Phytol 188:187–198

Li XM, Chao DY, Wu Y, Huang X, Chen K, Cui LG, Su L, Ye W-W, et al. (2015). Natural alleles of a proteasome $\alpha 2$ subunit gene contribute to thermotolerance and adaptation of African rice. Nat Genet 47:827–833

Lloyd J, Farquhar GD (2008) Effects of rising temperatures and $[CO_2]$ on the physiology of tropical forest trees. Phil Trans R Soc B 363:1811–1817

Loreto F, Fineschi S (2015) Reconciling functions and evolution of isoprene emission in higher plants. New Phytol 206:578–582

Loveys BR, Atkinson LJ, Sherlock DJ, Roberts RL, Fitter AH, Atkin OK (2003) Thermal acclimation of leaf and root respiration: an investigation comparing inherently fast-and slow-growing plant species. Glob Change Biol 9:895–910

Makino A, Nakano H, Mae T (1994) Effects of growth temperature on the responses of ribulose-1,5-bisphosphate carboxylase, electron-transport components, and sucrose synthesis enzymes to leaf nitrogen in rice, and their relationships to photosynthesis. Plant Physiol 105:1231–1238

Malhi Y (2012) The productivity, metabolism and carbon cycle of tropical forest vegetation. J Ecol 100:65–75

Malhi Y, Silman M, Salinas N, Bush M, Meir P, Saatchi S (2010) Introduction: elevation gradients in the tropics: laboratories for ecosystem ecology and global change research. Glob Change Biol 16:3171–3175

Meehl GA, Tebaldi C (2004) More intense, more frequent, and longer lasting heat waves in the 21st century. Science 305:994–997

Meir P, Grace J, Miranda AC (2001) Leaf respiration in two tropical rainforests: constraints on physiology by phosphorus, nitrogen and temperature. Funct Ecol 15:378–387

Menzel A, Fabian P (1999) Growing season extended in Europe. Nature 397:659

Menzel A, Sparks TH, Estrella N, Koch E, Aasa A, Ahas R, Alm-Kübler K, Bissolli P et al (2006) European phenological response to climate change matches the warming pattern. Glob Change Biol 12:1969–1976

Michaletz ST, Cheng D, Kerkhoff AJ, Enquist BJ (2014) Convergence of terrestrial plant production across global climate gradients. Nature 512:39–43

Monson RK, Jaeger CH, Adams WW, Driggers EM, Silver GM, Fall R (1992) Relationships among isoprene emission rate, photosynthesis, and isoprene synthase activity as influenced by temperature. Plant Physiol 98:1175–1180

Monson RK, Jones RT, Rosenstiel TN, Schnitzler J-P (2013) Why only some plants emit isoprene. Plant Cell Environ 36:503–516

Mora C, Frazier AG, Longman RJ, Dacks RS, Walton MM, Tong EJ, Sanchez TJ, Kaiser LR et al (2013) The projected timing of climate departure from recent variability. Nature 502:183–187

Murakami Y, Tsuyama M, Kobayashi Y, Kodama H, Iba K (2000) Trienoic fatty acids and plant tolerance of high temperature. Science 287:476–479

Nepstad DC, Tohver IM, Ray D, Moutinho P, Cardinot G (2007) Mortality of large trees and lianas following experimental drought in an Amazon forest. Ecology 88:2259–2269

O'Brien MJ, Leuzinger S, Philipson CD, Tay J, Hector A (2014) Drought survival of tropical tree seedlings enhanced by non-structural carbohydrate levels. Nat Clim Change 4:710–714

Olivares I, Svenning J-C, van Bodegom PM, Balslev H (2015) Effects of warming and drought on the vegetation and plant diversity in the Amazon basin. Bot Rev 81:42–69

Osmond CB, Björkman O (1972) Simultaneous measurements of oxygen effects on net photosynthesis and glycolate metabolism in C_3 and C_4 species of *Atriplex*. Carnegie Inst Wash Yearb 71:141–148

O'Sullivan OS, Weerasinghe KK, Evans JR, Egerton JJ, Tjoelker MG, Atkin OK (2013) High-resolution temperature responses of leaf respiration in snow gum (*Eucalyptus pauciflora*) reveal high-temperature limits to respiratory function. Plant Cell Environ 36:1268–1284

Pan Y, Birdsey RA, Phillips OL, Jackson RB (2013) The structure, distribution, and biomass of the world's forests. Annu Rev Ecol Evol Syst 44:593–622

Pau S, Wolkovich EM, Cook BI, Nytch CJ, Regetz J, Zimmerman JK, Wright SJ (2013) Clouds and temperature drive dynamic changes in tropical flower production. Nat Clim Change 3:838–842

Patiño S, Herre EA, Tyree MT (1994) Physiological determinants of *Ficus* fruit temperature and implications for survival of pollinator wasp species: comparative physiology through an energy budget approach. Oecologia 100:13–20

Peraudeau S, Lafarge T, Roques S, Quiñones CO, Clement-Vidal A, Ouwerkerk PB, van Rie J, Fabre D et al (2015) Effect of carbohydrates and night temperature on night respiration in rice. J Exp Bot 66:3931–3944

Phillips OL, Gentry AH (1994) Increasing turnover through time in tropical forests. Science 263:954–958

Phillips OL, Van der Heijden G, Lewis SL, López-González G, Aragão LE, Lloyd J. Malhi Y, Monteagudo A et al (2010) Drought–mortality relationships for tropical forests. New Phytol 187:631–646

Piao S, Sitch S, Ciais P, Friedlingstein P, Peylin P, Wang X, Ahlström A, Anav A et al (2013) Evaluation of terrestrial carbon cycle models for their response to climate variability and to CO_2 trends. Glob Change Biol 19:2117–2132

Pilkington SM, Encke B, Krohn N, Hoehne M, Stitt M, Pyl ET (2015) Relationship between starch degradation and carbon demand for maintenance and growth in *Arabidopsis thaliana* in different irradiance and temperature regimes. Plant Cell Environ 38:157–171

Pollastri S, Tsonev T, Loreto F (2014) Isoprene improves photochemical efficiency and enhances heat dissipation in plants at physiological temperatures. J Exp Bot 65:1565–1570

Portis AR Jr (2003) Rubisco activase–Rubisco's catalytic chaperone. Photosyn Res 75:11–27

Prasad PVV, Boote KJ, Allen LH Jr, Sheehy JE, Thomas JMG (2006) Species, ecotype and cultivar differences in spikelet fertility and harvest index of rice in response to high temperature stress. Field Crops Res 95:398–411

Prior LD, Bowman DM (2014) Big eucalypts grow more slowly in a warm climate: evidence of an interaction between tree size and temperature. Glob Change Biol 20:2793–2799

Raich JW, Schlesinger WH (1992) The global carbon dioxide flux in soil respiration and its relationship to vegetation and climate. Tellus B 44:81–99

Randerson JT (2013) Climate science: global warming and tropical carbon. Nature 494:319–320

Sage RF, Kubien DS (2007) The temperature response of C_3 and C_4 photosynthesis. Plant Cell Environ 30:1086–1106

Sage TL, Bagha S, Lundsgaard-Nielsen V, Branch HA, Sultmanis S, Sage R (2015) The effect of high temperature stress on male and female reproduction in plants. Field Crop Res 182:30–42

Salinas N, Malhi Y, Meir P, Silman M, Roman Cuesta R, Huaman J, Salinas D, Huaman V et al (2011) The sensitivity of tropical leaf litter decomposition to temperature: results from a large-scale leaf translocation experiment along an elevation gradient in Peruvian forests. New Phytol 189:967–977

Salvucci ME, Crafts-Brandner SJ (2004) Relationship between the heat tolerance of photosynthesis and the thermal stability of Rubisco activase in plants from contrasting thermal environments. Plant Physiol 134:1460–1470

Sasaki K, Saito T, Lämsä M, Oksman-Caldentey KM, Suzuki M, Ohyama K, Muranaka T, Ohara K, Yazaki K (2007) Plants utilize isoprene emission as a thermotolerance mechanism. Plant Cell Physiol 48:1254–1262

Saugier B, Roy J, Mooney HA (2001) Estimations of global terrestrial productivity: converging toward a single number? In: Roy J, Saugier B, Mooney HA (eds) Terrestrial global productivity. Academic Press, New York, pp 543–557

Sharkey TD, Monson RK (2014) The future of isoprene emission from leaves, canopies and landscapes. Plant Cell Environ 37:1727–1740

Sharkey TD, Yeh S (2001) Isoprene emission from plants. Annu Rev Plant Biol 52:407–436

Slot M, Poorter L (2007) Diversity of tropical tree seedling responses to drought. Biotropica 39:683–690

Slot M, Wright SJ, Kitajima K (2013) Foliar respiration and its temperature sensitivity in trees and lianas: in situ measurements in the upper canopy of a tropical forest. Tree Physiol 33:505–515

Slot M, Rey-Sánchez C, Gerber S, Lichstein JW, Winter K, Kitajima K (2014a) Thermal acclimation of leaf respiration of tropical trees and lianas: response to experimental canopy warming, and consequences for tropical forest carbon balance. Glob Change Biol 20:2915–2926

Slot M, Rey-Sánchez C, Winter K, Kitajima K (2014b) Trait-based scaling of temperature-dependent foliar respiration in a species-rich tropical forest canopy. Funct Ecol 28:1074–1086

Slot M, Kitajima K (2015) General patterns of thermal acclimation of leaf respiration across biomes and plant types. Oecologia 177:885–900

Slot M, Garcia MN, Winter K (2016) Temperature response of CO_2 exchange in three tropical tree species. Funct Plant Biol. In Press. doi:10.1071/FP15320

Smith NG, Dukes JS (2013) Plant respiration and photosynthesis in global-scale models: incorporating acclimation to temperature and CO_2. Glob Change Biol 19:45–63

Tambunan P, Baba S, Kuniyoshi A, Iwasaki H, Nakamura T, Yamasaki H, Oku H (2006) Isoprene emission from tropical trees in Okinawa Island, Japan. Chemosphere 65:2138–2144

Tingey DT, Manning M, Grothaus LC, Burns WF (1979) The influence of light and temperature on isoprene emission rates from live oak. Physiol Plant 47:112–118

Vanderwel MC, Slot M, Lichstein JW, Reich PB, Kattge J, Atkin OK, Bloomfield K, Tjoelker M, Kitajima K (2015) Global convergence in projected leaf respiration from estimates of thermal acclimation across time and space. New Phytol 207:1026–1037

Vargas GG, Cordero SR (2013) Photosynthetic responses to temperature of two tropical rainforest tree species from Costa Rica. Trees 27:1261–1270

Velikova V, Várkonyi Z, Szabó M, Maslenkova L, Nogues I, Kovács L, Peeva V, Busheva M et al (2011) Increased thermostability of thylakoid membranes in isoprene-emitting leaves probed with three biophysical techniques. Plant Physiol 157:905–916

Velikova V, Sharkey T, Loreto F (2012) Stabilization of thylakoid membranes in isoprene-emitting plants reduces formation of reactive oxygen species. Plant Signal Behav 7:139–141

Vlam M, Baker PJ, Bunyavejchewin S, Zuidema PA (2014) Temperature and rainfall strongly drive temporal growth variation in Asian tropical forest trees. Oecologia 174:1449–1461

Wahid A, Gelani S, Ashraf M, Foolad MR (2007) Heat tolerance in plants: an overview. Environ Exp Bot 61:199–223

Walck JL, Hidayati SN, Dixon KW, Thompson K, Poschlod P (2011) Climate change and plant regeneration from seed. Glob Change Biol 17:2145–2161

Wang D, Li XF, Zhou ZJ, Feng XP, Yang WJ, Jiang DA (2010) Two Rubisco activase isoforms may play different roles in photosynthetic heat acclimation in the rice plant. Physiol Plant 139:55–67

Way DA, Oren R (2010) Differential responses to changes in growth temperature between trees from different functional groups and biomes: a review and synthesis of data. Tree Physiol 30:669–688

Way DA, Yamori W (2014) Thermal acclimation of photosynthesis: on the importance of adjusting our definitions and accounting for thermal acclimation of respiration. Photosynth Res 119:89–100

Wright SJ, Muller-Landau HC, Schipper J (2009) The future of tropical species on a warmer planet. Conserv Biol 23:1418–1426

Zhang Y, Xu M, Chen H, Adams J (2009) Global pattern of NPP to GPP ratio derived from MODIS data: effects of ecosystem type, geographical location and climate. Glob Ecol Biogeogr 18:280–290

Zhang Y, Yu G, Yang J, Wimberly MC, Zhang X, Tao J, Jiang Y, Zhu J (2014) Climate-driven global changes in carbon use efficiency. Glob Ecol Biogeogr 23:144–155

Zotz G, Harris G, Königer M, Winter K (1995) High rates of photosynthesis in a tropical pioneer tree, *Ficus insipida*. Flora 190:265–272

Tree Biomechanics with Special Reference to Tropical Trees

Karl J. Niklas

Abstract All structures, whether engineered or natural, must obey the same physical laws and processes. Trees are particularly susceptible to these laws and processes because they are structures, composed mostly of wood, that begin and end their lives in the same location, which can experience dramatic changes in abiotic and biotic conditions (e.g., rainfall and epiphyte loads, respectively). The biomechanical behavior of trees is reviewed by presenting and discussing a few equations, with a particular emphasis on the effects of wind on branches, trunks, and roots. Limited space precludes a detailed review of these equations. Therefore, some basic references are listed to provide the necessary details. An important point is that unlike engineered objects and the engineering theory that deals with them, trees are growing biological entities that violate many of the assumptions of engineering theory. Consequently, the equations presented here provide only a first order approximation of how trees will respond to static (self) and dynamic (wind) loadings. Understanding the limits of these equations, therefore, is a critical first lesson in dealing with the biomechanical behavior of trees, regardless of whether they grow in tropical, temperate, or desert conditions. In the final analysis, every tree will ultimately fail. The challenge is to anticipate when and how.

Keywords Biomechanics · Buckling · Dynamic loading · Maximum height · Static loading · Trees · Wind

Introduction

The goal of this chapter is to provide an overview of the basic physical principles and phenomena that influence the mechanical behavior of trees. This overview revolves around eight biomechanical features that hold true for all trees, regardless

K.J. Niklas (✉)
Plant Biology Section, School of Integrative Plant
Science Cornell University, Ithaca, NY 14853, USA
e-mail: kjn2@cornell.edu

© Springer International Publishing Switzerland 2016
G. Goldstein and L.S. Santiago (eds.), *Tropical Tree Physiology*,
Tree Physiology 6, DOI 10.1007/978-3-319-27422-5_19

of their habitat: (1) trees sustain two general categories of mechanical forces (static loads and dynamic loads), (2) these forces are additive (stresses as well as strains are additive), (3) static loads increase slowly over time as trees increase in size and as epiphytes, if any, increase in number or size (therefore, tree growth patterns can compensate for these increasing loads), (4) dynamic loads can change dramatically over short periods of time (these loads are unpredictable and therefore potentially dangerous), (5) trees generally fail as a result of dynamic loads, (6) plant tissues resist bending more than twisting (eccentric loadings are potentially dangerous), (7) young parts of woody plants are more flexible than older parts, and (8) belowground growth generally does not keep pace with aboveground growth. Each of these claims will be explored in the following sections.

However, before exploring these topics, it is important to emphasize that trees grow and respond to their mechanical environment. This may seem trivial, but it is a vital fact. Fluid and mechanical engineering makes a number of assumptions that are consistently violated by anything that grows organically. For example, engineering theory by and large assumes that a structure is composed of materials that are homogeneous and elastic in behavior. No animal or plant tissue is homogeneous and every tissue is viscoelastic. Engineering practice deals with structures that are fabricated to meet specific specifications for a particular work-place environment. Natural selection has no agenda. Perhaps most important is the fact that an engineered structure cannot heal itself and cannot change its size, with the exception of corrosion. Every organism has the capacity to replace damaged parts and every organism changes size as it grows, reproduces, and ultimately dies. Trees grow in response to their local environment in ways that cannot be modeled easily nor comprehended fully. Engineering theory also typically deals with small deflections, whereas trees often experience extremely large deflections. For these reasons, many of the assumptions made in engineering theory and practice are naïve in the context of tree biology. Consequently, the principles presented in this chapter can provide, at best only guidance when predicting the mechanical behavior of trees.

As noted, the concepts presented in this chapter pertain generally to all trees, regardless of whether they grow in the tropics, deserts, or in temperate ecosystems. However, tropical trees flourish under some conditions that would seem to set them apart. For example, many tropical trees have buttressed root systems that have been interpreted to function as anchorage stabilizers in moist or soft soils. Tropical trees tend to have greater epiphyte loads than trees growing in most other types of ecosystems. They can also have large liana loads that additionally entangle the branches of neighboring trees in ways that can affect dynamic dampening. Likewise, arboreal animals and large fruits often distinguish tropical trees, which suggests greater canopy dynamics and seasonal variations in self-loading compared to trees in other ecosystems. Finally, dense stands of trees growing in low latitudes often experience limited light availability during the early and late daylight hours. These and other features seem to set tropical tree biomechanics apart from tree biomechanics in general. Nevertheless, with few exceptions, as for example buttressed root systems, which present a unique morphological and biomechanical context, these and other differences are a matter of degree and not of kind. For example, the deflection of a cantilevered uniformly loaded beam, which can be used to model tree branches, depends on the load per unit length (see Eq. 15), regardless of whether the load is solely due to self-loading or involves additional loads resulting from epiphytes.

Finally, space precludes giving the derivations for the equations presented here. The derivations of these equations are provided by Niklas (1992) and Niklas and Spatz (2012), and in additional literature, which is cited as required.

Static and Dynamic Loadings

In biomechanics, any internally or externally applied force is referred to as a load. There are two general types of loads—static loads and dynamic loads, denoted as P_s and P_d, respectively. Static loads (also called self-loads) are the result of gravity acting on the aboveground parts of a plant. They are the product of mass m and the acceleration of gravity g:

$$P_s = mg. \tag{1}$$

Although the numerical value of g differs as a function of altitude, this variation is irrelevant to dealing with tree mechanics. The conventional standard is $g = 9.80665$ m/s^2.

Dynamic loads are externally applied forces such as the pressure exerted by wind, the collision of stems or leaves with other stems or leaves, or the activities of animals in a canopy. The most pervasive dynamic load is wind pressure, which produces a drag force, denoted as D_f. This force is the product of air density ρ, wind speed U squared, projected (sail) area Sp, and a dimensionless parameter called the drag coefficient C_D:

$$D_{-f} = 0.5\,\rho\,U^2 Sp\,C_D \tag{2}$$

C_D is often assumed to be a constant for computational ease. However, it is not a constant for plants because leaves and stems reorient with respect to oncoming wind and thus reduce the numerical value of this "constant" (Vogel 1981). This reorientation also typically reduces Sp. Consequently, under some conditions, the drag force exerted by wind can be reduced even if wind speed increases. This phenomenology and the loss of leaves, twigs, and branches in strong winds can increase the safety factor of a tree (Niklas and Spatz 2000).

Mechanical loads are additive. Therefore, the total load experienced by a tree P_t is given by the formula

$$P_t = P_s + D_f = mg + 0.5\,\rho\,U^2\,Sp\,C_D. \tag{3}$$

Because loads are vectors (i.e., they have direction and magnitude), Eq. 3 is a vector equation.

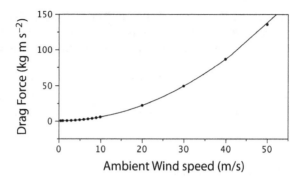

Fig. 1 The drag force exerted on a vertical cylinder (with a radius of 0.03 m and a length of 1.5 m) subjected to different uniform wind speeds. It is important to note that vertical wind speed profiles are never uniform. Under natural conditions ambient wind speeds increase from zero at ground-level to their maximum with increasing distance from ground level

With the aid of a few simplifying assumptions, Eqs. 1–3 can be used to evaluate whether static or dynamic loads contribute most to the total load. Consider a vertical cylinder with a radius of 0.03 m and a length (height) of 1.5 m subjected to a uniform wind speed of 11 m/s. Assuming that it is composed of wood with a density of 844 kg/m^3 (which is the average density of green conifer and angiosperm wood; see Niklas and Spatz 2010), Eq. 1 shows that the static load equals π $(0.03 \text{ m})^2 \times 1.5 \text{ m} \times 844 \text{ kg/m}^3 \times 9.80665 \text{ m/s}^2 \approx 35.09 \text{ kg m s}^{-2}$ or 35.09 N. With an ambient temperature of 20 °C, the density of air is 1.2041 kg/m^2. Assuming that the drag coefficient is approximately 1.0, Eq. 2 shows that the drag load equals $0.5 \times 1.2041 \text{ kg/m}^3 \times (11 \text{ m/s})^2 \times 0.06 \text{ m} \times 1.5 \text{ m} \times 1.0 \approx 6.56 \text{ kg m s}^{-2}$ or 6.56 N. Consequently, the drag load exerted on this hypothetical cylinder by air moving at 11 m/s is a small component of the total load. This situation changes as the ambient wind speed increases because the drag force scales as the square of wind speed (Fig. 1).

However, as we will see, what really matters are the stresses that develop at the base of the column when subjected to its own compressive load and to the stresses developed by the wind-induced drag. These stresses can be easily calculated, but this requires an understanding of other concepts and other formulas (see **Concluding Remarks**), which need to be presented first.

Stresses, Strains, and the Elastic and Shear Moduli

A load produces stresses and strains. A stress equals a load P divided by the area A over which the load is applied, i.e., stress is a force per unit area and thus has units of kg m^{-1} s^{-2} or N/m^2. There are three kinds of stresses: compressional stresses, tensile stresses, and shear stresses (denoted as σ_-, σ_+, and τ, respectively):

$$\sigma_-, \; \sigma_+, \; \text{and} \; \tau = P/A. \tag{4}$$

These stresses result when a material is subjected to pure compression, pure tension, or pure shear. They also result when a material is bent or twisted (Fig. 2).

A strain is a dimensionless measure of deformation. It is typically expressed as a decimal fraction or as a percentage, e.g., 0.01 or 1 %. Strains can be measured in different ways. For example, the Cauchy strain ε (also called the engineering strain) is the change in a reference dimension ΔL divided by the original non-deformed dimension L_0:

$$\varepsilon = \Delta L/L_0 = (L-L_0)/L_0. \tag{5}$$

The true strain $\varepsilon_{\text{true}}$ (also called the logarithmic or Henchy strain) is calculated by integrating the incremental Cauchy strain:

$$\varepsilon_{\text{true}} = \ln(1+\varepsilon). \tag{6}$$

For small strains (<5 %), Eqs. 5 and 6 yield similar results. For large strains, true strains should be used. Regardless of which equation is employed, it is always prudent to report how strains are calculated.

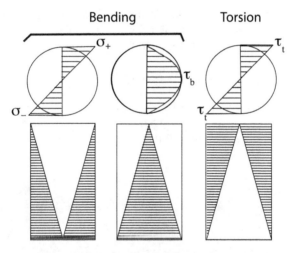

Fig. 2 Transverse and longitudinal distributions of tensile, compressive, and shear stresses (σ_+, σ_-, and τ, respectively) in a circular solid cylinder fixed at its base and subjected to bending and twisting. The magnitudes of each stress are denoted by the stippled lines in each figure. Note that σ_+ and σ_- reach their local maximum intensities at the surface of each transection and achieve their overall maximum the base of the cylinder, and that the bending shear stresses (τ_b) achieve their local maximum intensities at the center of each transection and achieve their overall maximum the base of the cylinder. In contrast, torsional shear stresses (τ_t) reach their local maximum intensities at the surface of the cylinder and their overall maximum at the free end

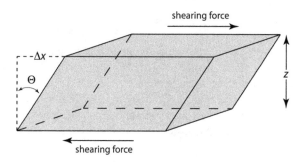

Fig. 3 Diagram of shear strains γ developing in a small volume of an isotropic material. The shear strain equals the deformation in the direction that parallels the direction of the shearing force divided by the distance over which the shearing force acts. In this illustration the direction of the shearing force is in the *x*–axis and the distance over which it acts is the height of the volume, which is *z*. Therefore, the shear strain γ equals Δ*x*/*z*, which in turn equals the tangent of Θ (see Eq. 7)

Shear strains γ are more complicated. Referring to Fig. 3, the shear strain equals the deformation in the direction that parallels the direction of the shearing force Δ*x* divided by the distance over which the shearing force acts *z*:

$$\gamma = \Delta x/z = \tan \Theta. \tag{7}$$

Equation 7 is convenient and simple. However, it is not appropriate when shearing forces produce substantial gradients of deformation within a structure. In this case, the shear strain must be calculated in its differential form:

$$\gamma = \partial x/\partial z. \tag{8}$$

The ability of a material to resist a load is measured by its elastic modulus *E* (also called Young's modulus) and its shear modulus *G*, which are given by the quotient of the appropriate stress and its corresponding strain:

$$E = \sigma/\varepsilon \tag{9}$$

and

$$G = \tau/\gamma. \tag{10}$$

Experimentally, these moduli are calculated by measuring the slope of the linear portion of bivariate plots of stress versus strain (Fig. 4).

Most materials are less capable of coping with shearing than with compression or tension because, for most materials, $G \ll E$. This fact does not imply that materials will fail more easily in shearing than in bending. It simply means that most materials shear more easily. The delamination of wood subjected to bending looks like a shearing failure, but it is not. It reflects the fact that upon bending wood

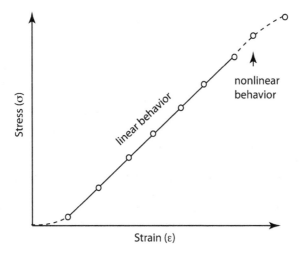

Fig. 4 A hypothetical bivariate plot of the increasing magnitudes of the stresses and strains (σ and ε, respectively) developing within a sample of a material subjected to either tension, compression, or bending. The slope of the linear portion of the stress versus strain plot equals the elastic modulus E. In this example, non-linear behavior is denoted by the *dashed lines*, which are curvilinear

fails more easily in the longitudinal direction than in the radial direction. It is worth noting further that engineering theory typically deals with what is called *pure* tension and *pure* compression (that is, compression and tension without shearing). However, this hardly ever happens in reality. Shearing always occurs in one form or another when a material is bent, compressed, or pulled. The fact that $G \ll E$ does help to explain why less force is required to twist a material than to bend it. However, it is important to draw a sharp distinction between a material and a structure. A structure can fail in a variety of ways depending upon the type and direction of loading. Indeed, as we will see, tall and slender structures can bend gracefully or catastrophically (see **Buckling**).

The Role of Geometry

Although the ability to resist loads depends on the physical properties of a material or structure, it also depends on size, shape, and geometry (note that shape and geometry are not the same thing). Specifically, the ability to resist bending is quantified as the product of the elastic modulus E and the second moment of area denoted by I, which is called flexural stiffness. In a similar manner, the ability to resist twisting is quantified as the product of the shear modulus G and the polar moment of area J, which is called the torsional stiffness:

$$\text{Flexural stiffness} = EI \qquad (11)$$

$$\text{Torsional stiffness} = GJ \qquad (12)$$

Formulas for I and J for different geometries are easily obtained from the literature. A few are provided in Fig. 5. The second moment of area and the polar moment of area quantify how the cross sectional geometry, shape, and size of an object contribute to the ability of the object to resist mechanical forces. For a circular cross section, $I = \pi\, r^4/4$ and $J = \pi\, r^4/2$, where r is radius. Notice that, in this particular case, $J = 2I$, and, second, both I and J increase dramatically as r increases. Although, J is twice the numerical value of I, the ability of a material or structure to resist twisting is still much less than the ability to resist bending because, as noted, the shear modulus G of virtually every material is many times lower than that of the

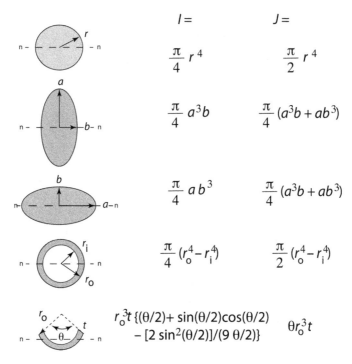

Fig. 5 Equations for computing the second moments of area (I) and the polar moments of area (J) of beams, columns, etc. characterized by different transverse geometries and shapes. Note that in each case $J > I$. Also note that, with the exception of the circular cross section, the numerical values of J and I depend on the plane of bending (i.e., the neutral plane), which is denoted in each figure by n – – – – n. The importance of the plane of bending is shown here by comparing the formulas for the two elliptical cross sections, which have different (orthogonal) orientations with respect to the neutral plane

elastic modulus E. For example, for structural steel, $G = 79$ and $E = 200$ GN/m^2, whereas, for Douglas Fir (*Pseudotsuga menziesii*) wood at a density of 625 kg/m^3, $G = 0.68$ and $E = 8.3$ GN/m^2.

A third feature is also important. With the exception of a circular cross section, the numerical value of I depends on the plane of bending. Consider the formulas for the second moment of area of a beam with an elliptical cross section (see Fig. 5). When the major axis of the ellipse is oriented in the vertical direction, $I = \pi a^3 b/4$. However, when the minor axis is oriented in the vertical direction, $I = \pi a b^3/4$. Consequently, the same object can resist bending differently depending on its orientation or the direction of an applied force. This phenomenon is evident when the cross sections of cantilevered branches or leaf petioles are inspected. Many of these structures have an elliptical cross section in which the major axis is oriented vertically. This orientation helps a branch or petiole to resist the pull of gravity, but it permits sideways deflection in the wind.

Bending and Twisting of Cantilevered Beams (Branches)

We are now in a position to consider bending and twisting moments. A moment, denoted as M, is the product of an applied load and the lever arm over which the force acts. This is illustrated for a cantilevered beam of length L fixed at one end and loaded at its free end (Fig. 6a). Assuming that the beam is uniform in its cross sectional area and homogenous in its composition, the bending moment at any point x from its fixed end given by the formula

$$M = P(L-x). \tag{13}$$

Thus, the bending moment increases linearly from the point-loaded free end toward the base of the beam where $x = 0$. The maximum deflection δ_{max} of such a beam occurs at the free end. It is given by the formula

$$\delta_{max} = PL^3/3EI. \tag{14}$$

In the case of a uniformly loaded beam, the maximum deflection is given by the formula

$$\delta_{max} = wL^4/8EI, \tag{15}$$

where w is the weight of the beam per unit length.

The situation with torsion is more complex. Consider a condition called pure shear in which a beam is twisted but not bent. Referring to Fig. 6b, the angle of twist θ results in strains γ that decrease toward the fixed end, and the circular arc between two points A and B has a length equal to $L\gamma$ and $r\theta$, where r and L are the

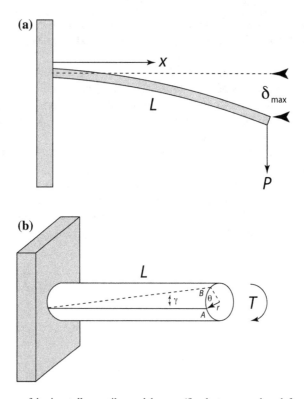

Fig. 6 Diagrams of horizontally cantilevered beams (fixed at one end and free at the other) subjected to bending (**a**) and to twisting (**b**). **a** The displacement of a beam (with length L subjected to a point-load P at the free end) from the vertical (δ) increases along the distance x from the fixed end toward the free end and reaches its maximum (δ_{max}) the free end of the beam (P). Likewise, the maximum bending moment (M_{max}) occurs at the fixed end of the beam where $x = 0$ (see Eqs. 13–14). **b** The material elements in a beam (with length L subjected to a clockwise torque T) located along line a will be displaced along the *dashed line b*. The magnitude of the displacement (the strain) of these material elements (γ) increases from the fixed end toward the free end. Note that angle of twist θ is equal to the strain γ (see Eqs. 16–17)

radius and the length of the beam, respectively. Assuming that the beam is homogeneous and elastic and that the distortions are not excessive, the relationships among the shear stress τ just beneath the surface of the beam, the shear strain γ, the shear modulus G, and the geometry of the beam (r and L) are given by the formula

$$\tau/\theta = G(r/L). \tag{16}$$

The total torque T experienced by the beam is given by the formula

$$T = GJ(\theta/L). \tag{17}$$

Buckling

Bending can result in buckling. This phenomenon results from the addition of a load that exceeds the maximum load that a column can sustain. Once this critical load is reached, the column undergoes mechanical instability that can lead to mechanical failure. Two forms of failure are possible, compressive (crushing) failure or buckling. Which of these occurs depends on the ratio of a column's radius r to its length L. The value of the quotient that characterizes the transition from one mode of failure to the other is given by the equation

$$r/L = (4/\pi) \left(\sigma_{comp}/E\right)^{1/2}, \tag{18}$$

where σ_{comp} is the maximum compressive stress a material (such as wood) can sustain. Equation 18 shows that very short and thick columns will undergo compressive failure, whereas tall and thin columns are more likely to bend under the applied load.

The load exerted on a tree trunk increases annually due to the addition of new leaves and branches and the loading condition of the trunk can be stylized as a point load at the top of a vertical column. The girth of a tree trunk also increases as a result of the annual addition of wood. However, different trees attain different r/L as they get older as a result of species-specific differences (e.g., *Populus* versus *Quercus*) and differences in local growing conditions (e.g., shaded habitats versus sunny habitats). Although it is conceivable that a trunk might undergo compressive (crushing) failure, a far more common response to applied loads is bending (Fig. 7). Assuming that a slender trunk has a solid cross section, the critical load is given by the formula

$$P_{crit} = \pi^2 EI/(2L)^2, \tag{19}$$

where L is the length (height) of the column. The equation describes what is called Euler buckling in honor of the great mathematician Leonhard Euler (1707–1783) who derived it. Euler's equation shows that the elasticity of the material (wood) used to construct a column (trunk) determines the load bearing capacity of the column. It also shows that the critical load is directly proportional to the second moment of area of the column. Thus, the load bearing capacity of the column can be increased by using a material with a higher elastic modulus, or by maximizing the second moment of area. In general, denser materials such as woods have higher elastic moduli, but a denser material contributes more weight per unit volume, which decreases the critical load because of increased self-loading.

An alternative to increasing the elastic modulus of a column is to distribute the material as far from the principal axis of the cross section as possible without incurring local buckling. This can be seen by dividing the critical load of a hollow column by the critical load of a solid column with the same material and the same

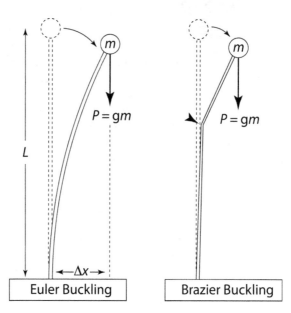

Fig. 7 Diagrams illustrating Euler (long-wave) buckling and Bazier (short-wave) buckling in a slender vertical thin-walled tube subjected to a point-load (P, which equals the point-mass m times the acceleration of gravity g) at the tube's free end. In each case the original undeformed tube and point-mass are indicated by *dashed outlines*. In Euler buckling, the column deflects under the point-load and the displacement along the *horizontal plane* (Δx) increases along the length of the column (L). In Brazier buckling, the tube deflects from the vertical (not shown) and its circular cross sections gradually ovalize until catastrophic failure occurs at a confined location (indicated by *short arrow*)

outside diameter, i.e., $P_{\text{crit=hollow}}/P_{\text{crit=solid}} = [\pi^2 EI/(2L)^2]_{\text{hollow}}/[\pi^2 EI/(2L)^2]_{\text{solid}} = 1 -(D_i/D_o)^4$, where D_i is the internal diameter and D_o is the external diameter. As D_i increases, $P_{\text{crit=hollow}}/P_{\text{crit=solid}}$ increases. Although this helps to explain why a tubular section is much more efficient than a solid section for the construction of a column, this tactic has limits because very thin tubes are susceptible to a phenomenon called Brazier buckling. This mode of failure, which is also called short-wave buckling, results when a bending load causes a cross section to ovalize. As the load increases, the ovalization of the cross section increases and eventually results in catastrophic localized crimping (Fig. 7). Hollow stems and leaves are susceptible to this mode of failure. It should be noted that Euler buckling and Brazier buckling are interrelated because a tall slender hollow column starts to undergo Euler buckling before it experiences Brazier buckling. The critical bending moment for this case is given by the equation

$$M_{\text{crit}} = \left(0.99E\, r_o t^2\right)/\left(1-v^2\right)^{1/2}, \tag{20}$$

where r_o is the outside radius, t is the thickness of the hollow column, and v is the Poisson ratio, i.e., the negative quotient of the transverse and the axial strains a material experiences. The equation holds true provided that $r_o > 10\ t$ and that the column is composed of an isotropic material (i.e., a material that mechanically responds the same regardless of the direction a load is applied). Whether a hollow trunk or branch undergoes compressive failure or Brazier buckling failure depends on many factors, including wall thickness, outside diameter, and length (a detailed treatment is provided by Spatz and Niklas 2013).

The Critical (Maximum) Buckling Height

Euler's formula for the critical buckling load (see Eq. 19) provides a way to calculate the critical height to which a column can be elevated before it begins to bend under it's own weight. This calculation begins by noting that the critical load P_{crit} has to be some fraction or multiple γ of the weight of a column. The weight of a column equals the column's density times the acceleration of gravity times its volume. The weight of an untapered circular cylinder with radius r and length L equals $\rho g \pi r^2 L$. Therefore, the critical buckling load equals $\gamma \rho g \pi r^2 L$. Inserting this expression into Euler's formula gives

$$\gamma \rho g \pi r^2 L = \pi^2 EI/(2L)^2. \tag{21}$$

Solving for the critical buckling length L_{crit}, we obtain

$$L_{crit} = [\pi^2 EI/(4\gamma \rho g r^2)]^{1/3}. \tag{22}$$

Since the column is circular, $I = \pi r^4/4$ (see Fig. 5). Inserting this expression into Eq. 22 gives

$$L_{crit} = (\pi^2/16\gamma)^{1/3}(E/\rho g)^{1/3} r^{2/3}, \tag{23a}$$

or

$$L_{crit} \approx (0.85/\gamma^{1/3})(E/\rho g)^{1/3} r^{2/3}. \tag{23b}$$

Unfortunately, we do not know the precise numerical value of γ. However, more sophisticated mathematics gives us the following relationships for columnar and conical stems:

$$\text{Columnar stems} \quad L_{crit} = 0.79(E/\rho g)^{1/3} D^{2/3}. \tag{24a}$$

$$\text{Conical stems} \quad L_{crit} = 1.25(E/\rho g)^{1/3} D^{2/3}, \tag{24b}$$

where D is basal stem diameter. Notice that a conical (tapered) trunk provides for greater maximum height than does a cylindrical (untapered) trunk.

However, the extent to which these formulas can be used to estimate the extent to which a tree has reached its maximum height becomes problematic, particularly when we examine the assumptions underlying Eq. 24a, b. Tree canopies are never completely symmetrical, trunks are never perfect cylinders or cones, and they almost always experience lateral loads due to moving air. For these and other reasons, Eq. 24a always over-estimates maximum tree height, which may help to explain why empirical observations indicate that trees growing in open habitats almost never reach their critical buckling heights. In contrast, trees growing in habitats sheltered from the wind, particularly, in dense stands, can approach their critical buckling heights in part because they are sheltered from wind by their neighbors and because they do not experience a phenomenon called wind-induced thigmomorphogenesis (see next section).

Before leaving the topic of critical buckling height, it is worth noting that the geometry of a tree and the material properties of a tree are equally important when considering safety factor analyses. This is no better illustrated than by the detailed study of six baobab (*Adansonia*) tree species by Chapotin et al. (2006). Despite the considerable girth with respect to the height of mature specimens, which might give the impression that these trees have very high factors of safety, Chapotin et al. (2006) report that safety factors based on estimated elastic buckling heights are rather low. Baobab trees are no more overbuilt than the majority of temperate or tropical trees. This feature is the result of baobab wood, which has a low elastic modulus owing to its high water content and parenchymatic volume fraction (Chapotin et al. 2006). It is also worthy noting that the elastic modulus of baobab wood can change as a function of the withdrawal of water, since the elastic modulus of this wood decreases with water content. This opens up the curious possibility that the second moment of area of the trunk and the elastic modulus of its wood are inversely correlated and thus possibly compensatory and altered by the addition or removal of water. In this context, it is interesting to return to Eq. (11), which shows that a stem's ability to resist bending depends on E and on I such that "fat but weak" can be just as affective as "slender and strong".

Thigmomorphogenesis

During their lifetimes, trees are constantly subjected to mechanical stresses resulting from the effects of bending and swaying under the influence of wind. Various workers have subjected woody as well as nonwoody plants to continuous swaying by mechanical devices and have found that, in general, the stems of woody plants tend to take on an elliptical cross section with the major axis aligned parallel to the major axis of the ellipse as a result of the differential growth of the vascular cambium (Knight 1811). Inspection of Fig. 5 reveals that a change from a circular cross section to an elliptical cross section results in a change in the second moment of area such that an elliptical cross section contributes more to the ability of a stem to resist bending if the major axis of the ellipse is aligned in the direction of the applied load. Jacobs (1954) provided additional insights on the affects of bending and swaying by guying the trunk of young Monterey pine trees (*Pinus radiata*) with wires approximately 20 ft (≈6.1 m) above ground so that trunks would sway above the attachment points of the wires but remain unperturbed below. The portion of the trunks above the guy wires was observed to grow in girth more rapidly that the portion of the restrained trunks. When the wires were removed, the lower portions of the trunks grew in girth at the same pace as the portions of trunks that were permitted to sway. These experiments indicate that wind-induced swaying stimulates the activity of the vascular cambium and results in thicker stems.

Experiments with nonwoody plants indicate that mechanical perturbation produces similar results. Mechanically stimulated stems grow more rapidly in girth and less rapidly in length compared to stems that are not permitted to move. In addition, mechanical tests reveal that the tissues of mechanically perturbed stems have lower elastic moduli compared to tissues that are not perturbed. Collectively these experiments indicate that mechanical stimulation increases the second moment of area but decreases the elastic modulus of stem tissues. These responses increase the contribution of geometry to bending and twisting but decrease the contribution made by the mechanical properties of tissues, presumably so that stems deflect more easily in the wind and thus reduce their projected area toward the oncoming wind.

This phenomenology is an example of thigmomorphogenesis, a term that refers to any growth response to mechanical perturbation (for a mechanistic review, see Telewski 2006). It is most often studied in terms of the changes in morphology or anatomy attending the application of external forces such as loading. Thigmomorphogenesis has been reported for over 80 % of all the species examined (Jaffe 1973). More recently, it has been studied in terms of the molecular events preceding changes in shape and size. Braam and Davis (1990) have shown that ten to twenty minutes after mechanical stimulation by handling, rain, or wind, the mRNA levels of mouse-ear cress (*Arabidopsis*) increase up to a hundredfold. Four touch-induced (TCH) genes are involved. These genes encode for calmodulin, suggesting that calcium ions are required for the transduction of mechanical signals, thereby enabling plants to sense and respond to dynamic as well as self-imposed mechanical forces.

Trunk-Root Interactions and Damping

The growth and development of woody roots are known to respond to chronic dynamic perturbations with a predictable directional component by adaptively altering cross-sectional morphology, patterns of wood deposition, and even the mechanical properties of tissues. The data reported from a number of case studies confirm that woody roots have the capacity to alter important physical and geometrical properties that are adaptive to stress and strain conditions (e.g., Knight 1811; Coutts 1983, 1986; Ennos 1993, 1994; Stokes et al. 1995, 1996; Niklas 1999).

Dynamic loadings typically actuate roots to bend and twist. Typically, the roots on the windward side of the tree will bend upward and the roots on the leeward side will bend downward as a result of the rotational pivoting of the trunk caused by the wind (Fig. 8). Roots oriented at an angle with respect to the direction of the oncoming wind will twist as well as bend to varying degrees. This bending and twisting can also cause the more distant parts of the root system to be pulled toward the base of the trunk. Gusts of wind coming in different directions can cause the root system to oscillate such that the direction of bending and twisting will change, e.g., root surfaces experiencing compression will experience tension and the direction of torsion will be reversed. Significant and sudden changes in the direction and magnitude of the wind force (drag) can result in root-wood fatigue-failure, some of which may not be visible until uproot occurs.

The extent to which roots bend and twist depends on many factors other than the force of the force and the strength of root-wood. For example, soil depth, compaction, and hydration can contribute to the ability of roots to resist bending, twisting and pulling by virtue of providing counteracting compressive forces near the trunk-root crown and by providing counteracting shearing or suction forces along the surfaces of more distal parts of the root system. Other important factors are (1) the general morphology of the root system, which influences how stresses are distributed along the length and breadth of individual roots (e.g., Crook and Ennos 1996; Stokes and Mattheck 1996; Crook et al. 1997), (2) the morphology and size of the tree canopy, which influences how drag forces are transmitted to the root system (Ennos 1993; Edelin and Atger 1994; Vogel 1996), and (3) soil-type and weather conditions, which influences the extent to which roots can remain anchored and the duration and magnitudes of wind-induced dynamic loadings (e.g., Casada et al. 1980; Marshall and Holmes 1988).

An important aspect of the transfer of energy from the wind to a tree, or to any large plant, is the damping of oscillations. Damping causes a decrease in the amplitudes of free oscillations and thus reduces the danger of a resonance catastrophe in dynamic winds. Oscillation and oscillation damping in trees have been widely studied (see Mayhead 1973; Milne 1991; Peltola et al. 1993; Moore and Maguire 2004; Jonsson et al. 2007). If friction among different trees, or among different branches, and dissipative mechanisms in the root-soil system are set aside, there are two principal sources of damping: fluid damping and viscous damping within the material. Fluid damping (dissipation of energy to the surrounding medium) depends on the square of the velocity of the object's movement relative to the surrounding medium. With estimates of the effective projected area and the drag coefficient (see Eq. 2), fluid damping can be calculated by iteration of the loss of energy during each cycle of the oscillation. Viscous damping in the material (conversion of mechanical energy into heat) is linearly related to the velocity of relative movements between adjoining branches or systems of branches that move in consort with one another. It can be determined in loading-unloading experiments by measuring the loss of energy in a hysteresis loop. In order to relate these measurements to a real tree, measurements should be performed using green wood, since the water in wood plays a pivotal role in viscous damping.

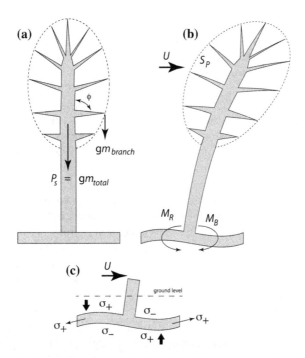

Fig. 8 Synoptic diagram of self-loading resulting from individual branches and the total above-ground mass of a tree (**a**), dynamic loading resulting from wind drag (**b**), and trunk-root interactions under conditions of wind-induced drag forces (**c**). **a** Self-loading for an individual branch with mass m_{branch} results in a bending moment M_B whose magnitude depends on tangent of the branch angle ϕ. In the simplest case, $M_B = g\, m_{branch} L \sin \phi$, where L is the length of the branch. Since $\sin \phi = 1.0$ when $\phi = 90°$, a horizontally oriented branch achieves the maximum bending moment (which occurs at its base). The maximum self load and the maximum bending moment, however, occur at the base of the trunk, which experiences the maximum compressive stresses, which is equal to the product of the acceleration of gravity and the total aboveground mass divided by the area of the base of the trunk, i.e., $\sigma_- = g\, m_{total}/A = P_S/A$. **b** A pictorial of a tree subjected to dynamic loading as a result of wind-pressure. U = wind speed. The drag force experienced by the tree canopy (with a projected area of S_p) is given by Eq. 2. The maximum bending moment M_B occurs at the trunk-root junction. M_B equals the drag force times the height of the tree. Mechanical stability requires a rotational (counter) moment M_R such that $M_R \geq M_B$. Flexure of the windward and leeward root system results in compression and tension within the root systems (see c). **c** Diagram of root-system flexure experiencing a wind-induced torque. Roots on the windward side bend upward and experience tensile stresses and compression stresses on their upper and lower surfaces (σ_+ and σ_-, respectively); more distal roots also experience tensile stresses as they are pulled toward the base of the trunk. Roots on the leeward side bend downward and experience compressive stresses and tensile stresses on their lower and upper surfaces (σ_+ and σ_-, respectively); more distal roots also experience tensile stresses as they are pulled toward the base of the trunk. Root flexure can be reduced if compacted soil resists root displacements and provides compression on the upper surfaces of upwind roots and compression on the lower surfaces of downwind roots (denoted by *upward* and *downward pointing arrows*)

In complex structures such as trees, certain processes can enhance damping.

In gusty winds, branches do not necessarily sway in line or in phase with their subtending stems. Rather, they can perform independent movements relative to one another. In this way, energy is distributed among branches and twigs and is dissipated more effectively than in a structure too stiff to allow relative movements between its elements. This phenomenon, which is referred to as structural damping (Niklas 1992), is not a different mode of damping. It merely emphasizes the enhancement of overall damping by the relative movements of structural elements (branches and twigs), which affect both fluid and viscous damping—processes that are most effective in the periphery of the tree canopy. Structural damping can be caused by the loose coupling mass (termed mass damping; James 2003) or by the distribution of mechanical energy through resonance phenomena within the tree (see James et al. 2006; Spatz et al. 2007), a phenomenon well known in the engineering sciences (Holmes 2001). As shown for a Douglas fir, a tree can react to dynamic wind loads like a system of coupled damped oscillators. This concept has been confirmed theoretically by Rodriguez et al. (2008). The interaction between the different elements leads to a higher damping ratio, and additionally to less strain on the stem as compared with a structure with much stiffer side branches. Multiple resonance damping is therefore essential for the survival of trees and other large plants growing in windy environments. An interesting but as poorly understood topic is the effect of epiphytes and lianas on the damping of the canopies of tropical trees. In theory, the eccentric loads created by epiphytes and the potential for lianas to entangle branches can increase damping ratios and produce coupled damped oscillators.

Buttressed Root Systems

As noted in the introduction, tropical trees manifest features that can set them apart from trees growing in other ecosystems. One of these features is the formation of buttressed root systems (Richards 1952). Various theories have been proposed to account for the formation of these triangular flanges joining the roots to the lower portions of trunks. Black and Harper (1979) suggested that buttresses prevent lianas from climbing trees, a hypothesis that was disproved by Boom and Mori (1982). Senn (1923) and Richards (1952) suggested that buttresses provide mechanical support, a hypothesis that has been supported by indirect observations that correlated tree development with environmental conditions (e.g., Richter 1984; Lewis 1988). For example, buttresses rarely if at all develop on the trunks of trees with well developed tap roots (e.g., Francis 1924; Corner 1988), whereas buttressing is correlated with emergent canopy trees (Richards 1952; Smith 1972) and with species growing in shallow waterlogged or weak silty soils (Richards 1952). An additional correlation that tends to support the biomechanical hypothesis is that buttresses on the upwind sides of trunks tend to be more well developed and extensive that those growing on the leeward sides of trunks (Senn 1923; Baker 1973; Lewis 1988; Warren et al. 1988). Perhaps for these and other reasons, Henwood (1973) proposed that buttresses act as tensile elements. Along similar lines, Mattheck (1991, 1993) noted that buttresses are often associated with bayonet-like sinker roots (Jenik 1978; Baillie and Mamit 1983) and suggested that sinker roots in tandem with buttresses provide a robust mechanism to resist wind throw (Fig. 9a).

The biomechanics of sinker roots attached to buttresses was examined in a seminal study provided by Crook et al. (1997) who investigated the anchorage mechanics of the buttressed root systems of *Aglaia* and *Nephelium* and compared it with the anchorage mechanics of the non-buttressed trunks of *Mallotus*. Using winches to simulate wind throw and strain gauges, Crook et al. (1997) exerted bending forces sufficient to rotate the root systems of trees and noted the following: (1) all trees failed in their root systems (and not by trunk failure), (2) despite differences among the species used in the study, windward buttressed laterals with our without sinker roots either pulled out of the ground or delaminated, while leeward buttresses pushed into the ground and broke near or at their ends (Fig. 9b–c), (3) buttresses with sinker roots resisted simulated wind throw better than those without sinker roots, and (4), in the non-buttressed root system, upwind roots extended in tension, uprooted, but rarely broke, leeward roots buckled appreciably, and taproots bent and compressed the leeward soil profile that often produced an upwind crevice (Fig. 9d).

Some of the results reported by Crook et al. (1997) are in disagreement with the Mattheck model (Fig. 9a). For example, the majority of buttresses examined in the study lacked sinker roots. Likewise, the Mattheck model posits that buttresses strengthen anchorage by preventing delamination at the junction of windward roots and the trunk, whereas windward buttresses with sinker roots delaminated rather than uprooted in the Crook et al. (1997) study. Nevertheless, trees with buttressed root systems had almost twice the anchorage strength as similar sized trees lacking buttressed systems, and leeward and windward buttresses appear to function in compression and tension, respectively.

The question as to why some species regularly produce buttressed root systems while others do not remains unanswered. It is possible that wood density may play a part in this, because tree species with higher wood densities may have thinner trunks and thinner roots that may be insufficient to provide anchorage when individual trees reach threshold critical heights. The scaling of stem and root diameters with respect to growth in height is likely to be another factor, because thicker stems and roots can compensate for lower wood densities or greater heights by virtue of magnifying second moments of areas. Clearly, many factors conspire to cope with anchorage requirements and each species "solves" these requirements using a multi-factorial approach based on its unique combination of functional traits.

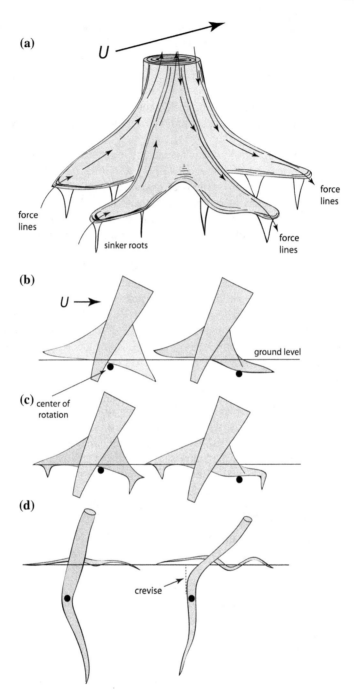

◀ **Fig. 9** Biomechanical behavior of buttressed root systems. **a** Mattheck's (1991, 1993) model for the distribution of mechanical forces running from sinker roots through buttresses. As the trunk is bent by wind (with speed *U*), the bending force is transmitted to upwind sinker roots that resist upward forces, whereas leeward roots and buttresses resist downward forces. Compare with Fig. 8. **b–d** Trunk, buttress, and sinker root displacements resulting from mechanical simulations of wind throw. Adapted from Crook et al. (1997). **b** Displacements of buttressed trunks without sinker roots at two different rotation centers (indicated by *black circles*). **c** Displacements of buttressed trunks with sinker roots at two different rotation centers (indicated by *black circles*). **d** Displacements of an unbuttressed trunk with superficial roots subjected to modest and large bending forces (*left* and *right*, respectively). For details, see text

Concluding Remarks

This chapter began with a treatment of the static and dynamic loadings experienced by a vertical cylinder with a radius of 0.03 m and a length (height) of 1.5 m subjected to a uniform wind speed of 11 m/s. It was noted that these loads are far less important than the stresses they produce. The treatment of biomechanical principles that followed allows us to calculate these stresses. As noted, the drag force exerted by a uniform wind speed of 11 m/s was 6.56 kg m s^{-2} or 6.56 N. The bending moment M_B resulting from this force equals the force times the length of the column, or 6.56 N × 1.5 m = 9.84 Nm. The maximum stress at the perimeter at the base of this column is given by the formula $\sigma_{max} = 4 \, M_B/\pi \, r^3 = [4 \, (9.84 \text{ Nm})]/[\pi (0.03 \text{ m})^3] = 4.64 \times 10^5$ N/m^2. The compressive stress σ_{comp} across the base of the column resulting from static loading is the mass of the column times the acceleration of gravity divided by the cross sectional area of the column. With a density of 844 kg/m^3, we see that $\sigma_{comp} = (\rho \, \pi r^2 \, L \, g)/(\pi \, r^2) = g \, \rho \, L = (9.806 \text{ m/s}^2) \, (844 \text{ kg/m}^3) (1.5 \text{ m}) = 1.24 \times 10^5$ kg m^{-1} s^{-2} or 1.24 × 10^5 N/m^2. Thus, the stresses produced by even a modest wind speed are more than 3.5 times the stresses produced by the static loading of the column.

This comparison shows that tree biomechanics is an extremely complex subject because much of what we think we know depends on the questions we ask and how we use equations to answer them. In addition, there are many aspects that are poorly understood, e.g., viscoelastic nonlinear behavior and the effects of lianas on damping. Thus, predicting the mechanical behavior of any particular tree poses many challenges because each tree has its own individual characteristics that reflect its particular genetic composition and growth responses to its particular habitat. The tropics is a huge and diverse ecosystem consisting of many different types of habitats. Likewise, the behavior of a tree observed at any particular time cannot be used to predict the behavior of the same tree a few years hence because trees grow in size and produce new branches and roots and loose branches and roots over the course of their lifetime. Nevertheless, the concepts and equations reviewed here provide some guidance in evaluating the potential mechanical behavior of trees. They provide boundary conditions as to what is possible and what is highly unlikely. However, in the final analysis, predicting the mechanical behavior of any tree requires experience and prudent judgment, attributes that thus far have eluded mathematical description.

Biblibography

Baillie IC, Mamit JD (1983) Observations on rooting in mixed dipterocarp forest, central Sarawak. Malays For 46:369–374

Baker HG (1973) A structural model of the forces in buttressed tropical rain forest trees (Appendix to Henwood 1973). Biotropica 5:89–93

Black HL, Harper KT (1979) The adaptive value of buttresses to tropical trees: additional hypothesis. Biotropica 11:240

Boom BM, Mori SA (1982) Falsification of two hypotheses on liana exclusion from tropical trees possessing buttresses and smooth bark. Bull Torrey Bot Club 4:447–450

Braam J, Davies RW (1990) Rain-, wind-, and touch-induced expression of calmodulin and calmodulin-related genes in *Arabidopsis*. Cell 60:357–364

Casada JH, Walton LR, Swetnam LD (1980) Wind resistance of Burely tobacco as influenced by depth of plants in soil. Trans Am Soc Agric Eng 23:1009–1011

Chapotin SM, Razanameharizaka JH, Holbrook NM (2006) A mechanical perspective on the role of large stem volume and high water content in baobab trees (*Adansonia* spp.; Bombacaceae). Am J Bot 93:1251–1264

Corner EJ (1988) Wayside trees of Malaya. Malaysian Nature Society, Malaysia

Coutts MP (1983) Root architecture and tree stability. Plant Soil 71:171–188

Coutts MP (1986) Components of tree stability in Sitka spruce on peaty gley soil. Forestry 59:171–197

Crook MJ, Ennos AR (1996) The anchorage mechanics of mature larch *Larix europea* x *L. japonica*. J Exp Bot 47:1507–1517

Crook MJ, Ennos A, Banks JR (1997) The function of buttress roots: a comparative study of the anchorage systems of buttressed (*Aglai* and *Nephileum ramboutan* species) and non-buttressed (*Mallotus wrayi*) tropical trees. J Exp Bot 48:1703–1716

Edelin C, Atger C (1994) Stem and root tree architecture: questions for plant biomechanics. Biomimetics 2:253–266

Ennos AR (1993) The scaling of root anchorage. J Theor Biol 161:61–75

Ennos AR (1994) The biomechanics of root anchorage. Biomimetics 2:129–137

Francis WD (1924) The development of buttresses in Queensland trees. Proc R Soc Queensland 36:21–37

Henwood K (1973) A structural model of forces in buttressed tropical rain forest trees. Biotropica 5:83–89

Holmes JD (2001) Wind loading of structures. Spon Press, London-England

Jacobs MR (1954) The effect of wind sway on the form and development of *Pinus radiate* D. Don Aust J Bot 2:35–51

Jaffe MJ (1973) Thigmomorphogenesis: The response of plant growth and development to mechanical stimulation. Planta 114:588–594

James KR (2003) Dynamic loading of trees. J Arboric 29:165–171

James KR, Haritos N, Ades P (2006) Mechanical stability of trees under dynamic loads. Am J Bot 93:1361–1369

Jenik J (1978) Roots and root systems in tropical trees: morphological and ecological aspects. In: Tomlinson PB, Simmerman MH (eds) Tropical trees as living systems. Cambridge University Press, Massachusetts, pp 323–348

Jonsson MJ, Froetzi A, Kalberer M, Lundström T, Ammann W, Stöckli V (2007) Natural frequencies and damping ratios of Norway spruce (*Picea abies* [L.] Karst) growing on subalpine forested slopes. Trees 21:541–548

Knight TA (1811) On the causes which influence the direction of the growth of roots. Philos Trans R Soc Lond 1811:209–219

Lewis AR (1988) Buttress arrangement in *Pterocarpus officinalis* (Fabaceae): effects of crown asymmetry and wind. Biotropica 20:280–285

Marshall TJ, Holmes JW (1988) Soil physics, 2nd edn. Cambridge University Press, Cambridge-England

Mattheck C (1991) Trees: the mechanical design. Springer Verlag

Mattheck C (1993) Design in der Natur. Der Baum als Lehrmeister. Rombach Verlag, Freiburg

Mayhead GJ (1973) Sway periods of forest trees. Scott For 27:19–23

Milne R (1991) Dynamics of swaying of *Picea sitchensis*. Tree Physiol 9:383–399

Moore JR, Maguire DA (2004) Natural sway frequencies and damping ratios of trees: concepts, review and synthesis of previous studies. Trees 18:195–203

Niklas KJ (1992) Plant biomechanics. University of Chicago Press, Chicago-Il

Niklas KJ (1999) Variation of the mechanical properties of *Acer saccharum* roots. J Exp Bot 50:193–200

Niklas KJ, Spatz H-C (2000) Wind-induced stresses in cherry trees: evidence against the hypothesis of constant stress levels. Trees 14:230–237

Niklas KJ, Spatz H-C (2010) Worldwide correlations of mechanical properties and green wood density. Am J Bot 97:1587–1594

Niklas KJ, Spatz H-C (2012) Plant physics. University of Chicago Press, Chicago-Il

Peltola H, Kellomöki S, Hassinen A, Lemittinnen M, Aho J (1993) Swaying of trees as caused by wind: analysis of field measurements. Silva Fenn 27:113–126

Richards PW (1952) The tropical rain forest. Cambridge University Press, Cambridge

Richter W (1984) A structural approach to the function of buttresses of *Quaranbea asterolepis*. Ecology 65:1429–1435

Rodriguez M, E. de Langre, Moulia B (2008). A scaling law for the effects of architecture and allometry on tree vibration modes suggests a biological tuning to modal compartmentalization. Am J Bot 95:1523–37

Senn G (1923) Uber die Ursachen der Brettwurzelbilding bei der Pyramiden-Pappel. Verhandlungen des Naturforschenden Gesellschaft in Basel 35:405–435

Smith AP (1972) Buttressing of tropical trees: a descriptive model and new hypothesis. Am Nat 106:32–46

Spatz H-C, Niklas KJ (2013) Modes of failure in tubular plant organs. Am J Bot 100:332–336

Spatz H-C, Brüchert F, Pfisterer J (2007) Multiple resonance damping or how do trees escape dangerously large oscillations? Am J Bot 94:1603–1611

Stokes A, Mattheck C (1996) Variation of wood strength in tree roots. J Exp Bot 47:693–699

Stokes A, Fitter AH, Coutts MP (1995) Responses of young trees to wind and shading: effects on root architecture. J Exp Bot 46:1139–1146

Stokes A, Ball J, Fitter AH, Brian P, Coutts MP (1996) An experimental investigation of the resistance of model root systems to uprooting. Ann Bot 78:415–421

Telewski FW (2006) A unified hypothesis of mechanoperception in plants. Am J Bot 93:1466–1476

Vogel S (1981) Life in moving fluids: the physical biology of flow. Willard Grant, Boston-MA

Vogel S (1996) Blowing in the wind: storm-resisting features of the design of trees. J Arbor 22:92–98

Warren SD, Black HL, Eastmond DA, Wtaaley WH (1988) Structural function of buttresses of *Tachigalia versicolor*. Ecology 62:532–536

Part VI
Ecophysiological Processes at Different Temporal Scales

Tree Rings in the Tropics: Insights into the Ecology and Climate Sensitivity of Tropical Trees

Roel J.W. Brienen, Jochen Schöngart and Pieter A. Zuidema

Abstract Tree-ring studies provide important contributions to understanding the climate sensitivity of tropical trees and the effects of global change on tropical forests. This chapter reviews recent advances in tropical tree-ring research. In tropical lowlands, tree ring formation is mainly driven by seasonal variation in precipitation or flooding, and not in temperature. Annual ring formation has now been confirmed for 230 tropical tree species across continents and climate zones. Tree-ring studies indicate that lifespans of tropical tree species average c. 200 years and only few species live >500 years; these values are considerably lower than those based on indirect age estimates. Size-age trajectories show large and persistent growth variation among trees of the same species, due to variation in light, water and nutrient availability. Climate-growth analyses suggest that tropical tree growth is moderately sensitive to rainfall (dry years reduce growth) and temperature (hot years reduce growth). Tree-ring studies can assist in evaluating the effects of gradual changes in climatic conditions on tree growth and physiology but this requires that sampling biases are dealt with and ontogenetic changes are disentangled from temporal changes. This remains challenging, but studies have reported increases in intrinsic water use efficiency based on $\delta^{13}C$ measurements in tree rings, most likely due to increasing atmospheric CO_2. We conclude that tree-ring studies offer important insights to global change effects on tropical trees and will increasingly do so as new techniques become available and research efforts intensify.

R.J.W. Brienen (✉)
School of Geography, Leeds University, Woodhouse Lane, Leeds LS2 9JT, UK
e-mail: r.brienen@leeds.ac.uk

J. Schöngart
Instituto Nacional de Pesquisas da Amazônia (INPA), Av. André Araújo 2936, P.O. Box 478, Manaus, AM 69011-970, Brazil

P.A. Zuidema
Forest Ecology and Forest Management, Centre for Ecosystems, Wageningen University, P.O. Box 47, 6700 AA Wageningen, The Netherlands

© Springer International Publishing Switzerland 2016   439
G. Goldstein and L.S. Santiago (eds.), *Tropical Tree Physiology*,
Tree Physiology 6, DOI 10.1007/978-3-319-27422-5_20

Introduction

Tropical trees are an important element of the global biosphere. Tropical forests and woody savannas cover 12–15 % of the Earth's terrestrial surface (FAO 2006), and host more than 40,000 tree species (Slik et al. 2015), they store about 60 % of total global forest biomass (Pan et al. 2011), and play a significant role in the water, carbon and nutrient cycles of the earth (Bonan 2008; Spracklen et al. 2012). Given these important roles, it is essential to understand how tropical forests respond to changing climatic conditions. This requires information about the climate sensitivity of tropical trees and the long-term dynamics of tropical forests. For instance, to project the effects of future climate change, it is pertinent to obtain insights into the responses of tropical trees to variation in climate and CO_2 (Zuidema et al. 2013), while long-term information on tree ages and forest disturbance history are important to infer residence times of carbon in forests as well as the interactive effects of disturbance and climate change on forest dynamics (Babst et al. 2014; Gebrekirstos et al. 2014).

Tree ring analysis (or dendrochronology) can importantly contribute to the study of tropical forest responses to changing climate. However, in the tropics it has lagged behind studies in temperate and boreal trees, where tree rings provided detailed insights into tree growth and functioning, and trees' responses to past climate (Fritts 1976; Speer 2010). While the occurrence of clearly defined annual rings in tropical trees is less common than in temperate trees, a substantial number of tropical species is known to form annual rings. One of the first documented observations on tree rings in a tropical tree species was by Brandis in 1856, who noticed that Teak (*Tectona grandis*) in Indonesia produced distinct rings which he assumed to be formed in response to an annually recurring dry season. This was confirmed in 1881 by Gamble by counting rings on plantation trees with known age, and much later by the publication of the first tropical tree ring chronology on Teak from Java by Berlage (1931). The research field of tropical dendrochronology has made some important advances since then. Two workshops specifically dedicated to this research field (1980 and 1989) summarized the advances and concluded that it has been clearly demonstrated that many tropical tree species in regions with seasonality in rainfall or flooding form distinct annual rings (Bormann and Berlyn 1981; Baas 1989). In 2002, Worbes evaluated 139 species from neotropical floodplains and terra firme, showing the occurrence of growth bands in at least a third of the listed species. In the last decade the number of species with proven annual rings has increased strongly (Zuidema et al. 2012) due to increased research efforts throughout the tropics. Despite these recent advances, even today the Berlage teak chronology (1514–1929) published in 1931 remains one of the longest tree ring chronologies in the tropics. The apparent lack of success in reconstructing long climate sensitive ring width chronologies, as witnessed for example in the paucity of tropical chronologies in the International Tree Ring Database (Grissino-Mayer and Fritts 1997), highlights undeniable problems in detecting tree rings for a large number of species, but is also partially due to insufficient efforts, and long-standing skepticism towards the occurrence of ring formation in tropical trees.

In this chapter, we present important findings obtained in tropical tree ring research relevant to the general theme of this book, the adaptation of tropical trees to changing climate. We first describe the different mechanisms of tree ring formation in tropical trees, relate this to tree phenology, and provide an overview of tropical tree species with proven annual rings. The following sections present results on ages, growth patterns and regeneration strategies of tropical trees, and a review of the sensitivity of tropical trees to inter-annual variation in climate and long-term tree responses to CO_2 increase. We end with a section on insights from isotope techniques and a summary of our main conclusions.

In our review, we included all studies from tropical latitudes (i.e., between 23.4° North and South) thus including both wet forests and savannas, very wet and dry areas, and flooded and non-flooded areas. We excluded studies in mountain systems (>1500 m a.s.l), and mangroves, where environmental conditions and cues are very different.

Tree Ring Formation in the Tropics

Tree-ring boundaries are formed when cambium cells in tree trunks are dormant. In mid- and high-latitude regions, the principal trigger behind cambial dormancy and the formation of anatomically distinct tree rings is seasonal variation in day length hours and temperature (Fritts 1976). In the tropics, there is no (or very limited) seasonality in these climatic variables. This notion was the basis for the long-held belief that tropical trees show relative constant growth throughout the year and would not form annual rings (cf. Lieberman and Lieberman 1985; Whitmore 1998). Nevertheless, many tropical trees do show annually recurring phenological patterns (van Schaik et al. 1993), which may slow down cambial activity or induce complete cambial dormancy, and result in an anatomically distinct layer of wood i.e., a tree ring boundary. Seasonality in leaf phenology in tropical trees is thought to be an adaptation to variation in external (abiotic) stress. Rainfall seasonality is by far the most common stressor for tropical trees (Worbes 1995, 1999). Over large areas of the tropics, especially further away from the equator, evapotranspiration exceeds rainfall for at least several weeks per year. This causes seasonal water-stress especially for shallow rooted species in dry sites (Borchert 1994), and may result in complete leaf shedding (deciduous), or exchange of leaves within a very short period (brevi-deciduous). When trees are leafless, cambial activity stops, and a distinct layer of wood is formed. While this behavior is most likely a plastic or evolutionary adaptation to reduce leaf cover during periods of prolonged water stress, the actual trigger to which trees respond may be different. For instance, in some deciduous species of seasonally dry forests, it has been observed that trees drop their leaves before water stress has developed and that bud break often occurs before the onset of the wet season and is triggered by variation in day length

(Rivera et al. 2002; Elliott et al. 2006). Another abiotic stress factor leading to the formation of annual rings is seasonal flooding. During flooding, anoxic conditions of the roots induce leaf fall, cambial dormancy and the formation of annual rings in many tree species (Worbes 1985; Schöngart et al. 2002).

Whether trees form annually distinct rings depends both on the species-specific physiology and wood anatomy, but also on exogenous factors like the seasonality of the rainfall or flooding. Trees showing strictly annual cycles of leaf fall and flushes are more likely to form annual rings, and the percentage of species with annual rings is expected to be higher in sites with strong annually recurring seasonality (Borchert 1999). Several studies demonstrated the importance of seasonality in environmental conditions by showing that some species may form clear annual rings in seasonal sites, but lack rings in very wet sites with low precipitation seasonality or in very dry sites with highly irregular precipitation (Geiger 1915; Coster 1927; Borchert 1999; Pearson et al. 2011). Hence, tree ring formation can be strictly annual, bi-annual (Jacoby 1989; Gourlay 1995) or irregular with false (non-annual) rings occurring every few years (Wils et al. 2011). Formation of annual rings can also differ between life stages with clear and annual rings in the adult phases, but absent, vague, or non-annual rings formed during the juvenile phases (Dünisch et al. 2002; Brienen and Zuidema 2005; Soliz-Gamboa et al. 2011). These issues call for cautious interpretation of tree-ring measurements in new species or in the same species at different sites.

A global dicot wood anatomy database with 5663 anatomical descriptions, indicates that ca. 23 % of the tropical woody dicots present distinct growth boundaries, somewhat lower than the global average of 34 % and the 76 % in temperate regions in the northern hemisphere (Wheeler et al. 2007). Considering the very high diversity of tropical tree species (Slik et al. 2015) and the proportionally low number of studied species, this lower occurrence of tree rings in the tropics still yields a potentially very large number of species forming anatomically distinct ring boundaries. As anatomic records do not indicate whether rings are formed annually, we provide a review of 130 studies on tree rings in tropical trees (see also Zuidema et al. 2012; Schöngart 2013) for which annual formation of tree rings has been verified using various methods (e.g. cambial marking, climate correlations, bomb-peak dating, trees of known age, etc., see Worbes 1995). This review shows that 230 different tropical tree species from 46 different families form annual rings (see summary Table 1). These studies cover all continents and climates (see Fig. 1), and also include some very wet sites (4000 mm annual precipitation) with limited seasonality in rainfall (Fichtler et al. 2003; Groenendijk et al. 2014). The majority of the studies however, are located in zones away from the equator that have at least one distinct dry season (cf. Fig. 1).

The existence of growth rings is closely related to the wood anatomical structure, and therefore—at least partially—genetically determined (Wheeler et al. 2007, but see; Fichtler and Worbes 2012). Variation in wood anatomy of tropical trees that defines ring boundaries is distinctly different from temperate northern hemisphere woods (Worbes 1995), thus requiring a separate classification. A useful and widely used anatomical classification of growth ring boundaries in tropical woods is that of

Table 1 Summary table of literature review of 130 studies showing the number of species with annual rings per vegetation type, the top-five species used for tree ring studies, and top five families with ring forming species

Vegetation type	# studies
Wet forest	76
Moist forest	117
Dry forest	18
Open savannah and desert	15
Floodplain	21
Top five species used in the tropics	
Tectona grandis	
Terminalia superba	
Cedrela odorata	
Triplochiton scleroxylon	
Macrolobium acaciifolium	

Top five families with ring forming species	# species
Fabaceae	95
Meliaceae	17
Malvaceae	11
Bignoniaceae	7
Combretaceae	7

A total of 230 species with annual formation of tree rings was found. The annual character of rings in these studies was verified using various methods (e.g. cambial marking, climate correlations, bomb-peak dating, trees of known age, etc.). A comprehensive list of these studies can be found in Zuidema et al. (2012) and Schöngart (2013). Studies in mountain systems (>1500 m a.s.l) or mangroves are excluded from this review

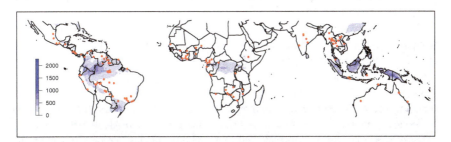

Fig. 1 Map with locations of tree-ring studies in the tropics. The *background color* shows the precipitation in the driest quarter (data: WorldClim, Hijmans et al. 2005) with *blue areas* marking areas that receive more than 300 mm during the driest quarter

Coster (1927) adopted by Worbes (1995, 2002). It distinguishes four different types of growth zones: (1) density variations; (2) marginal parenchyma bands; (3) repeated patterns of alternating parenchyma and fibre bands; (4) variations in vessel distribution and/or vessel size (i.e., ring-porosity) (Fig. 2). As currently no large-scale classification of wood anatomy of growth zones exists for tropical tree species, it is hard to make generalizations about relations between wood anatomy or the occurrence of growth zones and taxonomy. Nevertheless, the top families with ring-forming species are Fabaceae, Meliaceae and Malvaceae (see Table 1). Growth zones in Fabaceae and Meliaceae often show marginal parenchyma bands (Worbes 1989; Groenendijk et al. 2014), while alternating parenchyma and fibre bands are a common feature in Lecythidaceae, Sapotaceae, and Moraceae (Worbes 1989). Density variations and vessel size variation can occur in concurrence with the previous features, or delineate an annual growth zone by itself. The best example of a "ring-porous" species with wide vessels delineating the start of a new ring, is Teak (*Tectona grandis*), one of the most widely used tropical species for climate reconstruction in the tropics (cf. Fig. 2f). Examples of families with species with variations in wood-density include Annonaceae, Lauraceae and Euphorbiaceae (Worbes 1989).

Despite the advances in tropical tree ring analysis, tropical trees commonly present problems in the form of vague growth boundaries, false rings (Fig. 2h) and wedging rings (Fig. 2i) or discontinuous (locally absent) rings (Fig. 2g). For those interested in performing tree-ring studies in the tropics, please refer to Stahle (1999) and Worbes (1995, 2002) for practical guidance.

Inferences from Trees Rings on Tree Longevity, Growth Patterns and Regeneration

Tree rings allow detailed reconstruction of growth curves for individual trees over the full length of their life, thereby providing profound ecological insights that cannot be obtained using for example relative short-term growth measurements. We will highlight a few important ecological inferences that have been made from growth rings of tropical trees.

Tree rings provide accurate information on tree ages. The question of how old tropical trees become has occupied scientists for a long time, and has been the subject of fierce scientific debate. Ages in trees can be measured directly only by means of radiocarbon dating or tree ring analysis (Martinez-Ramos and Alvarez-Buylla 1998). Some radiocarbon dating methods revealed ages for tropical trees in excess of 1000 years (Chambers et al. 1998; Vieira et al. 2005), but given the controversy on these reported ages (Martinez-Ramos and Alvarez-Buylla 1998, 1999; Worbes and Junk 1999) and the focus of this chapter, we here report on the outcome of ages obtained by tree ring analysis only.

Figure 3 shows the range of maximum observed ages for 71 tropical tree species. These results provide an indication of tree longevities, although it should be noted that in most cases, the purpose was not to sample the largest individuals of the study species. The mean of the maximum observed age across all tropical tree species is 208 years (median = 200 years). In line with the high diversity in life-history strategies among tropical trees, ages vary from a few decades for pioneer species (Schöngart 2008; Brienen et al. 2009; Vlam 2014) to several centuries. *Taxodium mucronatum* trees show the highest longevity reaching ages of more than 1500 years

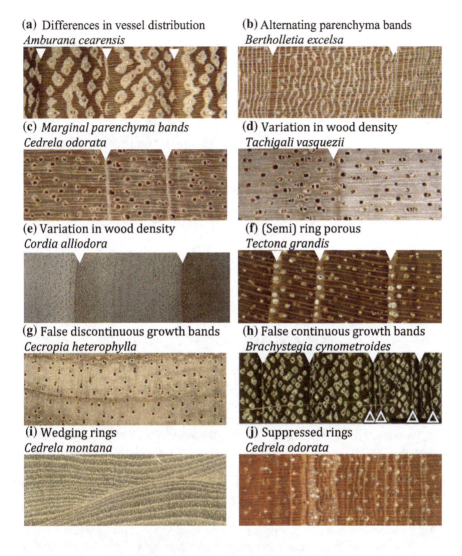

(a) Differences in vessel distribution
Amburana cearensis

(b) Alternating parenchyma bands
Bertholletia excelsa

(c) *Marginal parenchyma bands*
Cedrela odorata

(d) Variation in wood density
Tachigali vasquezii

(e) Variation in wood density
Cordia alliodora

(f) (Semi) ring porous
Tectona grandis

(g) False discontinuous growth bands
Cecropia heterophylla

(h) False continuous growth bands
Brachystegia cynometroides

(i) Wedging rings
Cedrela montana

(j) Suppressed rings
Cedrela odorata

Fig. 2 Wood anatomical features of tropical tree rings showing the most common ring boundaries (panels **a–f**) and some common problems encountered in tree ring studies (**g–j**). Growth direction in panels **a–f** is from *left* to *right*. *Filled white triangles* indicate annual growth boundaries; *open triangles* in panel **h** indicate false rings

in swamp forests in central Mexico (Stahle et al. 2012). However, such long-lived conifers can be regarded an exception, as most broadleaved trees in lowland areas are substantially younger. The oldest broadleaved tree with confirmed ages in excess of 1000 years using a combination of tree rings and radiocarbon dating are Baobab trees from Africa (Robertson et al. 2006; Patrut et al. 2007), while few studies report ages over 500 years old (Fichtler et al. 2003; Borgaonkar et al. 2010). This outcome

Fig. 3 Histogram of observed maximum ages of 71 tropical tree species by use of counting of annual tree rings

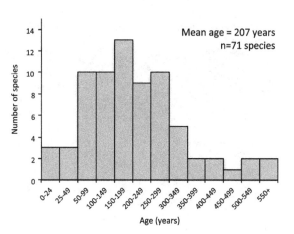

of the studies reviewed here are close to estimates of tree longevity derived from growth rates from rainforests of central Amazon (median of 296 years, Laurance et al. 2004) and Costa Rica (mean of 230 years from 10 cm DBH, Lieberman et al. 1985), and shows that life-spans of broadleaf tropical tree species are not different from those of temperate broadleaf tree species (cf. Loehle 1988). However, the results contrast with life-spans of several millennia for trees (mainly gymnosperms) growing in extreme environments (Brown 1996). Reliable estimates of tree longevities are an important basis for calculations of carbon residence times.

Reconstructions of age-size relationships show a large variation both within (Fig. 4a, b) and between species (Fig. 4d–f). Within species several magnitudes of variation in ages exist between trees of a similar size, even at the youngest stages. For example, within the same population the ages of *Cedrela odorata* trees of 10 cm in diameter range from 9 to 75 years (cf. Fig. 4a), and a similar magnitude of variation exists among *Macrolobium acaciifolium* trees from floodplain forests (cf. Fig. 4b). Hence, young juvenile trees may be as old as large canopy trees. Such large variation in growth rates between trees of the same species seems very common in tropical trees (Worbes et al. 2003; López et al. 2013; Groenendijk et al. 2014), reaffirming that size is a poor predictor for tree age. This variation between trees arises due to differences between individuals in light and water availability, and soil fertility (Schöngart et al. 2005; Baker and Bunyavejchewin 2006; Brienen et al. 2010a). The relative importance of these factors may vary between sites and species. Comparison of two *Cedrela* populations showed that in moist forest with a relatively dark understory, variation in juvenile growth was mainly governed by fluctuations in light availability, while at a dry site spatial variation in water availability was more important (Brienen et al. 2010a). In the Amazonian floodplain species *M. acaciifolium* the effect of differences in nutrient availability seemed to dominate variation in growth rates, with much slower growth in the nutrient poor, black-water floodplains (Igapó), compared to the nutrient rich white-water flood-plains (Várzea) (Schöngart et al. 2005). Remarkably, this growth variation is associated with tree longevities, with slow growers at the nutrient poor sites

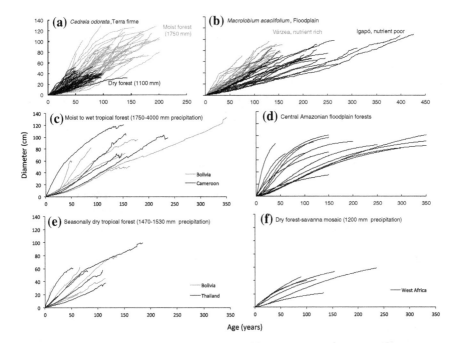

Fig. 4 Growth trajectories of tropical trees derived from tree ring analysis. Panels **a–b** show trajectories of individual trees for *Cedrela odorata* from two terra firme forests (Bolivia and Mexico; Brienen et al. 2010a) and for *Macrolobium acaciifolium* from Brazilian floodplains (Schöngart et al. 2005). Panels **c–f** show mean trajectories for different species from (**c**) moist to wet tropical forests (6 spp. northern Bolivia; Brienen and Zuidema 2006); 4 spp. Cameroon (Groenendijk et al. 2014), **d** Central Amazonian Várzea floodplain forests (14 spp.; Schöngart 2008), **e** seasonally dry tropical forests (5 spp. southern Bolivia; van der Sleen 2014), 5 spp. Thailand; (Vlam 2014), and **f** dry forest-savannah mosaics (6 spp. west Africa, Schöngart et al. 2006)

attaining much higher ages (i.e., 403 years) compared fast growers at the richer sites (i.e., 157 years). Such trade-offs between growth and longevity have been observed in temperate trees (Bigler and Veblen 2009) and may have important implications for future responses of forests to increased levels of CO_2 and temperature (cf. Bugmann and Bigler 2011).

Comparison of mean age-size relationships between species shows a comparable variation to that observed within a species (Fig. 4c–f) indicating that microsite and stochastic difference between individual trees may be as important as its taxonomy in shaping growth trajectories. Mean age of similarly-sized trees varies by a factor of four to five across species and across vegetation types (wet to dry terra firme and floodplain forests and savannas).

A final contribution of tree ages to tropical forest ecology is in evaluating changes in forest dynamics and reconstructing past disturbances. Distinct peaks in the age distribution of light-demanding tree species may indicate periods of disturbances if these have resulted in elevated recruitment rates of these species (Baker et al. 2005; Vlam et al. 2014b). The spatial configuration of tree ages of such light-demanding species may provide important additional information on the extent of these disturbances (Middendorp et al. 2013). Another indication of changes in forest dynamics can be obtained from the analysis of 'releases', periods of elevated growth rates, which can be caused by increased light levels. In forests where tree mortality is gradually increasing over time, more gaps are formed, which may lead to an increase in the incidence of such growth releases. The interpretation of the causes of growth releases is not straightforward though, as individuals and species may strongly differ in responses to gap formation (Brienen and Zuidema 2006; Soliz-Gamboa et al. 2012) and releases may be induced by climatic variability (Brienen et al. 2010a; Vlam 2014). So far, the few analyses of release frequency for tropical tree species have shown no or minor shifts in the rate of releases during the last century, suggesting that gap dynamics have remained rather stable over the last century (Rozendaal et al. 2011; Vlam 2014). It should be noted however, that retrospective tree ring studies miss a significant and possibly non-random proportion of the original historical tree population (Landis and Peart 2005; Rozendaal et al. 2010), which may lead to biases in reconstructions of historical growth rates and release frequencies (Brienen et al. 2012a).

Climate Sensitivity of Tree Rings

Temperature, water-availability, and incoming solar radiation all affect tree growth. The responses of tropical trees to these controls may be complex and non-linear, and may vary across species and ecosystems. Tree ring analysis is an ideal tool to evaluate growth responses to climatic fluctuations, as it yields long series of growth rates at annual resolution. Here we review growth responses of tropical trees to climatic variation based on 45 studies reporting climate growth relations (Table 2). The studies generally adopt standard dendrochronological techniques (Speer 2010) to develop a climate sensitive tree-ring chronology. A chronology is a time series of averaged growth for individuals that shows comparable growth fluctuations

Table 2 Summary table of the climate-growth relationships for different vegetation types (excluding floodplains)

Vegetation type	Correlation ring width-rainfall, mean (max)	n	Correlation ring width-temperature, mean (min)	n
Wet forest (>2000 mm)	0.49 (0.75)	6	–	–
Moist forest (1000–1500 mm)	0.44 (0.66)	39	−0.44 (−0.6)	8
Dry forest (<1000 mm)	0.57 (0.89)	7	−0.42 (−0.57)	5
Open savannah	0.49 (0.65)	7	−0.3 (−0.40)	2

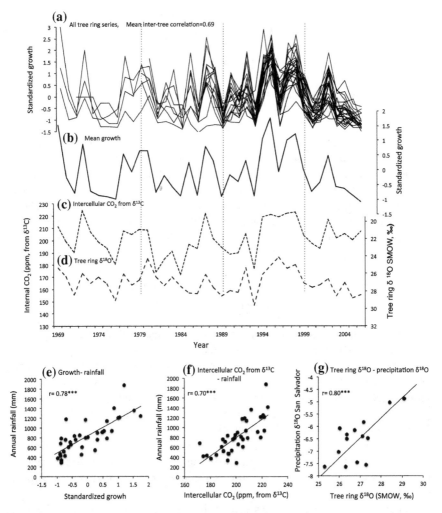

Fig. 5 Example of time series of growth (**a, b**), leaf intercellular CO_2 derived from tree ring carbon isotopes (**c**), and tree ring oxygen isotopes (**d**) for a dry forest tree species, *Mimosa acatholoba* from southern Mexico. Panels **e, f**, and **g** show the relationship of the ring width and isotope derived data to external variables. Growth and internal CO_2 (derived from $\delta^{13}C$ in tree rings) are most strongly related to annual rainfall, while oxygen isotopes related strongly to variation in isotopes in rainfall from a distant station in San Salvador and was also negatively correlated to local precipitation (r = −0.72). All relations are significant at p < 0.001. *Data-sources* Brienen et al. (2010a, 2011, 2013)

(cf. Fig. 5a). It should be noted that a straightforward comparison of the results of climate-growth relations is hampered by variation in the selection of trees to be included, the procedures to remove ontogenetic trends from the growth data, the length of the growth series and the statistical analyses applied.

Our analysis shows that annual variation in tropical tree growth is often associated with fluctuations in precipitation: in nearly 60 species, growth rates increased with rainfall (Table 2). This sensitivity to rainfall fluctuations was similar for dry, moist and wet forests (Table 2), although one would expect a higher sensitivity in drier forest sites. The lack of a clear relation of growth responses with underlying climate is at least partially due to methodological differences across studies, but also due to differences between species in their sensitivity to rainfall fluctuations, even when they occur in the same area (e.g., Schöngart et al. 2006; Mendivelso et al. 2013; Vlam et al. 2014a). In addition, some species are more susceptible to rainfall during the dry period, others to the wet period or transitional periods. And finally, species may also show growth responses to rainfall during the previous growing season (year). Such lagged responses are common, especially in strictly deciduous species (e.g., Teak; Buckley et al. 2007, and several Meliaceae; Dünisch et al. 2003; Brienen and Zuidema 2005; Heinrich et al. 2008; Vlam et al. 2014a), and may be due to use of stored reserves at the beginning of the rainy season to support wood formation (Dünisch and Puls 2003; Ohashi et al. 2009). These strong differences in sensitivity to rainfall reflect species' differences in stem water storage, phenology, rooting depth and use of reserves (Borchert 1994; Meinzer et al. 1999).

The strongest responses to precipitation are observed in dry sites in northern Columbia and Peru and in southern Mexico, where >60 % of variation in ring width is explained by year-to-year variation in rainfall totals (Rodríguez et al. 2005; Brienen et al. 2010b; Ramírez and del Valle 2011, 2012). In these dry to very dry sites (50–900 mm annual precipitation), rainfall can be extremely variable among years (partially due to ENSO), resulting in a strong control of precipitation on growth (see Fig. 5e). Despite these exceptions, the overall effect of precipitation on growth is relatively weak (mean correlation, r, across all species is 0.47, cf. Table 2), suggesting that rainfall fluctuations have a limited impact on tree growth. However, reserve storage in trees has a buffering effect and may reduce annual fluctuations in tree growth. Work in a dry tropical forest shows that trees may be relatively tolerant to annual fluctuations in rainfall, but sensitive to multi-annual droughts (Mendivelso et al. 2014).

Temperature negatively affects growth in tropical trees (cf. Table 2), but the number of studies reporting temperature growth relations is much lower than those reporting precipitation influences. This is at least partially because fewer studies included temperature explicitly in their climate growth relations, but also due to lower responsiveness of tropical tree growth to temperature fluctuations. Especially, in dry forests and savannas temperature influences were weaker than the influence of rainfall, and in some cases temperature influences simply resulted from co-linearity between inter-annual variation in rainfall and temperature (cf. Brienen et al. 2010b). Several studies in moist forests show temperature effects that are of comparable or slightly higher in magnitude compared to effects of precipitation (Buckley et al. 2007; López and Villalba 2011; Ramírez and del Valle 2012; Locosselli et al. 2013; Vlam et al. 2014a). These results along with findings of temperature influences on tree growth from repeated tree diameter measurements in several tropical forests (Clark et al. 2010; Dong et al. 2012) clearly show negative influences of inter-annual variation in temperature on tropical tree growth. This response is opposite to observed temperature responses of tree rings in cold climates (Fritts 1976), and consistent with negative effects of temperature on photosynthesis at high temperatures, above 30 °C (Doughty and Goulden 2008) and with effects of temperature on respiration and evaporative demand (Lloyd and Farquhar 2008).

While there is a large uncertainty in climate models with regard to rainfall predictions, there is a high degree of certainty for warming to occur throughout the tropics (Stocker et al. 2013). This predicted warming may negatively affect tree growth in the decades to come, but results of temperature sensitivity of tropical tree growth based on annual fluctuations in temperature cannot be easily translated into growth responses to a gradual rise in temperature, given possible acclimation of photosynthesis and respiration to gradual temperature changes (Lloyd and Farquhar 2008). Also, the negative temperature effects may be offset by increases in atmospheric CO_2 (Lloyd and Farquhar 2008). In all, our analysis indicates that tropical tree growth is moderately to highly sensitive to rainfall fluctuations and less so to temperature variation. Sensitivities to climatic fluctuations are highly species-specific, which may contribute to an overall resilience of diverse tropical forests to increased climatic variability in rainfall and temperature.

Global Change and Tropical Tree Growth

Future global climatic and atmospheric changes may affect the physiology and growth of tropical tree species in various ways. Major atmospheric and climatic drivers affecting tree growth include rising CO_2 levels, rising temperature, changes in precipitation regimes and frequency of droughts, and increased atmospheric deposition of nutrients. Tree-ring studies and analyses of stable isotopes in tree rings can be used to quantify the effects of past climatic and atmospheric changes on tree growth (Zuidema et al. 2013), and evaluate the performance of Earth system models and improve their projections (Babst et al. 2014). Most importantly, the effects of rising CO_2 levels on tree physiology and growth (CO_2 fertilization) can be studied using stable carbon isotopes (^{13}C) from tree rings and by changes in tree-ring width. Analyses of carbon isotope ratios ($\delta^{13}C$) in tree rings over periods of time during which atmospheric CO_2 levels have increased can be used to evaluate changes in the intrinsic water use efficiency (iWUE), the ratio of carbon fixed (photosynthesis) to water lost (stomatal conductance) (McCarroll and Loader 2004). In addition, stable oxygen isotopes (^{18}O) could potentially be used to infer changes in water fluxes (e.g. evapotranspiration from forests, Brienen et al. 2012a, b), and stable nitrogen isotopes (^{15}N) can be used to evaluate changes in nitrogen cycling (Hietz et al. 2011).

There are several advantages of using tree-ring analyses for studying global change effects on tropical forests, compared to analyses based on plots: tree rings cover longer periods of time, they yield growth rates at annual time resolution and additional measurements (stable isotopes, anatomy) can be obtained from tree rings (Zuidema et al. 2013; Babst et al. 2014). On the other hand, limitations of tree-ring analysis include that it cannot be conducted for full tree communities, and that it requires well-trained researchers and more sophisticated materials. In addition, there are a number of methodological issues (Brienen et al. 2012a) that need to be accounted for in tree-ring analysis. First, sampling issues arise due to the fact that alive trees sampled in tree-ring studies may be (and are likely) a non-random subset of the population, which may affect the outcome of analyses of growth trends over time (Briffa and Melvin 2011; Brienen et al. 2012a). Dealing with these biases requires the sampling of trees of all sizes (Brienen et al. 2012a; Nehrbass-Ahles et al. 2014). Some biases cannot be prevented by adjusting sampling schemes in the field, and need to be assessed by additional statistical analyses (Groenendijk et al. 2015) or simulations (Vlam 2014). Second, analyses of trends require that temporal trends in growth, isotope values or derived variables (e.g., iWUE) are separated from those occurring over the size range (or age range) of trees. Third, age distributions and recruitment waves of tree species may also affect the outcome of analyses of trends (Vlam 2014). There is a need for critical methodological evaluations and the development of statistical and simulation tools to evaluate the robustness of trends detected based on tree-ring analyses (Briffa and Melvin 2011; Brienen et al. 2012a; Vlam 2014; Peters et al. 2015). We therefore call for a cautious interpretation of published trends in growth, iWUE or isotope values for which such robustness checks are lacking.

So far, trends in iWUE and tree-ring width have been evaluated for a small number of tropical tree species. In several of these studies the abovementioned methodological issues have not been (sufficiently) taken care of, potentially affecting the sign and strength of trends in iWUE or growth. For instance, a number of studies on trends in iWUE over time have not or insufficiently separated the strong ontogenetic trend in iWUE from that occurring over time due to CO_2 rise (Hietz et al. 2005; Nock et al. 2010; Brienen et al. 2011; Locosselli et al. 2013). A recent study on 12 tree species across three tropical regions (Van der Sleen et al. 2014) evaluated trends in iWUE for fixed diameter categories (8 and 27 cm diameter), thus explicitly accounting for ontogenetic changes, and revealed a 30–35 % increase in iWUE over the last 150 years. A study on several Amazonian species used the same approach to account for ontogenetic growth patterns and reported increased growth rates for small individuals but not for large trees (Rozendaal et al. 2010). These trends however may be confounded by differences in survival of fast and slow growing trees (Brienen et al. 2012a). In a study on three Thai species, decreasing growth trends were found (Nock et al. 2010) which may partially have been generated by ontogenetic patterns and by effects of forest disturbances. And finally, in the abovementioned pan-tropical study (Van der Sleen et al. 2014) no growth increases were observed for 12 species over the past 100–150 years. While these studies have explicitly addressed or discussed the effect of several of the abovementioned biases, the reported trends may nonetheless have been affected by forest disturbance and absence of tree regeneration (cf. Vlam 2014), and other biases (cf. Groenendijk et al. 2015).

In all, tree-ring studies in the tropics revealed (modest) increases in iWUE that can be associated with the historical rise in atmospheric CO_2 levels. So far, the contribution of tree-ring and isotope analyses to generating insights on responses of tropical trees to global change has been small, in part due to the abovementioned methodological issues (biases), but also a limited research effort in the tropics. Nevertheless, this contribution can potentially be substantial if sampling designs and statistical analyses are appropriate, biases are taken into consideration and results are interpreted cautiously.

Advances in Stable Isotope Measurements in Tropical Tree Rings

Analysis of stable isotopes ($\delta^{18}O$, $\delta^{13}C$) is increasingly being used in tropical tree rings. Such measurements provide additional information: carbon isotopes mainly provide a measure for plant physiology (i.e., the magnitude of isotope discrimination is related to the ratio between carbon fixed per water lost; McCarroll and Loader 2004), while oxygen isotopes reflect variation in isotopic composition of the source water, tree transpiration rates and relative humidity (Sternberg 2009). In the section below, we will outline some recent advances on applications of isotope analysis in tropical tree rings, and the main insights obtained.

Carbon isotopes in tree rings provide a good drought signal in sites where moisture stress is limiting growth (Gebrekirstos et al. 2009; Fichtler et al. 2010; Brienen et al. 2011; Schollaen et al. 2013). During dry years stomatal aperture decreases, leading to reduced influx of CO_2 into leaf intercellular spaces, and thus a lower intercellular [CO_2] (ci) and lower isotope discrimination (Δ). An example of the effect of precipitation on intercellular CO_2 concentrations (derived from carbon isotopes in tree rings) for a dry forest species from southern Mexico is shown in Fig. 5f. At sites where trees experience less drought stress, the dominant factor controlling tree ring $\delta^{13}C$ may be the photosynthetic rate affected by irradiance (cf. McCarroll and Loader 2004). In line with this, a study on a moist tropical tree species comparing $\delta^{13}C$ in tree rings before and after gap formation shows a positive relationship between $\delta^{13}C$ and growth, thus suggesting variation in growth was most strongly driven by temporal changes in light availability (van der Sleen et al. 2014). These studies show the potential for carbon isotope measurements in tree rings to help interpret the causes of temporal growth rate variation in tropical trees.

Oxygen isotopes in tree rings reflect variation in source water $\delta^{18}O$ and plant physiological effects like leaf water enrichment due to transpiration (Sternberg 2009). Correlations of inter-annual variation in tree ring $\delta^{18}O$ of tropical trees with precipitation $\delta^{18}O$, suggest that plant physiological effects are not very pronounced and that tree rings in these species mainly record source water influences (Brienen et al. 2012b, 2013; Schollaen et al. 2013). Figure 5g shows the relationship between tree ring $\delta^{18}O$ from southern Mexico and precipitation $\delta^{18}O$ from San Salvador (ca. 700 km away). This shows that variation in source water $\delta^{18}O$ (even from distant stations) explains more than 60 % of the variation in tree ring $\delta^{18}O$, but the degree to which source water controls tree ring $\delta^{18}O$ may vary between sites and species. Variation in source water $\delta^{18}O$ depends on local precipitation intensity (i.e., the amount effect, cf. Dansgaard 1964), the origin of the water source, and rainout processes during water vapour transport. For which of these processes tree ring $\delta^{18}O$ provides a proxy, depends on the location and geography of the site. In the western Amazon, $\delta^{18}O$ in tree rings proved to be a strong indicator of total basin-wide precipitation and river discharge (Brienen et al. 2012b), while tree ring $\delta^{18}O$ at other (less continental) sites shows good correlations with more regional precipitation amounts (Poussart and Schrag 2005; Xu et al. 2011; Brienen et al. 2013; Schollaen et al. 2013).

Recent technical developments in isotope techniques allow for very precise dissection of small wood at high resolution (e.g., Schollaen et al. 2013). This has permitted the detection of annual cycles in species that lack anatomically distinct rings, which could subsequently be used to infer growth rates, determine tree ages, or relate the isotope or growth signals to climate (Poussart et al. 2004; Anchukaitis and Evans 2010; Schollaen et al. 2013; Xu et al. 2014). In addition, high-resolution isotope series of $\delta^{13}C$ has provided insights into differences between evergreen and deciduous species in allocation of photosynthates to reserves versus wood (Ohashi et al. 2009; Gulbranson and Ryberg 2013), and high-resolution oxygen isotopes may allow for more detailed seasonal reconstructions of historical rainfall regimes. For instance, it may allow studying differences in dry versus wet season precipitation (Schollaen et al. 2013), and can be used to detect short-term climate events caused by the El Niño-Southern Oscillation (ENSO) (Evans and Schrag 2004; Anchukaitis and Evans 2010) or tropical cyclones (Li et al. 2011).

Finally, a few other useful techniques include the measurement of wood density using gamma radiation, X-ray or high-frequency densitometry (Schinker et al. 2003; De Ridder et al. 2010). High-resolution densitometry measurements could greatly assist ring boundaries detection, for those species presenting growth boundaries defined by density variation (cf. Fig. 2d, e), and density variations themselves may contain climate information. For example, Worbes et al. (1995) found a significant relationship between density variations and the length of the terrestrial phase in floodplains of Central Amazonia.

Conclusions

While the study of tropical tree rings started over a century ago, most advances in this field have been made during recent decades. Important insights arising from these recent tropical tree-ring studies in relation to the theme of this book include:

- Tree-ring analyses of 71 tropical species shows that tree longevity is shorter than often believed (mean longevity ca. 200 years), suggesting relatively fast rates of turnover and carbon cycling in tropical biomes.

- Individual tropical trees show incredibly strong and persistent variation in long-term growth rates, resulting in a fourfold variation in the ages of similarly sized trees. Interestingly, this intraspecific growth variation exceeds the long-term growth variation between species.
- A review of the climate sensitivity of tropical trees shows that annual growth of tropical tree species is more sensitive to fluctuations in rainfall than temperature. While informative for tree sensitivity to climatic fluctuations, these results cannot be directly used to predict growth responses to long-term and gradual changes in temperature or rainfall.
- Long-term trends in water use and growth can be obtained from measurements of stable isotopes and tree-ring width, but such analyses need to take sampling biases and other methodological issues into account. So far, few long-term studies have revealed evidence for century-long increases in intrinsic water use efficiency of several tropical tree species. These results indicate that tropical tree physiology is changing, most likely due to rising CO_2.
- Combined studies on tree ring width and stable isotopes are promising to gain new insights into the response of tropical vegetation to climate change, validate coupled climate-vegetation models and diagnose large–scale changes in the climate system.

References

Anchukaitis KJ, Evans MN (2010) Tropical cloud forest climate variability and the demise of the Monteverde golden toad. Proc Natl Acad Sci USA 107:5036–5040

Baas P & Vetter RE (1989) Growth rings in tropical trees. IAWA Bulletin (Special Issue) 10: 95–174.

Babst F, Alexander MR, Szejner P, Bouriaud O, Klesse S, Roden J, Ciais P, Poulter B, Frank D, Moore DJ (2014) A tree-ring perspective on the terrestrial carbon cycle. Oecologia 176: 307–322

Baker PJ, Bunyavejchewin S (2006) Suppression, release and canopy recruitment in five tree species from a seasonal tropical forest in western Thailand. J Trop Ecol 22:521–529

Baker PJ, Bunyavejchewin S, Oliver CD, Ashton PS (2005) Disturbance history and historical stand dynamics of a seasonal tropical forest in western Thailand. Ecol Monogr 75:23

Berlage HP (1931) Over het verband tusschen de dikte der jaarringen van djatiboomen (*Tectona grandis* L.f.) en den regenval op Java. Tectona 24

Bigler C, Veblen TT (2009) Increased early growth rates decrease longevities of conifers in subalpine forests. Oikos 118:1130–1138

Bonan GB (2008) Forests and climate change: forcings, feedbacks, and the climate benefits of forests. Science 320:1444–1449

Borchert R (1994) Soil and stem water storage determine phenology and distribution of tropical dry forest trees. Ecology 75:1437–1449

Borchert R (1999) Climatic periodicity, phenology, and cambium activity in tropical dry forest trees. IAWA J 20:239–247

Borgaonkar H, Sikder A, Ram S, Pant G (2010) El Niño and related monsoon drought signals in 523-year-long ring width records of teak (*Tectona grandis* L.f.) trees from south India. Palaeogeogr Palaeoclimatol Palaeoecol 285:74–84

Bormann FH, Berlyn G, Borman FH & Berlyn G (1981) Age and growth rate of tropical trees: new directions for research. Yale university: School of Forestry and Environmental Studies. Bulletin No. 94, New Haven.

Brienen RJW, Zuidema PA (2005) Relating tree growth to rainfall in Bolivian rain forests: a test for six species using tree ring analysis. Oecologia 146:1–12

Brienen RJW, Zuidema PA (2006) Lifetime growth patterns and ages of Bolivian rain forest trees obtained by tree ring analysis. J Ecol 94:481–493

Brienen RJW, Lebrija-Trejos E, van Breugel M, Perez-Garcia EA, Bongers F, Meave JA, Martinez-Ramos M (2009) The potential of tree rings for the study of forest succession in Southern Mexico. Biotropica 41:186–195

Brienen RJW, Zuidema PA, Martinez-Ramos MM (2010a) Attaining the canopy in dry and moist tropical forests: strong differences in tree growth trajectories reflect variation in growing conditions. Oecologia 163:485–496

Brienen RJW, Lebrija-Trejos E, Zuidema PA, Martínez-Ramos MM (2010b) Climate-growth analysis for a Mexican dry forest tree shows strong impact of sea surface temperatures and predicts future growth declines. Glob Change Biol 16:2001–2012

Brienen RJW, Wanek W, Hietz P (2011) Stable carbon isotopes in tree rings indicate improved water use efficiency and drought responses of a tropical dry forest tree species. Trees (Berlin) 25:103–113

Brienen RJW, Gloor E, Zuidema PA (2012a) Detecting evidence for CO_2 fertilization from tree ring studies: the potential role of sampling biases. Global Biogeochem Cycles 26:GB1025

Brienen RJW, Helle G, Pons TL, Guyot J-L, Gloor M (2012b) Oxygen isotopes in tree rings are a good proxy for Amazon precipitation and El Niño-Southern Oscillation variability. Proc Natl Acad Sci 109:16957–16962

Brienen RJW, Hietz P, Wanek W, Gloor M (2013) Oxygen isotopes in tree rings record variation in precipitation $\delta^{18}O$ and amount effects in the south of Mexico. J Geophys Res: Biogeosci 118:1604–1615

Briffa K, Melvin TM (2011) A closer look at regional curve standardization of tree-ring records: justification of the need, a warning of some pitfalls, and suggested improvements in its application. In: Hughes MK, Swetnam TW, Diaz HF (eds) Dendroclimatology: progess and prospects. Springer, Dordrecht, pp 113–147

Brown PM (1996) OLDLIST: a database of maximum tree ages. In: Dean JS, Meko DM, Swetnam TW (eds) Tree rings, environment, and humanity. Department of Geosciences, The University of Arizona, Tucson, USA, pp 727–731

Buckley BM, Palakit K, Duangsathaporn K, Sanguantham P, Prasomsin P (2007) Decadal scale droughts over northwestern Thailand over the past 448 years: links to the tropical Pacific and Indian Ocean sectors. Clim Dyn 29:63–71

Bugmann H, Bigler C (2011) Will the CO_2 fertilization effect in forests be offset by reduced tree longevity? Oecologia 165:533–544

Chambers JQ, Higuchi N, Schimel JP (1998) Ancient trees in Amazonia. Nature 39:135–136

Clark DB, Clark DA, Oberbauer SF (2010) Annual wood production in a tropical rain forest in NE Costa Rica linked to climatic variation but not to increasing CO_2. Glob Change Biol 16: 747–759

Coster C (1927) Zur Anatomie und Physiologie der Zuwachszonen und Jahresbildung in den Tropen I. Annales Jardim Botanica Buitenzorg 37:47–161

Dansgaard W (1964) Stable isotopes in precipitation. Tellus 16:436

De Ridder M, Van den Bulcke J, Vansteenkiste D, Van Loo D, Dierick M, Masschaele B, De Witte Y, Mannes D, Lehmann E, Beeckman H (2010) High-resolution proxies for wood density variations in *Terminalia superba*. Ann Bot 107:293–302

Dong SX, Davies SJ, Ashton PS, Bunyavejchewin S, Supardi MN, Kassim AR, Tan S, Moorcroft PR (2012) Variability in solar radiation and temperature explains observed patterns and trends in tree growth rates across four tropical forests. Proc R Soc B: Biol Sci 279: 3923–3931

Doughty CE, Goulden ML (2008) Are tropical forests near a high temperature threshold? J Geophys Res Biogeosci 113:G00B07

Dünisch O, Puls J (2003) Changes in content of reserve materials in an evergreen, a semi-deciduous, and a deciduous Meliaceae species from the Amazon. J Appl Botany 77:10–16

Dünisch O, Bauch J, Gasparotto L (2002) Formation of increment zones and intraannual growth dynamics in the xylem of *Swietenia macrophylla, Carapa guianensis*, and *Cedrela odorata* (Meliaceae). IAWA J 23:101–119

Dünisch O, Montoia VR, Bauch J (2003) Dendroecological investigations on *Swietenia macrophylla* King and *Cedrela odorata* L. (Meliaceae) in the central Amazon. Trees-Struct Funct 17:244–250

Elliott S, Baker PJ, Borchert R (2006) Leaf flushing during the dry season: the paradox of Asian monsoon forests. Glob Ecol Biogeogr 15:248–257

Evans MN, Schrag DP (2004) A stable isotope-based approach to tropical dendroclimatology. Geochim Cosmochim Acta 68:3295–3305

FAO (2006) Global forest resources assessment 2005: progress towards sustainable forest management

Fichtler E, Worbes M (2012) Wood anatomical variables in tropical trees and their relation to site conditions and individual tree morphology. IAWA J 33:119–140

Fichtler E, Clark DA, Worbes M (2003) Age and long-term growth of trees in an old-growth tropical rain forest, based on analyses of tree rings and C-14. Biotropica 35:306–317

Fichtler E, Helle G, Worbes M (2010) Stable-carbon isotope time series from tropical tree rings indicate a precipitation signal. Tree-Ring Res 66:35–49

Fritts HC (1976) Tree rings and climate. Academic Press, London

Gebrekirstos A, Worbes M, Teketay D, Fetene M, Mitlohner R (2009) Stable carbon isotope ratios in tree rings of co-occurring species from semi-arid tropics in Africa: patterns and climatic signals. Global Planet Change 66:253–260

Gebrekirstos A, Bräuning A, Sass-Klassen U, Mbow C (2014) Opportunities and applications of dendrochronology in Africa. Curr Opin Environ Sustain 6:48–53

Geiger F (1915) Anatomische Untersuchungen uber die Jahresringbildung von *Tectona grandis*. Jahrbuch für Wissenschaftliche Botanik 55:521–607

Gourlay ID (1995) The definition of seasonal growth zones in some African Acacia species—a review. IAWA J 16:353–359

Grissino-Mayer HD, Fritts HC (1997) The International Tree-Ring Data Bank: an enhanced global database serving the global scientific community. The Holocene 7:235–238

Groenendijk P, Sass-Klaassen U, Bongers F, Zuidema PA (2014) Potential of tree-ring analysis in a wet tropical forest: a case study on 22 commercial tree species in Central Africa. For Ecol Manage 323:65–78

Groenendijk P, Sleen P, Vlam M, Bunyavejchewin S, Bongers F & Zuidema PA (2015) No evidence for consistent long-term growth stimulation of 13 tropical tree species: results from tree-ring analysis. Glob Change Biol 21:3762–3776.

Gulbranson EL, Ryberg PE (2013) Paleobotanical and geochemical approaches to studying fossil tree rings: quantitative interpretations of paleoenvironment and ecophysiology. Palaios 28:137–140

Heinrich I, Weidner K, Helle G, Vos H, Banks JC (2008) Hydroclimatic variation in Far North Queensland since 1860 inferred from tree rings. Palaeogeogr Palaeoclimatol Palaeoecol 270:116–127

Hietz P, Wanek W, Dunisch O (2005) Long-term trends in cellulose delta C-13 and water-use efficiency of tropical Cedrela and Swietenia from Brazil. Tree Physiol 25:745–752

Hietz P, Turner BL, Wanek W, Richter A, Nock CA, Wright SJ (2011) Long-term change in the nitrogen cycle of tropical forests. Science 334:664–666

Hijmans RJ, Cameron SE, Parra JL, Jones PG, Jarvis A (2005) Very high resolution interpolated climate surfaces for global land areas. Int J Climatol 25:1965–1978

Jacoby GC (1989) Overview of tree-ring analysis in tropical regions. IAWA Bull 10:99–108

Landis RM, Peart DR (2005) Early performance predicts canopy attainment across life histories in subalpine forest trees. Ecology 86:63–72

Laurance WF, Nascimento HEM, Laurance SG, Condit R, D'Angelo S, Andrade A (2004) Inferred longevity of Amazonian rainforest trees based on a long-term demographic study. For Ecol Manage 190:131–143

Li ZH, Labbé N, Driese SG, Grissino-Mayer HD (2011) Micro-scale analysis of tree-ring $\delta^{18}O$ and $\delta^{13}C$ on α-cellulose spline reveals high-resolution intra-annual climate variability and tropical cyclone activity. Chem Geol 284:138–147

Lieberman M, Lieberman D (1985) Simulation of growth curves from periodic increment data. Ecology 66:632–635

Lieberman D, Lieberman M, Hartshorn GS, Peralta R (1985) Growth rates and age-size relationships of tropical wet forest trees in Costa Rica. J Trop Ecol 1:97–109

Lloyd J, Farquhar GD (2008) Effects of rising temperatures and [CO2] on the physiology of tropical forest trees. Philos Trans R Soc B-Biol Sci 363:1811–1817

Locosselli GM, Buckeridge MS, Moreira MZ, Ceccantini G (2013) A multi-proxy dendroecological analysis of two tropical species (*Hymenaea* spp., Leguminosae) growing in a vegetation mosaic. Trees 27:25–36

Loehle C (1988) Tree life-history strategies—the role of defenses. Can J For Res 18:209–222

López L, Villalba R (2011) Climate influences on the radial growth of *Centrolobium microchaete*, a valuable timber species from the tropical dry forests in Bolivia. Biotropica 43:41–49

López L, Villalba R, Bravo F (2013) Cumulative diameter growth and biological rotation age for seven tree species in the Cerrado biogeographical province of Bolivia. For Ecol Manage 292:49–55

Martinez-Ramos M, Alvarez-Buylla ER (1998) How old are tropical rain forest trees? Trends Plant Sci 3:400–405

Martinez-Ramos M, Alvarez-Buylla ER (1999) Reply … Tropical rain forest tree life-history diversity calls for more than one aging method. Trends Plant Sci 4:386–387

McCarroll D, Loader NJ (2004) Stable isotopes in tree rings. Quatern Sci Rev 23:771–801

Meinzer FC, Andrade JL, Goldstein G, Holbrook NM, Cavelier J, Wright SJ (1999) Partitioning of soil water among canopy trees in a seasonally dry tropical forest. Oecologia 121:293–301

Mendivelso HA, Camarero JJ, Obregón OR, Gutiérrez E, Toledo M (2013) Differential growth responses to water balance of coexisting deciduous tree species are linked to wood density in a Bolivian tropical dry forest. PLoS ONE 8:e73855

Mendivelso HA, Camarero JJ, Gutiérrez E, Zuidema PA (2014) Time-dependent effects of climate and drought on tree growth in a Neotropical dry forest: short-term tolerance vs. long-term sensitivity. Agric For Meteorol 188:13–23

Middendorp RS, Vlam M, Rebel KT, Baker PJ, Bunyavejchewin S, Zuidema PA (2013) Disturbance history of a seasonal tropical forest in western Thailand: a spatial dendroecological analysis. Biotropica 45:578–586

Nehrbass-Ahles C, Babst F, Klesse S, Nötzli M, Bouriaud O, Neukom R, Dobbertin M, Frank D (2014) The influence of sampling design on tree-ring based quantification of forest growth. Glob Change Biol 20:2867–2885

Nock CA, Baker PJ, Wanek W, Leis A, Grabner M, Bunyavejchewin S, Hietz P (2010) Long-term increases in intrinsic water-use efficiency do not lead to increased stem growth in a tropical monsoon forest in western Thailand. Glob Change Biol 17:1049–1063

Ohashi S, Okada N, Nobuchi T, Siripatanadilok S, Veenin T (2009) Detecting invisible growth rings of trees in seasonally dry forests in Thailand: isotopic and wood anatomical approaches. Trees 23:813–822

Pan Y, Birdsey RA, Fang J, Houghton R, Kauppi PE, Kurz WA, Phillips OL, Shvidenko A, Lewis SL, Canadell JG, Ciais P, Jackson RB, Pacala SW, McGuire AD, Piao S, Rautiainen A, Sitch S, Hayes D (2011) A large and persistent carbon sink in the world's forests. Science 333:988–993

Patrut A, Von Reden KF, Lowy DA, Alberts AH, Pohlman JW, Wittmann R, Gerlach D, Xu L, Mitchell CS (2007) Radiocarbon dating of a very large African baobab. Tree Physiol 27:1569–1574

Pearson S, Hua Q, Allen K, Bowman DM (2011) Validating putatively cross-dated Callitris tree-ring chronologies using bomb-pulse radiocarbon analysis. Aust J Bot 59:7–17

Peters RL, Groenendijk P, Vlam M, Zuidema PA (2015) Detecting long-term growth trends using tree rings: a critical evaluation of methods. Glob Change Biol 21:2040–2054

Poussart PF, Schrag DP (2005) Seasonally resolved stable isotope chronologies from northern Thailand deciduous trees. Earth Planet Sci Lett 235:752–765

Poussart PF, Evans MN, Schrag DP (2004) Resolving seasonality in tropical trees: multi-decade, high-resolution oxygen and carbon isotope records from Indonesia and Thailand. Earth Planet Sci Lett 218:301–316

Ramírez JA, del Valle JI (2011) Paleoclima de La Guajira, Colombia; según los anillos de crecimiento de *Capparis odoratissima* (Capparidaceae). Revista de Biología Tropical 59:1389–1405

Ramírez JA, del Valle JI (2012) Local and global climate signals from tree rings of *Parkinsonia praecox* in La Guajira, Colombia. Int J Climatol 32:1077–1088

Rivera G, Elliott S, Caldas LS, Nicolossi G, Coradin VT, Borchert R (2002) Increasing day-length induces spring flushing of tropical dry forest trees in the absence of rain. Trees 16:445–456

Robertson I, Loader N, Froyd C, Zambatis N, Whyte I, Woodborne S (2006) The potential of the baobab (*Adansonia digitata* L.) as a proxy climate archive. Appl Geochem 21:1674–1680

Rodríguez R, Mabres A, Luckman B, Evans M, Masiokas M, Ektvedt TM (2005) "El Niño" events recorded in dry-forest species of the lowlands of northwest Peru. Dendrochronologia 22:181–186

Rozendaal DMA, Brienen RJW, Soliz-Gamboa CC, Zuidema PA (2010) Tropical tree rings reveal preferential survival of fast-growing juveniles and increased juvenile growth rates over time. New Phytol 185:759–769

Rozendaal DM, Soliz-Gamboa CC, Zuidema PA (2011) Assessing long-term changes in tropical forest dynamics: a first test using tree-ring analysis. Trees 25:115–124

Schinker MG, Hansen N, Spiecker H (2003) High-frequency densitometry-a new method for the rapic evaluation of wood density variations. IAWA J 24:231–240

Schollaen K, Heinrich I, Neuwirth B, Krusic PJ, D'Arrigo RD, Karyanto O, Helle G (2013) Multiple tree-ring chronologies (ring width, $\delta^{13}C$ and $\delta^{18}O$) reveal dry and rainy season signals of rainfall in Indonesia. Quatern Sci Rev 73:170–181

Schöngart J (2008) Growth-Oriented Logging (GOL): A new concept towards sustainable forest management in Central Amazonian varzea floodplains. For Ecol Manage 256:46–58

Schöngart J, Piedade MTF, Ludwigshausen S, Horna V, Worbes M (2002) Phenology and stem-growth periodicity of tree species in Amazonian floodplain forests. J Trop Ecol 18:581–597

Schöngart J, Piedade MTF, Wittmann F, Junk WJ, Worbes M (2005) Wood growth patterns of *Macrolobium acaciifolium* (Benth.) Benth. (Fabaceae) in Amazonian black-water and white-water floodplain forests. Oecologia 145:454–461

Schöngart J, Orthmann B, Hennenberg KJ, Porembski S, Worbes M (2006) Climate-growth relationships of tropical tree species in West Africa and their potential for climate reconstruction. Glob Change Biol 12:1139–1150

Schöngart J (2013) Dendroecological studies in tropical forests. Albert-Ludwigs-University Freiburg, Freiburg im Breisgau.

Slik JWF, Arroyo-Rodríguez V, Aiba S-I, Alvarez-Loayza P et al (2015) How many tropical forest tree species are there? Proc Natl Acad Sci 112(24):7472–7477

Soliz-Gamboa CC, Rozendaal DM, Ceccantini G, Angyalossy V, van der Borg K, Zuidema PA (2011) Evaluating the annual nature of juvenile rings in Bolivian tropical rainforest trees. Trees 25:17–27

Soliz-Gamboa CC, Sandbrink A, Zuidema PA (2012) Diameter growth of juvenile trees after gap formation in a Bolivian rain forest: responses are strongly species-specific and size-dependent. Biotropica 44:312–320

Speer JH (2010) Fundamentals of tree-ring research. University of Arizona Press

Spracklen DV, Arnold SR, Taylor CM (2012) Observations of increased tropical rainfall preceded by air passage over forests. Nature 489:282–285

Stahle DW (1999) Useful strategies for the development of tropical tree-ring chronologies. Iawa J 20:249–253

Stahle DW, Burnette DJ, Villanueva J, Cerano J, Fye FK, Griffin RD, Cleaveland MK, Stahle DK, Edmondson JR, Wolff KP (2012) Tree-ring analysis of ancient baldcypress trees and subfossil wood. Quatern Sci Rev 34:1–15

Sternberg L (2009) Oxygen stable isotope ratios of tree-ring cellulose: the next phase of understanding. New Phytol 181:553–562

Stocker TF, Qin D, Plattner G-K, Tignor M, Allen SK, Boschung J, Nauels A, Xia Y, Bex V, Midgley PM (2013) Climate change 2013: the physical science basis. In: Intergovernmental panel on climate change, working group I contribution to the IPCC fifth assessment report (AR5), Cambridge University Press, New York

van der Sleen P (2014) Environmental and physiological drivers of tree growth. A pan-tropical study of stable isotopes in tree rings. Wageningen University, Wageningen, NL

van der Sleen P, Soliz-Gamboa C, Helle G, Pons T, Anten N, Zuidema P (2014) Understanding causes of tree growth response to gap formation: $\Delta^{13}C$-values in tree rings reveal a predominant effect of light. Trees 28:439–448

van der Sleen P, Groenendijk P, Vlam M, Anten NP, Boom A, Bongers F, Pons TL, Terburg G & Zuidema PA (2014) No growth stimulation of tropical trees by 150 years of CO2 fertilization but water-use efficiency increased. Nature geoscience 8:24–28

van Schaik CP, Terborgh JW, Wright SJ (1993) The phenology of tropical forests: adaptive significance and consequences for primary consumers. Annu Rev Ecol Syst 24:353–377

Vieira S, Trumore SE, Camargo PBD, Selhorst D, Chambers JQ, Higuchi N, Martinelli LA (2005) Slow growth rates of Amazonian trees: consequences for carbon cycling. Proc Natl Acad Sci USA 102:18502–18507

Vlam M (2014) Forensic forest ecology. Unraveling the stand history of tropical forests. Wageningen University, Wageningen, NL

Vlam M, Baker PJ, Bunyavejchewin S, Zuidema PA (2014a) Temperature and rainfall strongly drive temporal growth variation in Asian tropical forest trees. Oecologia 174:1449–1461

Vlam M, Baker PJ, Bunyavejchewin S, Mohren GM, Zuidema PA (2014b) Understanding recruitment failure in tropical tree species: insights from a tree-ring study. For Ecol Manage 312:108–116

Wheeler E, Baas P, Rodgers S (2007) Variations in dicot wood anatomy: a global analysis based on the insidewood database. IAWA J 28:229–258

Whitmore TC (1998) An introduction to tropical rain forests. Oxford University Press, New York

Wils T, Sass-Klaassen U, Eshetu Z, Bräuning A, Gebrekirstos A, Couralet C, Robertson I, Touchan R, Koprowski M, Conway D (2011) Dendrochronology in the dry tropics: the Ethiopian case. Trees 25:345–354

Worbes M (1985) Structural and other adaptations to long-term flooding by trees in Central Amazonia. Amazonia 4:459–484

Worbes M (1989) Growth rings, increment and age of trees in inundation forests, savannas and a mountain forest in the neotropics. IAWA Bull 10:109–122

Worbes M (1995) How to measure growth dynamics in tropical trees—a review. IAWA J 16:337–351

Worbes M (1999) Annual growth rings, rainfall-dependent growth and long-term growth patterns of tropical trees from the Caparo Forest Reserve in Venezuela. J Ecol 87:391–403

Worbes M (2002) One hundred years of tree-ring research in the tropics—a brief history and an outlook to future challenges. Dendrochronologia 20:217–231

Worbes M, Junk WJ (1999) How old are tropical trees? The persistence of a myth. IAWA J 20:255–260

Worbes M, Klosa D, Lewark S (1995) Density fluctuation in annual rings of tropical timbers from central Amazonian inundation forests. Holz Als Roh-und Werkstoff 53:63–67

Worbes M, Staschel R, Roloff A, Junk WJ (2003) Tree ring analysis reveals age structure, dynamics and wood production of a natural forest stand in Cameroon. For Ecol Manage 173:105–123

Xu C, Sano M, Nakatsuka T (2011) Tree ring cellulose $\delta^{18}O$ of Fokienia hodginsii in northern Laos: A promising proxy to reconstruct ENSO? J Geophys Res 116:D24109

Xu C, Sano M, Yoshimura K, Nakatsuka T (2014) Oxygen isotopes as a valuable tool for measuring annual growth in tropical trees that lack distinct annual rings. Geochem J 48:371–378

Zuidema PA, Brienen RJW, Schöngart J (2012) Tropical forest warming: looking backwards for more insights. Trends Ecol Evol 27:193–194

Zuidema PA, Baker PJ, Groenendijk P, Schippers P, van der Sleen P, Vlam M, Sterck F (2013) Tropical forests and global change: filling knowledge gaps. Trends Plant Sci 18:413–419

Index

Wood density, 58, 110, 112, 207, 217, 218,
    221, 245, 246, 267, 281, 454

X

Xylem, 8, 9, 15, 53, 70, 72, 74–76, 110, 115,
    151, 185, 191, 210, 212, 229, 231, 239,
    243–247, 267

Y

Yucatan, 250

CPSIA information can be obtained
at www.ICGtesting.com
Printed in the USA
LVHW061122260319
611865LV00002B/30/P